INTELLIGENCE-BASED MEDICINE

INTELLIGENCE-BASED MEDICINE

Artificial Intelligence and Human Cognition in Clinical Medicine and Healthcare

ANTHONY C. CHANG

Medical Director, The Sharon Disney Lund Medical Intelligence and Innovation Institute (MI3);
Chief Intelligence and Innovation Officer, Children's Hospital of Orange County, Orange, CA, United States

AIMed
Artificial Intelligence
in Medicine

ELSEVIER

ACADEMIC PRESS

An imprint of Elsevier

Academic Press is an imprint of Elsevier
125 London Wall, London EC2Y 5AS, United Kingdom
525 B Street, Suite 1650, San Diego, CA 92101, United States
50 Hampshire Street, 5th Floor, Cambridge, MA 02139, United States
The Boulevard, Langford Lane, Kidlington, Oxford OX5 1GB, United Kingdom

Notices

Knowledge and best practice in this field are constantly changing. As new research and experience broaden our
understanding, changes in research methods, professional practices, or medical treatment may become necessary.

Practitioners and researchers must always rely on their own experience and knowledge in evaluating and using any
information, methods, compounds, or experiments described herein. In using such information or methods they should be
mindful of their own safety and the safety of others, including parties for whom they have a professional responsibility.

To the fullest extent of the law, neither the Publisher nor the authors, contributors, or editors, assume any liability for any
injury and/or damage to persons or property as a matter of products liability, negligence or otherwise, or from any use or
operation of any methods, products, instructions, or ideas contained in the material herein.

British Library Cataloguing-in-Publication Data
A catalogue record for this book is available from the British Library

Library of Congress Cataloging-in-Publication Data
A catalog record for this book is available from the Library of Congress

ISBN: 978-0-12-816462-4

For Information on all Academic Press publications
visit our website at https://www.elsevier.com/books-and-journals

Publisher: Stacy Masucci
Acquisitions Editor: Rafael E. Teixeira
Editorial Project Manager: Sara Pianavilla
Production Project Manager: Maria Bernard
Cover Designer: Christian Bilbow

Typeset by MPS Limited, Chennai, India

Working together
to grow libraries in
developing countries

www.elsevier.com • www.bookaid.org

Dedication

First and foremost, this book is dedicated to the many thousands of children and adults with their diseases whom I have had the pleasure to serve as their cardiologist (including a very special 9 year-old girl named Ilsa from Myanmar whose death motivated me to pursue this domain to make artificial intelligence (AI) in medicine available for everyone). This book is also dedicated to the millions of patients and families worldwide who are eternally dedicated to improve health care and to save lives. Their supreme fortitude and will to survive in their long and complicated medical journeys will continually inspire me to maintain my ardent passion to learn in this new and wondrous world of AI in clinical medicine and health care.

This book is also dedicated to the many clinicians who are open-minded in taking on this new and exciting domain, as difficult as it is and as challenging as it can be, as well as my great colleague and friend Dr. Nick Anas, whose presence I very much miss on a daily basis but who was always supremely supportive of my personal efforts in this nascent domain prior to his passing (and who probably has a huge smile on his face in heaven perusing this book).

Finally, I look forward to sharing this work with my beautiful daughters Emma and Olivia one day and thank them not only for their pure love and joy for me, but also for their utter affinity for sleep (including long afternoon naps), which allowed me just sufficient time to write most of this work (in front of the serene immensity of the Pacific Ocean while Mozart is playing in the background, which I would wholeheartedly recommend to any author).

Quote

Maybe that's enlightenment enough: to know that there is no final resting place of the mind; no moment of smug clarity. Perhaps wisdom... is realizing how small I am, and unwise, and how far I have yet to go. *Anthony Bourdaine, American cook/ author and global traveler*

Contents

III

The current era of artificial intelligence in medicine

7. Clinician Cognition and Artificial Intelligence in Medicine   193

8. Artificial Intelligence in Subspecialties   267

9. Implementation of Artificial Intelligence in Medicine   397

IV

The future of artificial intelligence and application in medicine

10. Key Concepts of the Future of Artificial Intelligence   415

11. The Future of Artificial Intelligence in Medicine   431

About the author

Anthony C. Chang MD, MBA, MPH, MS

Dr. Chang attended Johns Hopkins University for his BA in molecular biology prior to entering Georgetown University School of Medicine for his MD He then completed his pediatric residency at Children's Hospital National Medical Center and his pediatric cardiology fellowship at the Children's Hospital of Philadelphia. He then accepted a position as attending cardiologist in the cardiovascular intensive care unit of Boston Children's Hospital and as assistant professor at Harvard Medical School. He has been the medical director of several pediatric cardiac intensive care programs (including Children's Hospital of Los Angeles, Miami Children's Hospital, and Texas Children's Hospital). He served as the medical director of the Heart Institute at Children's Hospital of Orange County.

He is currently the chief intelligence and innovation officer and medical director of the Heart Failure Program at Children's Hospital of Orange County. He has also been named a Physician of Excellence by the Orange County Medical Association and Top Cardiologist, Top Doctor for many years, as well as one of the nation's Top Innovators in Health care.

He has completed a Masters in Business Administration in Health Care Administration at the University of Miami School of Business and graduated with the McCaw Award of Academic Excellence. He also completed a Masters in Public Health in Health Care Policy at the Jonathan Fielding School of Public Health of the University of California, Los Angeles and graduated with the Dean's Award for Academic Excellence. Finally, he graduated with his Masters of Science in Biomedical Data Science with a subarea focus in artificial intelligence from Stanford School of Medicine and has completed a certification on Artificial Intelligence from MIT. He is a computer scientist-in-residence and a member of the Dean's Scientific Council at Chapman University.

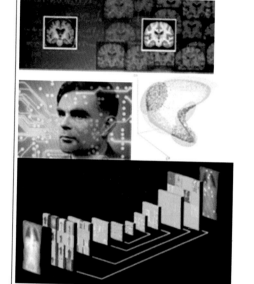

INTELLIGENCE-BASED MEDICINE

*PRINCIPLES AND APPLICATIONS OF
DATA SCIENCE, ARTIFICIAL INTELLIGENCE,
AND HUMAN COGNITION IN
CLINICAL MEDICINE AND HEALTH CARE*

ANTHONY CHANG, MD, MBA, MPH, MS

He has helped to build a successful cardiology practice as a start-up company and was able to complete a deal on Wall Street. He is known for several innovations in pediatric cardiac care, including introducing the cardiac drug milrinone and codesigning (with Dr. Michael DeBakey) an axial-type ventricular assist device in children. He is a committee member of the National Institute of Health pediatric grant review committee. He is the editor of several textbooks in pediatric cardiology and intensive care, including *Pediatric Cardiac Intensive Care, Heart Failure in Children and Young Adults*, and *Pediatric Cardiology Board Review*.

He is the founder of the Pediatric Cardiac Intensive Care Society that launched the multidisciplinary focus on cardiac intensive care for children. He is also the founder of the Asia-Pacific Pediatric Cardiac Society that united pediatric cardiologists and cardiac surgeons from 24 Asian countries and launched a biennial meeting in Asia that now draws over 1000 attendees.

He is the founder and medical director of the Medical Intelligence and Innovation Institute that is supported by the Sharon Disney Lund Foundation (since 2015). The institute is dedicated to implement data science and artificial intelligence in medicine and is the first institute of its kind in a hospital. The new institute is concomitantly dedicated to facilitate innovation in children and health care all over the world. He is the former organizing chair for the biennial *Pediatrics2040: Emerging Trends and Future Innovations* meeting as well as the founder and codirector of the Medical Intelligence and Innovation Summer Internship Program, which mentors close to 100 young physicians-to-be every summer. He is also the founder and chairman of the board of a pediatric innovation leadership group called the International Society for Pediatric Innovation.

He intends to build a clinician—computer scientist interface to enhance all aspects of data science and artificial intelligence in health and medicine. He currently lectures widely on artificial intelligence in medicine (he has been called "Dr. A.I." by the *Chicago Tribune*) and has been named one of the AI Influential Thinkers. He has given a TEDx talk and is a regular featured speaker at Singularity University's Exponential Medicine. He has published numerous review papers on big data and predictive analytics as well as machine learning and artificial intelligence in medicine. He is on the editorial board of the *Journal of Medical Artificial Intelligence*. He is currently completing a book project with Elsevier: *Intelligence-Based Medicine: Principles and Applications of Data Science, Artificial Intelligence, and Human Cognition in Medicine and Health care*. He is the founder and organizing chair of several *Artificial Intelligence in Medicine (AIMed)* meetings in the United States and abroad (Europe and Asia) that will focus on artificial intelligence in health care and medicine (www.ai-med.io). He intends to start a new group for clinicians with a special focus on data science and artificial intelligence as a nascent society (Medical Intelligence Group). He is the dean of the nascent American Board of Artificial Intelligence in Medicine (ABAIM).

He is the founder of three start-up companies in the artificial intelligence in medicine domain:

1. CardioGenomic Intelligence (CGI), LLC is a multifaceted company that focuses on artificial intelligence applications such as deep learning in clinical cardiology (cardiomyopathy and heart failure as well as other cardiovascular disease) and genomic medicine.
2. Artificial Intelligence in Medicine (AIMed), LLC is a multimedia and events company that organizes meetings and educational programs in artificial intelligence in medicine in local as well as global venues and across subspecialties.

3. Medical Intelligence 10 (MI10), LLC is an education and consulting/advising conglomerate for clinicians, executives, and leaders of health-care organizations and companies as well as investors for the evaluation and implementation of AI strategies in health-care organizations and health-care companies, and assessment and implementation of cybersecurity in health-care organizations. The proprietary MI10 assessment tool (MIQ) can evaluate any organization or company for AI readiness and quality and utilizes both deep learning and cognitive architecture for its AI-enabled strategy recommendations.

Foreword

The promise of artificial intelligence (AI) has become a dominant focus for many of today's tech companies and, in turn, for both the social and professional media that bring us our news and tout what lies in store for society. Topics such as machine learning, "big data" analytics, intelligent personal assistants, and self-driving cars are becoming commonplace notions for our society, all presumed to be just around the corner given recent progress and demonstrated early applications. But it is in the use of AI and machine learning for medicine and health care that our expectations have been raised to particularly high levels. Medical computing research that was once confined to technical journals in computer science or biomedical informatics is now frequently published in the most prominent clinical and health policy journals. The AI in Medicine era is upon us, after a half century of slow and steady progress that has exploded into the public's awareness, in large part because computing and communications technology has transformed what software innovations are practical and economically feasible in routine settings.

Today, we see major investments in medical AI, both in startup companies and major corporations. Health-care organizations and major medical centers are adjusting their budgets to incorporate substantial efforts to leverage AI methods for both business and clinical purposes. No one wants to be left behind, and the medical AI "buzz" is everywhere. Yet, for the average clinician, patient, or journalist, there is a great need to demystify this field and to understand something of the technology, its current status, and the barriers that need to be overcome if it is to reach its full potential.

In the nick of time, Dr. Anthony Chang has recognized the demand for a monograph that meets the need for an accessible source of comprehensive but nontechnical information on medical AI—its history, vocabulary, key concepts, current state, and promise for the future. Drawing on his formal training in medicine, public health, business, data science, and biomedical informatics, Chang has written a marvelous summary of the field—one that will meet the needs of the broad and diverse audience that he has sought to address. Although I have worked in the field for 50 years and presumably know much of what is covered in this volume, I have thoroughly enjoyed the logical and clear way in which he has addressed the key technologies and issues, often seeing connections between topics, or formulating definitions, that helped me with my own view of the field when viewed in its entirety. He brings the field together into a coherent whole, complementing it with a useful glossary and a summary of several forward-looking companies that are seeking to leverage AI in health care in innovative ways.

Any new and potentially revolutionary field needs its translators who can bridge communities, clarify terms or concepts that may be confusing to newcomers, and help to forge an energized movement that will help to assure that the potential is realized. Anthony

Chang is doing this for AI in Medicine, as this volume demonstrates to its readers. It is appropriately directed to the community that wants to both understand and advance the field as we seek to improve and potentially revolutionize the way we care for patients and keep individuals healthy.

Edward H. Shortliffe, MD, PhD
New York City
Editor, *Biomedical Informatics*

Foreword

Looking back in a century's time, we will shake our heads wondering how hard it must have been to deliver healthcare in an age before artificial intelligence (AI)—just in the same way we today find it hard to imagine the lives of those in the preantibiotic age. All those unnecessary deaths, all that avoidable hardship.

For better or worse, we have built a modern health-care system that is too complex for humans to manage or navigate.

The engines of science crunch out research findings at industrial scale and deposit them in digital warehouses where only a fraction ever gets looked at, and less is acted upon. Much of that research even today is considered to be "waste" because of methodological flaws, inherent biases, or simply because the question being asked has already been answered.

Our health-care delivery services for their part are less a system than a complex jumble of poorly interconnecting fragments. The "system" has never been designed but rather has accreted over time, using a hodgepodge of different technologies that are often not so much poorly interoperable, as they are fundamentally incompatible.

Citizens, for their part, dip in and out of this health system, and are fragmenting into different belief communities, aided by social media and informational manipulation. Fed on information diets of vaccine refusal, nonevidence-based alternative medicine, and making lifestyle decisions that welcome chronic diseases, people often put themselves out of reach of preventative health campaigns, to present late into the health-care system with significant but avoidable disease.

AI cannot cure all of these ills alone, but well-crafted technologies in the hands of effective individuals will make a big difference.

AI systems are perfect for the task of seeking out preexisting evidence, and summarizing it to answer a specific question about a patient's treatment, investigation or prognosis. Personalized answers can take the place of population-based answers, because AI will have the capacity to seek out evidence that best matches the circumstances of a given patient, in a way an unaided human never could. The precision of AI enabled diagnosis and therapy planning will soon far exceed what is possible in most health services by humans alone.

Smart health services will be better able to connect up with each other, and distributed AI systems will better navigate the complexities of the health-care system than unaided humans can. Finding out who to see next, or what clinical service to attend next, can be a guided and personalized journey when AI is employed as a personal guide.

People will always in the end believe what they want, but AI can also help citizens to access the best available evidence and explain it in a way that is meaningful and unbiased by preexisting beliefs. When we have personal AI assistants that we trust, our engagement with the research evidence is likely to be very different. Smart ways of helping citizens visualize the consequences of their decisions, for example, can make a difference their behaviors.

This book is emerging at a moment in history where we not only have the curiosity to develop AI for healthcare, but we now also have powerful motivation. We are seeing just how transformational AI is in other parts of our lives, and just how complex, expensive, and ultimately unsustainable modern pre-AI healthcare is.

The student of AI will find much to learn in these pages, and much of that has changed over the last three decades, when the earliest texts on AI in healthcare began to emerge. That pace of change, of innovation and discovery, is only going to continue. To manage these changes, it is more important to understand foundational principles than the specifics of a given technology, whose half-life is likely to be fleeting. It is also salutary to remember that many of the big challenges that preoccupied AI in health-care researchers three decades ago are still unsolved. How does a machine reason about cases it has never seen before? How do we interpret clinical findings in the presence of multiple and interacting diseases? How do we infer causality from associations in data? How can a human and a computer have a meaningful partnership?

We are at the cusp of two histories—medicine as it was practiced in the pre-AI era, and how it will be practiced in the era of AI-enabled healthcare. We know the old history very well. The new one is just about to be written.

Enrico Coiera
Editor, Guide to Health Informatics

Preface

Once you have tasted flight, you will forever walk the earth with your eyes turned skyward, for there you have been and there you will always long to return. **Leonardo Da Vinci**

Why a book on artificial intelligence (AI) in medicine and health care and why now?

As the world struggles in the grips of the coronavirus COVID-19, AI as a topic (as evidenced by the magazines on the stand) is equally hot: "Making Good on the Promise of AI" (*MIT Sloan Management Review*), "The AI-Powered Organization" (*Harvard Business Review*), and "Machine Intelligence" (*Nature*). Yet, with the exception of a few subspecialists in a few medical centers, the drumbeat of this AI revolution is eerily quiet in clinical medicine and our health-care ecosystem.

I was a very privileged student in the world of data science and AI during my years at Stanford almost a decade ago. The epiphanous 4-year journey was a personal transformation for me not only as a newly minted data scientist but also as a clinician with a much better appreciation for a balanced symmetry between the two disciplines of clinical medicine and AI.

The AI milieu can provide a sanctuary for all those who seek refuge and respite from the imbroglio and the perfect storm in biomedical sciences and clinical medicine in the present era. There are many daunting challenges ahead, but there can also be a myriad of invigorating discoveries and innovative solutions to the many problems we face in medicine and health care. This AI paradigm is a once-in-a-generation opportunity for all of us to transform medicine and health care in small and big ways, from the ivory tower specialized intensive care units to the malaria-stricken towns in sub-Saharan Africa.

This book with the accompanying compendium is designed for anyone who is interested in a comprehensive primer on the principles and application of data science, AI, and human cognition in health care and medicine, all toward an "intelligence-based medicine." The readers whom this book is written for include the curious but busy clinician, the interested data/computer scientist, the astute investor, the inquisitive hospital administrator and leader (from CEO to CIO and others), and any knowledge-seeking patient and family member. It is not a book full of mathematical formulae and esoteric data science topics so that the clinicians are not going to be engaged, nor is it a book filled with medical jargon and superficial descriptions of AI concepts, but with too little data science or AI substance. Like most elements in this domain, this book is a hybrid between the clinical, medical knowledge and the AI and data science. In short, it is written with everyone in mind in defining the rapidly growing interface between medicine and AI.

The sections of this book are designed to provide a comprehensive framework for anyone who is interested in understanding AI as well as the aspects of AI that would be relevant to biomedicine and health care. There are nearly 100 insightful commentaries by specially selected experts from diverse backgrounds covering many mini topics that pertain to this domain; these commentaries add a unique set of perspectives and expertise and are strategically interspersed throughout this book (Timothy Chou described these as "the ornaments for the tree"). There is an unavoidable overlap as I had asked each author to submit a commentary on their passion in this domain, but it is usually good to read more than one perspective. I believe that these "ornamental" commentaries comprise the singular strength of this work.

The first section introduces the many terms as well as elucidates the basic concepts of AI and also explores the relationship between neuroscience and AI. The section also delineates the early years of AI and the history of AI in medicine during this early era. The genesis of AI and its application in medicine is key to the understanding of state of the art today. The common misconceptions of AI in medicine with brief explanations are included in this section as a primer.

The second section details the current era or state of the art of biomedical data science and AI and covers basic elements of health-care data, databases, and biomedical data science. Machine and deep learning, by far the most well-used methodologies of AI in this current era, is the main part of this section. Additional key concepts such as cognitive computing, natural language processing, and robotics are also included in this section. A suggested strategy to learn basics of AI in medicine with a knowledge assessment (answers can be found in the compendium) are included in this section.

The third section of the book covers applications of AI in medicine in different areas and introduction to cognition aspects of clinicians. Following this orientation, AI in medicine as it relates to subspecialties is then separately discussed, particularly the subspecialties that are at the forefront of AI in medicine. Several useful guides (organizational assessment as well as obstacles and elements for successful implementation of AI in medicine) are included in this section.

The future of AI and applications in medicine is then covered in the fourth section of this book. The future of AI is discussed in terms of future elements such as augmented and virtual reality and Internet of everything. Additional key concepts such as virtual assistants and quantum computing are also briefly covered. The future of AI as it relates to medicine is separately discussed. Major takeaways of AI in medicine are included at the end of this section.

The last section of this book is a compendium of useful resources, including key references (books and journals), top 100 (and more) journal references, 100 important AI in medicine companies to know, and a comprehensive glossary of terms.

The author sincerely hopes that this work will inspire all of us to continually explore the mostly unfamiliar world of data science and AI in clinical medicine and health care and bring these capabilities to the myriad of domains in health care in order to help all patients. We often speak about patients as if patients are a separate human subspecies, but we are all patients (sooner or later). This Sisyphean task of deciphering AI and its many nuances and mysteries and deploying these tools in medicine and health care will be our greatest legacy for the next generations.

As I am finishing this book, I am myself a parent for my daughter with complex congenital heart disease having her heart surgery; it has been especially meaningful for me that this book was completed at the bedside. With an excellent multidisciplinary team taking extraordinary care of her, it was obvious to everyone that we still make many decisions without sufficient data or enough certainty, and that we do not naturally seek data science or AI as a resource in the medical domain. This theme of uncertainty in clinical medicine along with many other themes elucidated in this work are pervasive in medicine even at the premiere medical centers but can also be reduced greatly if some or all of the methodologies described in this book are put in practice. I certainly feel that I have become a better clinician with an embedded data science perspective.

My personal vision is to use AI (effective self-learning AI tools with expert clinicians providing oversight) to democratize "expert" opinions and perceived high-quality health care so that this will render rankings of hospitals (mainly a marketing ploy) much less relevant. In particular, we should "rank" diseases and conditions we need to eradicate rather than hospitals (as if these institutions are like college sports teams). The ranking criteria, however, should be retained as an adequate checklist and categories such as "A" or "B" (or "F") could be given instead of rankings.

It is perhaps more than a serendipity that Demis Hassabis of Google DeepMind so aptly stated that he compares AI (and his company DeepMind in particular) as the Apollo space program of our generation. It is also the wise words (that I will paraphrase) of the NHS Digital Chief Noel Gordon who commented at an AIMed Europe event that "AI is the accelerator for exit velocity we need" to escape the gravitational pull of the present healthcare conundrum. We are about the commemorate the semicentennial of the Apollo 11 lunar landing, so perhaps this event is inspiring for us to also think grand about AI as a once-in-a-generation dream that is reachable for those of us who want to make radical changes to improve health care. The opportunity to widely adopt AI as a paradigm in clinical medicine and health care is a wondrous one, and yet, similar to the vision of John F. Kennedy, very few among us think the grand vision is even remotely realizable. This AI journey will be even more difficult and far-reaching as Apollo 11 as there will be no obvious denouement as dramatic as the landing on the Moon 50 year ago, so we must persevere for the sake of all of our patients, and all of us.

Anthony C. Chang
Spring, 2020

Acknowledgments

I would like to thank my colleagues who were especially supportive of me during this project: Dr. Spyro Mousses, Dr. Sharief Taraman, Dr. William Feaster, Dr. Louis Ehwerhemuepha, and Dr. Terry Sanger. I would like to express my gratitude to the Sharon Disney Lund Foundation board members who supported my vision of having a dedicated institute for artificial intelligence, in particular Michelle Lund and Robert and Gloria Wilson. I would also like to thank Kimberly Cripe, Matt Gerlach, and John Henderson of Children's Hospital of Orange County, all exemplary in their professionalism. I like to express my thanks also to my Medical Intelligence and Innovation Institute (MI3) team members for their unwavering support: Deborah Beauregard, Tiffani Ghere, Julie Gillespie, Debbie Flint, Laura Beken, Seraya Martinez, Mijanou Pham, Dr. Addison Gearhart, Dr. Sharib Gaffar, Dr. Afnan Alqahtani as well as our past fellows in Artificial Intelligence in Medicine, Nathaniel Bischoff and Alex Barrett. My professional colleagues Dr. Chris Yoo, Dr. David Schneider, Sam King, Zhen Xu, Qingxin Zhang, Joe Kiani, Dr. Kevin Maher, Dr. Arlen Meyers, Dr. Jai Nahar, Dr. Vishal Nangalia, Sean Lane, Kevin Lyman, Joerg Aumueller, Dr. Annette ten Teije, Dr. Robert Hoyt, Dr. Daniel Kraft, Dr. Enrico Coiera, Dr. May Wang, Dr. Leo Celi, Dr. Arta Bakshandeh, Dr. Randall Wetzel, Dr. James Fackler, Dr. LinHua Tan, Dr. Uny Cao, Dr. Tony Young, Dr. John Lee, Dr. Peter Laussen, Dr. Uli Chettipally, Dr. Diane Nugent, Dr. Wyman Lai, Dr. Mustafa Kabeer, Dr. William Loudon, Dr. Hamilton Baker, Dr. Kathy Jenkins, Dr. Peter Holbrook, Dr. Peter Chang, Sylvia Trujillo, Vanessa Vu, Ria Banares, Jennylyn Gleave, Audrey He, Dr. Orest Boyko, Eric Smith, Matt Wilson, Kieran Anderson, Dr. Robert Brisk, Dr. Matthieu Komorowsky, Dr. John Lee, Dr. Piyush Mathur, Dr. William Norwood, Dr. Richard van Praagh, Angela Tripoli, Dr. Mitch Recto, Dr. Ioannis Kakadiaris, Brett McVicker, Sam Balcomb, Dr. Paul Lubinsky, Dr. Ken Grant, Dr. Amir Ashrafi, Dr. John Cleary, Dr. Afshin Aminian, Dr. Matthieu Komorowski, Steve Ardire, Steve Lund, Moe Levitt, Ken Collins, and Samras Phar have all been invaluable to me. For those names that I inadvertently have left out, please forgive me but your support was cherished.

I like to express my gratitude to the Elsevier staff: Rafael Teixiera, Sara Pianavilla, Justyna Kasprzycka, Maria Bernard for their utmost patience. I am grateful to my fellow leaders for the international Society of Pediatric Innovation (iSPI): Dawn Wolff, Dr. Claudia Hoyen, Sherry Farrugia, Leanne West, Dr. Alberto Tozzi, Dr. Srinivasan Suresh, and Dr. Todd Ponsky. I would like to thank my Stanford School of Medicine Biomedical Data Science program mentors (especially Drs. Ted Shortliffe, Russ Altman, Nigam Shah, Timothy Chou, and Dennis Wall) and classmates and teachers' assistants for their utmost patience and gracious encouragement during my 4-year sojourn as a curious and passionate student into the mysterious but fascinating world of biomedical data science and artificial intelligence. My computer science colleagues at Chapman University, Department of Computer Science, especially Drs. Michael Fahy and Cyril Rakovic who were always available for guidance and counsel are also acknowledged. I like to express my gratitude to the tireless and dedicated staff of the Artificial Intelligence in Medicine (AI-Med) project: Freddy White, Bansri Shah, Charlie Maloney, Kirsten Lane, Andrew Johnson, Suzy White, Andrew McDonald, Damian Doherty, Priya Samant, Hazel Tang, Ruki Rehman, Alexis May, Graham Wray, Laurie Griffiths, Ally Baker and the many colleagues and friends who have participated as faculty members as well as attendees at the Artificial Intelligence in Medicine (AI-Med)—related meetings around the world and now across subspecialties. Finally, I would like to thank all those esteemed authors and friends who have kindly contributed their insightful commentaries to this book and who have so kindly enlightened me now and even more in the future.

PART I

Introduction to Artificial Intelligence

On March 10, 2016, Google DeepMind's *AlphaGo* software made the game's 37th move as it competed against the best human Go champion Lee Sedol: this move was so astonishing in its ingenuity that Sedol felt compelled to leave the room to recover. This moment, in which the computer or machine intelligence may have created an entirely novel Go strategy, heralded the recent dawning of a new era in artificial intelligence (AI).

The recent impressive gains in sophistication of deep learning (DL) technology and utilization especially since 2012 have led to an escalating momentum for AI awareness and adoption. Major universities with AI departments (such as Stanford, MIT, and Carnegie Mellon) and technology giants [such as IBM, Apple, Facebook, and Microsoft in the United States as well as other large companies such as Baidu, Alibaba, and Tencent (BAT) in China] are all fervidly exploring real-life applications of AI. Even though the advent of data science and machine and DL has advanced information and analyses (such as financial interactions and sports performance) and promoted innovations (such as virtual assistants, autonomous cars, drones, and even a work of art completed by DL that has fetched a few hundred thousand dollars at Christie's), healthcare and medicine remain very much behind these other domains in leveraging this new AI paradigm. The recent major escalation of venture capital into healthcare and AI domain, however, promulgated over 100 companies in AI in healthcare with an expectant $50 billion to be spent on AI in healthcare by 2025 with more than $100 billion in savings. In early 2019 Google has announced its corporate direction in deploying its "AI-first strategy" into healthcare.

Since the first article published in the domain of AI in biomedicine in 1958 [1], there has been a relative paucity of published reports focused on AI in medical journals (perhaps about 100,000 total articles out of close to 50 million articles, or about 0.2%) and a congruent lack of serious interest amongst most clinicians in applications of AI in medicine. Even in 2018 there were only about 6000 reports on AI applications in medicine (under a myriad of AI-related search terms such as "artificial intelligence," "machine learning," "deep learning," "cognitive computing," and "natural-language processing") out of a total of close to 1.8 million articles in over 28,000 journals, or a mere 0.35% of total medical publications. Finally, there is publication activity only very recently in the more prestigious journals that have been relatively quiescent in this domain for a lengthy period [2–5].

We all face the imbroglio of healthcare with its complex ecosystem and data in disarray, and this has led to a significant rise in professional burnout amongst its caretakers. We have a once-in-a-generation opportunity to capture this robust AI resource for clinical medicine and healthcare, and potentially make the transformational change that is so direly needed in the coming decades.

From reactive "Sick" care to proactive healthcare: the future of digital health and artificial intelligence (AI)

Daniel Kraft[1,2]

Daniel Kraft, the founder and chair of *Exponential Medicine*, authored this commentary to provide a high-level and futuristic perspective of AI as the central part of a convergence of advanced technologies to transform health care.

[1]*Medicine, Singularity University, Santa Clara, CA, United States*
[2]*Exponential Medicine, Singularity University, Santa Clara, CA, United States*

As technology continues to advance, accelerate, and converge, massive new sources of data ranging from wearable devices, to personal genomics, to the information contained in our electronic medical records (EMRs) have manifested, including a wide array of "real-world" data increasingly sourced beyond the traditional four walls of the clinic or hospital bed. How and where we obtain, parse, and utilize this data when paired with the increasing capabilities of AI and machine learning have the potential to dramatically shift the practice of medicine—from one that is fundamentally a reactive "sick care" system, based on intermittent data historically only collected in the clinical environment, to one that is continuous, proactive, personalized, information-rich, and increasingly crowd-sourced and truly "healthcare" focused [1].

The first widely adopted consumer wearables only came to market in 2009 with the launch of FitBit, and 23andMe pioneered consumer genomics in 2007, democratizing access to genetic and wearable information and with the launch of Apple App Store in 2008 a rapidly growing app milieu, including 1000s of health-related apps were developed.

We have in the decade since seen an exponential growth of medical and health-related data, from ever higher resolution imaging platforms to emerging personal data sources, ranging from steps/activity and sleep data to consumer genome and microbiome sequencing, to data from Internet of Things, cameras, social media feeds, to now ECGs and blood pressure cuffs embedded in our clothing, smartwatches, and more.

The potential of our soon to be continuous data streams of our digital exhaust (coined the "digitome" (Fig. 1A) is enabling the measure of almost every component of physiology and behavior, brings the potential for a continuous, personalized, precise, and a proactive form of healthcare, moving to precision wellness … to diagnose/detect disease at earlier stages as well as help optimize and personalize therapy via iterative feedback loops.

> The Digitome: the summation of digital data gathered to capture an individual's current state of health, which might include genetic data, physiological parameters, medication status, diet, and lifestyle behaviors [1,2].

The "Quantified Self" movement has emerged, as many individuals track (and sometimes share) their personal data [3]. The opportunity has arrived to move beyond the individual Quantified Self, in which individuals recording, and analyze various aspects of their lives with data usually silo'd

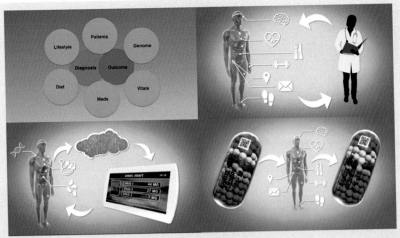

FIGURE 1 (A) The "digitome" includes an expanding array of digital data impacting health and disease; (B) flow of patient data and insights from patient to clinicians; (C) utilizing multiple data forms and AI to optimize personalized therapy, dosing, and combinations; and (D) feedback loop optimization and adjustment of therapy from leveraging various data streams.

in their own possession, to an era "Quantified Health," in which these various data streams can connect to the clinician and clinical care endeavor to "Quantified Health" enabling (1) improved individualized prevention based on objective measures, (2) earlier diagnosis leveraging algorithms to detect signs of problems at earlier stage, and (3) a more data and feedback-driven therapy utilizing everything from traditional drugs to digital interventions (Fig. 1B).

What is the clinician, already overwhelmed, to do with this wealth of new data (streaming real time, to historical), and ideally how might they leverage it to form actionable information that can be utilized across the healthcare continuum (wellness/prevention, diagnosis, and therapy to public health and clinical trials). Doctors "don't know what to do with data from wearables," nor is it often synthesized as meaningful, useful information integrated into the workflow of most medical record systems.

Recently, as application programing interfaces between various devices (both consumer and FDA grade) and massive consumer players from Apple, Samsung, Amazon, and Google move into healthcare, with platforms such as HealthKit, patient data (and the insights derived) are increasingly able to move from an individual's connected devices: blood pressure cuff, scale, glucometer, and other sources, via Bluetooth through their smartphone and (with appropriate permissions given) flow into the EMR of their provider [4] (Fig. 1B).

The utility of information from wearables is still at early stages, but showing promise in several trials [5].

Project Baseline [6] and the NIH's "All of Us" Trial [7] seek to collect, correlate, and glean actionable insights from crowd-sourced data and a diverse population of "data donors" to accelerate research and improve health outcomes. An early example of integrating large clinical data sets, guidelines, and other information to guide clinical care is Stanford Medicine's "Green Button" platform [8]. Given a specific case, Green Button provides a report summarizing similar patients in Stanford's clinical data warehouse, the common treatment choices made, and the observed outcomes.

In the future, based on the learning of these and similar trials, and with AI blended with decision support and effective user interfaces into clinical workflow, we can envision care models in which the provider can integrate and glean evidence-based, patient-specific guidance, which can lead to recommendations on both optimized prevention regimens and selection of therapy (Fig. 1C). For example, in common conditions such as hypercholesteremia and hypertension, the blending of patient data, guidelines, pharmacogenomics, and real-time measurement should help the clinician select an optimized and truly personalized set of medications that match the patient. As more individual and population-based data are analyzed and integrated, the promise of the "digital twin" will emerge in which each individual health and disease can be modeled and more individualized interventions prescribed [9]. As real-world clinical, behavioral, symptom, and lab data become more fluid, the therapies prescribed can become more integrated, from "personalized polypills" containing multiple medications, dosed, and combined to match the individual, which can be rapidly modified to adapt to measured values [10].

References

[1] Kraft D. 12 innovations that will revolutionize the future of medicine. Natl Geographic January 2019.
[2] Longmire M. Medable <http://MedableInc.com>.
[3] Fawcett T. Mining the quantified self: personal knowledge discovery as a challenge for data science. Big Data 2015;3(4):249−66.
[4] Apple reveals 39 hospitals to launch Apple Health Records. Healthcare IT News March 29, 2018. <https://www.healthcareitnews.com/news/apple-reveals-39-hospitals-launch-apple-health-records> and <https://www.apple.com/healthcare/health-records/>.
[5] Burnham JP, Lu C, Yaeger LH, Bailey TC, Kollef MH. Using wearable technology to predict health outcomes: a literature review. J Am Med Inform Assoc 2018;25(9):1221−7.
[6] Project Baseline. <https://www.projectbaseline.com/>.
[7] The NIH All of Us Trial. <https://allofus.nih.gov/>.
[8] Longhurst CA, Harrington RA, Shah NH. A 'green button' for using aggregate patient data at the point of care. Health Aff (Millwood) 2014;33(7):1229−35.
[9] Thotathil S. Digital twins: the future of healthcare delivery and improved patient experience. Beckers Hospital Rev March 6, 2019.
[10] <http://IntelliMedicine.com>.

Common sense advice to advance artificial intelligence (AI) in medicine: anecdotes from a layman

Charlie Moloney

Charlie Moloney a gifted journalist who is editor of the academic magazine AIMed, authored this commentary as someone who has had the opportunity to interview hundreds of stakeholders in the AI in medicine domain, and reminds us that we need to always put patients and families first.

The engagement of laypeople is essential to provide a diverse set of viewpoints to help keep the AI in medicine space robust and reduce the possibility of technology being designed that does not meet the needs of laypeople, many of whom are patients.

It has been well documented that homogenous teams of developers who do not actively seek out the views of people unlike themselves often design fundamentally flawed solutions [1].

I, a *bona fide* layman, have gained insights on laypeople in the AI medicine space from developing an academic journal on medical AI since 2017, conducting countless interviews, editing articles by world-renowned experts in this field, and attending some of the biggest industry events.

A key insight to share is to be careful in making assumptions about a person's ability to contribute to a solution you are building.

In an inspiring interview with Adriana Mallozzi, the CEO of Puffin Innovations, and a woman who was diagnosed with cerebral palsy as an infant, I learned about her work trying to educate clinicians about how patients experience consultations [2].

She described intervening on a patient's behalf when a therapist had recklessly prescribed him a wheelchair that did not fit his needs and was causing him discomfort. On investigating, Adriana discovered the therapist had not involved her patient in the decision-making process on the basis that he did not have sufficient mental capacity, which had led to him receiving the wrong treatment.

This clearly demonstrates the importance of patient engagement and why saying a patient did not have the ability to understand the treatment you are offering is no excuse to exclude them from the decision-making process.

Some people will argue that visionary entrepreneurs and coder geniuses can solve the big problems in healthcare from within the confines of Silicon Valley, but they are wrong. The utopianism of technology experts is more often than not extremely misplaced [3].

A common complaint encountered when conducting interviews with technology experts in this space is that their field is poorly understood and often solutions that are not "true" AI are being sent to market masquerading as just that.

And indeed, many vendors will brazenly tell you at AI conferences (if they recognize you are not a buyer) that they are utilizing the buzzword "AI" for marketing purposes.

But while the frustration these experts feel about the muddying of the waters by fake AI companies and spokespeople is genuine, they rarely perceive the onus is on them and not the confused members of the public to clear things up.

Most of the people you come across outside of the medical AI niche are laypeople who have no understanding of medical AI.

It is unlikely that a layperson will ever gain any detailed understanding of medical AI or learn to code.

Even if a layperson leaves an AI conference with a fistful of vouchers for a free starter course on programming neural networks, or a handwritten list of recommended reading material gifted to them by a well-meaning evangelist during the lunch break, in reality, laypeople have other interests and pursuits that take up their time.

A radiologist is unlikely to learn how to code, a business executive on the board of a hospital will not have read the most up-to-date journals on conceptual algorithms, and a patient who works a 9–5 job and has a family to feed cannot be expected to learn about AI at even a basic level.

Ultimately, the technology experts in the AI medicine space will have to acknowledge that it is incumbent upon them to reach out to the layman and explain to him what medical AI is all about.

Perhaps, a way to do this is by cutting down on the confusing jargon. It may be a step backward, for example, to change the term "artificial intelligence" and promote more synonyms such as "intelligently artificial" or "augmented intelligence" [4].

As a journalist covering AI medicine, I owe a lot to my mentors in the AIMed space who helped me to see that the important topics in healthcare AI are things such as quality of life, safety, ethics, diversity, and inevitably cost.

Once you can pinpoint those common themes, it allows you to find the signal amid the noise and ask the right questions that will allow you to extract the interesting story from the innovative start-up.

Similarly, if we continue to simplify the dialogue around medical AI in the public sphere, patients will be better able to ask the right questions of their doctors about how their patient history is being curated by algorithms, or why a certain routine procedure is now being automated, and what the benefits are [5].

References

[1] AIMed. <http://ai-med.io/ai-biases-ada-health-diversity-women/>.
[2] AIMed Magazine issue 05. <www.ai-med.io/magazine>.
[3] Frick W. The other digital divide. Harvard Bus Rev May 2017.
[4] Koulopoulos T. It's time to stop calling it artificial intelligence. Inc. May 2018. <https://www.inc.com/thomas-koulopoulos/its-time-to-stop-calling-it-artificial-intelligence.html>.
[5] AIMed Magazine issue 06. <www.ai-med.io/magazine>.

References

[1] Rosenblatt F. The perceptron: a probabilistic model for information storage and organization in the brain. Psychol Rev 1958;65(6):386−408.
[2] Beam AL, Kohane IS. Big data and machine learning in health care. JAMA 2018;319(13):1317−18.
[3] Rajkomar A, Dean J, Kohane I. Machine learning in medicine. N Eng J Med 2019;380:1347−58.
[4] Collins GS, Moons KG. Reporting of artificial intelligence prediction models. Lancet 2019;393(10181):1577−9.
[5] Faust K, Bala S, van Ommeren R, et al. Intelligent feature engineering and ontological mapping of brain tumour histomorphologies by deep learning. Nat Mach Intell 2019;1:316−19.

Basic Concepts of Artificial Intelligence

We tend to overestimate the effect of a technology in the short run and underestimate the effect in the long run. Roy Amara, Cofounder of the Institute for the Future

Definitions

The word "intelligence" is derived from the Latin root *legere*, which means "to collect, to gather, and to assemble" and its congener *intellegere* means "to know, to understand, to perceive, and to choose among." Intelligence can be defined as the ability to learn or understand, to deal with new situations, or to apply knowledge and skills to manipulate one's environment.

Intelligence is usually bundled with an interesting list of words such as data, information, knowledge, intelligence, and wisdom; these words form an information hierarchy but are often misunderstood (see Fig. 1.1). Data is the foundational layer of signals and facts that have little or no meaning without context. Information, then, is data in a more structured as well as more meaningful context and is often much better organized. If data are atoms of information, then information could be considered a molecule. When information becomes more contextual, this becomes knowledge. Knowledge can be explicit or tacit and involves understanding patterns and is also used to achieve goals. Intelligence is the ability to acquire and apply knowledge to achieve goals. Wisdom is an understanding of principles derived from intelligence and has embedded within it values and beliefs with self-reflection and futuristic vision. The difference between intelligence and wisdom is that the latter is informed decision powered by good intelligence using values and ethics, so it is more difficult to attain. There is a continuum from data to intelligence and with good intelligence, one can have wisdom; in health care, there should eventually be a bidirectional continuity from wisdom and intelligence directing how data, information, and knowledge be gathered, stored, and shared.

These definitions have interesting implications for artificial intelligence (AI). Perhaps the best definition of AI is the one conjured by the American cognitive scientist Marvin

Intelligence-Based Medicine
DOI: https://doi.org/10.1016/B978-0-12-816462-4.00001-7

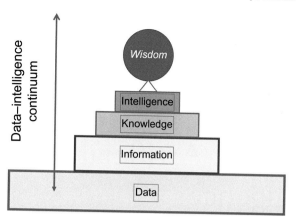

FIGURE 1.1 The data–intelligence continuum. The data to intelligence continuum can continue on to wisdom at the top of the hierarchy (not touching the rest of the continuum as it is not always a continuum). This continuum, from lighter to darker shade, should be bidirectional especially in health care.

Minsky: the science of making machines do things that would require intelligence if done by human. In a way, there is really nothing "artificial" about AI as humans are the progenitors of this discipline and any work, even if autonomous by machines, still has roots in earlier work by humans.

Types of Artificial Intelligence

AI can be categorized as weak or strong: weak (also termed "specific" or "narrow") AI pertains to AI technologies that are capable of performing specific tasks (such as playing chess or *Jeopardy!*) and strong (also termed "broad" or "general") AI is much more difficult to attain, it is also called artificial general intelligence (or AGI) or general AI (see Fig. 1.2). AGI relates to machines that are capable of performing intellectual tasks that involve human elements of senses and reason. The public's inaccurate perception of AI, however, continues to be that of the menacing robots that threaten mankind (such as HAL in *2001: A Space Odyssey* or the Terminator). Recently, this perception is modified to that of the more sophisticated and complex AI-inspired but still anthropomorphic robots or cyborgs seen in movies such as *Her* (2013) and *Ex Machina* (2015). The Swedish philosopher Nick Bostrom, in his enlightening book, cautioned the advent of a superintelligence that is essentially an intelligent agent that is superior to humans in intelligence (an intellect that is much smarter than the best human brains in practically every field, including scientific creativity, general wisdom, and social skills) [1]. The futurist Ray Kurzweil similarly described a technological singularity, a phenomenon in which the exponential increase in machine intelligence will supersede the human intelligence near the year 2045 [2]. In short, these AI intellects are concomitantly optimistic and cautious about the evolution of AI in the coming decades.

Artificial Intelligence and Data Science

Machine learning (ML) [and its more robust and specific type, deep learning (DL)] is not synonymous with AI but is often used interchangeably; ML as well as DL is AI methodology (see Fig. 1.3). AI, however, does overlap with data science and mathematics with statistics. Within data science are data analytics and data mining (in addition to some crossover to ML and AI). Data analytics is the discipline that starts with a hypothesis and then utilizes advanced algorithms on the dataset to answer the query; it is different than data science in that it deals more with descriptive and correlation-type predictive analytics than data science (which is more focused on causality-type predictive analytics as well as

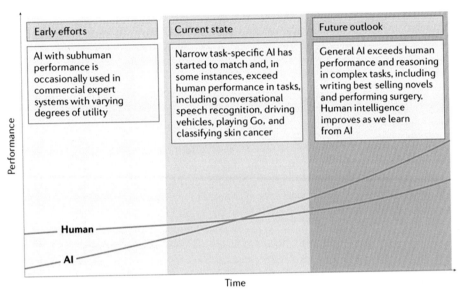

FIGURE 1.2 AI versus human performance. Early efforts involved expert systems, and AI did not perform at human level. In the current state, AI has reached human performance levels and in the future, general AI will exceed human performance level. *AI*, Artificial intelligence.

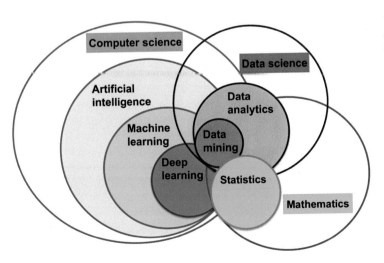

FIGURE 1.3 AI and data science. The circles show the spheres of computer science and AI, data science, and mathematics. Data science is at the intersection of computer science and mathematics. Deep learning and machine learning are within the domain of AI, whereas data analytics and data mining are inside the data science realm. *AI*, Artificial intelligence.

prescriptive analytics and ML). Data mining is the subdiscipline that discovers relationships or patterns from datasets to potentially engender questions and hypotheses. In short, a data scientist in the present era is expected to be an all-around data miner, data analyst, mathematician, and statistician as well as someone who is facile with AI (including machine and DL).

The Role of Mathematics in Artificial Intelligence in Health Care

Randall Moorman, Professor of Medicine and Biomedical Engineering, University of Virginia, VA, United States

Randall Moorman, a cardiologist as well as a cardiovascular engineer with a mathematics background, authored this commentary on the importance of clinicians learning the basics of mathematics in the algorithms of AI methodologies.

> There is also a rhythm and a pattern between the phenomena of nature which is not apparent to the eye, but only to the eye of analysis ... *RP Feynman, The Character of Physical Law, p 13.*

Twenty years ago, my coworkers and I set out to provide early detection of sepsis in premature infants based on continuous time series data from the bedside electrocardiogram (EKG) monitor. In hours and hours of looking at heart rate time series, we found a robust phenomenon—abnormal heart rate characteristics of reduced variability and transient decelerations—in the hours prior to clinical suspicion of illness [1]. We also found that this particular constellation of time series findings is resistant to detection by the then-canonical heart rate variability tools that are based on time- and frequency-domain analyses. Thus our decision was to either devise new and relevant applications of mathematics or to stop.

Then, there was no such thing as AI, or even ML (let alone DL), Big Data, or data science. So, while today we might hand the whole thing over to a computer to sort out, back then it was not an option. Instead, we devised a set of mathematical tools to quantify the degree of abnormal heart rate characteristics of reduced variability and transient decelerations. (One was sample entropy, which has gone on to a life of its own [2].) Some years of clinical study later, we demonstrated that the display of a risk estimate based on these mathematical time series analytics saved lives [3]. From this exercise, we emphatically learned the value of mathematics in the care of the individual patient.

Now we know about AI, and the promise that it will do all the work for us. One simply presents all the data to the computer, and it finds all the relationships, the obvious ones you knew about and all the ones you never dreamed of. It sounds too good to be true, and I am not saying it isn't. But I wonder—would an AI approach to neonatal sepsis detection have led to the same clinical tool and the same clinical benefit? Would AI have developed generally useful metrics such as sample entropy?

More recently, our colleagues at William and Mary developed elegant time-warping and wavelet transform-based methods for recognizing the major disorders of breathing in premature infants, neonatal apnea and periodic breathing [4]. The result—a quantitative breathing record for use in research and clinical care—may well change how doctors take care of babies [5]. Again, it is fair to ask whether AI approaches would yield the same results.

These are experiments we plan to perform. As we prepare, I read accounts of AI methodologies, and I am sometimes struck by the blindness of faith that the writers place in the algorithms. Thus a tutorial on, say, convolutional neural networks (CNN) might be confined to a qualitative description and some lines of code that call on library routines. This is very unsatisfactory. Had we approached time series analysis in the neonatal intensive care unit that way, I doubt we would have made any progress.

I suggest that the new generation might profit from the old in maximizing the good of AI and in minimizing the bad. I have some pointers.

1. If you are going to use AI, you need to understand every mathematical operation in the algorithms and where they came from. You may be daunted by the depth of what you need to know, or cheered by the age and solidity of the foundations. Suffice it to say that matrix algebra, the fundamental theorem of calculus, probability theory and random variables, and entropy estimation are all key—even if they all seem to have different names when used in AI. If you do not have an excellent working knowledge of the underlying mathematics of AI, you will never be as good at it as those who do.

2. If you know of useful features in, say, time series data, you should calculate them in advance and give them to the computer along with the raw data. That is to say, if you can tell the difference between two datasets with your eyes, nothing could be more valuable than to devise or adapt mathematical methods to quantify that difference.

Like any new science, AI has its skeptics. As a result, its practitioners bear responsibility for elevating the field for the future. In my view, this requires a bottom-up understanding of the relevant mathematics, both of the AI algorithms and the data features.

References

[1] Griffin MP, Moorman JR. Toward the early diagnosis of neonatal sepsis and sepsis-like illness using novel heart rate analysis. Pediatrics 2001;107:97–104.

[2] Richman JS, Moorman JR. Physiological time series analysis using approximate entropy and sample entropy. Am J Physiol 2000;278:H2039–49.

[3] Moorman JR, Carlo WA, Kattwinkel J, Schelonka RL, Porcelli PJ, Navarrete CT, et al. Mortality reduction by heart rate characteristic monitoring in very low birth weight neonates: a randomized trial. J Pediatr 2011;159:900–6 PMID 21864846.

[4] Lee H, Rusin CG, Lake DE, Clark MT, Guin LE, Smoot TJ, et al. A new algorithm for detecting central apnea in neonates. Physiol Meas 2012;33:1–17 PMID: 22156193.

[5] Dennery PA, DiFiore JM, Ambalavanan N, Bancalari E, Carroll JL, Claure N, et al. Pre-Vent: the prematurity related ventilatory control study. Pediatr Res 2019;. Available from: https://doi.org/10.1038/s41390-019-0317-8.

Other AI methodologies include cognitive computing and natural language processing (NLP) as well as computer vision, robotics, and autonomous systems (see Fig. 1.4). Cognitive computing (as exemplified by IBM's Watson cognitive computing platform) can involve a myriad of AI tools that simulates human thinking processes, while NLP involves connecting human language with computer programmed processing, understanding, and generation. Computer vision will not be separately discussed other than under DL and CNN. Robotics in its impressive panoply of forms is considered part of AI as well as its

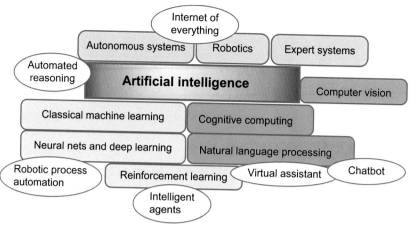

FIGURE 1.4   AI. The many domains of AI are shown with major ones shown in color. The figure shows one of many ways AI and its portfolio can be illustrated and is not at all meant to be inclusive of all the tools and methodologies. In addition, most of these domains have overlap and are not mutually exclusive. For example, cognitive computing is a portfolio of tools that includes natural language processing and machine learning. *AI*, Artificial intelligence.

related autonomous systems [in the context of AI not information technology (IT)]. It is perhaps reasonable to think about AI as a "symphony" of musical instruments, and you being the composer and/or conductor can put various music instruments together to realize the music you composed and envisioned. For example, there are many AI tools that are combinations of these elements, such as ML and NLP combining for robotic process automation or chatbots, or NLP aligning with ML for cognitive computing. All of these aforementioned AI methodologies will be covered in more detail later in the book.

The Human—Machine Intelligence Continuum

AI can be described in the context of a human—machine intelligence continuum [3] with three types of AI: assisted, augmented, and autonomous (see Table 1.1). The table illustrates these three types of AI as well as examples in the real and medical worlds. Assisted intelligence is when the machines perform tasks that are automated as the tasks

TABLE 1.1   The human-machine intelligence continuum.

Type of intelligence	Definition	Level of human involvement	Example	Example in health care
Assisted	System providing and automating repetitive tasks	Little or none	Industrial robots	UR robots for blood work (Copenhagen Hospital)
Augmented	Humans and machines collaboratively make decisions	Some or high	Business analytics	Watson for Oncology (Memorial Sloan Kettering)
Autonomous	Decisions made by adaptive intelligent systems autonomously	Little or none	Autonomous vehicle	IDx-DR for retinal images (University of Iowa)

UR, Universal Robots.

do not change and do not require an interaction with humans (such as certain automation tasks in a factory or the robot vacuum cleaner that are now ubiquitous). Augmented intelligence, on the other hand, implies that there is an active and ongoing interaction between the human and machine so that both humans and machines are informed and learned (such as ML). Some like to use this term in the context of avoiding the term "artificial" all together in hopes to raise the level of acceptance of AI in medicine and health care. Autonomous intelligence, as exemplified by the autonomously driven vehicle, involves an automated decision-making by the machine with the ML continuously.

There is an increasing number of examples of all types of intelligence, but the number of autonomously intelligent machines (such as autonomous vehicles and drones) has risen substantially in the last few years. The autonomous intelligence has already started in biomedicine: the recent the Food and Drug Administration approval of the first autonomously functioning diagnostic tool, a DL screening tool for diabetic retinopathy, does not require a physician's input and can return one of two results: more than mild diabetic retinopathy (referral to eye care professional) or negative for more than mild diabetic retinopathy (repeat screening in 12 months) [4].

The Analytics Continuum

In addition to the man—machine intelligence continuum, there is also an AI-inspired analytics continuum (descriptive, diagnostic, predictive, prescriptive, and cognitive analytics) that increases in intelligence and autonomous behavior from a data science perspective (see Table 1.2). Descriptive analytics (traditional business intelligence) is commonplace even in health care and uses well-established statistical methodologies and software packages for fulfilling mostly a reporting function. Methodologies used here include data visualization and data mining. Diagnostic analytics creates more value but is more difficult to achieve. Methodologies used here include queries and root cause analysis. Predictive analytics is less

TABLE 1.2 Analytic continuum.

Type of analytic	Focus	Tools	Question
Descriptive	Reporting	Statistical software	What happened?
		Data visualization	
Diagnostic	Insight	Statistical software	Why did it happen?
		Data visualization	
Predictive	Forecasting	Statistical models	What will happen?
		Predictive modeling	
Prescriptive	Optimization	Predictive modeling	What should I do?
		Machine learning	
Cognitive	Intelligence	Reinforcement learning	What is the best that could happen?
		Cognitive computing	

common than the two former types of analytics but provides insights by detecting patterns in data with the use of statistical methods (such as classification, regression, and clustering) that usually falls just short of ML. Prescriptive analytics is an even higher level of analytics that optimize human decision-making by prescribing recommendations with the utilization of machine and DL. This area is usually covered by data scientists more than data analysts. Cognitive analytics is the highest level of analytics (and therefore by far the most difficult to achieve) that is present when a project or an enterprise deploys AI methodologies (such as reinforcement learning, DL, and cognitive computing) to achieve a human-like cognition characterized by self-learning behavior with intelligence.

AI passion: clinical prescriptive analytics

John Frownfelter

John Frownfelter, a physician with an interest in prescriptive analytics, authored this commentary about the interesting entity called Eigen-space mapping, which can be associated with clinical vectors designed for risk prediction and intervention.

Health care, as it is delivered today in the United States, is unsustainable. While the costs continue to rise beyond GDP rates, quality languishes at the lower deciles for developed countries. This is now established with data, expert opinion, and is accepted as common knowledge. According to PricewaterhouseCoopers' Health Research Institute, $312B is wasted in clinical treatment that is either ineffective or not implemented. This waste is attributed to unnecessary procedures, preventable hospital readmissions, poorly managed chronic conditions, avoidable emergency room visits, hospital-acquired infections, treatment variation, and medical errors (http://www.oss.net/dynamaster/file_archive/080509/59f26a38c114f2295757bb6be522128a/The%20Price%20of%20Excess%20-%20Identifying%20Waste%20in%20Healthcare%20Spending%20-%20PWC.pdf). Tremendous energy is expended on both identifying root causes (and pointing fingers) and possible solutions. There will be no single answer to this complex challenge, but AI is destined to have a pivotal role.

For all the ways that AI may bring value in health care, I want to address one of the most difficult—the ability to deliver precise care that is particular to the individual patient is a goal of current health-care initiatives, but the bar is set far too low with the existing paradigm. To manage a patient's current conditions holistically, in a personalized way, that prevents gaps in care and avoidable harm is noble, but it will never be accomplished without the power of applied AI. Imagine a cancer patient, identified as at risk for mortality in the next 30 days, and the resulting interventions headed off the developing sepsis that would have rapidly led to her demise. What would the application of AI in this scenario even look like?

Let us consider how health care could change if we knew the events that would happen to a patient before they occurred and, furthermore, knew the specific interventions that would alter their natural history. This would transform health care from being reactionary to anticipatory, from pathway driven to data driven, and from a "checklist" approach to precise care. The power to identify risk and prevent adverse outcomes is within our grasp; imagine five diabetic patients who appear well today but are destined to be admitted to the hospital in the next 30 days. AI-driven prescriptive care will ensure that they are each individually managed with unique interventions that mitigate risk. The result: two, three, or all five are protected from that hospitalization.

Is this science fiction, or can AI actually help with this scenario? We must first accept that typical predictive modeling is limited in value and worse, which often contributes to the greater problems of increasing resources to manage and prevent complications. Predictive modeling does one thing relatively well: risk stratification allows us to identify patients at higher risk for a negative outcome—and this is better than not knowing the risk. The tradeoffs are that this approach neither identifies impactable patients nor does it offer interventions that are patient specific. Furthermore, utilization tends to increase as resources are deployed broadly toward the high-risk population.

There is a better way. In the Health Care Information and Management Systems Society (HIMSS) Analytics Adoption Model, the highest level of maturity is Stage 7: "Prescriptive Analytics" (https://www.himssanalytics.org/amam). While those levels (and analytics maturity) are important, the application of novel AI approaches allows an organization to leapfrog to Prescriptive Analytics without laying the foundation of a multiyear journey and the previous six stages. An example of one solution is through the use of spectral analysis and Eigen-space mapping, which is now being applied to patient care; similar to search engines, online purchasing, and social media, this approach allows the assimilation of diverse datasets without introducing bias that is typical in more traditional methods. This bottom-up approach using all available data reveals patients at risk that drop out of typical predictive models.

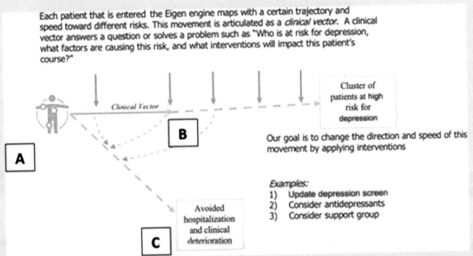

The combination of (A) Eigen-space mapping with AI techniques identifies (B) specific interventions that will impact individual patients and change their trajectory from a negative outcome to a (C) better one. The AI provides the ability to (1) identify impactable patients and (2) stack-rank interventions in order of potential benefit to the patient.

Through the use of these powerful techniques, there are already examples of dramatic reductions in complications such as falls, pressure injury, and even sepsis in the hospital setting. In the ambulatory setting, the application of this technology helps to reduce avoidable emergency department visits and hospitalizations and even anticipate and prevent comorbidities such as depression and opioid dependence. As predictive modeling ages out of style, clinical prescriptive analytics will usher in a new era that will drive health care closer than ever before to achieve the triple aim.

TABLE 1.3   Health Care Information and Management Systems Society adoption model for analytics maturity.

Stage	
7	Personalized medicine and prescriptive analytics
6	Clinical risk intervention and predictive analytics
5	Enhancing quality of care, population health, and understanding the economics of care
4	Measuring and managing evidence based care, care variability, and waste reduction
3	Efficient, consistent internal and external report production and agility
2	Core data warehouse workout: centralized database with an analytics competency center
1	Foundation building: data aggregation and initial data governance
0	Fragmented point solutions

In addition, HIMSS has promoted the use of an adoption model for analytics maturity cumulative capabilities, not to be confused with another seven-stage scoring system, electronic medical record adoption model, mentioned later (see Table 1.3).

Better health for everyone, everywhere with artificial intelligence (AI)

Steve Wretling and Shelley Price
HIMSS, Chicago, IL, United States
Steve Wretling and Shelley Price, both with a technology background and in leadership positions at HIMSS, authored this commentary on the HIMSS perspective of how a disruptive technology like AI can become a transformative tool for the global health ecosystem.

Like most disruptive innovations, AI can be deployed in a variety of ways. It also can be viewed as a positive or negative development. That is no surprise. In health care, we tend to (or should) view a change to the existing business/care model as potentially deleterious unless there is real return on investment (ROI) and it satisfies our mandate for safety, consistency, and outcomes. Even then, we have to ask: Is the risk-reward proposition for a new disruption worth it?

We are already witnessing that AI's potential to deliver better care is considerable. For example, as HIMSS's Chief Clinical Officer Dr. Charles Alessi has remarked [1], a mechanized approach to radiology, as long as it is as effective as current practice, offers significant advantage to the workforce as well as to the patient. The advantage to the patient is enormous because it improves access and convenience by eliminating the need to deploy clinicians on a 24-hour basis for routine radiology. In addition, it reduces the need for expensive schemes such as using offsite (and often remote) radiologists, which leads to savings in resourcing and reducing clinical risk in terms of governance.

In addition, as health-care workers become more and more stretched, AI holds out the promise to free staff from mundane and simple tasks, allowing them to further focus on the patient and more satisfying, complex, and dynamic work.

HIMSS INSIGHTS Editorial Director Philipp Grätzel von Grätz observed [2] that advancing AI in health care requires practical intelligence on the side of the innovators. This means identifying gaps instead of mimicking what already works reasonably well. Doctors know how to plan their tumor therapies, but they need assistance in identifying cancer patients who will benefit from immunotherapy. That is where AI can help.

Beyond advancing decision support, augmenting physician value, honing predictive analytics, and boosting population health management, AI can also help health care improve its *business and operations bottom lines*.

During and after the HIMSS 2018 Big Data and Healthcare Analytics Forum, *Healthcare IT News* Editor Mike Miliard explored [3] nonclinical AI uses with Sam Hanna, program director in health-care management at American University. In financial and operational areas—human resources, talent management, revenue cycle [4], etc.—AI and machine learning (ML) might be best positioned to deliver immediate ROI and tangible gains for hospitals and health systems. Hanna observed "(w)e always talk about the patient and the clinician. They are critical components, obviously, of our healthcare ecosystem. But patients and clinicians are supported by a number of professions: administrators, logistics people, IT people, finance people, HR people and many others."

For example, he said, AI can be used to sift through job applicants to determine those that best match the job description at hand. Another example is in strategic business planning, he said. Using AI in financial predictive models can help an organization understand the repercussions of their decisions and learn how to make better ones.

Given the potential ROI that AI can bring to bear both clinically and to the bottom line, health-care delivery systems are crafting strategies for AI, ML, and other emerging technologies. We explored this question in two 2018 HIMSS Media surveys.

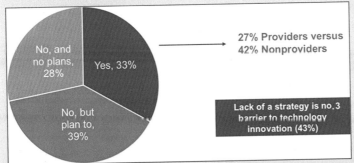

GRAPH 1.1 Are you crafting a strategy for emerging technologies, including AI and ML? *AI*, Artificial intelligence; *ML*, machine learning. *Source: Technology Innovation in Healthcare Survey, HIMSS Media.*

Even though 76% of the 142 surveyed professionals working in IT, business, informatics and clinical roles within hospitals, health insurers, and pharmaceutical companies reported a lack of clear understanding of how to best leverage AI/ML, 33% are strategizing for emerging technology, while 39% are planning to create strategies soon [5]. Meanwhile, 66% of the 180 surveyed professionals in another survey said they expect AI/ML to drive innovation in health care [6].

Though AI is still in its infancy, our findings, taken together with 2017 McKinsey research indicating [7] that 20% of respondents across multiple industries were using one of more AL technologies at scale, suggest AI as a disruptor is more than a "shiny new object." It also has the potential to transform business processes for the better. Among McKinsey's advice for organizations looking to take those tentative first steps: "Believe the hype that AI can potentially boost your top and bottom line."

A few concluding thoughts are as follows:

- Given the organizational, human, financial, and operational resources required to incorporate AI in clinical and business intelligence, will this disruption, at least in the short term, exacerbate existing health-care gaps in care, quality, and costs? This is a critical question as health care expands the use of this disruptive technology.
- Adoption of disruptive innovations such as AI also relies on there being a justified true belief from the user (e.g., the provider) and the recipient (e.g., the patient) of that knowledge. By its nature, AI is a particularly dark black box to many consumers. There will be a need for understandable health-care AI decision/ prediction explanation tools to provide providers with clinical decision support transparency rationale that can be shared with patients and reassure parties that the AI is accurate and justifiable.
- Countries from Finland to Germany are investing in national AI strategies, although most are lean on health-care applications [8]. Should there be larger strategies for more broadly applicable use of AI?

AI is a disruptor to reform the global health ecosystem by leveraging the power of information and technology. As we take these nascent steps with AI, beyond innovating at the level of a population or individual care settings, we must strive to leverage its potential to realize the full health of every human, everywhere.

Steve Wretling is Chief Technology and Innovation Officer of HIMSS. Wretling strategically guides enterprise-wide initiatives around digital innovations and technology.

Part of the Thought Advisory Team, Shelley Price leads the organization's efforts around payer, life sciences, and data and analytics—including population health, clinical and business intelligence, precision health, AI/ML—as a subject matter expert.

HIMSS is a global advisor and thought leader supporting the transformation of health through the application of information and technology. As a mission driven nonprofit, HIMSS provides thought leadership, community building, public policy, professional/workforce development, and engaging events to bring forward the voice of our members. Headquartered in Chicago, Illinois, HIMSS serves the global health information and technology communities with focused operations across North America, Europe, United Kingdom, Middle East, and Asia Pacific.

References

[1] Alessi C. AI and the physician — a blessing or a curse? In: Gratzel van Gratz P, editor. HIMSS insights 7.2: artificial intelligence. Chicago, IL: HIMSS Media; 2018. p. 31−3 [cited 21.05.19]. Available from: <https://pages.healthcareitnews.com/HIMSSInsights2.html>.

[2] Grätzel von Grätz P. The case for clinical intelligence. In: Gratzel van Gratz P, editor. HIMSS insights 7.2: artificial intelligence. Chicago IL: HIMSS Media; 2018. p. 2 [cited 21.05.19]. Available from: https://pages.healthcareitnews.com/HIMSSInsights2.html.

[3] Millard M. Making a persuasive business case for bigger AI investment. HITN [Internet] 2018; [cited 21.05.19]. Available from: <https://www.healthcareitnews.com/news/making-persuasive-business-case-bigger-ai-investment>.

[4] Sanborn BJ. Why the hospital revenue cycle is practically begging for artificial intelligence and machine learning. Health Care Finance [Internet] 2018; [cited 21.05.19]. Available from: <https://www.healthcarefinancenews.com/news/why-hospital-revenue-cycle-practically-begging-artificial-intelligence-and-machine-learning>.

[5] King J. Artificial intelligence and machine learning in healthcare. Chicago, IL: HIMSS Media; 2018. p. 17.

[6] Sullivan T. Artificial intelligence: 3 charts reveal what hospitals need in the near future. HITN [Internet] 2018; [cited 21.05.19]. Available from: <https://www.healthcareitnews.com/news/artificial-intelligence-3-charts-reveal-what-hospitals-need-near-future>.

[7] Bughin J, Chui M, McCarthy B. How to make AI work for your business. HBR [Internet] 2017; [cited 21.05.19]. Available from: <https://hbr.org/2017/08/a-survey-of-3000-executives-reveals-how-businesses-succeed-with-ai>.

[8] Grätzel von Grätz P. Old world new mission. In: Gratzel van Gratz P, editor. HIMSS insights 7.2: artificial intelligence. Chicago, IL: HIMSS Media; 2018. p. 11−5 [cited 21.05.19]. Available at <https://pages.healthcareitnews.com/HIMSSInsights2.html>.

Artificial Intelligence and the Neurosciences

In Greek mythology, Icarus yearned to fly by mimicking the bird with its trappings but failed; instead, man eventually learned to fly by building planes and spaceships after attaining a deep understanding of the principles of aerodynamics. Similarly, we can approach AI by building machine intelligence after a thorough comprehension of the brain and neuroscience [5]. Hassabis reviewed the past machine−brain interaction in the context of DL and reinforcement learning while projected the present state of mind−machine synergy in other contexts (attention, episodic memory, working memory, and continual learning). The exciting future of AI and neuroscience, in his opinion, resides in the bridging of the human intelligence with machines; areas of future interest are in intuitive understanding of the physical world, efficient learning, transfer learning, imagination and planning, and finally, virtual brain analytics. In short, innovative AI systems can be partly inspired by the brain just as the brain can be augmented by the machine.

Neuroscience and Artificial Intelligence (AI)

Sharief Taraman

Sharief Taraman, a pediatric neurologist with a passion for AI, authored this commentary on the increasingly close relationship between AI and the neurosciences, and the convergence of these two interrelated sciences toward a higher level of understanding of both of these domains.

The Evolution

Unlike other medical disciplines, neuroscience has a unique dynamic with AI. The human mind and our understanding of neuroscience has long inspired and advanced the field of AI and, conversely, advancements in AI have furthered our understanding of the human mind. Some of the earliest foundations of AI were fundamentally based in the neurosciences and many of those early scientists who created the first AI systems had cross-disciplinary expertise in neuroscience, psychology, mathematics, and/or computer science [1–4].

Over the past three quarters of a century, neural networks advanced through the intertwined history of neuroscience and AI as new modalities of supervised learning, unsupervised learning, reinforcement learning, recurrent neural networks, deep recurrent neural networks, CNN, recursive neural networks, and long–short-term memory networks, to name a few, emerged. Neuroscience has afforded a rich source of inspiration for these algorithmic approaches.

Simultaneously, the neurosciences and related medical disciplines such as neurology and psychiatry have readily adopted AI-based tools. Historically, AI was limited to primarily research applications; however, the implementation of AI into clinical health-care settings is steadily increasing with examples in neuroimaging [5], diagnosis and management of neurological disorders such as autism [6], Alzheimer's disease [7], and epilepsy [8], as well as in prognostication [9].

The Augmentation

The most successful use cases of AI in the neuroscience-related medical disciplines are those in which the technology is positioned to augment the clinician or health-care provider rather than replace them. It has been demonstrated repeatedly that a partnership of AI and the clinician outperforms an independent use of an algorithm versus the clinician performing independently. This synergistic partnership exists due to the dichotomous nature of the strengths and weakness of AI and of humans.

In health care specifically, AI is becoming more widely referred to as "augmented intelligence" or "symbiotic intelligence." This alternative conceptualization of AI focuses on the assistive role, emphasizing the ability to enhance humans rather than replace us.

Another emerging area of technology in the neurosciences has been brain–computer interfaces. Research in brain–computer interfaces began in the 1970s. The first neuroprosthetic devices emerged two decades later and the first implanted brain–computer interface in the early 2000s. ML for the development and application of brain–computer interfaces has proven to help solve some of the most complex challenges.

The Convergence

Despite the resistance to the notion that AI can replace human intelligence, we aspire to create an AGI, capable of full cognitive abilities, mimicking that of a human. This aspiration is within reach and is supported by advances in quantum and cognitive computing. Furthermore, to the extent in which the neurosciences and AI continue to cross-pollinate one other, the greater the likelihood that AGI will come to fruition within the lifetime of the author. Even at the level of AI in which AGI is a reality, it will serve as a tool, just as any other human creation. Our greatest advancements in both AI and neuroscience have and will continue to come from the integration of neuroscience and AI [10,11]. The future holds great promise for a new understanding of how the human mind works, unprecedented human cognitive abilities, and neural interfacing technology.

References

[1] McCulloch WS, Pitts W. A logical calculus of the ideas immanent in nervous activity. Bull Math Biophys [Internet] 1943;5:115–33 [cited 26.01.19]. Available from: <http://link.springer.com/10.1007/BF02478259>.

[2] Hebb DO. The organisation of behavior. New York: Wiley; 1949.

[3] Turing AM. I.—Computing machinery and intelligence. Mind [Internet] 1950;LIX:433–60 [cited 26.01.19]. Available from: <https://academic.oup.com/mind/article-lookup/doi/10.1093/mind/LIX.236.433>.

[4] Rosenblatt F. Perceptron simulation experiments. In: Proceedings of the IRE; 1960.

[5] Bagher-Ebadian H, Jafari-Khouzani K, Mitsias PD, Lu M, Soltanian-Zadeh H, Chopp M, et al. Predicting final extent of ischemic infarction using artificial neural network analysis of multi-parametric MRI in patients with stroke. PLoS One [Internet] 2011;6:e22626 [cited 26.01.19]. Available from: <https://doi.org/10.1371/journal.pone.0022626>.

[6] Abbas H, Garberson F, Glover E, Wall DP. Machine learning approach for early detection of autism by combining questionnaire and home video screening. J Am Med Inform Assoc [Internet] 2018;25:1000–7 [cited 19.12.18]. Available from: <http://www.ncbi.nlm.nih.gov/pubmed/29741630>.

[7] Zhang S, McClean SI, Nugent CD, Donnelly MP, Galway L, Scotney BW, et al. A predictive model for assistive technology adoption for people with dementia. IEEE J Biomed Health Inform [Internet] 2014;18:375–83 [cited 26.01.19]. Available from: <http://ieeexplore.ieee.org/document/6527964/>.

[8] Yang C, Deng Z, Choi K-S, Jiang Y, Wang S. Transductive domain adaptive learning for epileptic electroencephalogram recognition. Artif Intell Med [Internet] 2014;62:165–77 [cited 26.01.19]. Available from: <http://www.ncbi.nlm.nih.gov/pubmed/25455561>.

[9] Rughani AI, Dumont TM, Lu Z, Bongard J, Horgan MA, Penar PL, et al. Use of an artificial neural network to predict head injury outcome. J Neurosurg [Internet] 2010;113:585–90 [cited 26.01.19]. Available from: <https://thejns.org/view/journals/j-neurosurg/113/3/article-p585.xml>.

[10] Marblestone AH, Wayne G, Kording KP. Toward an integration of deep learning and neuroscience. Front Comput Neurosci [Internet] 2016;10:94 [cited 26.01.19]. Available from: <http://www.ncbi.nlm.nih.gov/pubmed/27683554>.

[11] Hassabis D, Kumaran D, Summerfield C, Botvinick M. Neuroscience-inspired artificial intelligence. Neuron [Internet] 2017;95:245–58 [cited 09.12.17]. Available from: <http://www.ncbi.nlm.nih.gov/pubmed/28728020>.

The Doctor's Brain and Machine Intelligence

The "doctor's brain" (see Fig. 1.5) for day-to-day clinical work can be conveniently deconstructed by its myriad of functions and matched to machine-equivalent capabilities. A cardiologist, for instance, will need to review medical images in the form of an EKG, echocardiogram, and perhaps a magnetic resonance imaging or CT scan. He/she also needs to think about a patient's case after hearing about the history from the patient and/or family members. Lastly, he/she needs to make complex diagnostic and therapeutic decisions.

The machines can perform some of these functions that clinicians routinely and usually effortlessly accomplish on a day-to-day basis. Image recognition by the retina and visual cortex of the occipital lobe can be done by computer vision with image interpretation done by DL, specifically CNN. Language comprehension and speech by the brain's Broca's and Wernicke's areas of the left hemisphere can be performed aptly by machine's NLP with also its understanding and generation. Thinking about the nuances of a particular clinical case based on existing data and past experiences can be partly represented by ML (although the processes are very different). The doctor's frontal lobe is a part of the brain that helps to make challenging decisions and this process is partly imitated by the computer's capabilities to have a decision support capability via cognitive computing, machine/DL, and reinforcement learning (including deep reinforcement learning).

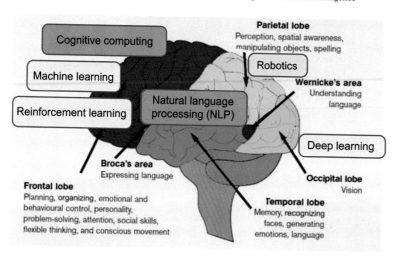

FIGURE 1.5   The doctor's brain. The various parts of the brain are mimicked by different types of machine intelligence. For instance, vision and medical image interpretation can be done by deep learning (and convolutional neural networks) (see text for more details).

References

[1] Bostrom N. Superintelligence: paths, dangers, strategies. London: Oxford University Press; 2014.
[2] Kurzweil R. The singularity is near. New York City: Viking Press; 2005.
[3] Quindazzi M. <bit.ly/2oyHqEw>.
[4] Personal communication with Dr. Michael Abramoff, 2018.
[5] Hassabis D, Kumaran D, Summerfield C, et al. Neuroscience-inspired artificial intelligence. Neuron Rev 2017;95(2):245−58.

2

History of Artificial Intelligence

The history of artificial intelligence (AI) can be traced back to the study of logic delineated by the Greek philosopher Aristotle as he formulated a system of syllogisms for proper reasoning as well as the ancient Greek Antikythera Mechanism, the oldest known analog computer in the world [1]. The statistician Thomas Bayes and his framework for probability and Bayesian inference, the mathematician George Boole and his Boolean algebra, and the polymath Charles Babbage and his early digital programmable computer (the Analytic Engine), all contributed to the underpinnings of present day AI (see Table 2.1).

Key People and Events

Alan Turing and the Turing Machine

It is the brilliant British mathematician and computer scientist Alan Turing, however, who would be considered the absolute progenitor of AI with his pioneering works that included his theory of computation and his work on computing machines [2,3]. His most valuable contribution was his deciphering of the German Enigma machine during the Second World War at Bletchley Park using machine intelligence (portrayed in the recent film The Imitation Game). The eponymous Turing Test is a test of machine AI's ability to pass itself as a human, as judged by humans blinded to the machine or human.

The Dartmouth Conference

In 1956 mathematicians and scientists gathered at the august seminal Dartmouth conference (organized by John McCarthy and others), and it is the proposal for this august gathering that the term "artificial intelligence" was coined by the Stanford computer scientist John McCarthy. This summer conference and its discussions is widely thought to be the birth of AI as an interdisciplinary field. McCarthy was also instrumental in designing the first AI programming language called LISP, which was the precursor to several important concepts such as tree data structure and object-oriented programming.

TABLE 2.1   A Brief History of Artificial Intelligence (AI).

Year	AI events	Key people
384–322 BCE	Syllogism	Aristotle
	Methods of logical argument and analytics	Greek philosopher
250–60 BCE	Antikythera mechanism	
	Oldest computer used to predict astronomical positions	Greek scientists and sailors
1763	Bayesian inference	Thomas Bayes
	Framework for reasoning about the probability of events	English statistician
1854	Logical reasoning	George Boole
	Framework for representation of logic in equations	English mathematician
1837	Analytical engine	Charles Babbage/Ada Lovelace
	First computer with general purpose computation	English mathematician/ programmer
1943	A logical calculus of the ideas immanent in nervous activity	Warren McCulloch and Walter Pitts
	Concept of artificial neurons and logical functions	American neuroscientist and logician
1945	ENIAC	Gladeon Barnes
	First electronic general-purpose computer	Chief of research and engineering
1949	Programming a computer for playing chess	Claude Shannon
	First reference on chess-playing computer program	American mathematician
1950	Computing machinery and intelligence	Alan Turing
	The "imitation game" that became the Turing test	English mathematician
1951	SNARC	Marvin Minsky
	First artificial neural network	American cognitive scientist
1952	First computer checkers program	Arthur Samuel
	Early demonstration of machine learning	American AI researcher
1955	The logical theorists	Allen Newell
	First AI program to mimic human problem solving	American computer scientist
1956	Dartmouth summer research project on AI	John McCarthy
	Term "artificial intelligence" coined and AI seminal event	American computer scientist
1957	The perceptron algorithm and machine	Frank Rosenblatt
	Precursor to neural network and deep learning	American psychologist

(Continued)

TABLE 2.1   (Continued)

Year	AI events	Key people
1958	LISP	John McCarthy
	Programming language for AI research	American computer scientist
1965	ELIZA	Joseph Weizenbaum
	Interactive NLP program with human–machine communication	German American computer scientist
1968	2001: A Space Odyssey	Arthur C. Clarke
	HAL, the sentient computer	English novelist and futurist
1989	Backpropagation algorithm	Yann LeCun (AT&T Bell Labs)
	Application in multilayer neural network	
	Modern AI era	
1997	Deep Blue (IBM)	Garry Kasparov
	Chess-playing program defeating world champion	Russian chess grandmaster
2011	Watson (IBM)	David Ferrucci
	DeepQA project defeating Jeopardy! champions	Principal scientist
2012	ImageNet	Geoff Hinton
	CNN with 650,000 neurons reduced error rate to 15.3%	English-Canadian computer scientist
2016	AlphaGo (DeepMind)	Demis Hassabis
	Reinforcement learning defeating Go champion Lee Sedol	English AI researcher and neuroscientist
2017	AlphaZero (DeepMind)	Demis Hassabis
	Superhuman play in multiple games and trained by self-play	English AI researcher and neuroscientist
2019	AlphaStar (DeepMind)	Demis Hassabis
	Deep reinforcement learning defeating StarCraft player	English AI researcher and neuroscientist

ENIAC, Electronic Numerical Integrator and Computer; *HAL*, hybrid assistive limb; *LISP*, list processing; *NLP*, natural language processing; *SNARC*, Stochastic Neural Analog Reinforcement Calculator.

Rosenblatt's Perceptron

Around this time, a significant contribution was made by the American psychologist Frank Rosenblatt in 1958 in the form of the perceptron (see Fig. 2.1), a biologically inspired, three-layer structure (with input, transfer function, and output) that was a simple but elegant supervised linear binary classifier (see section on machine learning). The perceptron became the early precursor of the artificial neural network and present day deep learning architecture that we are so familiar with today (see aforementioned reference).

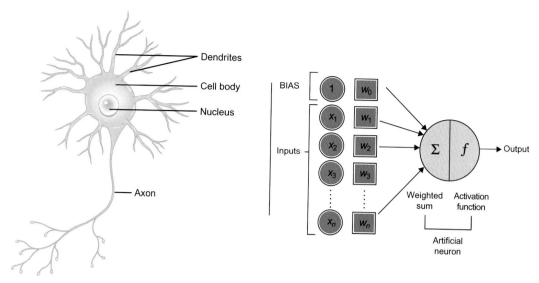

FIGURE 2.1 Biological neuron and computational perceptron. On the left, the biological neuron and its anatomy illustrates dendrites carry impulses toward the cell body and nucleus, and these impulses are processed and moved from the cell body via an axon and its connections and terminals. On the right is a schematic diagram of a perceptron. The inputs x are multiplied by their weights w and the resultant weighted sum then is all the multiplied values added together (not illustrated is an extra weight that helps to neutralize bias in the classifier). These inputs are equivalent to the dendrites carrying impulses toward the neuronal body. The activation function (or step function) is placed in the node and is linear or nonlinear depending on the data. The activation functions can be sigmoid, tanH (similar to a sigmoid), or ReLU (rectified linear unit) which has a slope of 1. After this function processes the sum, the output is delivered. *Source: Shiland BJ. Chapter 8-Introduction to Anatomy and Physiology. In: Zenith, editor. Medical Assistant: Introduction to Medical Assisting—MAIntro, Second Edition. 2nd ed. Elsevier; 2016. p. 204.*

Key Epochs and Movements

Good Old-fashioned Artificial Intelligence

Although current machine learning is a type of knowledge-acquisition type of AI, earlier knowledge-based systems (as represented by expert systems as well as knowledge representation and probabilistic reasoning) were considered relatively weak AI programs. Expert systems (see later) include rule-based reasoning and case-based reasoning (the latter finds solutions to new problems by taking previously good solutions to similar problems). These methodologies cannot scale nor accommodate larger and more complex problems. The symbolic AI in the 1950s to the late 1980s, also known as good old-fashioned AI (GOFAI) and the "neat" in neat versus scruffy AI, was mainly rooted in symbolic representations of problems and was considered the main school of thought during this epoch. In short, GOFAI was based on top-down approaches of logic, problem solving, and expert systems.

Knowledge representation (including the best known example, rule-based model) involves the difficult task of constructing a cognitive science and a general purpose ontology that covers many areas of knowledge domains; other examples of knowledge representation include semantic nets, frames, and conceptual graphs. Probabilistic reasoning involves handling of uncertainties in knowledge with various techniques, in particular Bayesian networks, which accounts for uncertainties by conditional probabilities. Automated reasoning, also known as automated deduction, is the art and science of having computers apply logical reasoning to solve problems such as prove theorems, solve puzzles, design circuits, and verify or synthesize computer programs. Automated reasoning is considered part of AI but does not deploy the typical AI methodologies.

Computational Intelligence

An alternative to GOFAI during this earlier period of AI was computational intelligence (the "scruffy" half of neat vs scruffy), which relied on heuristic algorithms such as the ones seen in fuzzy systems, evolutionary computation, and neural networks. Fuzzy system is based on fuzzy logic, a system that analyzes analog input values that take on continuous values between 0 and 1 (as opposed to digital 0 or 1 dichotomy); this is used in machine control and can be applicable in biomedicine and its physiological parameters. Evolutionary computation, exemplified by genetic algorithms, involves an optimization process to solutions that is based on evolutionary biology and its natural selection process. Neural network was also part of this bottom-up approach to AI and will be covered in much more detail in the next section.

The Artificial Intelligence Winters

Following this early epoch of machine intelligence, two AI "winters" (1974—80 and then 1987—93) occurred due to lofty expectations and suboptimal realities, resulting in an overall disappointing outlook on AI accompanied by decrease or even cessation of funding for AI projects. In between these two AI winters was a short period of AI boom during which expert systems and neural networks thrived for these years.

References

[1] Marchant J. Decoding the heavens: a 2,000 year-old computer and the century-long search to discover its secrets. Cambridge, MA: Da Capo Press; 2009.
[2] Turing AM. On computable numbers, with an application to the entscheidungsproblem. Proc London Math Soc 1936—37;Series 2, 42:230—65.
[3] Copeland J. The essential Turing. Oxford: Oxford University Press; 2004.

History of Artificial Intelligence in Medicine

Rule-based Expert Systems

Initial efforts in artificial intelligence (AI) and its application in medicine began in the 1960s with ruled-based domain-specific expert systems and focused mainly on diagnosis and therapy [1]. Among the best known early works on AI in medicine (AIME) was the Stanford physician and biomedical informatician Edward Shortliffe's innovative heuristic programming project MYCIN. This pioneering work was a rule-based expert system (written in the Lisp programming language) that had if−then rules; these rules yielded certainty values that mimicked a human's expertise (such as recommended selection of antibiotics for various infectious diseases) [2]. Other such expert systems in this era included *INTERNIST, AI/RHEUM,* and *ONCOCIN* [3].

The expert system worked as follows (see Fig. 3.1):

Knowledge is a repository of both factual and heuristic knowledge, and this knowledge from a human domain expert was entered into a knowledge base via a knowledge

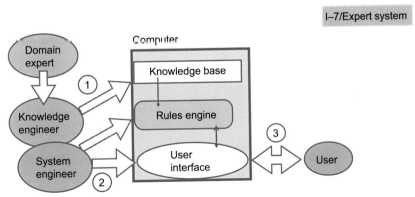

FIGURE 3.1 Expert system. The expert system has three major components: knowledge base, rules engine, and user interface (see text for more details).

TABLE 3.1   A Brief History of Artificial Intelligence in Medicine.

Year	Artificial intelligence in medicine events	Key people
1972	MYCIN	Ted Shortliffe
	Expert system for identifying bacteria for infection	American physician—computer scientist
1974	INTERNIST-1	Jack Myers
	Computer-assisted diagnostic tool for medicine	American internist
1982	AIME	Peter Szolovits
	First book on AIME	American computer scientist
1985	AIME	Mario Stefanelli
	First AIME meeting	Italian biomedical engineer
1986	DXplain	EP Hoffer
	Clinical decision support system with probabilistic algorithm	Massachusetts General Hospital
	MODERN AI ERA	
1998	AIME Journal	Mario Stefanelli
	First AIME journal	Italian biomedical engineer
2002	ISABEL (Isabel Healthcare)	Jason Maude
	Clinical decision support system for pediatrics	Founder
2017	CheXNet (Stanford)	Andrew Ng
	CNN algorithm for pneumonia detection at high level	American AI expert
	Cardio DL (Arterys)	Fabien Beckers
	First FDA-approved AI-assisted cardiac imaging in cloud	CEO
2018	ContaCT (Viz.AI)	Chris Mansi/David Golan
	First FDA-cleared AI-powered CDSS platform (stroke)	CEO/CTO
	IDx-DR (IDx)	Michael Abramoff
	First FDA-cleared autonomous AI device (diabetic retinopathy)	American ophthalmologist—scientist
2019	AI in Health-Care Regulation (FDA)	Scott Gottlieb
	FDA proposal for innovations in AI in health-care regulation	Chief, FDA

CNN, Convolutional neural network; *DL*, deep learning.

engineer (step 1). The knowledge base is organized, and the effectiveness of this knowledge base is proportional to the quality and accuracy of the knowledge entered. This knowledge base was in turn connected to a rules engine with its many if—then rules stored within it. A system engineer overlooks the rules engine (as well as the user interface) (step 2). The user then queries a user interface (step 3) that was coupled to the inference engine. The final reconciled advice was then given to the user via this user interface.

In essence, the expert system is a set of programmed rules that the computer can follow and output an answer based on these rules.

Of note, a precursor to MYCIN was the DENDRAL expert system that helped to identify unknown organic molecules. Although not successfully used in an actual clinical setting, MYCIN proved to be superior to human infectious disease experts when a comparison was performed. The ruled-based expert systems, however, were both tedious to construct and difficult to adapt and utilize in new and/or complex clinical scenarios. Another computer-assisted diagnostic tool, INTERNIST-1, was developed by Dr. Jack Myers, who was noted to state "The method used by physicians to arrive at diagnoses requires complex information processing which bears little resemblance to the statistical manipulations of most computer-based systems."

During this early era of AIME, several academic centers in the United States that focused on this burgeoning area included Stanford, MIT, Rutgers, and Carnegie Mellon University as well as a few centers in Europe. Overall, the main shortcomings during these early decades of AI tools in biomedicine included not only a lack of theory-to-use coupling but also an inadequate integration of the existing AI techniques into workflows to achieve user support (due to slowness of the tools and lack of adequate data or knowledge) [4] (see Table 3.1).

In 1970 a review article in the *New England Journal of Medicine* by Schwartz predicted that computers would have an entirely new role in medicine by the year 2000 as a powerful extension of the clinician [5]. The MIT professor of Computer Science and Engineering Peter Szolovits edited the book *Artificial Intelligence in Medicine* in 1982 that consisted of a collection of papers on various topics in this domain but mainly on expert systems [6] and also first reported the use of AI in making a medical diagnosis [7]. Around this time, Dr. Mario Stefanelli of Pavia, Italy, started the organization called Artificial Intelligence in Medicine (AIME) based in Europe. Szolovits organized a Medical Artificial Intelligence course at MIT in 2003 that was one of the first organized educational efforts on this topic but unfortunately did not continue after its first year [8].

Data science and knowledge engineering in medicine: building the evidence base

John Fox
Oxford University, Oxford, United Kingdom

John Fox, a leading artificial intelligence (AI) and cognitive science researcher, authored a commentary on a hybrid approach to AI in medicine, combining present era data science and machine learning with knowledge engineering of the earlier era.

The world has been astonished by the recent explosion of claims about artificial intelligence (AI). It is not that AI has suddenly changed from a field of academic research to one of commercial and economic significance (which seems to happen all the time in "tech") but that so many people believe that AI is set to outperform humans on a range of professional tasks, with predictions of a revolution in clinical services, offering better, safer care at lower cost than can be offered by health-care professionals who follow conventional practices alone.

The current wave of excitement about artificial intelligence in medicine: data science

The excitement was triggered several years ago by a small number of unexpected technical achievements, famously IBM Watson's performance in the *Jeopardy* general knowledge quiz and Deepmind Alphago's defeat of the world Go champion. Although the demonstrations were "mere games" they unleashed arguments that machine learning (ML), a key AI technique used in both systems has truly revolutionary significance.

Despite the excitement, many clinicians struggle with such claims. Countless variants of the revolution scenario have been promulgated by marketing departments, journalists, politicians, and some health-care professionals, though until recently the bulk of the medical community stayed largely silent; claims of breakthroughs come and go in medicine, with many technologies eventually falling by the wayside.

New technical concepts can surely lead to new thinking, but health-care professionals and clinical researchers are suspicious of innovations unless they are warranted by a clear medical need and convincing evidence of benefit yet the AI R&D programs that are being driven relentlessly forward in health care seem to ignore normal standards of *evidence-based medicine*. Other reasons for concern about AI are the much-discussed "black box" nature of complex algorithms, and the feeling that a mathematical view of decision-making based on population data engages poorly with clinicians' understanding of their expertise and individual patient-centered practice.

However, many clinical, research, funding, and other organizations now see an urgent need to develop a practical policy toward AI, with the aim of balancing the doubts, and fear about it with the need to be open-minded about new developments that might truly improve practice and outcomes at a time of acute pressure on human skills and services. A recent statement by the Royal College of Physicians of London[1] takes exactly this position

> The RCP should support all clinicians to critically appraise new technology, ask questions and engage in discussion about the accuracy and impact of its advice, efficacy and evidence base. This will ensure that no matter how technology progresses, doctors will be able to apply core professional principles in their approach and have the confidence to either agree or disagree with the recommendations made by AI technology.

The previous wave of artificial intelligence in medicine: knowledge engineering

Another reason for surprise about the current excitement is that medicine has been prominent in AI research for at least 40 years, and a far wider range of applications has been considered than is reflected in current excitement. The use of sophisticated algorithms that can process large collections of data and find statistical and other patterns to inform clinical decision-making is important, but AI is much wider and more ambitious than this; lessons learned from successfully developing and deploying AIME are in danger of being forgotten.

One important example is the use of knowledge engineering methods to formalize medical knowledge and practice starting from human expertise and human-readable *clinical guidelines* (text, flow sheets, etc.). This models medical knowledge and practice in an explicit symbolic or logical form rather than as a mathematical model or algorithm. Such models can still be executed on a computer to assist a clinician but are easier for health-care professionals to understand and critique than conventional software. "Expert systems" as they were called can capture and interpret patient data; assess risk and the plausibility of alternative diagnoses; make patient-specific recommendations for investigation, treatment, and follow-up; construct care plans; and manage clinical workflows and many other routine clinical tasks.

Clinical tasks	Trails and evaluations
Routine prescribing by GPs	Walton et al., *British Medical Journal* (1996)
Mammographic screening	Taylor et al., *Medical Image Analysis* (1999)
Genetic risk assessment	Emery et al., *British Medical Journal* (1999, 2000)
Genotyping and prescribing antiretrovirals	Tural et al., *AIDS* (2002)
Chemotherapy prescribing for ALL	Bury et al., *B Journal of Haematology* (2005)
Early referrals of suspected cancer	Bury et al., *Ph.D. Thesis* (2006)
Cancer risk assessment, investigation	Patkar et al., *British Journal of Cancer* (2006)
Hospitalisation decisions: asthma	Best Practice Advocacy Centre, NZ (2009)
Genetic risk assessment, treatment planning	Glasspool et al., *J Cancer Education* (2010)
Multidisciplinary decision making	Patkar et al., *BMJ Open* (2011)
Investigation, diagnosis of thyroid nodules	Peleg et al., *Endocrine Practice* (2015)
Diagnosis and treatment of stroke	Ranta et al., *Neurology* (2015)
Shared decision-making in chemotherapy	Miles et al., *BMJ Open* (2017)
Diagnosis of hyponatremia	Gonzales et al., *Int J Med Informatics* (2017)
Detection and Diagnosis of ophthalmic disease	Chandrasekaran *Ph.D. Thesis* (2017)
Kidney donor workup and eligibility	Knight et al., *Transplantation* (2018)

Peleg et al. [1] reviewed a number of approaches to guideline modeling including EON (United States), ASBRU (Israel), GUIDE (Italy), GLIF (United States), and PRO*forma* (United Kingdom). Peleg et al. systematically compared the ability of these different methods to capture practice guidelines, which often drew on ideas from AI but also incorporated important advances in medical informatics. Some of these efforts have tried to establish rigorous design methods and standards for modeling clinical practice and an evidence base to demonstrate clinical value; one of these (PRO*forma* [2]) has been successfully trialed in diverse applications summarized in the table mentioned earlier (details of these and other trials can be found in Fox 2018).[2]

The PRO*forma* approach to designing and deploying AI at the point of care is very different from the data science approach, in that it is grounded in cognitive science and long-term research goals of understanding human expertise and getting the best from a combination of human and machine intelligence [3]. This experience shows that knowledge engineering methods are highly versatile and that they can be successfully deployed at scale in a way that healthcare professionals can understand and critically engage with.

Although the current focus of public and business attention is on the use of data science and ML in medicine the knowledge engineering approach continues to advance in areas such as ontology design, autonomous agents for supporting clinical services, and so on. PRO*forma* and its peers are decades old and are not the last word in modern knowledge engineering or AI. Big data and ML can help to address some of its weaknesses; in return, a deeper understanding of medical knowledge and clinical expertise offers a humanly intuitive foundation for AI at the point of care; "hybrid" AI should now be the primary objective.

References

[1] Peleg M, et al. Comparing computer-interpretable guideline models: a case-study approach. J Am Med Inform Assoc 2003;10(1):52–68.
[2] Sutton D, Fox J. The syntax and semantics of the PROforma guideline modelling language. J Am Med Inform Assoc 2003;10(5):433–43.
[3] Fox J. Cognitive systems at the point of care: the CREDO program. J Biomed Inform 2017;68:83–95.

[1] In the UK reports on AI with this general goal of striking a balance are not only published by the RCP (https://www.rcplondon.ac.uk/projects/outputs/artificial-intelligence-ai-health) but also the Royal College of General Practitioners (https://sway.office.com/0tz3Q5xUQY5QnAD1?ref = Link), the Academy of Medical Royal Colleges (http://www.aomrc.org.uk/reports-guidance/artificial-intelligence-in-healthcare/) and many reports on AI ethics from the Wellcome Foundation, the Nuffield Trust, and others.

[2] Fox J. 2018 "PICO AI: A method for appraising AI systems in healthcare" (under revision, draft available from john.fox@eng.ox.ac.uk).

Combining data and knowledge in the medical domain

Annette ten Teije

Vrije Universiteit Amsterdam, Amsterdam, The Netherlands

Annette ten Teije, a computer engineer with expertise in the formalization of medical knowledge, authored this commentary on the combined model or knowledge- and data-driven approaches to medical knowledge acquisition.

The popular trend when developing intelligent support tools for medical decision-making is "data data data." However, we should not forget the usefulness of knowledge, in particular in the medical domain.

We all know that the last 10 years have seen a revolution in machine learning and data analytics, driven by large volumes of available data. But in the same period, a somewhat less visible revolution has happened in knowledge representation, through a smart combination of new logical formalisms and worldwide web technology, it has now become possible to construct very large knowledge bases, which contain highly curated facts and rules in very large quantities (easily up to 100s of millions, sometimes even billions of facts and rules). These so-called knowledge graphs are now in widespread use by major search engines (Google, Bing, Baidu), by major news corporations (BBC, New York Times, Reuters), by e-commerce platforms (Amazon, eBay, Uber, Airbnb) and governments (US, UK, EU), just to name few. An overview of the publically available knowledge graphs is available in the LOD (linked open data) cloud diagram at https://lod-cloud.net/ and reproduced next, but many more are being used inside organizations. Major knowledge graphs are also available in the medical domain, such as Drugbank, SNOMED, SIDER, Bio2RDF, MESH, Medline, LinkedCT, and, many others (the red segment of the LOD cloud diagram)

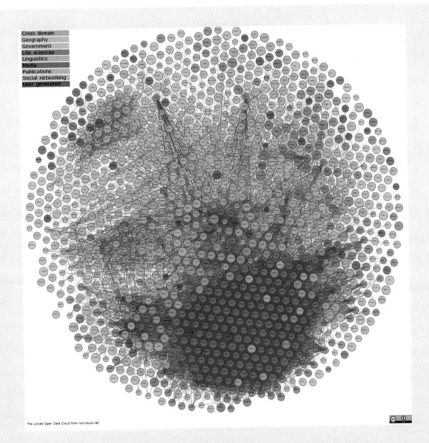

Furthermore, many techniques are available to model medical knowledge in such a way that it can be used in a variety of decision support systems in the medical field. Just a single example that we discuss briefly here is the use of computer models of medical guidelines.

Medical guidelines are deployed to ensure that clinical practice is maximally effective while risks and costs are being minimized. Such guidelines prescribe a process or routine to accomplish a medical task based on a set of assumptions, in particular the clinical evidence at the time of their creation.

The medical knowledge captured in a medical guideline can be modeled in such way that the guideline model is interpretable for a computer. The first benefit of such a computer-model is that inconsistencies, incompleteness, and other possible anomalies can be detected. Our work in the past has shown that even national standard guidelines contain inconsistencies that were only detected after computer analysis [1]. Furthermore, such a computer interpretable model of a medical guideline can be used for a variety of tasks, such as decision support in guideline execution, or for finding interactions among several guidelines. This latter is a serious problem given the rapidly rising number of multimorbid patients and the fact that a guideline typically covers only a single disease. In our work, we used public knowledge graphs such as SIDER and

Drugbank to detect interactions between the standard treatments of common chronic diseases such as hypertension, diabetes, and osteoporosis [2].

Besides their focus on a single disease, another problem with guidelines is their update frequency. Ideally, medical guidelines would stay in sync with the latest insights from the medical literature. However, new research is published at an astonishing rate—PubMed grows with 2000 papers per day—while it typically takes at least 3 years before a single guideline is updated. The question is how to enable a faster cycle of guideline updating to arrive at "living guidelines." We can exploit again the computer model of a guideline for the task of identifying new and relevant scientific publications (evidence) in PubMed to speed up the process [3]. Here we used evidence from the literature. Another way to update the guideline is to use evidence from practice by using patient records to improve the guideline. Evidence is automatically collected and interpreted to modify the preferred way of carrying out diagnosis and treatment activities based on experiences collected from patient data and workflows in hospitals, and in other words, guidelines stay in sync with the evidence collected from practice. Such combinations of knowledge and data are promising approaches to achieve high-quality guidelines that are subject to a faster the cycle of guideline update. The living guideline is then based on the original guideline model, the new available publications, and the evidences from practices (patient data).

The above are just a few examples of our research agenda that is in *combining* the model-driven (knowledge-driven) approach and data-driven approach. We believe that in particular in the medical domain, which is highly knowledge intensive, and where human expertise is essential, such a combination of knowledge-driven and data-driven approaches is a promising way to go beyond the current data-driven approaches.

References

[1] ten Teije A, Marcos M, Balser M, van Croonenborg J, Duelli C, van Harmelen F, et al. Improving medical protocols by formal methods. Artif Intell Med 2006;36(3):193—209.
[2] Zamborlini V, Da Silveira M, Pruski C, ten Teije A, Geleijn E, van der Leeden M, et al. Analyzing interactions on combining multiple clinical guidelines. Artif Intell Med 2017;81:78—93.
[3] Hu Q, Huang Z, ten Teije A, van Harmelen F. Detecting new evidences for evidence-based medical guidelines with journal filtering. In: Riano, D, Lenz R, and Reichert M. (Eds.), *Knowledge Representation for Health Care* KR4HC/ProHealth@HEC. 2016, Springer, 2016, pp. 120—132.

Other Artificial Intelligence Methodologies

In addition to the abovementioned expert systems, some other AI methodologies were used in medicine included fuzzy logic and neural networks [9]. Fuzzy logic, as discussed earlier, deals with degrees or continuum of truth (vs the dichotomous Boolean logic of true or false) and is therefore particularly well suited for biological systems with objective physiological parameters of continuous data (such as heart rate and blood pressure) [10]. A recent review concluded that fuzzy logic will need to be an essential part of AIME in the near future as algorithms alone are not feasible for solving enigmas and conundrums in biomedicine [11]. Neural network is a processing paradigm that is inspired by the brain; this methodology was applied to various clinical situations such as clinical diagnosis and medical images as well as the critical care setting [12].

Failure of Adoption

While this early period was initially characterized by knowledge engineering for medical expert systems, ML and data mining were AI methodologies only available in the later part of this early era but became the more dominant methods [13]. AI and its failed adoption in medicine during this early period were due to not only lack of favorable workflow logistics and slow speed of computation but also due to expectations that were unrealistically high. Both funding and enthusiasm dwindled and AIME became less popular.

In conclusion, the remarkable history of AI in the recent era started with Turing and his Turing Machine and the Dartmouth Conference. This AI naissance leads to the good old-fashioned AI period that was symbolized by expert systems. This recent modern era in AI is rooted in historical advances with AI competing with humans in various games: chess (expert systems), Jeopardy! (cognitive computing), and Go (deep reinforcement learning). The renaissance of AI is a convergence of three main forces: improved methodology, big data, and computational power with increased storage. In spite of these milestones, clinical medicine and health care have not adopted these AI tools to a large extent due to lack of trust and knowledge. It is hoped that clinicians will learn to leverage these AI tools in their practice in the near future.

Ten Common Misconceptions of Artificial Intelligence in Medicine

The following are common misconceptions (by both clinicians as well as data scientists) about AIME that are understandable and human (detailed explanations can be found in the upcoming sections in the book):

Clinicians will be replaced by AI. There is a fundamental lack of understanding of what clinicians do even amongst august data scientists and seasoned venture capitalists, and this deficiency renders it easy to think that computer vision and interpretation of medical images alone is sufficient to replace image-intensive subspecialties such as radiologists, pathologists, ophthalmologists, dermatologists, and cardiologists. The doctor's tasks can be divided into perception (visual image interpretation and integrative data analytics), cognition (creative problem solving and complex decision-making), and operation (procedures). The computer is much stronger in perception tasks but not yet facile with cognition nor operation parts of the clinician's tasks. It is also notable that all those who proclaim an end to radiologists are not radiologists (or even physicians) themselves. It would be prudent to end the AI versus MD debates and start discussing how these two health resources can work in synergy [14].

AI can be applied to every aspect of health care to bring value. While AI can improve workflow and increase accuracy of diagnoses, there are certain technologies that will not necessarily benefit from AI application. For example, applying AI to an older technology (such as auscultation for heart murmurs) may not increase the value proposition of such an application. There is a myriad of workflow deficiencies in health care, however, that can be improved with AI but these deficiencies are often neglected. It is therefore important to remember design thinking principles in applying AI (design AI) in delineating problems first.

AI, in conquering the game Go, will be successful for medicine and health care. AI was indeed successful in defeating the human champion in the ancient game of Go. Medicine and health care, especially in arenas such as a busy intensive care or emergency room setting or in domains such as chronic disease management and population health, is more akin to the real-time strategy game of *Starcraft* where real-time decisions on multiple fronts need to be made in a complex milieu that is different for each individual (so essentially it is even more akin to playing hundreds of *Starcraft* games simultaneously). It should be noted, however, that AI was recently successful in defeating a human real-time strategy player.

Deep learning (DL), especially convolutional neural network (CNN), will be the preferred AI tool for a long time. While DL is in its hype and is indeed very effective for computer vision and medical image interpretation as well as decision support, the future of these areas will need even more sophisticated tools involving cognitive architecture, which is the third wave of AI. Even DL gurus such as Geoff Hinton feels that cognitive elements such as capsular networks will be needed to improve DL performance [15]. DL with CNN will also need to be even more sophisticated in the future with tools such as recursive cortical network and transfer learning as there are limitations to the amount of available medical data. In addition, while CNN has been a major contribution in medical image interpretation, recurrent neural network (RNN) with robust natural language processing (NLP) can also be equally as valuable by extracting information and knowledge from hospital and clinic records. Variations of CNN with RNN can also be used to examine videos that are commonplace in medical imaging [16].

We need more biomedical data for DL in health care. There are several instances that big data in medicine and health care are not feasible. One situation involves patients with rare diseases as there is a limit in the sheer number of patients with the rare disorder with equal limit in the number of medical images. Another situation would involve a very sophisticated or invasive test or a test with excessive risks and/or costs; these tests would result in very few samples in a population. In all of these cases, creative uses of generative adversarial networks or low-shot learning to neutralize the lack of big data can obviate the absolute necessity of big data in biomedicine.

The area under the curve (AUC) of the receiver operating curve (ROC) is a good indicator of the performance of the algorithm. First, similar to parents having higher expectations of their children than they have of themselves, clinicians and data scientists can have relatively high (and perhaps unfair) expectations of AI. It is not uncommon for human clinicians to have not much better than 50% accuracy of certain medical diagnoses, and yet we have such higher (albeit understandable) expectations for machine intelligence [17]. To use the AUC of the ROC as the sole determinant of the accuracy of machine intelligence for a test, however, is problematic [18]. Some of this problem lies in the fact that the labels on images are often not entirely accurate to begin with (humans are labeling and are not infallible). Additional issues in the large datasets include lack of precise terminology (consolidation vs pneumonia), time element in diagnoses (early vs late manifestation of pneumonia), presence of multiple labels (diagnoses are often not exclusionary or dichotomous), and variability of the dataset (quality of image). Three key elements are necessary for there to be an accurate conclusion of performance derived from this ROC AUC assessment: accuracy, threshold, and prevalence of disease, the latter is a critical element for the analysis but often not included in the overall study description. The balance for this is the precision–recall curve that takes into account diseases with low incidence and therefore large true negative number (which artificially inflates accuracy and lower error or misclassification rate).

You have to be able to program to make a contribution to AIME. There are many ways other than actually programming and coding that anyone in medicine and health care can contribute to the overall paradigm shift of AIME. The most glaring deficiency in the AIME domain is not the lack of AI tools, but the quality and management of biomedical data. First, any clinician can provide domain expertise for an AI project or idea that can be misdirected. In addition, any health care worker can also delineate the workflow inadequacies so pervasive in health care for an AI project. Lastly, anyone in health care can also contribute to the foundation of the data–information–knowledge–intelligence pyramid by focusing more on the quality of data and integrity of data infrastructure.

AI is mainly for selected subspecialists such as radiologists and pathologists. While AI and DL have made significant contributions in these fields, the use of other AI methodologies such as cognitive computing as well as robotic process automation and NLP are helpful for almost all other subspecialists. In addition, these tools are essential in reducing the administrative burden of health-care systems irrespective of the clinical domain. AI, therefore, has much more to offer other than CNN and DL with medical image interpretation; the portfolio of AI tools will all provide a new resource to alleviate the burden of health-care delivery in all areas.

AI will make clinicians less human. With the appropriate application of AI, especially NLP and natural language understanding tools, clinicians will be in a position to be more human as the burden and distraction of electronic medical record would be lessened. It would be a laudable vision that the future sanctuary of the physician–patient setting will have no visible machines in the environment but will be the venue for only human-to-human bonding [19].

AI devices will be difficult to be understood or regulated. There is the possibility of a self-fulfiling prophecy if advanced AI tools lack explainability or interpretability [20]. Even if we treat AI and its panoply of tools as "software-as-a-device," how we can effectively and expediently approve all these upcoming AI tools as these emerge will be a challenge. Perhaps, we need to match this exponential paradigm shift in technology with a parallel trajectory in how we regulate. One potential solution is to not regulate AI devices per se but rather teams and/or individuals working on the AI tools in a specific program or institution. Another possible answer lies in the sagacious Turing philosophy of "machines to machines" and devises regulatory algorithms that will overlook algorithms on a continuous basis in addition to periodic checks by regulatory agencies.

AIME will be here in the future. As the computer scientist William Gibson so eloquently stated: "the future is already here, it is just not evenly distributed." The trajectory of medicine needs to change given the exponentially increasing amount of data and information to attain precision medicine and chronic disease management of our population. All stakeholders in both clinical medicine and data science have a special opportunity to create a special synergy for a once-in-a-generation transformative paradigm shift in medicine.

References

[1] Kulikowski CA. Artificial intelligence methods and systems for medical consultations. IEEE Trans Pattern Anal Mach Intell 1980;5:464–76.

[2] Shortliffe EH, David R, Axline SG, et al. Computer-based consultations in clinical therapeutics: explanation and rule acquisition capabilities of the MYCIN system. Comput Biomed Res 1975;8(4):303–20.

[3] Miller PL. The evaluation of artificial intelligence systems in medicine. Comput Methods Programs Biomed 1986;22:5–11.

[4] Personal communication with Dr. Shortliffe. 2014.

[5] Schwartz WB. Medicine and the computer: the promise and problems of change. N Engl J Med 1970;283:1257–64.

[6] Szolovits P. Artificial intelligence in medicine. Boulder, CO: Westview Press Inc; 1982.

[7] Szolovits P, Patil RS, Schwartz W. Artificial intelligence in medical diagnosis. Ann Intern Med 1988;108:80–7.

[8] Personal communication with Dr. Szolovits. 2015.

[9] Hanson CW, Marshall BE. Artificial intelligence applications in the intensive care unit. Crit Care Med 2001;29:427–35.

[10] Ramesh AN, Kambhampati C, Monson JR, et al. Artificial intelligence in medicine. Ann R Coll Surg Engl 2004;86(5):334–8.

[11] Thukral S, Singh Bal J. Medical applications on fuzzy logic inference system: a review. Int J Adv Networking Appl 2019;10(4):3944–50.

[12] Yardimci A. A survey on the use of soft computing methods in medicine. In: Proceedings of the 17th international conference on artificial neural networks, Porto, Portugal. p. 69–79.

[13] Peek N, Combi C, Marin R, et al. Thirty years of artificial intelligence in medicine (AIME) conferences: a review of research themes. Artif Intell Med 2015;65(1):61–73.

[14] Goldhahn J, Rampton V, Spinas GA. Could artificial intelligence make doctors obsolete? BMJ 2018;363:k4563.

[15] Hinton G. Deep learning—a technology with the potential to transform health care. JAMA 2018;320 (11):1101–2.

[16] Yu F, Silva Croso G, Kim TS, et al. Assessment of automated identification of phases in videos of cataract surgery using machine learning and deep learning techniques. JAMA Netw Open 2019;2(4):e191860.

[17] Hill AC, Miyake CY, Grady S, et al. Accuracy of interpretation of pre-participation screening electrocardiograms. J Pediatr 2011;159(5):783–8.

[18] Mallett S, Halligan S, Collins GS, et al. Exploration of analysis methods for diagnostic imaging tests: problems with ROC AUC and confidence scores in CT colonography. PLoS One 2014;9(10):e107633.

[19] Verghese A, Shah NH, Harrington RA. What this computer needs is a physician: humanism and artificial intelligence. JAMA 2018;319(1):19–20.

[20] Vellido A. The importance of interpretability and visualization in machine learning for applications in medicine and health care. In: Neural computing and applications. 2019. https://doi.org/10.1007/s00521-019-04051-w.

Key Concepts

- The recent impressive gains in sophistication of deep learning (DL) technology and utilization especially since 2012 have led to an escalating momentum for artificial intelligence (AI) awareness and adoption.
- Even though the advent of data science and machine and DL has advanced information and analyses and promoted innovations, health care and medicine remain very much behind these other domains in leveraging this new AI paradigm.
- There is a continuum from data to intelligence and with good intelligence, one can have wisdom; in health care, there should eventually be a bidirectional continuity from wisdom and intelligence directing how data, information, and knowledge be gathered, stored, and shared.
- Perhaps the best definition of AI is the one conjured by the American cognitive scientist Marvin Minsky: the science of making machines does things that would require intelligence if done by human.
- AI can be categorized as weak or strong: weak (also termed "specific" or "narrow") AI pertains to AI technologies that are capable of performing-specific tasks (such as playing chess or *Jeopardy!*) and strong (also termed "broad" or "general") AI is much more difficult to attain; it is also called artificial general intelligence or general AI.
- Machine learning (ML) and its more robust and specific type, deep learning (DL), are not synonymous with AI but are often used interchangeably; ML as well as DL are AI methodologies.
- Other AI methodologies include cognitive computing and natural language processing as well as computer vision, robotics, and autonomous systems.
- AI can be described in the context of a human–machine intelligence continuum with three types of AI: assisted, augmented, and autonomous.
- In addition to the man–machine intelligence continuum, there is also an AI-inspired analytics continuum (descriptive, diagnostic, predictive, prescriptive, and cognitive analytics) that increases in intelligence and autonomous behavior from a data science perspective.
- Innovative AI systems can be partly inspired by the brain just as the brain can be augmented by the machine.
- The "doctor's brain" for day-to-day clinical work can be conveniently deconstructed by its myriad of functions and matched to machine-equivalent capabilities.
- It is the brilliant British mathematician and computer scientist Alan Turing, however, who would be considered the absolute progenitor of AI with his pioneering works that included his theory of computation and his work on computing machines.
- The symbolic AI in the 1950s to the late 1980s, also known as good old-fashioned AI (GOFAI) and the "neat" in neat versus scruffy AI, was mainly rooted in symbolic representations of problems and was considered the main school of thought during this epoch.
- An alternative to GOFAI during this earlier period of AI was computational intelligence (the "scruffy" half of neat vs scruffy), which relied on heuristic algorithms such as ones seen in fuzzy systems, evolutionary computation, and neural networks.

- Initial efforts in AI and its application in medicine began in the 1960s with ruled-based domain-specific expert systems and focused mainly on diagnosis and therapy.
- AI and its failed adoption in medicine during this early period were due to not only lack of favorable workflow logistics and slow speed of computation but also due to expectations that were unrealistically high.

Data Science and Artificial Intelligence in the Current Era

It is a capital mistake to theorize in advance of the facts. Insensibly one begins to twist the facts to suit the theories, instead of theories to suit the facts. Sherlock Holmes

The data mining and machine learning focus in the 1990s slowly revived the field of artificial intelligence (AI) and this era (see prior table on history of AI) was best symbolized by IBM's iconic supercomputer Deep Blue, which defeated the reigning world chess champion Gary Kasparov in 1997 (on the second attempt). Another IBM supercomputer, Watson (named after its first CEO Thomas Watson and not as some believe, Sherlock Holmes' sidekick) demonstrated the prowess of cognitive computing (with its access to over 200 million pages of content and its immense capability from the DeepQA project) by easily defeating the human champions Ken Jennings and Brad Rutter on February 14, 2011 on the game show *Jeopardy!* In a similarly dominant fashion, the AlphaGo program of DeepMind (now with Google) easily defeated the human Go champion Lee Sedol in March 2016, thus heralding a new era of AI with deep learning (more specifically, deep reinforcement learning)

The recent advent of an AI "triad" (conveniently "ABC") consisted of (1) the emergence of sophisticated algorithms, particularly machine and deep learning with all its variants, (2) the increasingly large volumes of available data that requires new computational methodologies (or simply "Big Data"), and (3) the escalating capability of computational power (with faster, cheaper, and more powerful parallel processing that defied Moore's Law) with coupling to the widely available cloud computing (with nearly infinite storage). These elements have converged to engender this new resurgence of AI [1] (see Fig. 1).

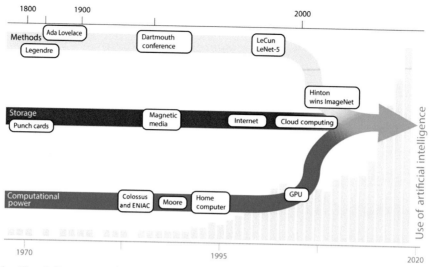

FIGURE 1   Use of AI. Among the significant forces that have increased the use of AI. First, methods improved with more sophisticated algorithms and deep learning. In addition, there is increased storage capacity with cloud computing. Third, computational power has also escalated with the advent of GPUs. Lastly, the availability of data in large amounts also rendered AI methodologies much more relevant. *AI*, Artificial intelligence; *GPU*, graphics processing unit.

The role of graphics processing units (GPUs) in artificial intelligence for healthcare and biomedicine

Abdul Hamid Halabi

AI & Healthcare at NVIDIA, Santa Clara, CA, United States

Abdul Hamid Halabi, with decades of experience in health-care technology, authored this commentary on the advent of GPUs and its immediate impact on a myriad of health-care venues, from medical images to intensive care setting.

AI in healthcare presents a life-changing opportunity in the service of humanity. From improving the quality of life for patients to streamlining routine clinical operations, AI can assist with better diagnostic accuracy, delivery of treatments, and lower cost of care.

At the core of the technology behind AI advancements in healthcare is the graphics processing unit (GPU). Its parallel processing power is a key tool for accelerating high-performance computing, deep learning, and graphics. NVIDIA GPUs are programmable, based on a singular architecture, and deployable from the network edge to the cloud. This flexibility and widespread accessibility help one to solve critical information technology infrastructure challenges in healthcare.

Today, high-performance computing and parallelization are enabling doctors to derive meaning from a growing amount of health-care data much more quickly and efficiently.

This explosion of computing power is also leading to a generational leap in the capabilities of medical instruments. The future of medical imaging is faster and uses a lower dose of radiation and contrast. Some examples include the reconstruction of high-quality magnetic resonance images from undersampled and noisy projection data and the conversion of 2D ultrasound scans into 3D images using a series of AI algorithms, developed on NVIDIA GPUs.

NVIDIA in health-care artificial intelligence

During a landmark year for AI in 2012, the GPU was a key ingredient in accelerating the deep learning creation and training process from hours to minutes [1]. However, no single company or organization can bring AI to health-care systems on its own. NVIDIA works with medical institutions, emerging startups, and instrument vendors to facilitate breakthroughs in care across all populations and medical specialties.

In pediatric intensive care units, children arrive with complex, often chronic conditions and many others come as emergency cases. To find the best treatments for children, a team of data scientists at the Children's Hospital Los Angeles used GPUs to train deep learning models to identify patterns from a decade's worth of pediatric ICU health records [2].

Through GPU-powered neural networks, researchers were able to predict the probability of survival in parallel with the patient's physiology through time, aiding their understanding of the key relationships between the patient's vitals and the interventions performed in the unit [2].

In March 2019 the Food and Drug Administration granted breakthrough device designation to Paige.AI, a computational pathology start-up focused on increasing diagnostic accuracy and improving patient outcomes, starting with prostate and breast cancer [3]. Paige.AI uses an AI supercomputer made up of 10 interconnected NVIDIA DGX-1 systems to assess pathology data derived from one of the preeminent cancer research facilities in the United States, Memorial Sloan Kettering. The supercomputer's enormous computing capabilities have the potential to develop a clinical-grade model for pathology, bridging the gap from research to a clinical setting to the benefit of future patients.

In the area of biological analysis, Oxford Nanopore Technologies is democratizing genome sequencing with MinIT, a portable rapid analysis, and device control accessory to run their pocket-sized, real-time MinION DNA/RNA sequencer. Powered by the NVIDIA AGX system for high-throughput, real-time analysis, MinIT will enable DNA/RNA sequencing by anyone, anywhere.

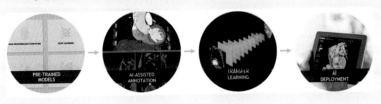

Clara AI is a system for radiologists to deliver AI-assisted annotation, adapt AI for their patients, and deploy it in the hospital.

Bringing artificial intelligence to the clinic: NVIDIA Clara artificial intelligence

To usher deep learning into hospital workflows, it is vital for physicians to be involved in the process. It is critical to give them tools that work within their own data and practice at institutions. This begins at the front lines of many care pathways with radiology, which informs physicians and patients of the next steps in the care process.

Approximately 70% of medical-imaging research is now based on deep learning on GPUs, but few algorithms have made their way into clinical deployment [4]. Medical imaging AI tends to be sensitive to patient demographics, the type of the imaging instruments, and settings of those instruments when the images are acquired. The data needed to create models as well as building deep learning is expensive, time consuming, and demands expertise.

To spur deployment, NVIDIA created Clara AI, a toolkit that includes state-of-the-art classification and segmentation AIs and software, which can help radiologists integrate AI in their unique clinical tools and hospital workflows. The Clara AI-assisted annotation capability, for example, speeds up the creation of structured datasets, enabling annotations in minutes instead of hours [5].

Another NVIDIA Clara AI tool, transfer learning, customizes deep learning algorithms to data that includes local demographics and imaging devices, without compromising patient data privacy. As a result, doctors can create models for their own patients with $10\times$ less data than starting from scratch [5]. The toolkit facilitates the integration of AI models into existing radiology workflows using industry standards.

GPUs accelerate processes at the heart of health-care workflows. A vast and growing ecosystem of partners is harnessing them to lead cutting-edge AI innovations in healthcare and power the next generation of intelligent medical applications.

References

[1] Huang H. Accelerating AI with GPUs: a new computing model. NVIDIA Blog January 2016; [cited 05.04.19]. Available from: <https://blogs.nvidia.com/blog/2016/01/12/accelerating-ai-artificial-intelligence-gpus/>.
[2] Zee S. How deep learning can determine drug treatments for better patient outcomes. NVIDIA Blog May 2016; [cited 05.04.19]. Available from: <https://blogs.nvidia.com/blog/2016/05/17/deep-learning-5/>.
[3] Healio. Cancer statistics reports for the UK, FDA grants breakthrough device designation to artificial intelligence technology for cancer diagnosis. Healio March 2019; [cited 05.04.19]. Available from: <https://www.healio.com/hematology-oncology/prostate-cancer/news/online/%7B4dfae387-2af3-445b-894f-03205456a617%7D/fda-grants-breakthrough-device-designation-to-artificial-intelligence-technology-for-cancer-diagnosis>.
[4] Powell K. How healthcare industry is using NVIDIA AI to better meet patients' needs. NVIDIA Blog November 2018; [cited 05.04.19]. Available from: <https://blogs.nvidia.com/blog/2018/11/20/healthcare-industry-uses-nvidia-ai/>.
[5] Halabi AH. Clara AI lets every radiologist teach their own AI. NVIDIA Blog March 2019; [cited 05.04.19]. Available from: <https://blogs.nvidia.com/blog/2019/03/18/clara-ai-gtc/>.

The discussion of AI in medicine, however, needs to begin with a thorough review of the complex nature of biomedical data and the myriad of issues with data and databases in healthcare that was delineated even two decades ago in the context of AI in medicine [2]. Much of the future of AI in medicine and its success will be rooted in the quality and integrity of biomedical data and databases.

References

[1] Chang AC. Big data in medicine: the upcoming artificial intelligence. Prog Pediatr Cardiol 2016;43:91−4.
[2] Altman R. AI in medicine: the spectrum of challenges from managed care to molecular medicine. AI Mag. 1999;20(30):67−77.

Health-care Data and Databases

Health-care Data

The complex portfolio of health-care data includes not only electronic medical records (EMRs) (patient encounters, vital signs, laboratory results, prescriptions, etc.) but also advanced imaging studies (such as MRI, CT scans, and echocardiograms and angiograms) [1]. Structured data are usually presented in an organized table or relational database format (integers, strings, etc., in a spread sheet, for example) and requires less storage, while unstructured data are without this table format (text, medical image, audio file, video, etc.) and requires more storage. It is estimated that about 80% of health-care data are unstructured [2] (see Fig. 4.1).

International Statistical Classification of Diseases and Related Health Problems (ICD) is a medical classification list put forth by the World Health Organization. The current ICD-10 is the 10th version of this list and has codes for diseases (including signs and symptoms of these disease) as well as abnormal findings, injuries, and social situations in great detail (with over 69,000 billable codes, many more than its previous version ICD-9). The 3—7 digits alphanumeric diagnosis characters and procedure codes (e.g., H91.21 for sudden idiopathic hearing loss of the right ear) support interoperability and HIE exchange internationally. Another important aspect of medical records is the Current Procedural Terminology (CPT) code that is a list of medical, surgical, and diagnostic services organized by the American Medical Association. The 5-digit CPT codes (e.g., 93303 or 93306 for transthoracic echocardiogram, the latter code with Medi-Cal insurance) relate to services rendered (rather than diagnoses as in ICD-10).

SNOMED Clinical Terms (CT) is an ontological collection of medical terminologies used in clinical documentation that includes clinical findings, symptoms, diagnoses, procedures, and anatomical structures. The four main core components of SNOMED CT are concept codes, descriptions, relationships, and reference sets. In addition, the Unified Medical Language System (UMLS) from the NIH National Library of Medicine integrates and distributes key terminology and standards to provide interoperable biomedical information systems, including electronic health record (EHR). A common "dictionary" is very useful since there can be many ways to describe the exact same medical condition. Logical Observation Identifiers Names and Codes (LOINC) is a database specific for medical laboratory observations. The full LOINC name has several components—property:timing:specimen:scale (e.g.,

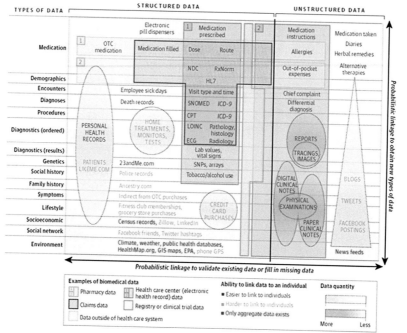

FIGURE 4.1   Biomedical data. This complicated schema illustrates the very challenging data landscape in health care. The types of data are listed in the left column. As complicated as this diagram is, it is still not complete (wearable technology data not included) nor realistic (as most of health-care data are unstructured). In addition, most of the health-care data are actually outside the health-care system.

2951-2 SOIDUM:SCNC:PT:SER/PLAS:QN). Lastly, Digital Imaging and Communications in Medicine (DICOM) for medical imaging is an accepted standard for both communication and management of this image data, while Picture Archive and Communication System (PACS) is the computer network for digitized medical images and reports. PACS include imaging data, secure network for transmitting information, workstations for reviewing the images, and archives for storing the images.

Big Data

Data that have escalated in a myriad of ways to the point that traditional data processing applications are no longer adequate is termed "Big Data." The four "V"s of big data often mentioned are (1) volume (more than 40 ZB, or the equivalent of 40 trillion gigabytes, are expected to be in existence by 2020 with Internet of Things accelerating this growth), (2) variety (videos, wearable technology, images, and structured vs unstructured types of data created a digital "tsunami"), (3) velocity (speed data are accessed such as with streaming data and over 20 billion network connections in the near future powered by 5G), and (4) veracity (uncertainty of data is not only costly but also leads to inaccurate conclusions) [3]. Additional "V"s in big data can include value, visualization, and variability.

The current imbroglio in health-care data is highlighted by an escalating volume of unstructured, heterogeneous medical data with little embedded predictive analytics or machine learning [4,5]. In regards to the size of biomedical data, while EMRs are in the range of 5−10 MB, radiological and cardiac imaging studies can be 10−100-fold more at 50−100 MB; these data still are not nearly as big as clinical genomic data (which can be in the range of 20 GB or more). Lastly, current estimate of the entire health-care data volume is above 150 EB in volume and escalating rapidly [6].

Self-driving autos, latter day EHR's, and monsters under the bed: operationalizing data for outcomes which matter to patients

Richard A. Frank

Siemens Healthineers, Erlangen, Germany

Richard A. Frank, an internist with decades of experience in industry, authored this commentary on AI and its capability of transforming health care data into actionable elements to fulfill the Quadruple Aims, especially improved patient outcomes.

Patients and physicians alike feel they have reason to fear augmented intelligence. Among the "Triple Aims" of improved outcomes, enhanced patient experience, and reduced costs, patients reasonably anticipate suffering from an efficiency overdrive rather than improvements in the outcomes which matter to patients. This is analogous to autonomous vehicles, which prioritize the smoothness of ride over the safety of pedestrians; sensors detected a pedestrian walking a bike, but emergency braking maneuvers were not enabled while the vehicle was under computer control in order to reduce the potential for erratic vehicle behavior [1]. The car carried on to impact without pause—albeit with fatal consequences.

Physicians rightly fear a repeat of the overpromise of EHRs and the shortcomings of noninteroperability. Physicians are persecuted by alerts and burdened by the administrative requirements of EHR compliance. The eventual mandates to adopt EHRs engendered resentment. Until the advent of prior authorization, EHRs were the principal cause of physician burn-out and suicidal ideation. This led physicians to propose the Quadruple Aim [2], adding "improving the work life of health care providers, including clinicians and staff."

The good news is that artificial intelligence (AI) is not EHR, or at least it need not be if we can learn from experience. Patients can be reassured that clinical decision-making is augmented, not dictated, by health-care AI. To keep up with the medical literature, a physician would have to read 13 articles daily. Over the last few decades this cognitive burden has, along with revenue incentives, driven physicians from primary to specialty care. Today, even specialists suffer this cognitive burden, having become victims of their own success in generating so much high quality, actionable knowledge. This is complicated further by the growing shortage of physicians, which shortage is particularly acute in some geographic areas.

Patients will be reassured that the physician who consults AI will bring the greatest discernment and the most medical knowledge to bear on their problem. The physician will be fulfilled by providing better care.

Physicians can be reassured that, at least from traditional innovators of medical devices, there will be interoperability (witness DICOM for imaging), ease of workflow, and availability at the point of care [3] as we operationalize the value of data. The key to achieving the savings and outcomes promised by AI [4] will be incentives in the form of payment warranted by the value added by AI—rather than mandates to override resistance to the administrative burdens of EHRs.

Physicians welcome improvements in outcome and can ascribe their Seal of Approval, "standard of care" to that AI which qualifies for wide adoption into clinical practice [5], a de facto mandate to better care. Health-care professionals may anticipate a reduction in liability claims attributable to better outcomes—not only reductions in medication errors.

Incentives will enable equitable access geographically and socioeconomically, from tertiary-level academic medical centers and world-class municipalities to community clinics and rural settings. Access to care, particularly among the underserved, is complicated by a growing shortage of physicians which can be alleviated by AI as a companion along the care pathway, freeing up physicians from mundane and routine tasks while relieving both administrative and cognitive burdens.

Beyond the streamlined interoperability of DICOM, mature innovators are building a digital ecosystem which not only respects the importance of cybersecurity, patient privacy, and workflow, but also facilitates integration consistent with the multidisciplinary nature of most diseases.

The monsters under the bed—overdrive to cost-efficiency and mandate to comply—may yet devour the beneficent improvements in patient outcome and physician satisfaction. Instead, let us vanquish the monsters by valuing, investing in, and paying for the work done by machines in augmenting clinical decision-making, thereby reaping the benefits of new standards of care based on operationalizing data for outcomes which matter to patients.

References

[1] NTSB. Preliminary report highway HWY18MH010. Available from <https://www.ntsb.gov/investigations/AccidentReports/Reports/HWY18MH010-prelim.pdf>; 2015 [Accessed 05.04.19].

[2] Bodenheimer T, Sinsky C. From triple to quadruple aim: care of the patient requires care of the provider. Ann Fam Med 2014;12(6):573–6. Available from <http://www.annfammed.org/content/12/6/573.full> [Accessed 05.04.19].

[3] Daniel G, Silcox C, Sharma I, Wright MB, Blake K, Frank R, et al. Current state and near-term priorities for AI-enabled diagnostic support software in health care. Available from <https://healthpolicy.duke.edu/news/white-paper-release-current-state-and-near-term-priorities-ai-enabled-diagnostic-support>; 2019 [Accessed 05.04.19].

[4] Collier M, Fu R. Technology; ten promising AI applications in health care. Harvard Business Review May 10, 2018. Available from <https://hbr.org/2018/05/10-promising-ai-applications-in-health-care> [Accessed 05.04.19].

[5] JASON report. Artificial intelligence for health and health care. Dec 2017. Section 2.3. Available from <https://fas.org/irp/agency/dod/jason/ai-health.pdf> [Accessed 05.0419].

Despite the large volume, variety, velocity, and veracity of big data in biomedicine, there is little dividend in the form of information from this health-care big data [7,8]. Yet, there are sizable opportunities for utilizing health-care big data to reduce costs, reduce readmissions, improve triage, predict decompensation, prevent adverse events, and introduce treatment optimization [9]. The current situation will soon be far more complex and daunting with the advent of data "tsunamis": genomic data (as a result of the high throughput next generation sequencing) [10] and physiologic data (from home monitoring and wearable physiologic devices) [11].

Health-care Data Conundrum

Health-care data are unique in several ways that renders the data challenging for AI applications:

1. Data size—The size of health-care data, especially genomic data and some imaging data as well as future wearable technology data, is getting larger and more difficult to manage and store.
2. Data location—The data are in various formats (clinical data vs claims data) and are often stored in several repositories such as clinics, hospitals, and other departments (radiology, laboratories, etc.).
3. Data structure—Most (over 80%) of the health-care data, from handwritten doctors' notes to echocardiograms, remain unstructured and, therefore, it is difficult to handle as a data bundle.
4. Data integrity—It is not uncommon to have some or most of the data in health care missing and/or inaccurate, but this aspect can be partly neutralized with data mining and data-analytics strategies.
5. Data consistency—Data are often inconsistently recorded as diagnoses, and conditions are frequently defined differently by clinicians as there is often no universal definition for even the simplest of medical terms.

Finding the relevant data

Pieter Vorenkamp

Pieter Vorenkamp, an electric engineer and the father of someone special who succumbed to cancer, authored this commentary about his difficult saga to get and manage data from his son's medical records in order to improve his changes for surviving his cancer.

> *How Patient Registries can be a valuable tool to progress treatment outcomes for rare diseases*
> "Tim's Dad"—The Live for Others Foundation, Laguna Beach CA (www.L4OF.org)
> COO—Syntiant Corp., Irvine CA (www.syntiant.com)

Pediatric cancers are unique and extremely rare compared to adult cancers. Based on the data from the CI5-X database [1] as illustrated in Fig. 1, more than 90% of all cancer patients are diagnosed at the age of 45 years or older. Only 1% of all cancer patients are diagnosed at the age of 20 years and younger.

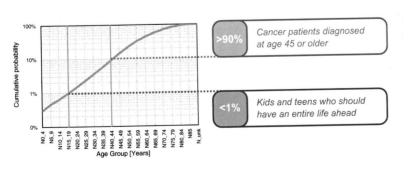

FIGURE 1 Cumulative and normalized size of the total cancer population.

On top of this the types of cancers found in adult patients is fundamentally different from pediatric cancer patients. As illustrated in Fig. 2, there is a complete lack of correlation between the top-5 adult cancers (which include lung, breast, prostate, colon, and bladder) and the top-10 pediatric cancers (which include lymphoid leukemia, brain, Non-Hodgkin's lymphoma, Hodgkin's disease, bone, myeloid leukemia, connective and soft tissue, kidney, thyroid, and testicular).

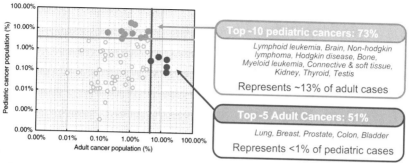

FIGURE 2   Correlation between adult and pediatric cancers.

A focus on synovial sarcoma

Synovial sarcoma is a subset of connective and soft tissue cancer, and primary diagnosis is typically with teens and young adults. The initial tumor is mostly found at or near the joints of extremities. Synovial sarcoma is characterized by a unique chromosomal translocation $t(X; 18)$ [2,3]. Metastasis is commonly reported in the lungs and drops the average survival rate to less than 5 years from diagnosis.

Because of these factors, research and drug development through the pharmaceutical industry is rare and limited. In addition, most research in the field of pediatric cancers is usually focused on the "success stories," where, based on novel and experimental treatments, the observed median overall survival rate has shown improvement, usually on a statistically insignificant sample size [4]. Unfortunately, the "nonsuccess stories" are forgotten and not considered. Often, the records and treatment history of patients who do not survive get lost or disregarded. The important data that is hidden in these patient records is difficult to find. This data are typically unstructured and often include handwritten notes. Unfortunately, there is no tool available today to extract a complete patient history from this potentially valuable data. Furthermore, this data do not get shared with clinicians and research teams. However, these records contain essential information. The question is "How do we access and leverage this information?" Or perhaps the even more pertinent question is "How are future Synovial Sarcoma patients receiving the benefit of what we can learn from past patients, their medical history and the treatment protocols they received?"

Expanding the data set

We strongly believe that by including and sharing the information associated with the few "success stories" and these "nonsuccess stories," and leveraging DNA and RNA breakthroughs, combined with strong collaborations across research and clinical teams supported by focused foundation organizations we can find patterns and commonalities leading to the development of relevant statistical models and advancement toward a cure—even for the rarest diseases.

In 2017 Live for Others Foundation (L4OF) established a Patient Registry, a component of the ShareMD collaborative initiative. The concept which is based on collecting data from synovial sarcoma (SS) patients has been received with significant enthusiasm not only by patients and their parents but also by researchers and oncologists. To date, L4OF has registered over 80 patients, and the number is growing rapidly.

About the live for others foundation—L4OF.org

Founded by Tim Vorenkamp in 2015, the L4OF is a nonprofit organization that was created to help raise awareness for and fund the research needed to find a cure for SS. During Tim's 5-year battle with this extremely rare cancer, he became a beacon of awareness and hope. Even though he was only 18 years old, Tim was known for his positive spirit and never-ending zest for life, working hard until his final days to get the Foundation in place so he could help others and keep his mission alive.

In order to make progress in fulfilling Tim's wish and also inspired by the urgent desire of patients and families with this disease, we strongly believe that the sharing and analyzing of medical records combined with genetic information can enable a crucial step forward.

L4OF has created the Synovial Sarcoma Patient Registry, a component of the *ShareMD* collaboration, dedicated to SS patients, an IRB (Institutional Review Board) approved patient registry. Patient information will be anonymous (fully compliant with HPPA requirements) and will not be shared with anyone without proper consent. The long-term vision of the foundation is to expand this platform beyond SS patients to include a wider range of medical professionals and researchers in the future.

Life is a Journey, live for the people you experience it with. —*Tim Vorenkamp*

References

[1] Forman D, Bray F, Brewster DH, Gombe Mbalawa C, Kohler B, Piñeros M, et al., editors. Cancer incidence in five continents, Vol. X, IARC scientific publication no. 164. Lyon: International Agency for Research on Cancer; 2014.
[2] Ren T, Lu Q, Guo W, Lou Z, Peng X, Jiao G, et al. The clinical implication of SS18-SSX fusion gene in synovial sarcoma. Br J Cancer 2013;109:2279—85.
[3] Coindre JM, Pelmus M, Hostein I, Lussan C, Bui BN, Guillou L. "Should molecular testing be required for diagnosing synovial sarcoma?" A prospective study of 204 cases. Cancer 2003;98(12):2700—7.
[4] Krieg AH, Hefti F, Speth BM, Jundt G, Guillou L, Exner UG, et al. Synovial sarcomas usually metastasize after >5 years: a multicenter retrospective analysis with minimum follow-up of 10 years for survivors. Ann Oncol 2011;22(2):458—67.

Health-care Data Management

Data Processing and Storage

An ETL (extract, transform, and load) process is employed in order to extract data out of the system and configure the data for the data warehouse that is favored by business professionals as the data is usually structured (but storage usually more costly). A data lake is a lower cost data storage repository preferred by data scientists and can hold large amounts of raw data, including unstructured data, for later analytic use. The data in the data lake are also somewhat more agile for flexible configuration, whereas the data in the data warehouse is usually less agile. There are important distinctions between data warehouse and data lake, but perhaps a hybrid "data reservoir" would be the best data repository to take advantage of both types of data storage. Health-care data can be stored in a database-management system (DBMS) either on a server or a distributed computing storage platform for both access and analysis (Hadoop is such a system). Vendors such as Amazon, Google, and IBM now have data storage and analytics available in the cloud.

Electronic Medical Record Adoption and Interoperability

In order to facilitate clinical and administrative data being transferred between software applications, especially with AI-related work, health level-7 (HL7) denotes the seventh (application) level of the International Organization for Standardization seven-layer communications model for open systems interconnection. The HL7 vision is "a world in which everyone can securely access and use the right health data when and where they need it," so AI work in a health-care organization mandates HL7 as it promotes interoperability of EHRs. Interoperability, according to Health care Information and Management Systems Society (HIMSS), is "the ability of different information systems, devices, or applications to connect, in a coordinated manner, within and across organizational boundaries to access, exchange, and cooperatively use data amongst stakeholders with the goal of optimizing the health of individuals and populations." This interoperability can be foundational, structural, semantic, or organizational. This aspect of health-care data is vital to multiinstitutional collaborations in AI projects. The Fast Health care Interoperability Resources, developed by HL7, is an application programming interface that functions as a standard for data formats for exchanging EHR to promote interoperability.

HL7 is sometimes confused with the HIMSS EMR Adoption Model and its stage designations (see Table 4.1): stage 7 is an environment where paper charts are no longer used (complete EHR), whereas stage 6 is its precursor when health-care organizations are at the forefront of EHR adoption with interpretable EHR and just prior to stage 7.

Health-care Databases

Database-management Systems

The types of DBMSs include hierarchical, network, relational, and object-oriented DBMS. Medical databases have traditionally been primitive flat files with little or no database management and have not advanced far due to EMRs only having been recently implemented [12]. The health-care data have, therefore, been static with indirect sharing mostly through hyperlinks. In short, most of the present health-care data remain embedded in flat files or at best, in relatively simplistic hierarchical or relational DBMS with most of the data centralized and locked into local operating systems that reside in hospitals or offices. There is a paucity of literature on object-oriented approaches in the biomedical area [13].

TABLE 4.1   Health Care Information and Management Systems Society Electronic Medical Record (EMR) Adoption Model (EMRAM).

Stage	
7	Complete EMR, data analytics to improve care
6	Physician documentation (templates), full CDSS, closed loop medication administration
5	Full R-PACS
4	CPOE; clinical decision support (clinical protocols)
3	Clinical documentation, CDSS (error checking)
2	CDR, controlled medical vocabulary, CDS, HIE capable
1	All three ancillaries installed—lab, radiology, pharmacy
0	All three ancillaries not installed

CDR, Clinical data repository; CDS, Clinical decision support; CDSS, clinical decision support systems; CPOE, computerized physician order entry; HIE, health information exchange; PACS, Picture Archive and Communication System.

Relational Database

As the most common health-care database is a relational database, its management system is called a relational DBMS (relational DBMS or simply RDBMS). Oracle and Structured Query Language (SQL) Server are examples of relational DBMS, while Not Only SQL (NoSQL) databases such as MongoDB are nonrelational DBMS. An online transaction processing database is the predominant use case for RDBMS. A major disadvantage of such a database is that the data is often sequestered. Often, an online analytical processing database is used as an enterprise data warehouse solution to solve the problem. There are limitations to relational DBMS for health-care data: these lack sufficient infrastructural support for the larger health-care data (such as time-series data, large text documents, and image/videos). In addition, queries are difficult due to the structure of relational DBMS.

Object-oriented Database

While this type of DBMS is more efficient and flexible, it lacks the practical functionalities of a relational DBMS especially for search and query functions. A hybrid object-relational DBMS, therefore, can take advantage of strengths from both relational and object-oriented DBMS; it can, therefore, accommodate the larger, more complex health-care data elements while retaining the relational table structure for query purposes (using Hadoop, Oracle, or SQL). This object-relational DBMS does, however, require more expertise to use, as it is more complex to configure. The NoSQL or next generation databases represent databases that are characterized by large data volumes, scalable replication and distribution, and efficient queries; these databases are exemplified by document-based systems (such as MongoDB) or graph databases and are the future of health-care databases.

Graph Database

A graph DBMS (used in LinkedIn and Twitter as well as Zephyr Health and Doximity and visualized by Neo4j) can store data in the nonlinear form of graph elements (nodes and edges): a node (also called a vertex) is an entity and an edge is the relationship between the nodes (see Fig. 4.2). This type of database is more "three-dimensional" and has distinct advantages over the traditional relational database (see Table 4.2). This central tenet of delineating connectedness and relationships in a rapid-changing world is often very much needed in biomedicine as quality of care, overall efficiency, and innovation direction become the new paradigm in health care.

In a graph DBMS, each data element in the graph will need to be described in the universal language. Resource description framework (RDF) [14] as a "triple" (<Subject> <Predicate> <Object>) is then stored in a semantic database that can be queried using the semantic version of SQL, SPARQL (Simple Protocol and RDF Query Language). Ontologies and accompanying inference rules can then be embedded in the data to enrich the database.

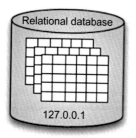 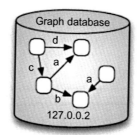

FIGURE 4.2   Relational versus graph database. The figure on the left shows the traditional relational database in table format, whereas the figure on the right shows nodes (*rounded squares*) and edges (*arrows*).

TABLE 4.2   Comparing relational versus graph databases.

	Relational database	Graph database
Format	Table	Graph
Relationship	Row and column	Any-to-any relationship
Data	Structured	Structured and unstructured
Data Type	Simple to moderate	Complex
Number of relationships	Hundreds	Thousands to billions
Quality of relationship	Low and sparse	High and rich
Data model type	Collection of interlinked tables	Multirelational graph
Data model	Infrequently changed	Constantly changing
Agility	Static	Dynamic
Schema	Rigid	Flexible
Deep analytic performance	Poor	Good
Human and machine	Labor intensive	Machine assisted

In short, if relationships are a high priority and data are constantly evolving (such as medicine and health care), a graph DBMS is much more accommodating than traditional relational databases. The graph DBMS with these search algorithms is especially well designed for complex queries in health care such as chronic disease management, acute epidemiological crises, and health-care resource allocation. The location of a similar patient to an index patient can also be performed using this strategy. The major limitation of graph DBMS is that it is relatively large and complex, but this limitation can now be partly neutralized with large storage capacity, semantic storage improvements, and superior search algorithms. A graph or even its more advanced version, hyper graph, is perhaps an essential element for AI in medicine and health-care to advance to the next level.

The Data-to-intelligence Continuum and Artificial Intelligence

The continuum from data to intelligence starts with good data and database management; this is particularly important in health care as often the health-care data are inaccurate and/or incomplete. To have the best AI or intelligence in health-care, one will need to start with good health-care data (the foundation layer of the data-intelligence pyramid) (see the previous figure). As discussed more philosophically at the beginning of the book, data are processed and interpreted to induce meaning, and this leads to the next level, information. While computers need data, humans need information. The recent surge in medical image interpretation and deep learning reminds us just how important this data aspect of AI in biomedicine is. For instance, to simply take a large dataset such as the CXR14 from the NIH with its hundreds of thousands of images and labels (interpretations), to perform feature selection and extraction for deep learning, and expect that the diagnostic tool is near perfect is overly optimistic and naive. There is a long list of issues with data even with large datasets from respected institutions; these include insufficient variability, poor labeling method, inaccurate labeling, inconsistent level structure, hidden stratification (from unlabeled findings), poor documentation, and poor image quality [15].

From information, one attains knowledge, which is derived from one's experience and analysis. Data science is instrumental in the conversion of information to useful knowledge and intelligence. Intelligence is, therefore, the ability as well as the velocity to apply this knowledge. Wisdom is thought to be a quality of "knowing" without necessarily having the logic to affirm the observation or decision. Present day innovative AI methodologies, especially deep reinforcement learning, recursive cortical network, and cognitive architectures, are changing the human role as well as the machine expectations in this data-to-intelligence continuum.

How to Properly Feed Data to an ML Model

Adapted from "Machine Intelligence Primer for Clinicians: No math or programming required"
Alexander Scarlat
https://www.amazon.com/Machine-Intelligence-Primer-Clinicians-Programming/dp/
1794256067/ref = sr_1_1?ie = UTF8&qid = 1549684006&refinements = p_27%
3AAlexander + Scarlat + MD&s = Books&sr = 1-1&text = Alexander + Scarlat + MD

While the other chapters in this book will give you an idea about what AI/ML models can do for us, in this chapter, I will try to sketch what we must do for the machines—before asking them to perform magic. Specifically, the data preparation before it can be fed into an ML model.

For the moment, assume the raw data is arranged in a table with samples as rows and features as columns. These raw features/columns may contain free text, categorical text, discrete data such as ethnicity, integers such as heart rate, floating point numbers such as 12.58, various ICD, DRG, CPT codes, as well as images, voice recordings, videos, and waveforms.

What are the dietary restrictions of an artificial intelligence agent? ML models love their diet to consist of only floating point numbers, preferably small values, centered and scaled/normalized around their means $+/-$ their standard deviations.

No Relational Data

If we have a relational database management system (RDBMS), we must first flatten the one-to-many relationships and summarize them, so one sample or instance fed into the model is truly a good representative summary of that instance.

For example, one patient may have many hemoglobin lab results, so we need to decide what to feed the ML model:

- An aggregator such as the minimum Hb, maximum, Hb daily average, count only abnormal Hb results per day, or . . ..
- Use all the data available as a time series: with each combination of a unique patient-encounter being assigned a time stamp for each entity (Hb), value (12.2), and unit (g/dl).

No Missing Values

There can be no missing values, as it is similar to swallowing air while eating. 0 and n/a are not considered missing values. Null is definitely a missing value. The most common methods of imputing missing values are:

- numbers—the mean, median, 0, etc.
- categorical data—the most frequent value, n/a or 0

No Text

We all know by now that the genetic code is made of raw text with only four letters (A, C, T, and G). Before you run to feed your ML model some raw DNA data and ask it questions about the meaning of life, remember that one cannot feed an ML model raw text, unless you want to see an AI entity, burp and barf.

There are various methods to transform words or characters into numbers. All of them start with a process of tokenization, in which a larger unit of language is broken into smaller tokens. Usually, it suffices to break a document into words and stop there:

- Document into sentences.
- Sentence into words.

- Sentence into *n*-grams, word structures that try to maintain the same semantic meaning (three-word *n*-grams will assume that chronic atrial fibrillation, atrial chronic fibrillation, and fibrillation atrial chronic are all the same concept).
- Words into characters.

Once the text is tokenized, there are two main approaches of text-to-number transformations so text will become more palatable to the ML model:

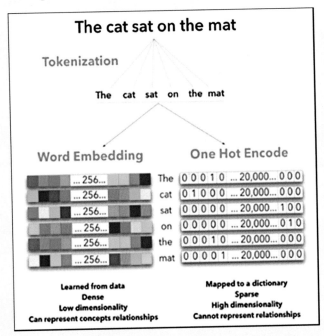

One Hot Encode (right side of the above figure)

Using a dictionary of the 20,000 most commonly used words in the English language, we create a large table with 20,000 columns. Each word becomes a row of 20,000 columns. The word "cat" in the above figure is encoded as 0,1,0,0, ..., 20,000 columns, all 0s except one column with 1. One Hot Encoder—as only one column gets the 1, all the others get 0.

This is a widely used simple transformation that has several limitations:

- The table created will be mostly sparse, as most of the values will be 0 across a row. Sparse tables with high dimensionality (20,000) have their own issues, which may cause a severe indigestion to an ML model, named the curse of dimensionality (see next).
- In addition, one cannot represent the order of the words in a sentence with a One Hot Encoder.

In many cases, such as sentiment analysis of a document, it seems the order of the words does not really matter. Words such as "superb," "perfectly" versus "awful," "horrible" pretty much give away the document sentiment, disregarding where exactly in the document they actually appear:

From "Does sentiment analysis work? A tidy analysis of Yelp reviews" by David Robinson [1].

On the other hand, one can think about a medical document in which a term is negated, such as "no signs of meningitis." In a model where the order of the words is not important, one can foresee a problem with the algorithm not truly understanding the meaning of the negation at the beginning of the sentence.

The semantic relationship between the words mother−father, king−queen, France−Paris, and Starbucks−coffee will be missed by such an encoding process.

Plurals such as child−children will be missed by the One Hot Encoder and will be considered as unrelated terms.

Word Embedding/Vectorization

A different approach is to encode words into multidimensional arrays of floating point numbers (tensors) that are either learned on the fly for a specific job or using an existing pretrained model such as:

- word2vec, which is offered by Google and trained mostly on Google news and
- GloVe from Stanford trained on a Wikipedia, Tweeter, or other crawlers.

Basically an ML model will try to figure the best word vectors—as related to a specific context—and then encode the data to tensors in many dimensions so another model may use it down the pipeline.

This approach does not use a fixed dictionary with the top 20,000 most used words in the English language. It will learn the vectors from the specific context of the documents being fed and create its own multidimensional tensors "dictionary."

An Argentinian start-up generates legal papers without lawyers and suggests a ruling, which in 33 out of 33 cases has been accepted by a human judge [2].

Word vectorization is context sensitive. A great set of vectorized legal words (like the Argentinian start-up may have used) will fail when presented with medical terms and vice versa.

In the previous figure, I have used many colors, instead of 0 and 1, in each cell of the word embedding example to give an idea of about 256 dimensions and their capability of storing information in a much denser format. Please do not try to feed colors directly to an ML model as it may void your warranty.

Consider an example where words are vectors in only 2 dimensions (not 256 dimensions). Each word is an arrow starting at 0,0 and ending on some X,Y coordinates:

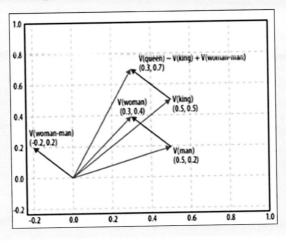

From Deep Learning Cookbook by Douwe Osinga [3].

The interesting part about words as vectors is that we can now visualize, in a limited 2D space, how the conceptual distance between the terms man–woman is being translated by the word vectorization algorithm into a physical geometrical distance, which is quite similar to the distance between the terms king–queen. If in only two dimensions the algorithm can generalize from man–woman to king–queen, what can it learn about more complex semantic relationships and hundreds of dimensions?

We can ask such an ML model interesting questions and get answers that are already beyond human level performance:

- Q: Paris is to France as Berlin is to? A: Germany.
- Q: Starbucks is to coffee as Apple is to? A: IPhone.
- Q: What are the capitals of all the European countries? A: UK—London, France—Paris, Romania—Bucharest, etc.

The above are real examples using a model trained on Google news.

One can train an ML model with relevant vectorized medical text and see if it can answer questions such as:

- Q: Acute pulmonary edema is to CHF as ketoacidosis is to? A: diabetes.
- Q: What are the three complications a cochlear implant is related to? A: flap necrosis, improper electrode placement, and facial nerve problems.

The word vectorization allows other ML models to deal with text (as tensors)—models that do care about the order of the words, such as algorithms that can deal with time series or text sequences.

Discrete Categories

Consider a drop-down with the following mutually exclusive drugs:

1. Viadur
2. Viagra
3. Vibramycin
4. Vicodin

As the previous text seems already encoded (e.g., Vicodin = 4), you may be tempted to eliminate the text and leave the numbers as the encoded values for these drugs. That is not a good idea. The algorithm will erroneously deduce that there is a conceptual similarity between the abovementioned drugs just because of their similar range of numbers. After all, two and three are really close from a machine's perspective, especially if it is a 10,000-drug list.

The list of drugs being ordered alphabetically by their brand names does not imply that there is any conceptual or pharmacological relationship between Viagra and Vibramycin.

Mutually exclusive categories are transformed to numbers with the One Hot Encoder technique detailed previously. The result will be a table with the columns: Viadur, Viagra, Vibramycin, and Viocodin (similar to the words tokenized previously: "the," "cat," etc.).

Each instance (row) will have one and only one of the abovementioned columns encoded with a 1, while all the others will be encoded to 0. In this arrangement the algorithm is not induced into error and the model will not find conceptual relationships where there are none.

Normalization

When an algorithm is comparing numerical values such as creatinine = 3.8, age = 1, heparin = 5000, the ML model will give a disproportionate importance and incorrect interpretation to the heparin parameter, just because heparin has a high raw value when compared to all the other numbers.

One of the most common solutions is to normalize each column:

- Calculate the mean and standard deviation.
- Replace the raw values with the new normalized ones.

When normalized, the algorithm will correctly interpret the creatinine and the age of the patient to be the important, deviant from the average kind of features in this sample, while the heparin will be regarded as normal.

Curse of Dimensionality

If you have a table with 10,000 features (columns), you may think that is great as it is feature-rich. But if this table has fewer than 10,000 samples (examples), you should expect ML models that would vehemently refuse to digest your data set or just produce really weird outputs.

This is called the curse of dimensionality. As the number of dimensions increases, the "volume" of the hyperspace created increases much faster, to a point where the data available becomes sparse. That interferes with achieving any statistical significance on any metric and will also prevent an ML model from finding clusters since the data is too sparse.

Preferably the number of samples should be at least three orders of magnitude larger than the number of features. A 10,000-column table better be garnished by at least 1,000,000 rows (samples).

Tensors

After all the effort invested in the data preparation previously, what kind of tensors can we offer now as food for thought to a machine?

- 2D—table: samples, features
- 3D—time sequences: samples, features, time
- 4D—images: samples, height, width, RGB (color)
- 5D—videos: samples, frames, height, width, RGB (color)

Note that samples is the first dimension in all cases.

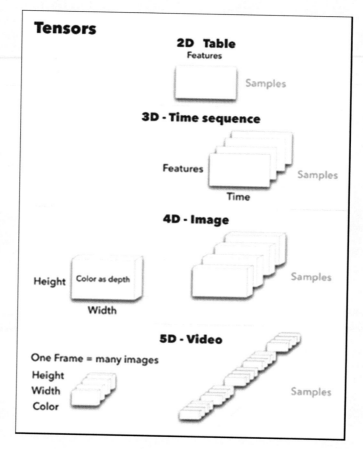

Hopefully this chapter will cause no indigestion to any human or artificial entity.

References

[1] http://varianceexplained.org/r/yelp-sentiment/

[2] https://www.bloomberg.com/news/articles/2018-10-26/this-ai-startup-generates-legal-papers-without-lawyers-and-suggests-a-ruling?srnd = businessweek-v2

[3] https://www.amazon.com/Deep-Learning-Cookbook-Practical-Recipes-ebook/dp/B07DK1ZZXT?keywords = douwe + osinga&qid = 1540440172&s = Books&sr = 1-1-fkmrnull&ref = sr_1_fkmrnull_1

References

[1] Weil AR. Big data in health: a new era for research and patient care. Health Aff 2014;33:1110.

[2] Healthcare Content Management White Paper. Unstructured data in electronic health record (HER) systems: challenges and solutions. <www.datamark.net>; 2013.

[3] Chang AC. Big data in medicine: the upcoming artificial intelligence. Prog Pediatr Cardiol 2016;43:91—4.

[4] Chang AC, et al. Artificial intelligence in pediatric cardiology: an innovative transformation in patient care, clinical research, and medical education. Cong Card Today 2012;10:1−12.

[5] Roski J, et al. Creating value in health care through big data: opportunities and policy implications. Health Aff 2014;33(7):1115−22.

[6] Hughes G. How big is "big data" in health care? SAS Blogs October 11, 2011.

[7] Jee K, et al. Potentiality of big data in the medical sector: focus on how to reshape the health care system. Healthc Infrom Res 2013;19(2):79−85.

[8] Schneeweiss S. Learning from big health care data. N Eng J Med 2014;370:2161−3.

[9] Bates DW, et al. Big data in health care: using analytics to identify and manage high-risk and high-cost patients. Health Aff 2014;7(2014):1123−31.

[10] Feero WG, et al. Review article: genomic medicine—an updated primer. N Engl J Med 2010;362:2001−11.

[11] Chan M, et al. Smart wearable systems: current status and future challenges. Artif Intell Med 2012;56 (3):137−56.

[12] Mandl KD, et al. Escaping the HER trap—the future of health IT. New Engl J Med 2012;366:2240−2.

[13] Gu H, et al. Benefits of an object-oriented database representation for controlled medical terminologies. J Am Med Inform Assoc 1999;6(4):283−303.

[14] Anguita A, et al. Toward a view-oriented approach for aligning RDF-based biomedical repositories. Methods Inf Med 2015;53(4):50−5.

[15] Blog from Luke Oaken-Rayner, < www.lukeoakdenrayner.wordpress.com >; February 25, 2019.

5

Machine and Deep Learning

Introduction to Machine Learning

The following are introductory concepts in machine learning (ML).

Data Mining and Knowledge Discovery

Data science encompasses data mining as well as ML, and these two areas have some overlap and are therefore sometimes can be confusing to the reader. While a data scientist can use data mining with its databases and statistical methodologies to draw information from the data (such as emerging patterns), a data scientist utilizes ML to provide the machine the opportunity to learn on its own. In other words, data mining is more a source of information for ML to benefit from (with its capability to learn automatically and to make predictions). Two data-mining methods are association rule mining that describes relationships among the attributes but is limited by its binary nature and sequential pattern discovery that is designed for sequential or temporal data.

History and Current State of Machine Learning

ML, defined as the ability of the machine to learn from its experience going through tasks, is widely used in our society (from search engines to spam filtering). ML, a term initially coined by Arthur Samuel in 1959, is an increasingly popular subdiscipline of AI and is the art of computer programing that enables the computer to learn and improve its performance without an external program instructing it to do so. In other words, in ML, the algorithms self-improve and "learn" from trial and error just as humans do from experience.

Pedro Domingos, in his book *The Master Algorithm*, described the five schools or "tribes" of ML that share one paradigm in common: discovery of the knowledge hidden in the data. He proposed that a master algorithm would need elements from each of the five schools described in Table 5.1.

Whereas the symbolists clearly were the main force in the earlier AI era and the Bayesians also had a good run during this time (see the previous section), it is evident that the current era is mostly dominated by the connectionists (especially since the popular

TABLE 5.1 Schools in machine learning.

School	Representation	Origin and influence	Methodologies	Key algorithm
Symbolists	Logic	Philosophy Computer science	Production rule system Inverse deduction Decision trees	Inverse deduction
Connectionists	Neural networks	Neurosciences	Back propagation Deep learning Deep reinforcement learning	Back propagation
Evolutionaries	Genetic programs	Evolutionary biology	Genetic algorithms Evolutionary programing Evolutionary game theory	Genetic programing
Bayesians	Graphical models	Statistics	HMM Graphical model Causal inference	Probabilistic inference
Analogizers	Support vectors	Psychology	k-NN SVM	Kernel machines

HMM, Hidden Markov model; k-NN, k-nearest neighbor; SVM, support vector machines.
Adapted from The Master Algorithm.

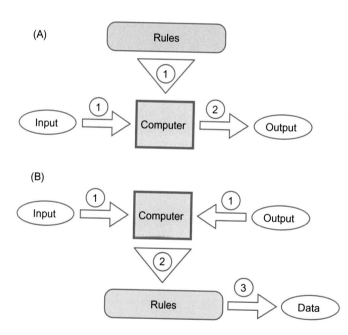

FIGURE 5.1 (A) Traditional programing and (B) machine learning. The diagrams illustrate the major difference between traditional programing and machine learning. In traditional programing (and statistical analysis), a top-down approach provides rules for the input data and output is derived. In machine learning, both the input data as well as output data (labeled by humans) are entered into the computer and the rules are derived from the data. The new rules are then applied to the new set of data (see the text for more details).

deep-learning algorithm surged in popularity since 2012). It is hoped that someday there can be a unifying effort to incorporate all five schools into one "master algorithm."

Machine Learning versus Conventional Programing

In traditional programing (see Fig. 5.1A), data (Input) along with a computer program (Rules) (step 1) are entered into the computer and then the answers (Output) are derived (step 2). In ML (see Fig. 5.1B), however, a computer uses both the data (Input) and the corresponding answers (Output) (step 1) to derive the "Rules" by examining the patterns in the data (step 2) (and thus "learns" from data). These new rules are then applied to new data to get answers (predictions) (step 3). This latter "bottom-up" approach is distinctly different than the traditional programing (such as the aforementioned rule-based expert system) in which the human is imprinting instructions to the computer to follow (therefor a "top-down" approach). ML is also different than traditional statistical analysis (which is also "top down" like programing in that the data is analyzed by statistical rules to yield output, or answers). In short, ML "learns" from both the input and the output data and then finds the patterns that relate the input data to the output data; these patterns (in the form of a model) learned from ML are then applied to new data to see how this model would fit the data.

The advent of complex and efficient algorithms (sets of steps to accomplish certain tasks) that are available for not only calculations and data processing but also automated reasoning has advanced the capabilities of machine intelligence. Examples of complex algorithms that are in current use include Pixar's coloring of 3D characters in virtual space (rendering algorithm) and NASA's operations of the solar panels on the international space station (optimization algorithm). Even the recent first-ever picture of the black hole of the galaxy Messier 87 was helped by data scientists and a new algorithm (called Continuous High-Resolution Image Reconstruction using Patch Priors, or CHIRP) that put together data from a virtual consortium of telescopes that rendered this into an Earth-sized giant telescope.

Machine Learning Workflow

The following is a brief description of the steps of the ML workflow for supervised learning (but can also be applied to unsupervised learning with a small alteration) (see Fig. 5.2):

Data collection (step 1): Prior to data collection, the team as a group should discuss the expected goals of the project to have a clear direction for what data to collect. When consensus is reached, then members can help to gather the relevant data for the important next step, which prepares the data. Data collection is sometimes challenging depending on the data acquisition and storage capabilities of the institution. In health care, data is mostly unstructured and stored in many different formats (current procedural terminology and ICD-10 codes, encounter information, demographics, medications, nurses and doctors' notes, vital signs, etc.) as well as venues (hospital, clinic, and now wearable devices), and this makes the process of data collection particularly tedious (see the previous section for details)

Data processing (step 2): This step after collection of data involves utilizing data processing tools to have the data cleaned, readied, and organized into a more structured format for ML methodologies to follow. The data preparation involves measures such as missing value imputation, imbalanced data processing, outlier detection, and normalization. There is also data wrangling, or data munging, which are terms describing the process of converting as well as mapping data from the raw form from step 1 to another format that would be more friendly to the ensuing steps. This step can entail both data transformation (to tables, comma-separated value, etc.) and data platform (using Hadoop, MapReduce, Cassandra, etc.). Data curation involves the grouping of data (procedures, complications, comorbidity, etc.).

The steps of collecting and processing the data can easily make up the majority of the effort and time needed to do a project for the data scientist, especially in a clinical setting.

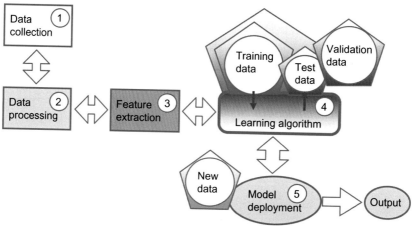

FIGURE 5.2   Machine-learning workflow. The entire machine-learning workflow is illustrated, from data collection all the way to deployment of the model to real-world data at the very end. Most of the workflow has feedback to the prior step(s) and also the model is evaluated and reassessed even during the initial deployment phase so that needed adjustments can be made. The double arrows in the diagram signify the fluidity of these steps so that one can return to previous steps to add/refine/change the data, features, or even model. This is not possible for deep learning as the intermediate steps are "compressed" so that once labeled samples go in, the feature extraction and classification steps are combined in deep learning (machines are performing the feature extraction instead of humans). In other words, steps 3 and 4 are combined in deep learning. For unsupervised learning, the algorithm yields grouping of objects instead of the predictive model in the aforementioned machine-learning workflow.

It is hoped that these two steps will be much more efficient in the future as projects in AI in medicine become more commonplace and at a more sophisticated level.

Feature extraction (step 3): Features, also known as variables or parameters, are the "column names" of the data (such as blood pressure, patient names, and medications), whereas examples are the actual numbers or data. This step of feature extraction first involves feature selection, which is a strategic selection of only relevant features that will create a good predictive model. Feature selection is then followed by feature extraction, which takes the existing features and transforms this into a new set of features that will most likely lead to a good model. Feature engineering is the sophistication of feature extraction using domain expertise. Of note, representation learning, a type of ML in which no feature engineering is involved, is a methodology that renders data easier to extract useful information when building the classifier.

Learning algorithm (step 4): This step involves the algorithm learning to come up with the best predictive model to yield good-to-excellent predictions of future new data. The substeps involve using a dataset and dividing it into two smaller datasets: the training data (usually larger of the two and around 70%) is used to train (or "fit") the model or algorithm and the smaller subset of test data is sequestered and then used for assessing the generalization error of the final chosen model in an unbiased way at the end. Depending on the availability and size of the labeled dataset, a third smaller dataset can be allocated as validation data (also smaller than the training dataset and similar in size to the test dataset) which is used to estimate prediction error for model selection. Overfitting (see later underfitting for a detailed discussion) occurs when the data fit the training data too closely but ends up not generalizing well for other sets of data.

Model deployment (step 5): This final step is to deploy and refine the predictive model for real-life applications with new data. The model's performance is evaluated (see the section next) and its algorithm refined if necessary for new data or projects. The ML "learns" and self-improves by being able to improve its prediction capability or performance via measures such as optimization or loss function built into the ML algorithm (these concepts will be discussed later in this section).

Design thinking and biomedical artificial intelligence

Jules Sherman

Jules Sherman, a strong proponent of design thinking in biomedicine, authored this commentary on the concept of design science, and how this can impact on data science in health care by addressing the problems.

Design thinking is in many ways a distinctly human process. It requires designers to empathize with users to understand what users really want, and it takes a "people first" approach to tackling problems by imagining desirable solutions that meet explicit or latent needs [1]. While I enjoy being a researcher and medical device designer with the tools I have now, I can imagine myriad ways in which my job could be optimized through artificial intelligence (AI). Namely, AI will enhance the iteration and test phases of the design thinking process.

AI is already changing the biomedical sector in many ways. To name a few, it helps gather real-time, real-world data for clinical trials; it can test for millions of complex interactions in seconds to evaluate drug efficacy; and it has been shown to assess vast amounts of data to provide highly accurate diagnoses, sometimes surpassing the efforts of trained specialists. Prosthetics design is a great example where the employment of AI has led to better outcomes [2]. Machines take measurements in real-time to help manufacturers understand how a user's gait has changed, or how their body is compensating with the addition of an artificial limb. These data allow manufacturers to personalize these wearable products far beyond the initial design.

At its core, AI will offer personalized solutions. Once AI becomes a regular part of the design thinking process, it will make biomedical products work better for each individual. Since AI is able to dynamically synthesize consistent user feedback, it will allow designers, engineers, and manufacturers to iterate more precisely at break-neck speed.

Imagine another scenario—a skincare line with a companion smartphone application. The app could remind the consumer to take photos of her skin as she uses the products so that AI could scan the images to measure improvements in tone, acne, or fine lines. The app could then send information back to the manufacturer of the products so that the manufacturer can tweak the formula to give the consumer better and better results as she reorders the product. The product would never be the same. It would change as the consumer's skin changes so that results constantly improve. This is the future of biomedical AI: rapid iterative improvements based on measuring outcomes, a pillar of the design thinking process.

We are at the beginning of a revolution. Humans are now merging with computers in a way that has never occurred previously. Innovations in data aggregation and advances in additive manufacturing processes represent an incredible convergence of bytes and atoms. As a result, machines will gradually take over many of the steps of design thinking, including research and synthesis, in addition to ideation, testing, and refinement.

In a world that is becoming ever more automated, humans' primary role will be more "meta." Designers, who were once tasked with drawing up concepts from scratch and passing over specifications to developers or manufacturers, will transition to become "design scientists." These design scientists will calibrate values, select parameters, and apply relevant aesthetics models for AI embedded into a software system. They will be tapped to conceptualize programs and tools that will orchestrate a complex series of steps in order to simulate what we have been trained to do manually.

Design scientists will work alongside programers, mathematicians, and statisticians in order to create practical design process tools that utilize data from successful innovation in various fields and apply that knowledge to innovation in medicine and medical product design. This new model will indeed automate the design thinking process; still, machines may not be able to improve upon the initial step of need finding, which often requires human-to-human empathy and long-term vision. With the help of AI, however, humans will be able to use synthesized data (i.e., data with meaning) to accelerate their development of empathy for the user.

"Design science" is being born now. In order to reach its full potential, interdisciplinary collaboration between doctors, computer scientists, patients, and designers will be key.

References

[1] Brown T. Design thinking. Harv Bus Rev, June 2008;84—95.
[2] Berboucha M. Artificial intelligence and prosthetics join forces to create new generation bionic hand. Forbes Jul 13, 2018.

Data Science in Biomedicine

As we discussed earlier, data science is considered to be the intersection of mathematics and statistics (including modeling and biostatistics) and computer science (programing, data concepts, and data mining). The current paradigm of AI in medicine for biomedical data science is adding another domain of knowledge to computer science and mathematics: the domain knowledge of biomedicine (bioinformatics and clinical informatics as well as biology, genetics and genomics, medicine, and health sciences). Bioinformatics is a field that focuses on collecting and analyzing complex biological data (especially genetic information), whereas clinical informatics (or biomedical informatics) is the study and practice of data and information approach to health care. Both are germane to not only biomedical data science but AI in medicine and health care.

Some programs and hospitals offer a fellowship in biomedical informatics, and there is now a board certification in clinical informatics. At Children's Hospital of Orange County, there is a pilot program to have a senior fellow in data science and AI in medicine and health care. In the near future, there will be many such positions for specialized training and education as a subarea for most subspecialties.

Agile data science: an AI-driven approach to clinical research and informatics

Ryan O'Connell

Ryan O'Connell, a pathologist and informatician, authored this commentary on the futuristic notion of agile data science, a dimension of medicine in which one can query large databases and answer questions in a very short amount of time, minutes to hours rather than weeks and months.

The idea of agile data science pertains to the ever-increasing ability to

1. acquire large data sets,
2. rapidly parse data,
3. actively pivot and evolve queries in real time, and
4. use established platforms and libraries rather than develop tools de novo.

The availability of health information databases for research is slowly increasing despite a multitude of regulatory and financial roadblocks. One such database is the Medical Information Mart for Intensive Care III (MIMIC-III) which contains the deidentified data generated by over 40,000 patients who received care in the ICU between 2001 and 2012 at Beth Israel Deaconess Medical Center. There are several options for downloading the MIMIC-III Database, including flat file copies of the database tables. Databases for imaging studies also exist, such as the Chest X-ray 14 Dataset released by the NIH, which contains over 100,000 chest X-ray images.

Once these data sets have been obtained, storage and compute power are easily purchased from an ever-increasing number of vendors within the space, with or without HIPAA compliance. The compute power needed for analyzing large datasets can often be met using cloud computing resources with Amazon Web Services and Microsoft Azure being the largest providers of cloud services. The need for cloud-computing tools rests mainly on the availability of specialized elastic compute instances. The elasticity implies that more computing resources can be assessed in real time or scaled down as need. This shift to elastic cloud resources has seen one of the major electronic medical records providers, Cerner Corporation, to develop agile data science tools that use cloud computing resources as the underlying computing engine. These agile cloud computing tools for data science often use Jupyter as the underlying front-end programing interface. Using an open-source computational environment such as Jupyter Notebook supports programing languages such as Python and R required for parsing the data because Jupyter is programing language agnostic and supports collaboration between multiple individuals. This also facilitates the ability for investigators to both parse data, manipulate queries, and visualize results in real time. Rapidly cycling through this process of asking a question and visualizing the data allows for iterative evaluation and improvement of the clinical question or theory. Finally, once a clinically impactful project is identified, the use of open-source machine—learning (ML) libraries such as Pytorch and TensorFlow drastically abbreviate a process that could have taken months of development a decade ago. These cloud computing tools also provide the state of the art in the analysis of big data using Apache Spark. Spark is a parallel distributed cloud computing programing paradigm that provides a faster analysis, preprocessing, and model development on datasets often too large to fit in a single computer. It is the defacto tool for real-time model development, update, and implementation.

The recommended test to illustrate the concept of agile data science is to acquire a freely available data set and team up at least one clinical expert and one data scientist. Then have the clinical expert identify a patient cohort of interest to them and propose an outcome that concerns them the most in that cohort. The team then reviews the elements available in the data set and determines which variables potentially contribute to the identified outcome. Patient counts and simple descriptive statistics can serve as a "sanity checks" along the way. When cohorts become undesirably small or the data fails a "sanity check" the team can quickly pivot their inquiry or dive into why there might be errors in a field (i.e., missing or double counting). This stage of the analysis is part of data preprocessing that usually takes the greatest chunk of time from data scientists or other analysts. But with agile data science tools, incredibly fast preprocessing can be conducted to determine the feasibility of a study and to build initial models as a measure of the expected performance of a predictive model. The combination of the data scientist, domain experts (such as a provider), and agile data science tools will help revolutionize the whole data science process and accelerate discoveries in medicine as well as other application domains.

The data science team for biomedicine consists of the following members:

A data scientist is very well-rounded and is someone who can usually take a data science project starting from data collection through ML and then finishing with data visualization. The data scientist has a skill set from mathematics and statistical analysis, database warehousing, and engineering as well as programing skills [particularly with R, Python, and structured query language (SQL)]. A data engineer (also known as database administrators or data architects) is mainly focused on processing and managing large datasets with software engineering so that the data scientists can work on these datasets. A data engineer works mainly with Hadoop, NoSQL, and Python (but not usually R as ML is usually not in his/her skill set). Finally, a data analyst is typically someone who is more business-focused and utilizes tools such as Excel, Tableau, and SQL for data visual presentation and communication for the enterprise. Data analysts are usually not working directly with data science projects [so machine and deep learning (DL) are usually not in their portfolio of skills] but are focusing on data warehousing, Hadoop-based analytics, and are familiar with data architecture and extract, transform, and load tools. These personnel may report to the same person in the C-suite or different leaders depending on the institution. There is considerable overlap with these job descriptions based on the individual's skills and experiences as well as preferences.

This data science team usually works with a leadership team in the health care organization or the chief executives in a company. The chief information officer (or information technology director) is usually the most senior executive in the information technology sector of the organization. The chief medical information (or informatics) officer (or chief health information officer) is usually a physician executive in charge of the health informatics sector of the organization and is the liaison between the clinical and the IT domains. A chief technology officer (more common in companies rather than health care organizations) is usually someone who knows how to monetize software and deals with software engineering issues. Finally, chief intelligence officer or chief AI officer (very few at present especially in a health care organization) is someone who has a comprehensive knowledge about AI and the know-how about evaluation and deployment of AI projects.

Data scientists are from Mars, clinicians are from Venus

David Ledbetter

David Ledbetter, a senior data scientist at a children's hospital who has spent an inordinate amount of time with clinicians in an intensive care setting, authored this commentary to elucidate the cultural differences between clinicians and data scientists as well as the collaborative process.

Virtual Pediatric Intensive Care Unit, Children's Hospital Los Angeles, Los Angeles, CA, United States

Data scientists are now an essential resource within a healthcare organization. They bring a multitude of skills that can augment the capabilities of an institution including helping to analyze and visualize data, build data infrastructure, and most iconically, train machine learning (ML) models to make specific predictions. That being said the vast majority of data scientists lack the necessary clinical foundation making it difficult to properly understand clinical contexts. This makes it difficult for data scientists to know which problems are clinically relevant and impedes their ability to provide actionable intelligence (AI).

Clinicians, on the other hand, as the boots in the trenches treating patients day in and day out, are intimately familiar with the clinical settings and terminology which may be opaque or worse for data scientists. They understand what problems they face, what information they need to make more informed decisions, and at what point in the clinical workflow such information would be actionable. Nevertheless, most clinicians are not mathematicians and are not comfortable with datasets that are too large to analyze in Excel. Clinicians are typically trained to think about problems in terms of P-values and R-values, rather than a more pragmatic out-of-sample performance assessment.

There are astronomical distances dividing the clinical and data science teams. Most importantly, the languages spoken by either group is barely recognizable to the other. Pressors or fluids or Lasix, oh my. Clinicians think about solving problems in terms of how solutions fit into a clinical workflow and ease of use (how many clicks do I need to make in the electronic medical record?), whereas data scientists are often concerned about esoteric error functions such as receiver operating characteristics or the technical novelty of an algorithm (will I be able to submit this to NeurIPS?). Without the ability to communicate with each other, there is a world of problems that cannot be overcome. Approaching these problems will require both mindsets working together.

To help bridge that divides, it is important to provide as much exposure to the clinical world as possible for data scientists. A great method to provide clinical exposure is to have data scientists observe morning rounds. This provides a great perspective on the machinations of a clinical unit and the context of real urgent life and death problems. Data scientists can see firsthand how information is transferred among nurses, doctors, parents, and patients and glean insight into what information the clinical team uses and how decisions are made within the clinical workflow. They can begin to understand the relevance of their solutions.

It is also important to team data scientists up with clinicians to work on specific problems. This ensures that problems are analyzed from both a clinical and a data science perspective so that an actionable solution can be synergistically achieved that fits within the clinical workflow. Data scientists are able to talk through issues with a clinician and make that critical leap from merely optimizing a cost function to understanding the clinical implications of deploying a model at a selected threshold and the impact of all four quadrants of the cost analysis (true positives, true negatives, false positives, and false negatives).

Collaboration involves every step of the process:

Conception → I have this problem in the ICU.
Design → What information at what time would help?
Munging → What do these values actually mean?
Assessment → Can this be used at the bedside?

Being able to analyze model failures from a clinical perspective will often shed substantial insights into what went wrong in the model; finding common threads and systemic issues will provide an additional avenue of model refinement and improvement.

It cannot be stressed enough how important it is to build a common culture between the data science and clinical teams. Reminding both groups that data scientists are not just robots sitting behind computer screens and doctors are not just... doctors. Ancillary activities such as going out to happy hour together or going to karaoke as a team enables both groups to see the humanity on either side.

It is also important to exposure clinicians to the data science world. Have clinicians sit in on data science lectures and conceptually walk through how data scientists think about problems. Clinicians will benefit from striving to understand and demystify what it is data scientists do and how and what the limitations are. Nothing data scientists do is much more complicated than logistic regression. Random forests are just lots of random binary splits; neural networks are just stacked logistic regressions. At the end of the day, data scientists are just trying to fit a line through data.

The most valuable characteristic of both a good data scientist and a good clinician is the ability to communicate, listen, and learn. Unfortunately, not everyone is a natural communicator, but communication is a skill, and like any other skill, it can be practiced and honed. Without the ability to communicate there will always be projects that are outside the scope of what is possible.

It is important to keep in mind that both groups want to solve problems in order to help improve patient care. Together, Mars and Venus can understand what are the real problems, how are decisions made, and what actionable information (AI) can be provided within the clinical workflow.

Transition of a machine learner into the clinical domain

Anna Goldenberg

Anna Goldenberg, who entered the clinical venue of a children's hospital to be a valuable asset as a data scientist, authored this commentary to provide a personal perspective on the cultural differences between the two domains and how to effectively implement AI in a health care setting.

Learning to converse. What is a feature? One dimension of the input space. What is input space? All the data that you want to use to make the prediction of the outcome. How do you actually learn a model? We optimize the objective function. What is an objective function? A mathematical construct, often representing the error between prediction of the outcome and the actual outcome, the quantity we aim to reduce. And on and on. Does the clinician need to know all of these terms? No. Will a machine learner inadvertently use these or similar terms in their discussion with a clinician at some point in time? Likely yes. Before starting to work with the clinicians, I did not realize just how technical machine learning (ML) speak really is. It took years of practice and conversing with clinicians and biologists to understand that what our collaborators want to know is what my model will do for them, not how it really works.

Embracing uncertainty. An important part of ML is uncertainty. We often develop probabilistic models to estimate whether Patient A has disease subtype X or Y rather than predicting diagnosis directly. Given the imperfect data that we use for learning, it is more satisfying to express the result of a model in terms of probabilities rather than certainties. In clinical practice, however, early on I was told that my clinical collaborators have to give a diagnosis to the patient and they want to see what diagnosis my algorithm is predicting, not have to deal with probabilities. As it happens, though, the medical opinions differ. On the scale of most uncertainty in emergency departments to most certainty in the critical care units, uncertainty is omnipresent. It is important to acknowledge limitations of our knowledge and embrace probability. I believe that in the clinical world, understanding uncertainty will help to make better use of the ML tools in the future. As machine learners, we have to keep working on making our probabilistic results more amenable to clinical interpretation.

Interpreting black-box models. Or not. Another interesting, heated, ever-present discussion is around the noninterpretability of AI, so-called black-box models. Many clinicians claim that they could never trust such a system. At the same time, numerous clinicians acknowledge that they cannot always explain decisions that they are making, rather they make decisions based on the gestalt of their previous experience, an internal classifier that they synthesized over the years of observing patients. If even humans are creating internal complex models, explaining them post hoc, why are the standards for AI algorithms so much higher? Another question arises when clinicians are asking to design models that they understand—who said that human reasoning, with its obvious inability to make inference in high dimensions is the best way to go? Perhaps, the requests for model explainability are due to the recency and power of these tools or perhaps clinicians are simply asking for a clear and trustworthy way to interact with the systems. Though ML practitioners are split with respect to the reasons and purpose of the explainability, and even the definition of what explainability really means in this context, the quest for building trustworthy and reliable models as well as models that users can easily interact with is for now at the heart of many attempts to build AI models for healthcare. I have great hope that by working together clinicians and computer scientists converge on a path that leads to accurate trustworthy models.

Coming to the clinical domain with the background of computer science is akin to moving to a new country requiring to learn a new language, immersing oneself in an unfamiliar system of values, rules and regulations, and of course, idiosyncrasies. It is an amazing opportunity to learn and immerse oneself in expert knowledge and at the same time it is a constant source of frustration at the inability to access relevant data, the many missing values and records even when it is available. Despite the fact that working with clinical data is hard due to its many imperfections, much can be filled through the many interactions with clinical experts. Overall, it is an invigorating feeling to be able to get closer to helping a vulnerable population with the spectrum of disease and conditions to get a more precise diagnosis, prognosis, and treatment. As my graduate advisor said, it would be nice to use a phrase "Let me through, I'm a computer scientist." I believe I'm living that dream.

Programing Languages in Biomedical Data Science

There are several programing languages that are particularly useful for anyone interested in AI and data science in biomedicine:

Python (named after the comedy series *Monty Python's Flying Circus*) is a very flexible and relatively simple programing language and can be used for a myriad of purposes; it is arguably the most popular language in data science. In addition, Python has numerous special libraries (including NumPy and Pybrain for scientific computations and ML, respectively).

R is a common programing language for statistical learning and data analytics and is particularly strong in visual presentation of data in plots. R tends to be more popular in academic institutions. Like Python, R is also open source and has many libraries for ML (such as Gmodels, Class, and Tm) in Comprehensive R Archive Network (CRAN).

MATLAB is a high-level programing language by MathWorks that has an interactive environment for numerical computation as well as visualization and programing. It is widely used in science and engineering.

Statistical Analysis System (SAS) is a relatively expensive commercial analytics software that offers a large portfolio of statistical functions.

Most of the ongoing debate in biomedical data science, however, is between Python and R as the preferred language. Here is a table to help illustrate the strengths and weaknesses of each (see Table 5.2).

For those working in predictive analytics as a domain, R is more popular than Python (with SAS close to R as a programing preference). For data scientists, Python has a slight edge over R in terms of popularity (and this difference seems to be widening). Overall, if one is more interested in efficient deployment, Python would have an edge; if one is more focused on statistical analysis, especially graphical presentations of the data, R would be a slightly better choice.

Both programing languages are excellent for ML and AI. One data scientist astutely draws the analogy that R is more like Batman, who is intelligent and more brain than brawn, whereas Python is more like Superman, who is strong and more muscle than brain (compared to Batman) [1]. The author similarly draws the analogy that Python is more like the everyday car that you take to work (functional and gets the job done), but R is your fancy weekend sports car (good for showing off). Perhaps, one can have a combined visual: Superman in a Prius (for Python) and Batman in a Porsche (for R).

TABLE 5.2   Programing languages in data science and AI: Python versus R.

	python	R
Purpose (year of inception)	General purpose (1991)	Statistical analysis (1993)
Users	Programers and developers	Researchers and scholars
Popularity	+++	++
Ease of learning	+++	++
Usage and application	+++	+++
Data handling	++	+
Speed	++	+
Community support	++	+++
Presentation	++	+++
Advantages	Jupyter notebook for sharing Code readability and syntax simplicity Agile development	Superior data visualization Large data analysis library (CRAN)
Disadvantages	Not as many libraries (as R) Visualization less flexible (as R)	Slow learning curve for especially novices Losing popularity Less deep-learning support

CRAN, Comprehensive R archive network.

There are also several frameworks and libraries frequently used in AI projects. Hadoop (Apache Software Foundation) is not a programing language but an open-source framework (usually written in Java but not always) with a set of tools (such as its distributed file system and Map Reduce programing model) to store and processes large volumes of data on a cluster of commodity hardware. TensorFlow (Google Brain) is perhaps the best known AI library (written in C++, Python, or CUDA) used for DL. The flexible architecture accommodates a variety of platforms [not only central processing unit (CPU) and graphical processing unit (GPU) but also TPU, Google's tensor processing unit]. Lastly, Keras is an open-source library (written in Python) that can run on top of TensorFlow, but is more of a high-level application programming interface for training deep-learning models. Other deep-learning libraries and frameworks include Caffe (BAIR), Theano, and MXNet.

Making machine learning (ML) more available and easier to understand

Robert Hoyt

Robert Hoyt, a clinical informatics physician who has passionately taught clinical informatics as well as data science to clinicians, authored this commentary to offer the novice data scientist a short introduction as well as practical resources for machine learning.

Editor: Health Informatics: Practical Guide

Artificial Intelligence (AI) is pervasive in all industries, to include healthcare, and is an important skill set for biomedical data scientists. One of the most important components of AI is ML that uses algorithms, from simple decision trees, to complex neural networks to make predictions and solve problems. ML is the backbone of clinical decision support and risk prediction today.

Because ML has its roots in computer science, it is frequently not taught in noncomputer science departments. When it is taught, it often demands advanced math and/or programing experience. As a result, most graduate-level informatics students are not taught ML. Similarly, clinical and nurse informaticists infrequently are introduced to ML. While they do not necessarily need to have extensive experience in ML, they should be conversant in ML core concepts and be comfortable with the available software. As the field of medicine embraces biomedical data science and AI, we need to look for ways to improve the educational approach and look for software that is built for the task.

Newer tools (many open source) are being developed that simplify statistics, programing languages, and ML, so noncomputer science students and healthcare workers can get more involved in biomedical data science. In addition, most of these newer tools are affordable and do not require an expensive site license. For example, Jamovi is a free statistics program based on the R programing language [1]. Rattle is a free graphical user interface for R that is geared toward predictive analytics and is associated with a textbook [2].

There is also affordable and intuitive ML software. Programs such as WEKA, KNIME, Orange, and RapidMiner are some of the leading open-source software choices for ML [3–6]. WEKA does not require manipulating visual operators to create a workflow, unlike KNIME and Orange. Using WEKA is fairly intuitive and is associated with extensive algorithms, a textbook, and many free online courses [7,8].

One of the easiest and most modern ML software platforms to learn and use is RapidMiner that makes its platform free to faculty and students. It includes tools such as TurboPrep that expedite early data preparation (exploratory data analysis). Auto-Model is an intuitive feature that performs supervised and unsupervised learning. Multiple appropriate algorithms are automatically chosen and run simultaneously, for example, for classification or regression. Performance results are automatically generated, using standard parameters such as area under the curve and they are compared. A dynamic model is also created with sliders so you can adjust the predictor (independent) variables and see the effect on the outcome (dependent) variable.

Unsupervised learning is similarly intuitive. You have the choice to use k-means so you have to set the number of clusters and/or you can select x-means where it determines the number of clusters in the dataset.

With WEKA you can adjust or tweak the algorithms but with RapidMiner you cannot. For most ML users, this is ok while they learn the basics and get comfortable with ML.

The simplicity of these ML programs will make it easier for faculty to teach basic ML. With today's ML software choices, there is no longer a reason not to be comfortable with ML, so the potential is there for ML to be understood and adopted by many more people.

References

[1] Jamovi. <https://www.jamovi.org>.
[2] Rattle. <https://rattle.togaware.com/>.

[3] WEKA. <https://www.cs.waikato.ac.nz/ml/weka/>.

[4] KNIME. <https://www.knime.com/>.

[5] Orange. <https://orange.biolab.si/>.

[6] RapidMiner. <https://rapidminer.com/>.

[7] Witten I, Eibe F, Hall M. Data mining: practical tools and techniques. 4th ed. Morgan-Kaufmann; 2017. <https://www.cs.waikato.ac.nz/ml/weka/book.html>.

[8] Free online courses on data mining with machine learning techniques in WEKA. <https://www.cs.waikato.ac.nz/ml/weka/courses.html>.

Trusting artificial intelligence: a clinical programer's perspective

Rob Brisk

Rob Brisk, a cardiologist with a data science background, authored this commentary as someone who has the dual perspective of a clinician as well as a data scientist and provided his insight into the interpretability of ML/DL as not exceedingly difficult to attain.

Department of Cardiology, Craigavon Hospital/School of Computer Science, Ulster University, Coleraine, Northern Ireland

Machine learning: a steppingstone to the future of medicine

Artificial intelligence (AI) is a broad technological church, but today's conversation regarding its use in healthcare is driven particularly by the explosion in ML technology. In very broad terms, "classical" programing requires a human software engineer to expound a precise and exhaustive set of language-based rules that allow a computer to transform some input data into some desired output data, whereas ML allows us to produce computer programs by obtaining (usually large sets of) matched of input data and output data, then tasking an ML algorithm with discerning a set of logic-based rules that transform the former into the latter.

This becomes advantageous in two settings: first, in automating tasks that humans perform at a partially subconscious level and, consequently, cannot elucidate all the required steps; and second, in situations where one has access to a set of input data that is likely to have some causal link to a set of output data but where human experts are unable to discern the patterns within the former that map to the latter. Two examples in the clinical domain are, respectively, detecting rhythm abnormalities from ambulatory electrocardiogram (ECG) signals and detecting incipient Alzheimer disease from brain imaging that plausibly contains diagnostic clues but where radiologists cannot detect any [1,2].

The "black box effect": can there be trust without transparency?

When a software engineer writes the internal logic for a computer program, they generally do so in a language-based format that can be interrogated by others who understand that language. The rules governing the behavior of ML systems, conversely, are often encoded within complex mathematical architectures that the conscious mind is incapable of interpreting. A visualization of a rudimentary artificial neural network (ANN) algorithm (the most sophisticated genre of ML algorithm, which underpins deep-learning applications) is shown in Fig. 1. There have been many attempts to explain the rules encoded within such algorithms using saliency maps,

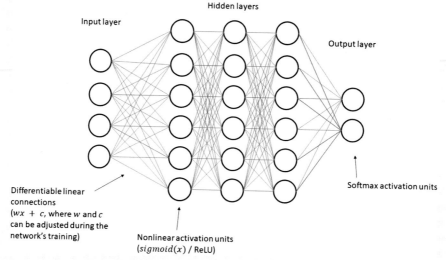

FIGURE 1   A rudimentary artificial neural network.

deconvolution of convolutional networks, etc. Such approaches can offer useful insights into certain aspects of a system's functionality but the wider internal logic remains overwhelmingly opaque.

The importance of this in the clinical setting is that even if an ML system has been extensively informed by and validated upon high-quality data, it is extremely difficult to anticipate when and how it will defy an operator's expectations. Any parent is well placed to grasp this concept: one may develop and validate a predictive model of a child's behavior based on enormous quantities of observational data, and the model may be reliable the vast majority of the time. However, the opaque nature of the internal logic that governs a child's behavioral impulses means it is impossible to foresee when and how a child will confound expectations. How many of us, for example, have anticipated that our children will continue to play nicely if we leave the room for a few moments—based, in all likelihood, on this pattern having been observed countless times before—only to find the walls covered in crayon on our return?

For real-world examples, consider the Google algorithm that labeled a photograph of an Afro-Caribbean couple as gorillas or the Tesla whose autopilot caused a fatal collision by failing to detect a truck pulling out on to the road, despite the accident report concluding that the truck would have been visible to the car for a full 7 seconds prior to impact [3,4]. The former example is sobering because if Google cannot exert enough control over their ML algorithms to avoid a public relations disaster, one must wonder how smaller companies in the digital health space will fare. The latter is even more concerning from a healthcare perspective, as it exemplifies the potentially grave consequences of reliance on opaque technology in a high stakes setting.

Moving forwards

The need for transparency in AI is becoming widely acknowledged, though how this is achieved remains an open question for regulatory bodies [5,6]. A notable success in this respect came from a collaboration between DeepMind and Moorfields Eye Hospital, where the interpretation of optical coherence tomography scans was broken down into several parts and a human-readable visualization of the diagnostic process generated at each step [7]. Similar approaches are being developed in other areas of medicine, though the application of such frameworks to tasks less intuitive to the human mind such as mutliomics analysis may prove challenging.

Ultimately, the responsibility for the quality of care delivered to patients sits with us as clinicians. It is therefore paramount that we collectively inform ourselves regarding emerging technologies and begin to discuss mechanisms that will allow us to leverage the enormous potential of ML in a safe and responsible way.

References

[1] Rajpurkar P, Hannun AY, Haghpanahi M, Bourn C, Ng AY. Cardiologist-level arrhythmia detection with convolutional neural networks. July 6, 2017. arXiv: 1707.01836.

[2] Ding Y, Sohn JH, Kawczynski MG, et al. A deep learning model to predict a diagnosis of Alzheimer disease by using F-FDG PET of the brain. Radiology 2019;290(2):456—64.

[3] Hern A. Google's solution to accidental algorithmic racism: ban gorillas. The Guardian Jan 12, 2018. <https://www.theguardian.com/technology/2018/jan/12/google-racism-ban-gorilla-black-people> [accessed 09.06.19].

[4] Stewart J. Tesla's autopilot was involved in another deadly car crash. Wired Mar 30, 2018. <https://www.wired.com/story/tesla-autopilot-self-driving-crash-california/> [accessed 09.06.19].

[5] The European Commission High Level Expert Group on Artificial Intelligence. Ethics guidelines for trustworthy AI. <https://ec.europa.eu/digital-single-market/en/news/ethics-guidelines-trustworthy-ai>; 2019 [accessed 09.06.19].

[6] The US Food & Drug Administration. Proposed regulatory framework for modifications to artificial intelligence/machine learning (AI/ML)-based software as a medical device (SaMD)—discussion paper and request for feedback. <https://www.fda.gov/media/122535/download>; 2019 [accessed 09.06.19].

[7] De Fauw J, Ledsam JR, Romera-Paredes B, et al. Clinically applicable deep learning for diagnosis and referral in retinal disease. Nat Med 2018;24(9):1342—50.

Classical Machine Learning

ML, or more accurately, classical ML, is better suited for smaller and less complicated datasets and clinical scenarios with less features. Classical ML is categorized into two types of learning: (1) supervised learning and (2) unsupervised learning (see Fig. 5.3). Additional discussions on semisupervised and ensemble learning as well as DL will follow.

Supervised Learning

Supervised learning takes refined raw data and uses an algorithm to predict the outcome (with the algorithm having derived from previously studying the labeled data in a training set of data in the process described earlier). In other words, the training data help to guide the machine to the right prediction or output via an algorithm so that the model can be used to make predictions of the new data. Active learning is a supervised learning that involves a learning algorithm that can query the user for labels to avoid manually labeling a large number of samples.

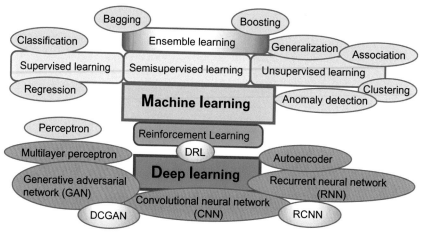

FIGURE 5.3   Machine (classical) and deep learning. Classical machine learning is divided into supervised and unsupervised learning. Ensemble learning as well as semisupervised learning are also variations of machine learning. Deep learning is divided into the various types of deep learning, and reinforcement learning is considered a different type of learning (although deep reinforcement learning, or DRL, combines aspects of both). Other types of deep learning also have similar hybridizations: GAN and CNN (DCGAN) and CNN and RNN (RCNN) (see text for details). *CNN,* Convolutional neural network; *DCGAN,* deep convolutional generative adversarial network; *GAN,* generative adversarial network; *RCNN,* recurrent convolutional neural network; *RNN,* recurrent neural network.

In short, supervised learning develops a predictive model from both input and output data (the later labeled by humans) and this model is then used to make predictions on a new set of data. These supervised learning methodologies lead to classification (dichotomous or categorical) or regression (to a continuous variable). For classification, the popular methodologies are support vector machines (SVMs), naive Bayes classifier, *k*-nearest neighbor, and decision trees (with boosting or bagging); logistic regression is a misnomer and is in fact a classification methodology. For regression, linear and polynomial regression methods are most commonly used, but other types [such as ridge and least absolute shrinkage and selection operator (LASSO) regression] may become more popular in the future.

Classification

This methodology leads to predicting or assigning a label or category for an unlabeled sample [e.g., "tumor" vs "not tumor" on magnetic resonance imaging (MRI)]. Classification strategy is good for fraud detection and facial recognition. In medicine and health care, classification is good for medical images, phenotyping, and cohort identification.

The supervised classification methodologies are listed in Table 5.3 but only the more popular methods will be described in more detail. Logistic regression (which is a misnomer as it is not a true regression but rather a classification methodology) is included in this classification section.

TABLE 5.3   Supervised learning: classification methodologies.

Classification methodologies	Main tool(s)
Decision trees	Nodes and branches
Discriminate analysis	Discriminant function
k-NN	Decision boundary
Logistic regression	Logistic function
Naive Bayes classifier	Decision boundary
SVM	Hyperplane and kernels

k-NN, k-nearest neighbor; SVM, support vector machines.

Support vector machine

This "maximum-distance" classification methodology is achieved by creating a line or an optimal hyperplane (a decision surface) in a high-dimensional space that represents the largest separation between the two classes (see Fig. 5.4). In general, the larger this separation, the better the SVM performance. The points that are closest to the border are termed "support vectors." SVM can be in two forms: (1) linear SVM: use of linear optimization for linear separation (if possible) and this methodology is relatively fast (and resembles logistic regression) and (2) kernel SVM: use of many different kinds of dividing structures called kernels (Note. The so-called kernel trick is a mathematical trick that adds an extra dimension so that what is not possible in certain number of dimensions can now be possible with more dimensions.) In short, kernels delineate the boundary in a nonlinear fashion (either curved line or an optimal plane).

Common usages are spam filtering, sentiment analysis, image classification and segmentation, recognition of handwritten characters, and fraud detection.

Advantages and disadvantages

This popular but relatively complex ML methodology can be good for complex nonlinear relationships between input features and output. This methodology is often considered one of the most accurate algorithms for classification, especially when there is a high number of features (compared to the number of data points) and is therefore effective in high-dimensional data. SVM is also robust and thus can avoid noise as well as minimize overfitting. SVM is frequently used when one has a small training dataset as it does not scale well to larger datasets (larger training sets may be better suited for other classification methodologies or even DL). While linear SVM is considered relatively fast (but less accurate), kernel SVM is perhaps more accurate (but slower) for classification.

Disadvantages include SVM needing high levels of both memory and processing power and difficulty in interpreting the precise relationship between input features and output.

An example in the recent biomedical literature is the usage of SVM in CT-based radiomic features to effectively identify high and low grades of clear cell renal cell carcinoma in only 227 patients with area under the curve (AUC) values of 0.88–0.91 [2].

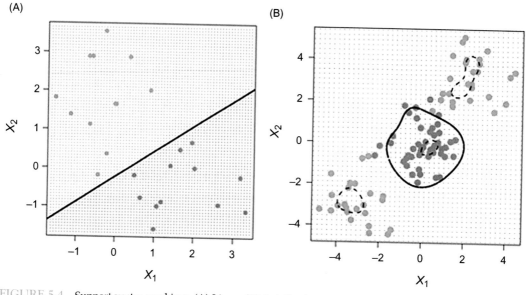

FIGURE 5.4   Support vector machines. (A) Linear SVM. A line is separating the two variables seen in blue and in red with the maximal distance (best separation of all the possible lines) seen in the line drawn. (B) Kernal SVM. An SVM with radial type of kernel captures the decision boundary as a near circle. *SVM*, Support vector machine. *Source: James G, Witten D, Hastie T et al. An Introduction to Statistical Learning with Applications in R. Springer, New York, 2013. Left: Figure 9.2; pg. 340. Right: Figure 9.9; pg. 353.*

Naive Bayes classifiers

This supervised learning methodology uses probability to divide or categorize the data and is based on Bayes' theorem (with its concept of prior probability which selects the outcome with the highest probability). The presumption is that the predictors are independent of each other (hence the term "naive") so the model becomes essentially a probability table where the presence of a certain feature is not dependent on any other feature. In short, the probability of a certain outcome is the product of probabilities given by the features. The dividing line, or Bayes' decision boundary (see Fig. 5.5), is used to separate the samples into the two populations.

This methodology is well suited for real-time prediction, text classification/spam filtering, and recommendation system.

Advantages and disadvantages

This supervised learning methodology utilizes statistical modeling and has relatively fast speed (given parallel process) and is also good when there is a high dimensionality of the inputs. Another advantage of the Bayesian probabilistic approach is that it does not require large training sets and is relatively simple to implement as well as to interpret. For its relatively fast speed, however, there is compromise in terms of accuracy (compared to kernel-type SVM that was just discussed).

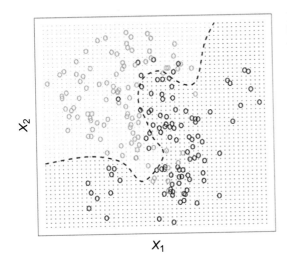

FIGURE 5.5   Naive Bayes classifier. The orange and blue background designate the two regions divided by the Bayes decision boundary (*dashed line*). Source: *James G, Witten D, Hastie T et al. An Introduction to Statistical Learning with Applications in R. Springer, New York, 2013. Figure 2.13; pg. 38.*

Disadvantages include the underlying principle that there is total independence of features (uncommon situation) and relatively lower level of performance in lower dimensional data.

An example in the recent biomedical literature is the application of Bayesian network modeling to pathology informatics by constructing several Bayesian network models to assess individual patient-specific risk for subsequent specific histopathologic diagnoses and their related prognosis in gynecological cytopathology and breast pathology [3].

k-Nearest neighbor

This supervised learning algorithm is used for both classification as well as regression and identifies the number of nearest neighbors of any element, hence the appropriate name (see Fig. 5.6). *k* is the integer of neighbors that is in the feature space closest to a designated point and the class is determined by majority observed in that space (e.g., if *k* is 3 and 2 out of 3 neighbors is a certain class, this class "wins"). Similar to the Bayes' classifier just discussed, there is a decision boundary called the *k*-nearest neighbor (*k*-NN) decision boundary.

k-NN is well suited for text mining or categorization as well as stock market trends and forecasting.

Advantages and disadvantages

This methodology is relatively simple to interpret and is therefore considered a "lazy" learning that is instance-based (vs model-based). *k*-NN is well suited for datasets in which there is no prior knowledge about the distribution of the data. Like some of the aforementioned supervised methodologies, it is relatively robust to noisy training data.

Disadvantages include *k*-NN not performing well with either high-dimensional data or massive amounts of data that may contain noise and nuances such as missing data, but this algorithm itself can be utilized to perform missing values imputation, noise filtering,

(A)

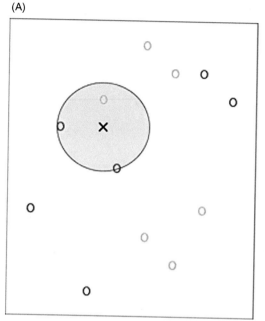

(B)

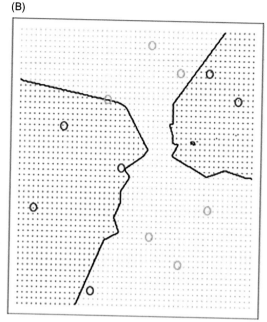

FIGURE 5.6 *k*-NN. (A) The *k*-NN process. Test observation (designated with a cross) is shown with its $k = 3$ nearest neighbors showing 2 blue and 1 red. Since the test observation location belongs to the more common class (blue vs red), this area belongs to the blue class. (B) The *k*-NN decision boundary. The black lines delineate the decision boundary and the blue and orange background designate each respective blue and orange regions. *k*-NN, *k*-nearest neighbor. *Source: James G, Witten D, Hastie T et al. An Introduction to Statistical Learning with Applications in R. Springer, New York, 2013. Figure 2.14; pg. 40.*

and data reduction [4]. *k*-NN is also susceptible to overfitting from the curse of dimensionality. Lastly, the *k* value can have a large effect on the performance of this model.

An example in the recent biomedical literature is the application of a modified *k*-NN algorithm (enhanced with instance weights) to patients with diabetic retinopathy for more accurate diagnosis [5].

Decision trees

The decision tree methodology is the most straight forward (also known as classification and regression trees, or CART) with the trees drawn upside down; the decision points are termed nodes and the segments of the tree that interconnect the nodes are called branches (Fig. 5.7). The leaves are therefore outcomes. To continue the tree terminology, a computational "pruning" process can reduce the size of the tree without sacrificing the accuracy of the model. This supervised methodology uses branches in decisions to achieve classification (but can also be used for regression); decision trees, however, are not considered as accurate as other aforementioned methodologies. Three strategies, therefore, are necessary to build more powerful prediction models

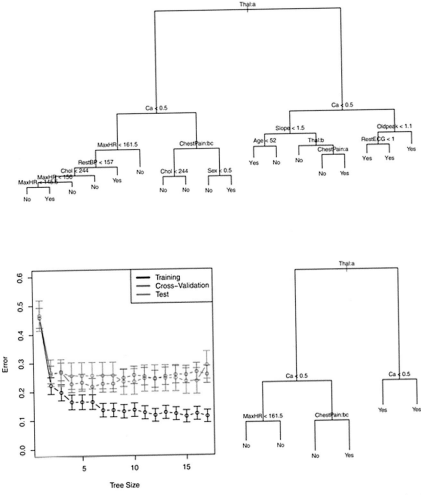

FIGURE 5.7 Decision trees. (A) Unpruned tree and (B) Pruned tree. The pruned tree is a result of decreasing the total number of nodes in the decisions without decreasing accuracy. *Source: James G, Witten D, Hastie T et al. An Introduction to Statistical Learning with Applications in R. Springer, New York, 2013. Figure 8.6; pg 313.*

with decision trees: bagging, boosting, and stacking (see later under the Ensemble Learning).

Advantages and disadvantages

This methodology has higher explainability as it is much more relatable (compared to the other classification methodologies discussed earlier) and can be easily visualized as it has a good graphical representation. Decision trees are relatively robust to noise and incomplete data. Decision trees are also very accommodating for nonlinear relationships as well as outliers. Lastly, it is also relatively fast compared to other methodologies.

Disadvantages of decision trees include that these are not as accurate as other methods amongst supervised learning methods (but are relatively fast and efficient like logistic regression). These trees would usually need to be in an ensemble format, however, to be more accurate and be less prone to overfitting. Lastly, decision trees tend not to work well with small training data sets.

An example in the recent biomedical literature is the application of decision trees (vs risk scores) for predicting drug-resistant infections that revealed the decision tree (with five predictors) was more user-friendly with fewer variables for the end user in a study of over 1200 patients [6].

Regression

The supervised regression methodologies lead to numerical representation of output variables in order to predict a number. Regression is good for market forecasting, growth prediction, and life expectancy calculation. In medicine and health care, regression is good for risk prediction and outcome prediction.

Regression methodologies are listed in Table 5.4 but only linear and polynomial regression will be discussed in detail.

Linear regression

This regression is probably the methodology that is most familiar to clinicians. This ML method (derived from statistics) delineates the strength of the relationship between two continuous variables (x being the independent variable, whereas y is the dependent variable). The method for fitting a regression line in linear regression is the method of least squares with a correlation coefficient r. This regression is termed "simple" when there is a single input variable and "multiple" when there are multiple input variables (see Fig. 5.8). There is also polynomial (or nonlinear) regression in which the relationship is not linear. In addition, generalized linear model (GLM) is an ordinary linear regression with a generalization that accommodates variables that do not have a normal distribution. Finally, LASSO regression increases its prediction accuracy by performing both regularization (a process that decreases overfitting) and variable selection.

TABLE 5.4   Supervised learning: regression methodologies.

Regression methodologies	
LASSO regression	Model is based on shrinkage and simple sparse models
Linear regression	See text
Polynomial regression	See text
Ridge regression	Model is a regularized linear regression
Support vector regression	Model is based on maximum margin but output is a number

LASSO, Least absolute shrinkage and selection operator.

(A)

(B)

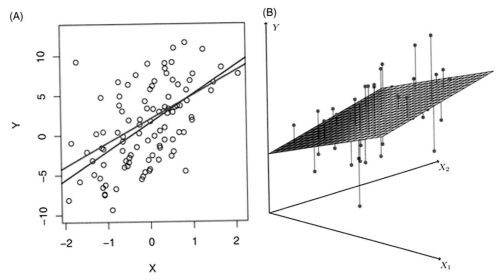

FIGURE 5.8   Linear regression. (A) Simple linear regression shows the population regression line (in red) and the least squares line (in blue). (B) Multiple linear regression shows that the least squares line becomes a plane, which minimizes the sum of the squared vertical distances between the observations (in *red dots*) to the plane. *Source: James G, Witten D, Hastie T et al. An Introduction to Statistical Learning with Applications in R. Springer, New York, 2013. A: Figure 3.3; pg. 64. B: Figure 3.4; pg. 73.*

Advantages and disadvantages

Linear regression is perhaps the most familiar regression methodology to all those who are not formally educated in data science. Linear regression is considered fast but not perhaps as accurate compared to other methodologies that perform regression. Linear regression performs poorly when the relationship is nonlinear. Simple linear regression is usually not as accurate also given that LASSO and Ridge regressions are regularized (penalizing large coefficients) and therefore observed to exhibit less overfitting.

An example in the recent biomedical literature is the use of a hierarchical linear regression analysis of patients with chronic nonspecific low back pain and how this condition correlated with emotional distress [7].

Logistic regression

This is the adaptation of the aforementioned linear regression to a binary classification (via a logistic function to yield maximum likelihood); it is not, therefore, a true regression like linear regression. Multiple logistic regression utilizes multiple predictors and is commonly used as a statistical tool for patient studies. A closely related classifier to logistic regression is linear discriminant analysis (LDA), which is more stable than logistic regression in certain situations.

Advantages and disadvantages

Logistic regression is relatively fast compared to other supervised classification techniques such as kernel SVM or ensemble methods (see later in the book) but suffers to some degree in its accuracy. It also has the same problems as linear regression as both techniques are far too simplistic for complex relationships between variables. Finally, logistic regression tends to underperform when the decision boundary is nonlinear.

An example in the recent biomedical literature is the use of logistic regression (compared with three other ML methodologies) in a 2-year mortality prognostication study of a small number of patients (76 patients) with heterogenous glioma with highly dimensional datasets [8].

Unsupervised Learning

Unsupervised learning takes unlabeled data and uses algorithms to predict patterns or groupings in the data set without any human intervention. It is more challenging than supervised learning as there are no "answers" but can be coupled with supervised learning. This type of learning is more for exploratory purposes (such as to discover market segmentation) or to analyze and label new data. In medicine and health care, there is use for this unsupervised learning in subgroups of various cancers based on their gene expressions.

These unsupervised learning methodologies lead to clustering, generalization, association, or anomaly detection.

Clustering

These methodologies group data by similar characteristics without any human intervention. Clustering can be used for customer segmentation and recommender systems. In medicine and health care, clustering is good for biological hypothesis generation, identifying new populations or therapies, and novel phenotype identification.

Clustering methodologies are listed in Table 5.5 with a brief description but only *k*-means clustering will be described in detail.

TABLE 5.5   Unsupervised learning: clustering methodologies.

Clustering methodologies	
Affinity propagation	Clusters based on graph distances between points
DBSCAN	Density-based algorithm for clusters of dense regions
Fuzzy C-means clustering	Model based on each data can belong to >1 cluster
Gaussian mixture models	Probabilistic model-based on Gaussian distributions
HMM	Probabilistic model-based approach to sequences
Hierarchical or agglomerative clustering	Model based on a hierarchy of clusters
k-Means clustering	See text
Mean-shift clustering	Model based on kernel density estimation

DBSCAN, Density-based spatial clustering of applications with noise; *HMM*, hidden Markov model.

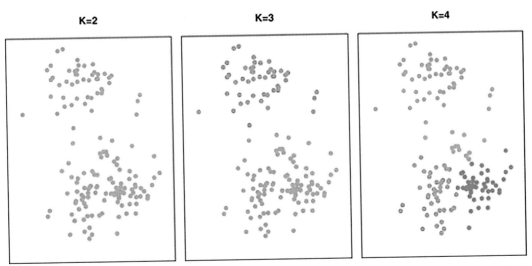

FIGURE 5.9   *k*-Means clustering. The panel shows 150 observations with results of different values of *k* (*k* = 2, 3, and 4) with different colors designating the groups for the clustering methodology. *Source: James G, Witten D, Hastie T et al. An Introduction to Statistical Learning with Applications in R. Springer, New York, 2013. Figure 10.5; pg. 387.*

k-means clustering

This is a commonly used simple unsupervised learning that has an algorithm used to find clusters or groups in the data with *k* number of groups that form organically based on similarity (see Fig. 5.9). The end result of the process of using distance formulas yields *k* feature vectors called centroids of clusters. When *k* is not designated, the classifier will determine the best *k* based on two techniques, reconstruction error (sum of the mean squared error between all the points and their centroid) or peakedness (best *k* is where the largest cluster is the one with the highest peak in comparison with others).

Advantages and disadvantages

One advantage of *k*-means clustering is that it is relatively easy and simple to implement. It is also faster than the other clustering methodologies especially with larger number of variables. Lastly, this methodology also produces relatively tight clusters. The weakness of this methodology is that sometimes it is difficult to predict the *k* value. In addition, the clusters may lack the hierarchical significance and consistency that other clustering methodologies may yield.

An example in the recent biomedical literature is the effective use of *k*-means clustering for a cluster-based classification strategy (based on severity) of the heterogenous disorder of bipolar disorder in 224 subjects [9].

Generalization (or dimension reduction). Generalization is a methodology that reduces the dimensionality of the data, usually by merging features. The advantages of having such abstracted models are that these models can then be more efficient and use less features. This methodology is also used for data visualization and compression of data. In

TABLE 5.6   Unsupervised learning: generalization methodologies.

Generalization methodologies	
Laplacian eigenmaps	Technique for nonlinear dimensionality reduction that is computationally efficient
LSA	Technique for creating a vector representation of a document in NLP
p-SNE	Model designed for visualization of high-dimensional datasets
Principal components analysis	See text
Random projection	Model for representing high-dimensional data in Euclidean space to a low-dimensional feature space
SVD	Model for decomposing or factorizing a matrix into its constituent elements
t-SNE	Model designed for visualization of high-dimensional datasets

LSA, Latent semantic analysis; NLP, natural language processing; p-SNE, power-law stochastic neighbor embedding; SVD, singular value decomposition; t-SNE, t-Distributed stochastic neighbor embedding.

medicine and health care, generalization is good for data visualization, data compression, and variable selection.

These methodologies include the popular principal component analysis (PCA) and other generalization methodologies that are listed in Table 5.6 with a brief description.

Principal component analysis

PCA identifies the features that are most significant in classification, and therefore these selected features are then chosen to be used for computation; it is considered a dimension reduction method as it reduces a large set of variables into a low-dimensional representation of the dataset with this feature extraction. Each principal component is a linear combination of variables that are compressed. PCA can also be used as a dimension reduction technique for the purpose of regression. PCA is often used for data visualization or data preprocessing prior to supervised methodologies are used.

Advantages and disadvantages

The advantages are several: low noise sensitivity, decreased requirement for capacity and memory, and increased efficiency. The disadvantages of this methodology are centered around its assumptions: linearity and principal components being orthogonal. In addition, the new principals are not interpretable.

An example in the recent biomedical literature is a study that used PCA to eliminate unwanted low-frequency signal drift as well as spontaneous high-frequency global signal fluctuations in 4D functional MRI so that these artifacts as part of a more sophisticated preprocessing step in the study acquisition [10].

Other categories under unsupervised learning include association rules, or pattern search or recognition and identify sequences or relationships in data. The associative unsupervised methodologies include Apriori, FP-Growth, and Equivalence CLAss

Transformation (ECLAT) algorithms and are used in sales and marketing strategies as these algorithms can predict buyer behavior.

In addition, there is anomaly detection (also called outlier detection) that can be performed from unsupervised learning, although supervised as well as semisupervised techniques can also be applied to anomaly detection. In addition to fraud detection in finance and structural defects in industry, this last category of unsupervised learning can also be very useful in biomedicine for detecting medical problems or errors. It is precisely this population of anomalies or outliers in biomedicine that can be an important source of new knowledge.

An example in the recent biomedical literature is the use of outlier detection in preventing medication errors using an unsupervised methodology named density—distance—centrality to detect potential outlier prescriptions in a dataset of over 560,000 prescribed medications [11].

Finally, a Boltzmann machine, a network of symmetrically connected nodes (each node is connected to every other node), is an unsupervised ML algorithm that can discover latent features in the dataset. Restricted Boltzmann machine (RBM) will be discussed later.

Semisupervised Learning

A hybrid technique of supervised and unsupervised learning is semisupervised learning, which uses a small amount of labeled data and then a relatively large amount of unlabeled data. Semisupervised learning can also produce proxy labels on unlabeled data.

Advantages and disadvantages

These methodologies can therefore be trained on a mixture of small amount of labeled and larger amount of unlabeled data, which will be more efficient as it saves human time and effort. The introduction of unlabeled data may actually reduce human bias and improve the accuracy of the final model.

An example in the recent biomedical literature is a report on the semisupervised learning approach [combined with generative adversarial networks (GANs)] in providing a small number of labeled medical data to build a platform for IoT-based medical data and interpretation as part of decision support for this new source of medical data [12].

Ensemble Learning

This ensemble learning strategy (bagging, boosting, and stacking) involves training a large number of models that together will surpass the performance of a single model; in short, it is the creation of a metamodel that has better prediction and more stability. This ensemble of models reduces noise, bias, and variance. A common scenario is the utilization of decision tree algorithms to achieve this ensemble to improve accuracy (although generally these ensemble learning methodologies can be slower than others).

There are three ensemble learning strategies that can improve performance (see Fig. 5.10).

First, bagging (also called bootstrap aggregating) involves the creation of many duplicates or different sets of the training data followed by training with the same model or

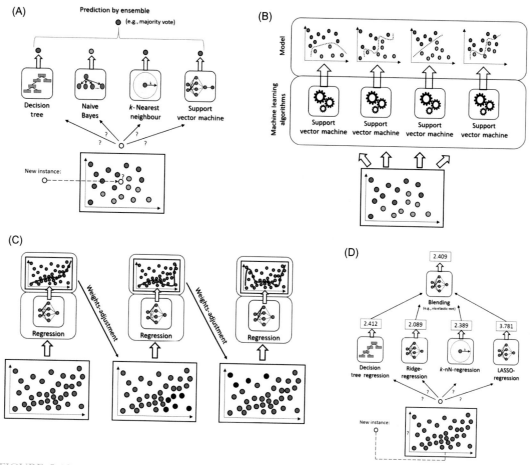

FIGURE 5.10 Ensemble learning: bagging, boosting, and stacking. In (A), the general concept of ensemble learning is illustrated with several models assembled to surpass the performance of a single model. In (B), bagging with several SVM models is seen to reduce the model's variance. In (C), boosting (Adaboost) with several models to enable the weaker models into a stronger one by giving more weight to the models with better performance; bias is also reduced. Lastly, in (D), stacking is seen with several different regression algorithms collectively involved in increasing the predictive force of the classifier. *SVM*, Support vector machine.

algorithm and yielding eventually the average of all the models; the variance is reduced. This process of building models is done in parallel. Random forest is a popular supervised learning algorithm that consists of an ensemble of many decision trees that collectively yield an accurate prediction and minimizes overfitting. One major limitation of random forest is its slow nature (with its large number of trees) so it is not ideally suited for real-time predictions. Random forest is a methodology that is used for detecting fraud and stock projections.

Second, boosting involves the creation of many models with the training data to enable each new algorithm to correct the errors of the previous algorithm and, therefore, enable the weak models into a stronger one; the bias is reduced. This process of building models is done in sequence. While equal weight is given to all models in bagging, the performance of a model determines the weight of the model in boosting. Gradient boosting is one such ensemble algorithm that typically uses decision trees, and another one is adaptive boosting (or AdaBoost). Other tools for this boosting function include XGBoost, LightGBM, and CatBoost.

Third, stacking involves having several different algorithms to deliver the output to one last algorithm as an arbitrating algorithm for a final decision; the predictive accuracy is increased as a result.

Advantages and disadvantages

This strategy of combining algorithms is that the collective strength of several or many models is a good improvement on individual models, particularly decision trees. In addition, less stable or more fragile algorithms can actually (paradoxically) add a higher quality to the ensemble of models. Finally, while bagging reduces variance and thus overfitting, boosting reduces bias (but may increase overfitting) so both ensemble methods have its own advantage.

An example in the recent biomedical literature of ensemble methods is the recent study of an ensemble ML model for trauma risk prediction that proved to be superior to three established risk prediction models (including one that used Bayesian logic) [13].

Two additional categories of ML that are not usually considered "classical" or traditional ML include (1) reinforcement learning (RL) and (2) neural networks and DL. These more elaborate types of ML are better suited for large data sets or complicated data types (like images or videos); the corollary to this generalization is that using these advanced methodologies for relatively straight forward datasets with few features may be excessive and unwarranted (in other words, classical ML methodologies discussed previously may be sufficient).

Reinforcement Learning

In addition to the aforementioned supervised (task-driven with classification or regression) and unsupervised (data-driven with clustering) learning, another type of learning is RL. Although RL is often described as a third or additional type of ML along with supervised and unsupervised learning, it is distinctly different than the former two types of ML.

RL has its origin over 100 years ago with the psychologist Edward Thorndike and his cat experiments in which the cats learned to "reinforce" their positive outcome with the appropriate behavior of pushing a lever. In 1951 Harvard's Marvin Minsky devised an equipment to emulate this nature's RL; this equipment, called Stochastic Neural Analogy Reinforcement Computer (SNARC), consisted of motors, tubes, and clutches that functioned as dozens of neurons and synapses that favored behavior that lead to a positive outcome.

RL and its wondrous capability (along with a Monte Carlo tree search algorithm) was best demonstrated in the Google DeepMind's *AlphaGo* program and its recent successful

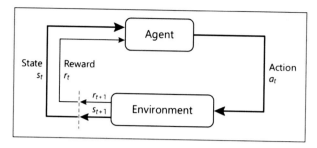

FIGURE 5.11   Reinforcement learning. An agent (which represents the algorithm) interacts with the environment via an action and has knowledge of the state of the environment. With the action the agent receives a reward (or a penalty). Reinforcement learning enables the agent to maximize the reward via learning from its experience with all the interactions with the environment.

defeat of the Go champion Lee Sedol. DeepMind (based in London and founded in 2010) and its founder Demis Hassabis aim to be the Apollo program of the AI domain with a large number of highly trained AI and ML scientists and with a singular focus on achieving general-purpose, self-learning AI with deep RL. With an innovative form of RL, *AlphaGo Zero* was able to teach itself the game Go entirely on its own and was able to soundly defeat its precursor *AlphaGo* in just 40 days. DeepMind was acquired by Google in 2014 for $500 million and DeepMind Health focuses on AI applications in health care.

In RL, the model is not relating itself to data but rather finding the optimal method via exploration to achieve the most desirable outcome while receiving input data in a dynamic environment (analogous to humans attempting to attain the highest score in a game) (see Fig. 5.11). In other words, there is a positive and negative feedback to the solution of the algorithm so the goal of RL is to learn a policy (defined as a function that maximizes the reward in a long term setting, or reward maximization). RL is therefore well suited for a sequential decision process needed for video games, automated trading, and robotic navigation.

An intelligent agent, or simply "agent" in AI parlance, is an autonomous entity that can perform a task based on an input (perceptions) and intelligent processing to lead to an output (action). These agents are different from traditional software programs in that these agents have characteristics of perception, autonomy, learning, and communication. It is essentially a self-contained software program with goal orientation embodied with knowledge and overall represents someone.

On reinforcement learning in medicine—a data scientist perspective

Louis Ehwerhemuepha

Louis Ehwerhemuepha, a PhD in data science, authored this commentary on his perspective as a data scientist in a hospital on reinforcement learning and how this type of learning can help clinicians in decision-making and knowledge discovery.

CHOC Children's Hospital, Orange, CA, United States

Reinforcement learning (RL) is a branch of artificial intelligence (AI) where the process of autonomously understanding trends in data and making predictions of the future is achieved through learning algorithms that adapt base on incentivization of correct decision and penalization of inaccurate ones. RL is developed on a set of formalism that includes the concept of a virtual agent and environment [1], and actions, rewards, and observations [2]. The virtual agent learns by making autonomous decisions on a course of action while observing the consequence of the action from its environment. This consequence could be a reward or penalty. Through trial and error, the virtual agent learns an optimal path or set of decisions to be made to achieve the highest reward based on the state of its environment [3,4]. The underlying theoretical framework is a stochastic process called Markov Decision Process [2,5,6].

The application of RL in medicine is nascent. RL has been applied to the problem of predicting sepsis, a leading cause of death, in adults [7]. This application to sepsis was achieved through an "AI Clinician" that was developed using ICU data from the Medical Information Mart for Intensive Care III (MIMIC III) [8] and the eICU Research Institute ICU databases [9]. The RL system was built on multidimensional discrete time series clinical data from the databases such that rewards were associated with survival and penalties for death. The authors showed that the AI Clinician's selected treatment plans were on average better than human clinicians [1], although the goal of the development of such "AI Clinician" should be the ensemble of the human and AI clinician in the development of treatment plans.

Other applications of RL in medicine includes prediction of readiness for extubation [10], heparin dosage [11], ventilator support in the ICU [10], and prediction of optimal treatment regimens for nonsmall cell lung cancer [12]. RL has also been shown to be applicable to the optimization of medication delivery such as the delivery of anesthesia [13]. Multiagent deep RL models may also be applied toward the discovery of de novo drugs and cheaper alternatives to existing drugs that are expensive or difficult to administer [14].

These examples of the application of RL in medicine indicate that there is a great opportunity for real-time AI decision support and interaction engine for physicians. Integration of AI decision support tools will provide substantial improvement in the quality of care delivered by clinicians. In combination with traditional supervised and unsupervised learning, future RL models would be able to provide physicians with multiple treatment options for patients with complex medical conditions. Also, RL models may become invaluable in cases where providers have difficulty in providing a diagnosis or in safely treating a patient condition. Success will depend largely on participation of clinicians toward the determination of appropriate reward mechanisms for training of RL models. The goal is not to replace clinicians, as the title of recent applications of deep convolutional neural network studies may portend, but to provide information from an RL model as additional data points/information throughout the development or modification of treatment plans for patients. Indeed, data scientists and machine learning (ML) engineers can only be successful in building AI systems that are safe and effective through active collaboration with care providers such as physicians and registered nurses. There may be a rise in the training of physician-data scientists who would hold and support bridges in knowledge gaps/difference between the highly technical and vast fields of data science and medicine.

References

[1] Komorowski M, Celi LA, Badawi O, Gordon AC, Faisal AA. The artificial intelligence clinician learns optimal treatment strategies for sepsis in intensive care. Nat Med. 2018;24(11):1716.
[2] Lapan M. Deep reinforcement learning hands-on: apply modern RL methods, with deep Q-networks, value iteration, policy gradients, TRPO, AlphaGo zero and more. Packt Publishing Ltd; 2018.
[3] Henderson P, Islam R, Bachman P, et al. Deep reinforcement learning that matters. arXiv:1709.06560.aaai.org. <https://www.aaai.org/ocs/index.php/AAAI/AAAI18/paper/viewPaper/16669> [accessed 09.02.19].

[4] Kaelbling LP, Littman ML, Moore AW, et al. Intelligence AM. J Artificial Intelligence Res 1996;4:237–85.

[5] Karlin S. A first course in stochastic processes. Academic Press; 2014.

[6] Puterman ML. Markov decision processes. Handbooks Oper Res Manag Sci 1990;2:331–434.

[7] Singer M, Deutschman CS, Seymour CW, et al. The third international consensus definitions for sepsis and septic shock (Sepsis-3). JAMA 2016;315(8):801–10.

[8] Johnson AEW, Pollard TJ, Shen L, et al. MIMIC-III, a freely accessible critical care database. Sci Data 2016;3:160035.

[9] Pollard TJ, Johnson AEW, Raffa JD, Celi LA, Mark RG, Badawi O. The eICU Collaborative Research Database, a freely available multi-center database for critical care research. Sci Data 2018;5:180178.

[10] Prasad N, Cheng L-F, Chivers C, Draugelis M, Engelhardt BE. A reinforcement learning approach to weaning of mechanical ventilation in intensive care units. 2017. arXiv:1704.06300.

[11] Nemati S, Zeng D, Ghassemi MM, Clifford GD. Optimal medication dosing from suboptimal clinical examples: a deep reinforcement learning approach. Conf Proc IEEE Eng Med Biol Soc 2016;2016:2978–81.

[12] Zhao Y, Zeng D, Socinski MA, Kosorok MR. Reinforcement learning strategies for clinical trials in nonsmall cell lung cancer. Biometrics 2011;67(4):1422–33.

[13] Padmanabhan R, Meskin N, Haddad WM. Closed-loop control of anesthesia and mean arterial pressure using reinforcement learning. Biomed Signal Process Control 2015;22:54–64.

[14] Popova M, Isayev O, Tropsha A. Deep reinforcement learning for de novo drug design. Sci Adv 2018;4(7): eaap7885.

The real dividend in reinforcement learning is in its potent combination with DL, which uses a large neural network for pattern recognition, in the form of deep RL. *AlphaGo* combines RL with DL (deep RL) and is ideally designed for the myriad of nuanced humanlike decision-making aspects gaming since it accommodates for sequences of better decisions by a combination of recognition of complex patterns, long-term planning, and "intelligent" decision-making (see Fig. 5.12) [14]. *AlphaGo* has three

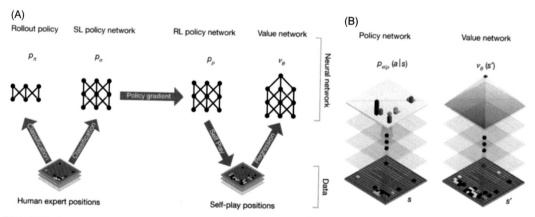

FIGURE 5.12   Deep reinforcement learning in AlphaGo. (A) Reinforcement learning policy network. The human expert positions indicate rollout policy as well as supervised learning policy network. A reinforcement learning policy improves by policy gradient learning to maximize outcome. (B) Neural network architecture of AlphaGo. The policy network and the value network work in conjunction to determine the best next move. The policy network takes board position *s* as its input and passes it through many convolutional layers with parameters from the policy networks, and outputs a probability distribution of moves. The value network also uses many convolutional layers that predict the expected outcome in the proposed position.

components: (1) Policy network—this element evaluates the current situation and predicts the next step; (2) Fast rollout—this component improves the speed of the decision; and (3) Value network—this element evaluates the situation and predicts which side will win.

RL and its AI congener deep RL is particularly valuable assets for biomedicine as these methodologies are well designed to make sequential decisions in an uncertain environment toward a long term goal that can be to minimize error (leading to morbidity and/or mortality). In medicine and health care, RL is ideal for process optimization and decision sequence optimization. This RL, especially coupled with DL (deep RL), is one step closer to what some would consider *real* AI.

RL methodologies are listed in Table 5.7 and briefly described with Deep Q-network (DQN) being the first large-scale application of RL with deep neural network [15]. In DQN, the synergistic combination of RL with a novel artificial agent named DQN can learn successful policies directly from high-dimensional sensory inputs using end-to-end RL.

Advantages and disadvantages

The advantage of RL is the type of learning is considered more human learning. In addition, RL is not task nor data driven like supervised and unsupervised learning. RL is also capable of functioning in a changing environment. A limitation with RL, however, is that some biomedical problems have multiple simultaneous interactions and are real-time without time delay (continuous and not discrete). In addition, RL is data hungry and relatively opaque as well as narrow and brittle. Finally, RL has to balance exploration and exploitation.

A brief review of not only RL but also DRL is useful (but relatively esoteric on the mathematics) to appreciate basic principles of RL and DRL and their applications in medicine [16]. An example in the recent biomedical literature is the use of deep RL (in the form

TABLE 5.7 Reinforcement learning methodologies.

Reinforcement learning methodologies	
A3C	Algorithm obsoleting DQN and leverages deep learning for continuous action spaces
DDPG	Model relies on actor critic architecture with experience replay and separate target network
DQN	Model that leverages neural network to estimate the Q-value function
Genetic algorithm	Use mutations and crossovers to converge to local optima
Q (Quality)-learning	An off-policy algorithm that aims to maximize the Q-value
SARSA	An on-policy algorithm that learns the Q-value based on the action by current policy
Temporal difference	Model-free methodology in which learning is by bootstrapping from current value function

A3C, Asynchronous advantage actor critic; *DDPG,* deep deterministic policy gradient; *DQN,* deep Q-network; *SARSA,* state-action-reward-state-action.

of double DQNs) for optimal pain management in the intensive care setting that is more effective than conventional methodologies [17].

Neural Networks and Deep Learning

A type of ML that was inspired (perhaps in a slightly exaggerated manner) by the brain with its neurons and intricate synaptic interconnections is termed ANN, or neural networks (also neural nets for short). The aforementioned perceptron, also called a node, is a computational model of a biological neuron.

Compared to the ML techniques just discussed, the more sophisticated neural networks and deep-learning techniques (with sometimes hundreds of layers of neurons) are particularly well suited for nonlinear and complex relationships, which are not uncommon in biomedicine and health care. This category of neural networks includes perceptrons (including both the simple perceptron as well as perceptrons in multiple layers called multilayer perceptrons, or MLP), autoencoders, GAN, convolutional neural network (CNN), and recurrent neural network (RNN) (see Table 5.8).

Perceptron and Multilayer Perceptrons

A perceptron is essentially a single layer neural network and is the simplest of all the neural computational models. It is essentially a binary linear classifier. This node or neuron model, with its biologically inspired structure, takes in inputs in the form of data, performs designated computational function(s) (also termed activation function, see later), and then gives out an output. In short, the node or neuron is where computational function(s) occur. Even though the neuron and the perceptron have similarities in their architectures (dendrites and axons are equivalent to inputs and outputs respectively), a major difference between these two models is the number of connections: while the perceptrons can be connected to a few other perceptrons, biological neurons can be connected to as many as 10,000 other neurons.

TABLE 5.8   Neural networks.

Type of neural network	Features	Function
MLP	Hidden layer	Adaptive learning
Antoencoder neural network	Encoder-decoder	Dimension reduction
		Data denoising
GAN	Generator-discriminator	New data generation
CNN	Convolutional layer and ReLU	Computer vision
RNN	LSTM	Sequential data

CNN, Convolutional neural network; GAN, generative adversarial network; LSTM, long–short-term memory; MLP, multilayer perceptron; ReLU, rectified linear unit; RNN, recurrent neural network.

These nodes are connected via their connections, which are respectively modulated by parameters called weights. These weights determine the relative strength of the transmitted signal for each connection and can be positive or negative. In other words, this weight influences the effect that the input will have on the neuron and is adjusted in order for the neural network to "learn." There is also a bias, which is a number that tells one how high the weighted sum needs to be before the neuron is activated. In summary, the sum of the inputs is multiplied by the weights and then the bias is considered.

In addition, there is an activation function of the neural network that learns via connection between the input and the output (and can limit the breadth of the neuron's output). The activation function determines the value of the perceptron's output. The activation function can therefore improve the performance of the neural network; the activation function of a neuron is usually in a sigmoid shape as it is the balance between linear and nonlinear behavior.

Training of the neural network occurs in two different stages: forward and back propagations. Forward propagation is essentially the weighted sum of the inputs and the predicted label (compared to the true label). Backward propagation (short for backward propagation of errors) then is a mechanism in which the neural network can learn (or fine-tune). The difference between the actual output and the desired output is used to calculate a modification called the loss or cost function (see also next) to achieve the desired outcome.

The MLP is therefore a multilayer feedforward network with one or more hidden layers. Each of these neural networks has an input layer, a hidden layer, and an output layer (see Fig. 5.13). The hidden layer performs computations received from the input layer and passes the results to the output layer. Learning, therefore, is for the network to find not just the right weights but also the best biases (usually many more the former than the latter) to perform correctly.

Loss or cost function is an assessment of all the weights and biases of the network and provides one number as a performance "grade" of the network; the lower the number, the better the network's performance. [Note. Some authors do differentiate loss function (or error) as an element for a single training example whereas cost function is for the entire training set]. In essence, "learning" in a network is reflected by minimizing this loss or cost function in an iterative fashion.

A gradient descent is an optimization algorithm to improve the parameter of the ML model (the coefficients in linear regression and the weights in neural networks) so that these models can be at the lowest cost function (or at the highest accuracy). This strategy is performed by iteratively decreasing the gradient until a global minimum is reached (as a local minimum is usually not the lowest) of the cost function (see Fig. 5.14).

Advantages and disadvantages

The advantages of MLPs include its adaptive learning capability and its backpropagation algorithm can perform mapping between input and output well. Among the disadvantages are its relatively slow speed of learning and that it requires labeled data. In addition, there is also its possibility of overfitting with MLP.

An example in the recent biomedical literature is a recent report using radiomic features and multilayer perception network classifier for a relatively robust MRI classification strategy (compared to GLM or even CNN) for distinguishing glioblastoma from primary central nervous system lymphoma [18].

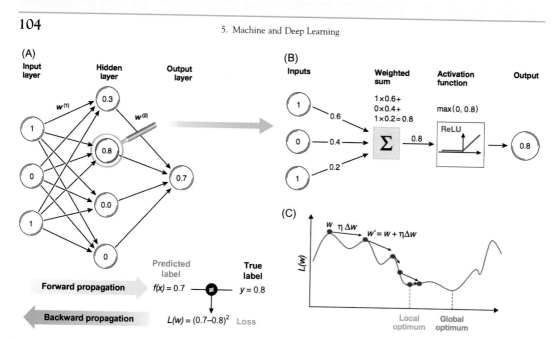

FIGURE 5.13   The ANN is seen in (A), with the input layer pushing data through the hidden layers and then end up in the output layer. Weights, $w^{(1)}$, are parameters that learn from input and output comparisons. Learning occurs by minimizing the loss function $L(w)$ that measures the fit of the output from the model to the actual sample. In (B), data and mathematical functions of the neuron are seen. Each neuron computes the inputs with a weighted summation (seen in blue) and then applies an activation function (ReLU in this case) so that an output is calculated. The loss is backward propagated through the network in order to generate the gradients of the $L(w)$, which is optimized using the gradient descent approach for a global optimum [seen in (C)]. ANN, Artificial neural network; ReLU, rectified linear unit.

Deep Learning

DL promulgated from three separate influences. First, the evolution of methods from early mathematicians to LeCun's CNN in 1989 and culminating in Hinton's work later on with ImageNet in 2012. Second, the storage capacity of computers increased dramatically from punch cards in earlier periods to Internet, and finally, the current cloud form of storage. Finally, computational power also increased from its origin in the form of the ENIAC computer to the present-day GPUs, which are specialized type of microprocessors with memory that are parts of the graphics cards and have more processing cores than CPUs (see the previous figure).

In 2012 the team from the University of Toronto lead by Geoff Hinton used a deep-learning algorithm with 650,000 neurons and 5 convolutional layers to reduce the error rate in half during a computer vision challenge called ImageNet [19]. Following this milestone, Andrew Ng of Stanford and Google as well as others then synthesized huge neural networks by increasing the number of layers and neurons to enable larger and larger data sets to be trained to promulgate DL [20−22]. With robust open-source software tools (such as TensorFlow, PyTorch, and Keras), powerful supercomputers (such as the NVIDIA DGX-1) as well as the abundance of many types of data, DL is an exciting but also enigmatic new extension of ML. Current applications of DL include speech recognition and

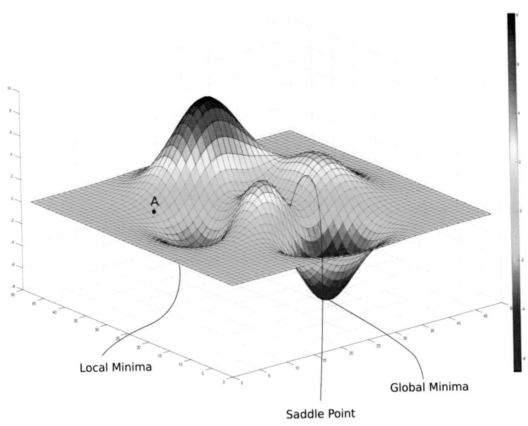

Local Minima

Global Minima

Saddle Point

FIGURE 5.14 Gradient descent and global minimum. Optimization with gradient descent to a global minimum point (global minima). A local minimum is seen but it is not the lowest point. The complicated nature of the image reflects the nonlinear nature of neural networks. *Source: Mechelli A, Vieira S, Lopez Pinaya WH, Garcia-Dias R. Chapter 9 - Deep neural networks. In: Vieira S, Mechelli A, editors. Machine Learning. Methods and Applications to Brain Disorders. Elsevier Inc; p. 162.*

natural language processing, computer vision with visual object recognition and detection, speech recognition, and autonomous vehicle driving.

An excellent review on DL in medicine discusses DL in computer vision, natural language processing, RL, and generalized methods [23]. The following types of DL will now be discussed: autoencoder neural network, GANs, CNN, RNN, and others.

Autoencoder Neural Network

This is a relatively simple three-layer (or more) neural network that is an unsupervised learning tool that "encodes" input data (as a vector) into a more compressed representation (see Fig. 5.15); it is therefore a form of dimension reduction.

An autoencoder has an intentional "bottleneck" in the network so that this area forces a compressed knowledge representation (or exact copy) of the original input; a decoder is

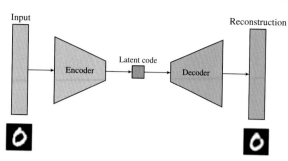

Input   Reconstruction

Latent code

Encoder

Decoder

FIGURE 5.15 Autoencoder. The autoencoder transforms a higher level input into a lower-dimensional representation (Z) via an encoder and decoder. *Source: Mechelli A, Vieira S, Lopez Pinaya WH, Garcia-Dias R. Chapter 11 - Autoencoders. In: Vieira S, Mechelli A, editors. Machine Learning: Methods and Applications to Brain Disorders. Elsevier Inc; p. 195.*

usually paired with this encoder in order to reconstruct the input data that was compressed. An autoencoder is also used for data denoising. Unlike PCA and its capability for dimension reduction, autoencoder performs dimension reduction in a nonlinear fashion. An interesting application of autoencoders is the variational autoencoder (VAE), which not only compresses input data but is also generative—it synthesizes new, similar data of the type that the autoencoder has observed (essentially new images from old images).

An autoencoder can be applied to computer vision, anomaly detection, and information retrieval.

Advantages and Disadvantages

An autoencoder is easier to use for dimension reduction than PCA and is considered to be more accurate as well. In addition, as autoencoder is a neural network, it is well suited for image and audio data. One drawback of an autoencoder is that one can miss theoretical insights and information about the input data, especially with VAEs. Another potential weakness is that it requires a relatively high volume of data to train.

An example in the recent biomedical literature of an autoencoder is the use of a novel prediction method for Parkinson's disease gene prediction using a three-part strategy: extracting features of genes based on network, reducing the dimension using deep neural network in the form of an autoencoder, and predicting Parkinson's disease genes using a ML method (SVM) [24].

Generative Adversarial Network

Introduced by Ian Goodfellow in 2014, GANs are deep neural network architectures that consist of two networks that are competing against each other but can generate data from scratch [25]. In addition, according to Yann LeCun, director of Facebook AI, GAN is "the most interesting idea in the last 10 years in ML." In GANs, two adversarial models (called "generator" and "discriminator") can cotrain through back propagation as a form of unsupervised learning (thus giving the computer the capability of "imagination"). There have been many applications of GANs in computer vision and images with training semisupervised classifiers and generating higher resolution images from originals with lower resolution.

The deep-learning concept is the following (see Fig. 5.16): the generator, a neural network that is generating new data instances, is a dyad with another neural network called the discriminator, which is assessing the created data instances from the generator for authenticity. In essence, the discriminator acts as a "judge" to force the generator to produce more authentic images so these two neural nets are simultaneously being trained. GANs have been coupled with CNN to create an unsupervised learning deep convolutional GAN, or deep convolutional GAN (DCGAN) [26]. DCGAN differs from CNN in that it has only the convolutional layers but not the pooling nor the fully connected layers.

Advantages and disadvantages

GANs are good at training classifiers in a semisupervised way as it does not require labeled data; it is relatively easy to for this deep-learning methodology to generate new (artificial but good) data. While GANs are good at generating image data, it is not easy for these neural networks to generate text data. In addition, some experts feel that it is not easy to train GANs as these are relatively huge computations. Lastly, "mode collapse" occurs in GANs when the generator produces samples of extremely little variety and is considered a weakness of GANs.

An example in the recent biomedical literature of GANs is the recent work by Guan on using both GANs and transfer learning for breast cancer detection by CNN as innovative dual solutions for lack of training images [27]. In short, the authors applied GANs for image augmentation and transfer learning in CNN for breast cancer detection.

A RBM is a shallow (two-layer) ANN that is essentially an unsupervised generative deep-learning algorithm; the first layer is called the visible or input layer and the other layer is the hidden layer. The "restricted" term from no two nodes in the same layer has a connection. An RBM can also be considered the building blocks of deep belief networks (DBM) and can be used for classification and dimensionality reduction. Overall, RBM is not used as frequently now as most users have changed to GANs or VAEs (as discussed earlier).

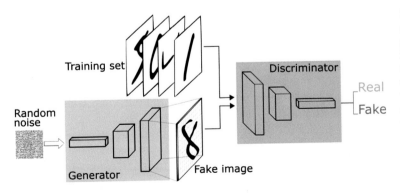

FIGURE 5.16 Generative adversarial network. The generator attempts to increase the probability of making the discriminator mistakes its inputs as real whereas the discriminator allows the generator to produce more images (see the text for details).

Convolutional Neural Network

This is a very popular deep neural network that consists of the characteristic three-dimensional convolutional layers or blocks inspired by cognitive neuroscience and the visual cortex functions. The biological construct of vision as it relates to computer vision is reviewed in a comprehensive fashion by Cox; this review also discussed nuances such as moving images and other elements [28]. CNN is particularly good for computer vision with hierarchical or spatial data (usually images or characters) as well as natural language processing.

CNN can be applied to medical images in three ways: (1) classification—determination of a category (absence/presence of malignancy or type of malignancy); (2) segmentation—identification of pixels/voxels that constitute an area of interest (such as an organ or bleed); and (3) detection—prediction of an area of interest.

Note. Computer vision is usually considered a branch of AI and includes image processing, object recognition, optical mark recognition, and other areas but will not be separately covered as the discussion later pertains to computer vision as it relates to medicine.

First, it is important to introduce the concept that the computer "sees" a matrix of numbers representing pixel brightness (rather than the shades of gray that the human eye sees) (see Fig. 5.17).

The building blocks of the CNN architecture consists of convolution layers, pooling layers, and fully connected layers as well as rectified linear unit (ReLU) (see Fig. 5.18) [29]. These layers are constructed to enable CNN to learn spatial hierarchies of features and are discussed as follows:

1. A convolutional layer (for feature map extraction) involves a convolutional "filter" or "kernel" that is placed over the source pixel or input image. This convolutional filter (or kernel) transforms the source pixel or input image into a new pixel value, and it becomes the destination pixel or output feature map. In other words, this layer integrates and transforms the collection of pixels in the images into significant feature characteristics on the output feature map. Essentially, convolution signifies "filtering" and the filter matrix is applied over the image matrix to produce a "convolved" feature map or matrix. In short, a third function (feature map) is derived from two functions (input data and convolutional kernel).

   A stride is the number of pixels to shift or slide over the input matrix (a stride of one signifies a movement of one pixel). There cannot be a stride of 0.

   The ReLU is an activation function, a key component of CNN that ensures a nonlinearity from linear operations; it improves CNN by speeding up its training. The ReLU is unlike the other possible functions (sigmoidal or hyperbolic tangent); in that, it is perfectly linear for positive inputs while blocking negative inputs and therefore better suited for CNN.

2. A pooling layer (for feature aggregation) usually follows the convolutional layer described previously and receives the output of the convolutional layer as its input but reduces the number of parameters (in order to reduce overfitting). This process also enables faster computation as the number of input parameters is reduced. Therefore this "pooling" (also called "downsampling") process reduces the dimensions while retaining important information.

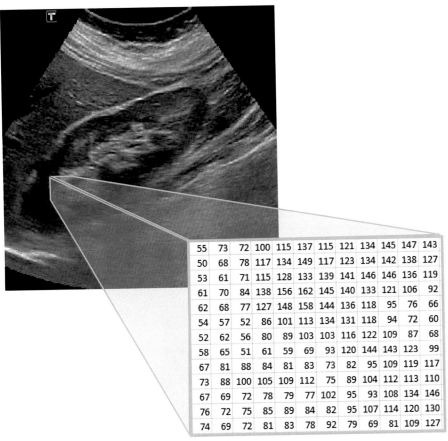

FIGURE 5.17   Human and computer vision. Whereas a human sees the image as a kidney, the computer sees a matrix of numbers representing pixel brightness. Computer vision involves computing the matrix number patterns (so-called features) and then applying machine learning algorithms to these images.

A process of max-pooling (usually preferred over min, average, or sum pooling) (see Fig. 5.19) condenses the more essential information from previous layers into a smaller tensor (a tensor is a term for a multidimensional array more complicated than either a simple vector or a higher dimensional matrix). The pooling layer increases the accuracy as well as the velocity of training of the model. Overall, the resolution necessarily decreases as the pooling layers progress due to the reduction of dimensions but the information of the images (per space) become more and more rich and relevant.

Overall, the first layers of the CNN are more involved with basic image features (such as edges or shapes), while subsequent layers are more focused on abstract features. It is these later layers that contain the robustness as well as the lack of full explainability of DL. In other words, CNN features are more generic in the earlier layers and more complex in the later layers.

II. Data science and artificial intelligence in the current era

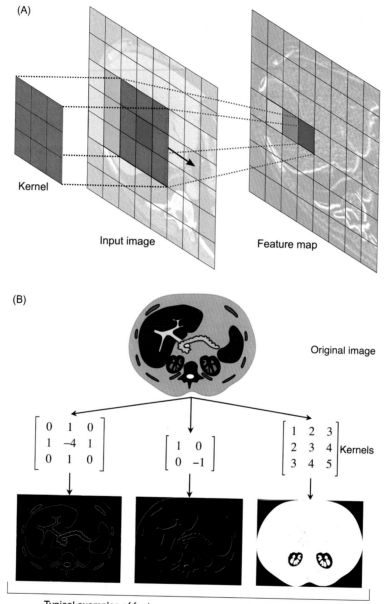

FIGURE 5.18 Convolution. A convolution kernel (or filter matrix) is placed over the source pixel or image, and this convolving process leads to a new destination pixel with a new value on the feature map.

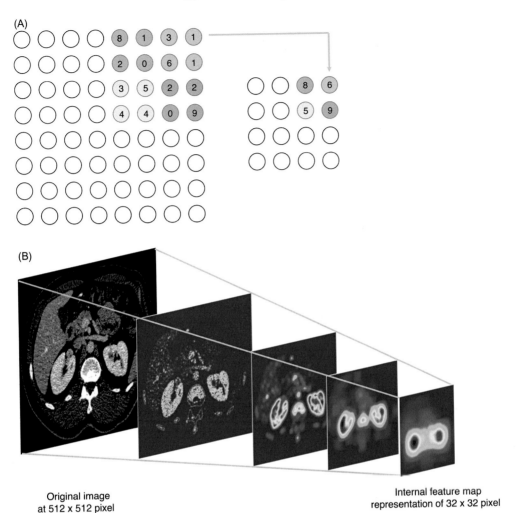

FIGURE 5.19 The CNN pooling process. In (A), the max-pooling process is illustrated with only the maximum activation propagating to the next layer. For example, the green numbers of 1, 1, 3, and 6 are downsampled to a single number 6. In (B), the downsampled representation of the original 512×512 kidney CT image to the internal feature map of 32×32 pixel is shown. This process reduces the memory requirements while retaining the most important information. *CNN*, Convolutional neural network.

3. The fully connected layer (for classification) maps the features into the final output. A process called upsampling in the latter layers increases the resolution of the CNN by reducing the storage and transmission requirements for the images. The last output of the CNN is a topographic display of areas of interest via a process called flattening. This is converting the output of the convolutional part of the CNN to a feature vector so that an ANN classifier can be applied. A softmax activation is applied to the very

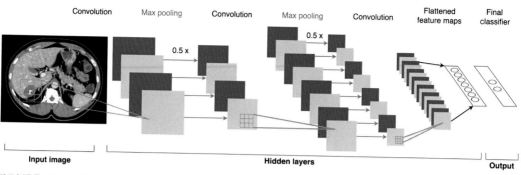

FIGURE 5.20 Convolutional neural network. The image on the left is the input image and goes through a series of convolutions (with convolution kernels) with a stack of feature maps with low-level features (such as edges and corners). The feature maps are then "downsampled" or "pooled" by a max-pooling layer that propagates only the maximum activation to the next layer of learned convolutions with higher features (such as parts of organs). These convolutional and max-pooling layers are stacked in an alternating pattern for the deep network. In the end, there is a flattened feature map (a single vector) that will perform the final classification or regression for the target task.

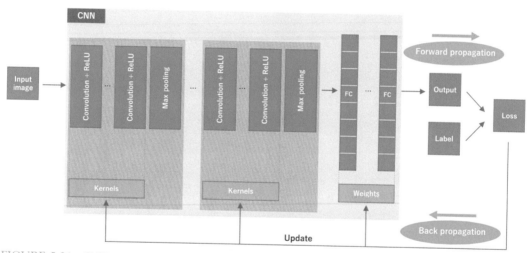

FIGURE 5.21 CNN architecture. This more detailed CNN architecture consists of three main types of layers or blocks: convolutional layers (paired with ReLU), pooling layers (max-pooling), and FC layers. Kernels and weights are part of the model's performance enhancement tools for both forward as well as back propagation. CNN, Convolutional neural network; FC, fully connected; ReLU, rectified linear unit.

last layer (instead of ReLU) so that the output can be converted into a probability distribution (see Figs. 5.20 and 5.21).

The entire convolution and pooling (downsampling or subsampling) process can be summarized in Fig. 5.22.

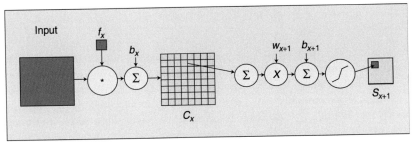

FIGURE 5.22   Convolution and pooling (downsampling or subsampling). The convolution process (linear) started with convolving (transforming) an input (image at the early stages of CNN or feature map at the later stages of CNN) with a trainable filter (f_x) or kernel and subsequently with a trainable bias (b_x) to produce the convolution layer C_x. The subsampling process (nonlinear) then consists of summing a group of pixels (four in this case, outlined in red) and weighting by scalar w_{x+1} (with trainable bias b_{x+1}). Finally, this is passed through a sigmoid function to produce a smaller feature map S_{x+1}. In essence, these two processes convert an input (either an image or a feature map) into something that is reduced in size but retained the information. CNN, Convolutional neural network.

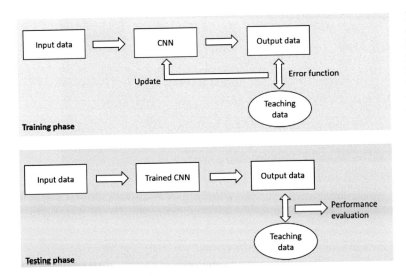

FIGURE 5.23   Overview of deep-learning training and testing process. During the training phase, output data from the CNN are analyzed via an error function with the teaching data. The parameters within the CNN are adjusted to minimize the error. During the testing phase, input data (from training data) are fed to the trained CNN, and the performance is evaluated by using the output data and the teaching data. CNN, Convolutional neural network.

The deep-learning process of training and testing is shown next (see Fig. 5.23).

Convolutional neural networks in medical image processing

Birgi Tamersoy, Brian Teixeira and Tobias Heimann
Birgi Tamersoy, Brian Teixeira, and Tobias Heimann, members of an outstanding engineering team focused on convolutional neural network (CNN) development, authored this commentary

on CNN with an engineering perspective with discussion of the nuances of CNN, from the encoding to decoding stages.

In recent years, CNNs have helped us reach or surpass human performance in a variety of medical image processing tasks. There are multiple factors contributing to this success.

Like other deep neural networks, CNNs also build on the concept of learning increasingly higher levels of abstraction. For example, a CNN that is trained to localize anatomical landmarks in topogram images [1] may detect low-level image features such as "edges" or "blobs" in the earlier layers, process organ-specific structures in the intermediate layers and then build on this information for retrieving the actual landmark locations in the final layers. Such a CNN is illustrated in Fig. 1.

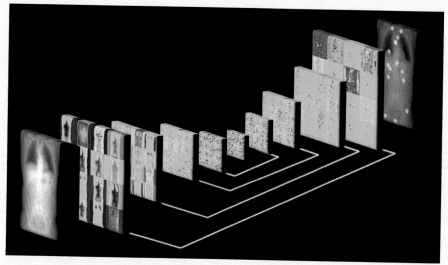

FIGURE 1   Localizing anatomical landmarks in topogram images. From left to right, the image shows how the input information is processed through different layers of a butterfly-shaped network until the final output (a heatmap of landmark probabilities) is produced. The white lines represent skip connections between different layers of the network.

Butterfly-shaped CNNs, similar to the one illustrated in Fig. 1, have been especially prominent in the medical image processing domain. These CNNs (commonly referred to as "U-Nets" [2]) have two primary stages: the encoding stage, where input information is summarized in an intelligent and task-specific compression; and the decoding stage, where the compressed information is extracted to obtain the desired results.

Image resolution is progressively reduced through the encoding stage of a U-Net. This is traditionally achieved by an operation called "max-pooling," where spatial subregions of the output of a layer are summarized before being used as input in the following layer. Through multiple stages of max-pooling, the effective "receptive fields" of the filters in subsequent layers are expanded, or in other words, these filters are exposed to larger regions of the original input image. Having this global view of the input is crucial for generating globally consistent results, which is very important in medical image processing.

The decoding stage of a U-Net takes this heavily compressed information and progressively increases its resolution by an operation called "up-sampling." It is common to use "skip connections" during the decoding stage, where information from the encoding stage is incorporated back into the processing. Skip connections effectively combine higher resolution local context with lower resolution global context and allow U-Nets to have globally consistent and locally accurate results.

CNNs in medical image processing domain are used for modeling very complex input/output relationships. In order to achieve this, utilized networks not only have to be very deep, but they also need to employ nonlinear activation functions. Traditionally, squeezing functions such as sigmoid or hyperbolic tangent have been used for this purpose. However, for larger absolute input values, the gradients of these functions can become so small that numerical limits are hit when trying to back-propagate them through the network during training. This problem of vanishing gradients has limited the sensible depth of neural networks for a long time. Rectified linear units (ReLU) have a derivative of 0 for negative values (nonactivation) and a derivative of 1 for positive values (activation). Given this, gradients can be propagated over active nodes through an arbitrary number of layers [3]. Hence, the use of ReLU activation functions has helped with the training of deeper networks in medical image processing domain.

CNNs have also enabled the successful application of deep reinforcement learning concepts in the medical image processing domain. One recent example is the efficient parsing of large medical volumes using artificial agents [4], where the agents employ CNNs for determining their next moves given their limited local contexts. With the huge popularity CNNs have been gaining in recent years and the large amount of research efforts that is spent on this technology, we expect CNNs to stay relevant for a long time in medical image processing.

References

[1] Teixeira B, Singh V, Chen T, Ma K, Tamersoy B, Wu Y, et al. Generating synthetic X-ray images of a person from the surface geometry. Proceedings of the conference on computer vision and pattern recognition, CVPR 2018. June 18–23, 2018, Salt Lake City, UT. IEEE; 2018. p. 9059–67.

[2] Ronneberger O, Fischer P, Brox T. U Net: convolutional networks for biomedical image segmentation. Proceedings of the international conference on medical image computing and computer-assisted intervention, MICCAI 2015. October 5–9, 2015, Munich, Germany. Springer LNCS 9351; 2015. p. 234–41.

[3] Maas AL, Hannun AY, Ng AY. Rectifier nonlinearities improve neural network acoustic models. In: Proceedings of the ICML workshop on deep learning for audio, speech, and language processing, WDLASL 2013. June 16, 2013, Atlanta, GA

[4] Ghesu FC, Georgescu B, Zheng Y, Grbic S, Maier AK, Hornegger J, et al. Multi scale deep reinforcement learning for real-time 3D-landmark detection in CT scans. IEEE Trans Pattern Anal Mach Intell 2019;41 (1):176–89.

Advantages and disadvantages

CNN is different from conventional ML in that CNN requires large amounts of data for model training; CNN, on the other hand, does not require manual (human-derived) feature extraction nor image segmentation (see Fig. 5.24). While CNN is particularly good at image recognition and classification, it has difficulty if images have alterations (like rotation or any orientation that is different from previously presented). Other issues with CNN and medical images are its limitation with small datasets as well as its problem

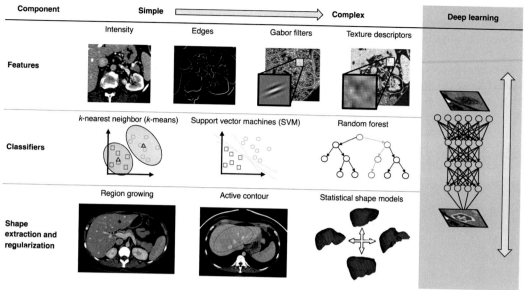

FIGURE 5.24　Deep learning and features. For each component such as features, classifiers, and shape extraction and regularization, the schematic shows elements from simple to complex. For instance, features get more complex as one goes from intensity to texture descriptors. In deep learning the paradigm shift is that features are no longer handcrafted (as in machine learning) but learned in an end-to-end manner to maximize the classifier's performance.

with overfitting. Finally, DL is difficult to explain and its lack of transparency is a serious issue for CNN adoption.

There are numerous good-to-excellent review papers on DL and medical images (see under Radiology and Cardiology) for in depth discussions with helpful diagrams. An example in the recent biomedical literature of a CNN is the work from Stanford's group on interpretation of ECG across a wide variety of diagnostic classes in 12 rhythm classes using single-lead ECGs from over 50,000 patients with an AUC of the receiver operating characteristic (ROC) at 0.97 and an average F1 score (harmonic mean of the positive predictive value and sensitivity) of 0.84 [30].

How AI can bring imaging to billions of people

Esteban Rubens

Esteban Rubens, an expert in AI and data storage, authored this commentary on the grand vision of convolutional neural network (CNN) and AI for medical image interpretation coupled with federal learning to bring medical image interpretation and education around the world.

According to the Pan American Health Organization (part of the World Health Organization), two-thirds of the world's population have no access to diagnostic imaging. Furthermore, between 70% and 80% of diagnostic problems can be resolved through the basic use of X-rays or ultrasound examinations [1]. In the healthcare and development world, doing more with less is the mantra, and finding "low-hanging fruit" a priority. Therefore bringing diagnostic imaging to the roughly 5 billion people who lack access to it seems like a winning proposition. Unfortunately, the hurdles have been significant: imaging modalities are expensive, and radiologists are scarce where most of these underserved populations reside [2].

Evolutionary advances in imaging modalities have not resulted in even the most basic ones (X-ray and ultrasound) becoming more commonly available, especially in remote parts of the developing world. Fortunately, recent advances in solid-state technology have triggered a revolutionary leap in ultrasound imaging. Whereas traditional ultrasound modalities require expensive and fragile transducers based on piezoelectric crystals, it is now possible to leapfrog that paradigm by leveraging technology used to make computer chips in order to produce micromachines called "capacitive micromachined ultrasound transducers" [3]. These new solid-state transducers can be produced cheaply, are sturdy enough to use in the field, and, crucially, can be connected to a smartphone instead of requiring a bulky proprietary computer console. The significant computing and networking capabilities of even the most basic smartphones provide a strong platform for these new transducers and the software required to use them.

We can imagine health workers with little more than basic skills bringing these new portable point-of-care ultrasound (POCUS) devices to places where no ultrasound imaging has been available before. They can be the feet on the ground to acquire images, even benefitting from embedded AI to guide them for optimal probe positioning and hence the best possible image acquisition. Bringing imaging technology to remote parts of the world, however, is only the first step. What good are good images if there are no radiologists available to interpret them?

Perhaps unsurprisingly, AI can bridge that gap. It is well known that a subset of AI, deep learning, in the form of CNNs and other similar classes of algorithms, is perfectly matched to computer-vision problems such as those encountered in radiology, cardiology, and other types of medical imaging [4]. CNNs can be trained to be very accurate in the usual sense of sensitivity and specificity, with the accepted standard being that they should be at least as good as a subspecialty-trained radiologist. The key requirement of achieving that level of accuracy is to have large amounts of labeled data. In the context of medical imaging, "labeled data" means the images themselves together with the radiology reports as well as all the available metadata such as annotations, segmentation data, and anything else that the radiologist (or other physician) reading the images generated as part of the interpretation process.

Once the training dataset with labeled data is available, the deep-learning model (CNN) can be trained. This training process is iterative, requiring multiple runs and the adjustment of the neural-network parameters at each stage in order to achieve the lowest possible error. The ultimate goal is to have a trained model that delivers accurate inferences or predictions when exposed to unlabeled data. It would be reasonable, then, to envision such a trained CNN being embedded in a low-cost, durable, portable POCUS that could be deployed globally to improve the health outcomes of billions of people that had until then been marginalized from the health-care system. These AI-powered POCUS (AI-POCUS) would significantly augment the capabilities of human radiologists by quickly triaging normal images, keeping them out of PACS reading worklists, automatically interpreting clear-cut cases (either normal or with findings), and transmitting only nonobvious images with relevant pathology to a PACS where a radiologist could interpret them.

Alas, the solution is not as simple. It is also known that training CNNs with data from a single population results in models that do not generalize well to global populations [5,6]. Thus it would be necessary to compile vast global training datasets with labeled images in order to train CNNs that could be used effectively across large swathes of the world. This large-scale health-data collection exercise is impractical to the point of impossibility. Private health data, including imaging data, is tightly regulated across the world for good and appropriate reasons. In addition, many countries limit the mobility of health data so that it cannot leave its borders, what is known as data-residency requirements [7].

How, then, can it be possible to secure the benefits of AI-POCUS for the billions of underserved people around the world without compromising the privacy of their health data and staying within the guidelines of the myriad different legal frameworks protecting it? Enter Federated Learning. First proposed by Google researchers in 2017, Federated Learning [8] works by making a CNN model available to a multitude of portable devices. Each device is used to train that CNN further with its own local data, which never leaves the device. The retrained model is uploaded to a central repository, where additional software merges all the newly received models into a newly trained CNN that incorporates all the improvements contributed by each local device. The process continues iteratively until the desired level of accuracy is reached, or on an ongoing basis for continuous improvement.

Given that no training data leaves each local device, that the models are encrypted, and that the methodology ensures that no single update contributed by a local device can be inspected before being merged, Federated Learning, initially in conjunction with radiologists, can be the keystone to enabling a global model-training collaboration that could lead to the desired outcome of making medical imaging available for the first time to billions of people who have never been able to benefit from it. It is impossible not to feel hopeful for the massively significant positive impact that this combination of new technologies (POCUS and Federated Learning) can have on the lives of billions of people who through no fault of their own have been unable to reap the benefits of modern medical imaging.

References

[1] Pan American Health Organization. World radiography day: two-thirds of the world's population has no access to diagnostic imaging, <https://www.paho.org/hq/index.php?
option=com_content&view=article&id=7410:2012-dia-radiografia-dos-tercios-poblacion-mundial-no-tiene-acceso-diagnostico-imagen&Itemid=1926&lang=en>; 2012 [accessed 30.04.19].
[2] Lungren MP, Hussain S. Global radiology: the case for a new subspecialty. J Glob Radiol 2016;2(1) Article 4.
[3] IEEE Spectrum. New "ultrasound on a chip" tool could revolutionize medical imaging, <https://spectrum.ieee.org/the-human-os/biomedical/imaging/new-ultrasound-on-a-chip-tool-could-revolutionize-medical-imaging>; 2017 [accessed 30.04.19].
[4] Chartrand G, Cheng PM, et al. Deep learning: a primer for radiologists. RadioGraphics 2017;37:2113–31 <https://doi.org/10.1148/rg.2017170077>; [accessed 30.04.19].
[5] Saria S, Butte A, Sheikh A. Better medicine through machine learning: what's real, and what's artificial. PLoS Med 2018;15(12):e1002721 <https://doi.org/10.1371/journal.pmed.1002721> [accessed 30.04.19].
[6] Zech JR, Badgeley MA, Liu M, Costa AB, Titano JJ, Oermann EK. Variable generalization performance of a deep learning model to detect pneumonia in chest radiographs: a cross-sectional study. PLoS Med 2018;15(11): e1002683 <https://doi.org/10.1371/journal.pmed.1002683> [accessed 30.04.19].
[7] Information Technology & Innovation Foundation. Cross-border data flows: where are the barriers, and what do they cost?, <https://www.itif.org/publications/2017/05/01/cross-border-data-flows-where-are-barriers-and-what-do-they-cost>; 2017 [accessed 30.04.19].
[8] Federated Learning. Collaborative machine learning without centralized training data, <https://ai.googleblog.com/2017/04/federated-learning-collaborative.html>; 2017 [accessed 30.04.19].

Recurrent Neural Network

This is another type of deep neural network that consists of a feedback loop (next state depends on the prior state) (Fig. 5.25) [31]. In short, each neuron has an embedded memory of an element. RNN, therefore, is capable of having an active data memory known as long–short-term memory, or LSTM. A gated recurrent unit, or GRU, is a variation of LSTM that is structurally similar to LSTM but simpler (2 "gates" vs 3) and does not possess internal memory. RNN is therefore able to recall a memory as there are longer term dependencies (compared to CNN) so it is good for temporal or sequential data (financial transaction data, musical passages, or speech patterns as well as serial biometric measurements such as blood pressure and heart rate).

Even a brief discussion of RNN should include two entities: hidden Markov model (HMM) and neural Turing machine (NTM). HMM is more of a stochastic process based on a Markov chain and makes the Markovian assumption (future state is only dependent on the present state, but not before; therefore "memoryless"). HMM, like RNN, deals with sequential data but is a simpler and linear model, whereas RNN is more complex (and adaptive) and nonlinear. In addition, RNNs do not have Markovian property so they can accommodate long-distant dependencies. NTM is an RNN that extends the neural network concept and couple this with logical flow and external (infinite) memory sources. The NTM has four components: the controller or neural network; the memory; and the read and write heads.

Advantages and disadvantages

RNN therefore has two advantages: it can store information by using the feedback connections and it can learn sequential data. This type of DL is used for repeating the same task for sequential information that are dependent on each other (like time series in the ICU setting or stock market analyses as well as language generation or translation). The disadvantages of RNN include its slow speed for recurrent computation and often it is difficult to access information in the past.

An example in the recent biomedical literature is the use of the independently RNN to classify seizures (against nonseizures) by EEG with a new approach to expand the time scales with superior results for diagnosis [32]. Another illustrative example is the RNN

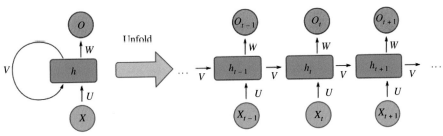

FIGURE 5.25 Recurrent neural network. This architecture shows a feedback mechanism at recurrent layers. Note that the maroon arrows demonstrate a feedback into the nodes of the hidden layer and it is therefore the recurrent layer of the RNN. *Source: Settisara Janney S, Chakravarty S. 33 - Deep learning in medical and surgical instruments. In: Pal K, Kraatz H-B, Khasnobish A, Bag S, Banerjee I, Kuruganti U, editors. Bioelectronics and Medical Devices: From Materials to Devices—Fabrication, Applications, and Reliability. Elsevier Ltd.; 2019. p. 842.*

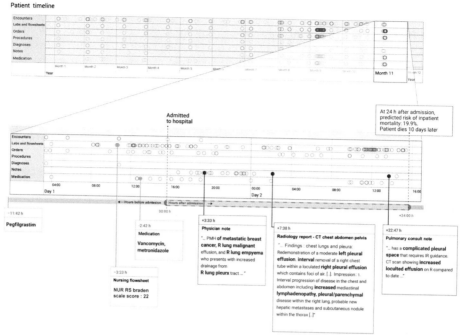

FIGURE 5.26   Patient record and mortality prediction. This patient is a woman with metastatic breast cancer with pleural effusions. The top graph shows the timeline with data types and circles for every time-step. The close-up view illustrates the most recent data-points with the relevant tokens highlighted in red. The prediction in terms of percentage probability of inpatient mortality is seen to the right.

work on EHR that showed accurate predictive models can be built directly from EHR data with the FHIR standard for a variety of clinical scenarios (see Fig. 5.26) [33].

In short, CNN is good for spatial data while RNN is designed for sequential or temporal data. There is, however, a hybrid "CNN–RNN" model (also called recurrent CNN, or recurrent convolutional neural network (RCNN) but not to be confused with regional CNN, R-CNN) that has some potential in biomedical data, such as multilabel image classification and serial complex biomedical data.

An example in the recent biomedical literature of RCNN is the ingestible wireless capsule endoscopy technology with the use of RCNN for a reliable real-time monocular visual odometer method for endoscopic capsule robot operations [34]. The novel RCNN architecture models sequential dependence and complex motion dynamics across endoscopic video frames.

Assessment of Model Performance

Assessment Methods

There are several methodologies to assess the prediction model performance but it is good to have the understanding that a perfect score (classification problem with

100% accuracy and regression problem with 0% error) is simply an unrealistic expectation. There will be inevitable error from data issues (inaccurate or incomplete data) or algorithm limitations. A good overall strategy prior to evaluation of the prediction model is to have a good baseline prediction model (good predictive models to start with include random forest or gradient boosting, see under ensemble learning) or simply try all the prediction models that you are familiar with (and then select the best one or few simple ones).

The evaluation of a model with a test set of data (that the model has not seen before) can follow two methods: cross-validation or holdout method. In cross-validation, the original dataset is divided into k equal-sized subsets called folds, so that $k-1$ subsets are used as the training dataset. In the holdout method, the master dataset is divided into training set, validation set (not always used), and test set (see the ML workflow section).

Evaluation of Regression Models

For regression models, the performance of the model can be assessed with the coefficient of determination (also called R^2 with range 0–1, with 1 being the best), which may need to be adjusted for additional independent variables which can increase value of R^2 (without increase in accuracy). In addition, mean absolute error (MAE) as well as root mean square error (RMSE) are also used: in the former, MAE is the mean of the absolute differences between predictions and actual values, and in the latter, RMSE measures the average magnitude of the error by taking the square root of the average of squared differences between again, the prediction and the actual values.

Model parameters are configuration variables that are "internal" to the model and are properties that will learn on their own during the training by the model (not set by humans). Examples of these parameters include weights in neural networks, coefficients in linear or logistic regression, and support vectors in SVMs. A hyperparameter, on the other hand, is a parameter that is prior belief so these are initialized before training a model and are "external" to the model. Examples of these hyperparameters include k in k-nearest neighbor and learning rate for neural network. In essence, hyperparameters are settings of a model that are adjusted to optimize performance of that model. Finally, the process of an automatic optimization of hyperparameters is termed hyperparameter optimization or tuning; two examples are grid search and random search.

Evaluation of Classification Models

For a binary classification model, the performance is measured by the confusion matrix, the AUC in a ROC curve, and AUC in a precision–recall curve (PRC).

Confusion matrix

The confusion matrix (aptly named perhaps according to some readers) is a 2×2 table (see later) of predicted versus actual values for the model (most clinicians are

TABLE 5.9 Confusion matrix.

	Actual positive (for disease)	Actual negative (for disease)	
Predicted positive (based on test)	True positive	False positive (type 1 error)	*Total predicted positives*
Predicted negative (based on test)	False negative (type 2 error)	True negative	*Total predicted negatives*
	Total actual positives	*Total actual negatives*	*Total population*

familiar with this table from epidemiology and biostatistics) (see Table 5.9):

$$\text{Accuracy} = \frac{\text{True positive} + \text{true negative}}{\text{Total population}}$$

$$\text{Precision} = \frac{\text{True positive}}{\text{True positive} + \text{false positive}}$$
(predicted positive)
(Precision also called positive predictive value)

$$\text{Specificity} = \frac{\text{True negative}}{\text{False positive} + \text{true negative}}$$
(actual negative)

$$\text{Sensitivity} = \frac{\text{True positive}}{\text{True positive} + \text{false negative}}$$
(actual positive)
(Sensitivity also called recall)

$$\text{Error rate} = \frac{\text{False positive} + \text{false negative}}{\text{Total population}}$$

(Error rate also called misclassification rate)

$$F_1 \text{ score} = \frac{2 \times (\text{Recall} \times \text{precision})}{\text{Recall} + \text{precision}}$$

The F_1-score, also called F-measure or balanced F-score, is probably the least familiar to most readers; it is the harmonic mean between precision and recall and can be used to assess binary or multiclass classification models for accuracy. The F_1-score conveniently ranges from 0 to 1 (with higher scores being higher in accuracy). Two other related F measures place more emphasis on either recall or precision and include (1) F_2-score or measure, which places more emphasis on recall (high recall model) and therefore more focus on false negatives (false negatives are not acceptable but false positives are acceptable, as in some medical situations) and (2) $F_{0.5}$-score or measure, which places more emphasis on precision (high precision model) and therefore more focus on false positives (false positives are not acceptable but false negatives are acceptable, as in spam situations). Lastly, the G-measure is the geometric mean of recall and precision (as opposed to F-measure being the harmonic mean).

TABLE 5.10 Confusion matrix for a low-incidence disease.

	Actual positive for disease	Actual negative for disease	Low incidence for disease (4/1000)
Predicted positive based on test	3	1	4
Predicted negative based on test	1	995	996
	4	996	1000

Limitations

In the above confusion matrix, both accuracy and error rate (or misclassification rate) can appear better than the true underlying performance of the model if the incidence of the disease is very low and therefore the true negatives of the entire population is relatively high (see later for an illustrative example). In addition, recall (or sensitivity, quantity, completeness) may need to be higher or lower depending on the disease: for instance, one certainly would prefer to have recall be as high as possible for cancer (vs a benign rash) since recall reflects people that were missed in the screening process for cancer. Concomitantly, one would also like precision (quality or exactness) to be appropriate for certain disease states: again for cancer, precision reflects that a person who is tested positive actually has the disease so a false positive could be catastrophic for a diagnosis of a brain tumor (but much less significance for mild myopia, for instance). In other words, a disease in which there is no significant penalty for a false-positive diagnosis would have less demand on a higher precision. Lastly, the strength of the F_1 score is perhaps also its potential weakness: precision and recall are balanced but sometimes this is less than ideal. Depending on the classification prediction model, the F_1 score may need to be weighted so that either precision or recall is given relatively more importance (as discussed earlier).

We can calculate F_1 for an imaginary test for imaging of cancer (which is a popular classification methodology for CNN) for a cancer that has a low incidence (see Table 5.10).

$$\text{Accuracy} = \frac{\text{True positive} + \text{true negative}}{\text{Total population}} = \frac{3 + 995}{1000} = 0.998$$

$$\text{Precision} = \frac{\text{True positive}}{\text{True positive} + \text{false positive}} = \frac{3}{3 + 1} = 0.75$$

$$\text{Specificity} = \frac{\text{True negative}}{\text{False positive} + \text{true negative}} = \frac{995}{1 + 995} = 0.999$$

$$\text{Sensitivity} = \frac{\text{True positive}}{\text{True positive} + \text{false negative}} = \frac{3}{3 + 1} = 0.75$$

$$\text{Error rate} = \frac{\text{False positive} + \text{false negative}}{\text{Total population}} = \frac{2}{1000} = 0.002$$

$$F_1 \text{ score} = \frac{2 \times (\text{Recall} \times \text{precision})}{\text{Recall} + \text{precision}} = \frac{2 \times 0.75 \times 0.75}{0.75 + 0.75} = 0.75$$

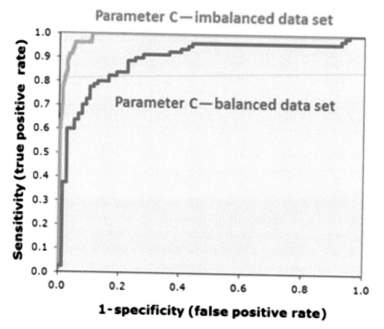

FIGURE 5.27 ROC and AUC. The ROC plots sensitivity (y axis) against the false-positive rate (1 − specificity). ROC is shown for two different data sets (for an imagined parameter C). One is balanced (in blue) and the other one is imbalanced (in orange). A perfect model will be close to the upper left corner. In this case the AUC of the imbalanced data set appears to be greater than the one for the balanced data set. *AUC, Area under the curve; ROC, receiver operating characteristic.*

So one can surmise from this specific confusion matrix example that it is relatively easy to have a higher accuracy as well as a lower error or misclassification rate (real and perceived) when the incidence of disease is *low* because the true negatives are a relatively high number in the calculations for both accuracy and error rate. In short, accuracy and error rate are not good indicators of performance especially when the incidence of disease is very low (like for cancer) because the true negatives, a majority of the cases usually in low incidence disease states, is a relatively high number to make accuracy and error rate look more favorable. This is a good example of how precision and the F_1 score will be more realistic reflection of the classification model prediction performance, especially when there is an imbalance in the classes as in the case of true negatives being a very large number. When the true negatives are a large number, one can also consider the PRC (see next).

Receiver operating characteristic

The most familiar curve and metric for assessing classification models is the ROC and its accompanying AUC (see Fig. 5.27). Similar to the aforementioned PRC, one can compare different predictors by estimating their AUC with a major difference: in PRCs, the upper right is near perfection with both precision and recall close to 1, but in ROC, it is the upper left part of the graph that is considered better (since the x axis is 1 − specificity to derive the false-positive rate).

The AUC measures the area under the curve in the ROC plot and is often used to measure the performance of a classification model as it reflects both true positive rate or sensitivity (y axis) and false-positive rate or 1 − specificity (x axis). True positive rate is also called recall, and false positive rate reflects data that have been misclassified.

Performance is not outcomes: safety in medical artificial intelligence

Luke Oakden-Rayner

Luke Oakden-Rayner, a radiologist and an AI advocate as well as probably the brightest mind in radiology and AI, authored this commentary on a very important caveat in looking at performances of AI tools in radiology.

Performance is not outcomes.

This should be the mantra of anyone who is building medical AI systems.

To explain why, let me introduce some terminology.

Performance testing is what we have seen in medical AI research papers and regulatory approvals so far. We take a set of patients, define how we will measure the performance of our model, and identify what "good" performance will be (usually a comparison against current practice). We then analyze the results with some sort of statistical test to estimate how reliable they are.

This is like doing an experiment in a laboratory, which is why it is often also called laboratory testing (despite the severe lack of laboratories in most AI research). The point is that in this type of experiment, we control for all factors other than the AI model.

Clinical testing has the goal of not controlling the experiments. Unlike in performance testing, we want to see how the system operates in the context of real healthcare. We want to see that good performance actually leads to better clinical outcomes.

Clinical outcomes are what happens in practice. The two types of outcomes we care about are patient outcomes, that is, the rates of death and disability for patients who have a specific condition and healthcare system outcomes, such as the amount of money spent per patient.

So the key components of clinical testing are

- real clinical environments,
- real patients, and
- real outcomes that really matter.

It is easy to assume that high performance in experiments should result in good clinical outcomes. If we look at recent papers, we often see experiments that directly compare the performance of AI systems to those of doctors, with favorable results.

But we have to consider these results with caution, because performance is not outcomes.

The experience we have had in computer-aided diagnosis (CAD) over the last few decades is instructive. CAD is the term we used for 1990s computer vision techniques as they applied to screening mammography. The methods used were mostly expert systems using handcrafted rules and support vector machines with hand-crafted features.

While this technology was frequently unsuccessful in computer vision tasks more broadly (unlike modern AI), the US government decided to pay radiologists $8 more to report a screening mammogram if they used CAD. Unsurprisingly, by 2010 it was estimated that 74% of mammograms in the United States were read by CAD [1].

The early experiments were promising. A performance study of CAD undertaken in 1990 [2] showed humans combined with CAD outperformed humans alone. Many more studies followed, with similar results. The first FDA approval of mammography CAD was in 1998, and Medicare in the United States started to reimburse the use of CAD in 2001.

But in practice, many radiologists felt that these systems did not work very well, and that using them could be frustrating. This feeling was born out as outcomes-based clinical trials were performed over the following decades.

In 2007 Fenton et al. [3] showed that in a cohort of 222,000 women undergoing 430,000 mammograms, across 4 years and three states, implementing CAD was associated with a reduction in specificity from 90.2% to 87.2%. The rate of biopsy increased by 19.7%, but the change in the cancer detection rate (from 4.15 to 4.20 per thousand) was not significant.

In 2015 an even larger study by Lehman et al. [4] looked at 630,000 mammograms from 320,000 women across a 6-year period. They found that sensitivity, specificity, and cancer detection rates were not any different between radiologists that used CAD, and those that did not. They also found that for the radiologists who had practiced both with and without CAD during the study period, their sensitivity dropped from 89.6% to 83.3%.

Similar results were seen in other trials. A systematic review in 2008 showed that CAD did not change detection rates, but increased recall rates. It also showed that double reading (which is often considered an alternative to using CAD) increased detection rates and decreased recall rates.

The cause of the discrepancy between laboratory studies and clinical trials has been widely debated, but it is likely that the humans using the systems played an important role. Several studies have shown [5,6], that when doctors take part in controlled experiments they behave differently than when they are treating real patients, a phenomenon called the laboratory effect [7].

Another issue may be that humans tend to overrely on cues from computers and undervalue other evidence, often called automation bias or automation induced complacency.

This effect has also been directly cited as the cause of several deaths in self-driving cars, as well as being a possible reason for the failure of mammography CAD. One particularly interesting study showed that using CAD resulted in worse sensitivity (less cancers picked up) when the CAD feedback contained more inaccuracies [8], suggesting that humans became less accurate because they trusted the computer system.

Whatever the cause of the failure of CAD, it is clear that the reliance on performance testing in regulatory approvals and Medicare reimbursement decisions in the United States lead to negative patient and system outcomes. We need to learn from this experience and make sure we do not make the same mistakes in the rush to bring exciting new technologies into our clinics [9].

Modern AI has the capacity to affect patients on an unprecedented scale, so as we approve these systems and apply them to patients, we need to remember that performance is not outcomes.

References

[1] Rao VM, Levin DC, Parker L, Cavanaugh B, Frangos AJ, Sunshine JH. How widely is computer-aided detection used in screening and diagnostic mammography? J Am Coll Radiol 2010;7(10):802—5.
[2] Chan HP, Charles E, Doi K, Vyborny CJ, et al. Improvement in radiologists' detection of clustered microcalcifications on mammograms. The potential of computer-aided diagnosis. Invest Radiol 1990;25 (10):1102—10.
[3] Fenton JJ, Taplin SH, Carney PA, Abraham L, Sickles EA, D'Orsi C, et al. Influence of computer-aided detection on performance of screening mammography. N Engl J Med 2007;356(14):1399—409.

[4] Lehman CD, Wellman RD, Buist DS, Kerlikowske K, Tosteson AN, Miglioretti DL. Diagnostic accuracy of digital screening mammography with and without computer-aided detection. JAMA Intern Med 2015;175(11):1828–37.

[5] Rutter CM, Taplin S. Assessing mammographers' accuracy: a comparison of clinical and test performance. J Clin Epidemiol 2000;53(5):443–50.

[6] Gur D, Bandos AI, Cohen CS, Hakim CM, Hardesty LA, Ganott MA, et al. The "laboratory" effect: comparing radiologists' performance and variability during prospective clinical and laboratory mammography interpretations. Radiology 2008;249(1):47–53.

[7] Mosier KL, Skitka LJ. Chapter 10 Human decision makers and automated decision aids: made for each other? In R. Parasuraman and M. Mouloua (Eds.). Automation and Human Performance: Theory and applications 1996;10:201–20.

[8] Alberdi E, Povyakalo A, Strigini L, Ayton P. Effects of incorrect computer-aided detection (CAD) output on human decision-making in mammography. Acad Radiol 2004;11(8):909–18.

[9] Posso M, Carles M, Rué M, Puig T, Bonfill X. Cost-effectiveness of double reading versus single reading of mammograms in a breast cancer screening programme. PLoS One 2016;11(7):e0159806.

The ROC is, then, a plot of true- and false-positive rates at various classification threshold so that these thresholds will provide a different pair of true positive and false-positive rates. An AUC of 0.5 is essentially a random classifier, whereas an AUC of 1 is perfect (conveniently again, like the R^2 coefficient for regression models and F_1 score for classification models, 1 is a perfect score). Usually, an AUC of 0.80 or higher is considered good and 0.90 or higher is considered excellent. The higher the AUC, the better the classifier with a few caveats.

A Gini coefficient is the ratio between the area between the ROC curve and the diagonal line and the area above the triangle:

$$\text{Gini} = 2 \times \text{AUC} - 1 \text{ (where index above 0.60 is considered a good model)}$$

Limitations

Just as accuracy and error rate are vulnerable to a potentially large true negative number, the ROC and AUC are as well since the false-positive rate is (1 − specificity), which has true negatives in the denominator [false positives/(false positives + true negatives)]. A large true negatives number will lower the false-positive rate (x axis), which will push the performance toward perfect classification, or the upper left portion of the ROC curve. In other words, an imbalanced dataset will falsely increase the performance of the classifier when if fact there is no change in the performance. In addition, the ROC does not accommodate changes in prevalence. This is very important as there are rare or unusual diseases in medicine and this does affect the classifier as the false positives will have a high level of impact.

Precision−recall curve

As the name implies, this is a curve plotting precision and recall for various threshold values (in a similar fashion to ROC) with both precision and recall values plotted from 0 to 1. Therefore the upper right part of the graph is precision−recall "nirvana": both values are near 1 (see Fig. 5.28). The previous discussion on the impact of true negatives on interpretation of metrics and curves is significant in that the much less familiar PRC is more appropriate as an assessment tool when there is an imbalance in

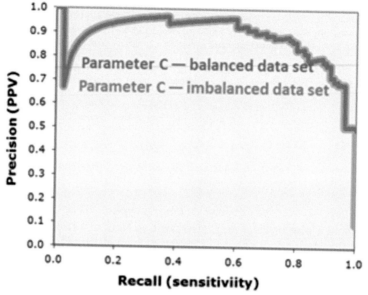

FIGURE 5.28 PRC and AUC. The PRC plots precision (or positive predictive value) on the *y* axis against recall (or sensitivity) on the *x* axis. PRC is also shown for two different data sets (for an imagined parameter C). One is balanced (in blue) and the other one is imbalanced (in orange). See the text for details. A perfect model will be close to the upper right corner. In this case the AUC of the imbalanced data set appears to be about the same as the one for the balanced data set. *AUC,* Area under the curve; *PRC,* precision–recall curve.

datasets; one study even showed that PRC was superior to ROC and even other curves such as concentrated ROC (CROC) and cost curve (CC) in the presence of an imbalanced dataset [35]. In other words, "stacking" a lot of people who are true negatives for disease (no disease and no positive test) will inflate the ROC curve favorably but without true improvement in sensitivity nor positive predictive value [36]. This is called the imbalanced classification problem.

Limitations

The PRC is more appropriate for situations in which there is an imbalanced dataset (usually due to large true negatives) but the area under a PRC takes the arithmetic mean of precision values so it is more difficult (vs ROC) to interpret or visualize. In addition, the RPC does not take into account that precision and recall may not have the same significance depending on the situation (but it is considered to be equal in significance).

Fundamental Issues in Machine and Deep Learning

Interpretability and Explainability

A common issue with ML resides in its "black box" characteristic: for those who are not data scientists or those who have not had education in this domain, it is difficult to

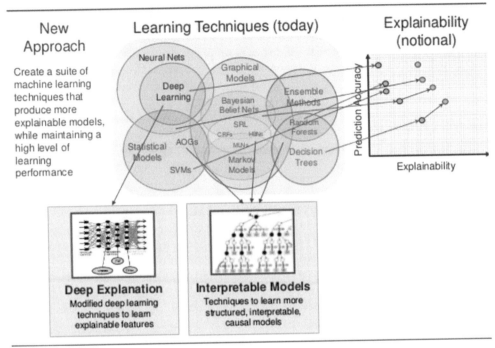

FIGURE 5.29 Machine and deep learning and explainability. The figure shows that the learning techniques such as deep learning have relatively higher prediction accuracy (*y* axis) but also relatively lower explainability. On the other hand, learning techniques such as decision trees are more interpretable models and more easily understood and explained. There is a strategy to increase the explainability for all of these techniques (*orange circles* with *arrow* to *green circles*). *AOG*, Stochastic and-or-graphs; *CRF*, conditional random fields; *HBN*, hierarchical Bayesian networks; *MLN*, Markov logic network; *SRL*, statistical relational learning.

understand the many learning techniques that exist today, especially sophisticated ML and DL (see Fig. 5.29) [37]. While there is a difference between interpretability and explainability, AI methods need at least interpretability for it to be widely adopted by clinicians. Interpretability is the capability to observe a cause-and-effect while explainability is the understanding of the inner workings of a system or technology. For example, a cardiologist can program a pacemaker and see the consequences of an action (programing the pacemaker and see pacing at the rate that was set), but not necessarily fully understand the engineering aspects of the pacemaker itself (it is helpful to understand the latter but arguably less essential).

The natural black box versus the artificial black box

Jonathan D. Lima and Joao A.C. Lima
Jonathan and Joao Lima, father–son AI enthusiasts and father a cardiologist, authored this commentary on the somewhat unfair black-box moniker for AI, while humans often have complex cognitive processes that are symmetrically enigmatic.

Over the course of the next several decades, artificial intelligence (AI) will disrupt every significant technical endeavor in society. This prediction is easy to make, because all of these endeavors will depend heavily on intelligence, and natural intelligence (NI) is comparatively hard to commodify. While these changes face resistance in many forms, with varying levels of validity, this paper seeks to reframe a particular type of concern, "Should I trust a physician or a black box?"

There is a fundamental misconception that underlies this question. Namely, that a physician's judgment is *not* the end result of a complex algorithm which no one can quite explain. From matters as straightforward as dismissing ordinary vital signs to determining whether or not to biopsy a tumor, physicians execute algorithms that have been designed by their experience in training and practice. More and more frequently, these algorithms are becoming standardized across the medical industry where they can be. We wish therefore to reframe the question, "Should I trust a natural black box or an artificial black box?"

Of course, the answer is not straightforward, so let us consider the advantages of the artificial black box. One major advantage is processing speed. For many learning tasks, humans have better learning curves than machines, but a human simply cannot play 4.9 million games of Go in three days. Another major advantage is the scalability/portability of AI. While a skilled cardiologist could detect the arrhythmia proceeding a cardiac episode, we cannot assign a physician to every individual at risk, but maybe we can assign them a watch. The last major advantage is centrality. Because one algorithm can be used universally, entire groups of people can be dedicated to its development and evaluation. This practice tends to result in rapid improvement over time.

While these advantages are significant, let us not underestimate the human condition. The fundamental advantages of NI are reason and communication. In the medical arena, this manifests itself in the Theory of Medicine, which is codified in journals and textbooks, and administered in hospitals and universities. While imperfect, this theory equips physicians with a causal dynamical model for the entire complex system of the human body. This theory is essential in dealing with one of the greatest challenges in medical practice, that no two high-level medical problems are truly the same.

Patient histories are complexly structured data sources with a lot of data that isn't relevant to the task at hand. In machine learning (ML), adding uninformative input data hurts your learning curve. Moreover, it is hard to create models that capture the structure, which again, hurts your learning curve. Equipped with theory, physicians can focus on the essential information and understand the complex relationships between its components. They use that information to execute a delicate procedure: making diagnoses with some degree of uncertainty and making decisions about treatment, with the present option of querying for more data, at some cost.

Because of the usefulness of theory in making higher level decisions, this is where NI is most impactful. The jumping off point of any successful ML endeavor is a well-defined task with tightly coupled inputs and responses. In these settings, we should expect that AI will eventually prove more effective, less expensive, and faster than NI. The job of a diagnostician does not fit that description, but with the use of medical theory, the patient history, and reason, one can factor the job into tasks that do.

Perhaps, it is best for physicians to consider a thought experiment. Imagine you have an assistant, not the most clever, but fast and reliable. It will take great effort to train them in a task, but once trained, they will do it better than you. Also, once trained, for low cost, you can have 1 million of them. You can send them to your colleagues to assist in their labs, or even send them home with your patients to execute their task in the field. What would you train them to do? What would save you the most time? What would save the most money? What would save the most lives? Microsoft chose hospital readmission.

Understanding the relative advantages of natural and human intelligence is essential to answering the most important questions related to the integration of AI into medicine. What kind of AI software should we be developing in our labs? What kind of NI physicians should we be developing in our schools? Perhaps, most importantly, what is the data that we as a society should begin collecting to further enable this revolution? (The vast quantities of data collected throughout medical practice are observational in nature, whereas programs like annual imaging would produce highly useful data for training AI. This is really step zero.)

As a final note, the healthcare industry is relatively immune to the problem of automating away jobs. Better health outcomes lead to longer lives and more healthcare needs. This is partially why solving the major health crises of the 20th century (infectious disease, trauma care, malnutrition, maternal and early childhood mortality) has landed us in a shortage of healthcare professionals. And in the end, while machines are much faster and more precise thinkers than humans, this makes them terrible at learning. Learning is rooted in possibility and uncertainty, concepts that are foreign to machines, and must therefore be elements of the program's design. The human brain was naturally designed for this purpose.

Some of the higher prediction accuracy ML methodologies (DL, random forest, SVMs, etc.) have the least explainability, whereas others (Bayesian belief nets, decision trees) have more explainability (but relatively lower prediction accuracy). There is an ongoing effort to elevate explainability in the form of "explainable AI or xAI" while maintaining (or even increasing) prediction accuracy with a new suite of techniques. The overall strategy to increase explainability is to raise awareness and education of ML and other techniques is to generalize algorithms, understand features, and utilize available support tools for explanations such as the Local Interpretable Model-Agnostic Explanations (LIME).

Interpretable deep learning (DL) for advanced and transparent clinical decision support

Ying Sha and May D. Wang
Ying Sha and May D. Wang, both bioinformaticians, authored this commentary to offer an intelligent strategy to render DL more interpretable by focusing on feature scoring and data synthesis (DS).

As a result of the development of advanced biotechnologies and related instruments, biomedical big data have accumulated exponentially, providing us with rich opportunities for developing data analytics for clinical decision support (CDS). Traditionally, we would develop a clinical data analysis pipeline that consists of missing data imputation, feature engineering and selection, and training of a machine learning (ML) model. However, the pipeline has several nonnegligible shortcomings. Specifically, the absence of information in a clinical data set may be indicative of a physician's judgment of a patient's health status [1]. For example, physicians tend to measure a patient with a worse condition more often than one with a more stable condition. In addition, feature engineering and selection may restrict us from discovering novel feature representation, and a shallow ML model, such as logistic regression, might not be able to capture the sophisticated temporal dependencies and interfeature relationships in clinical data. To address these challenges, we could develop DL-based models to characterize the sophisticated patterns of clinical data, as in recent decades, DL models have achieved significant results in various tasks including nature image classification [2,3], object detection [4,5], natural image captioning [6,7], machine translation [8], or videogame artificial intelligence (AI) [9]. However, DL models have been criticized for their black-box properties. That is, because of the stacked nonlinear transformation of typical DL models, researchers find them difficult to interpret compared to traditional ML models such as logistic regression, for which one feature is associated with only one coefficient. Therefore in addition to applying DL models, we also need to develop and evaluate approaches for interpreting DL models, for ultimately accurate and transparent clinical decision support.

We could approach the problem of interpreting DL models from two perspectives, feature scoring (FS) and DS. First of all, we will need to build and train a DL model, such as long–short-term memory [10], to represent clinical data and their corresponding missing pattern information. After we get a reasonably good DL model, we will fix its parameters and then interpret it in a post hoc way. Specifically, FS refers to approximating a feature importance score for every input feature using typical gradient-based approaches and their variants [11–13]. On the other hand, DS refers to synthesizing an input sample that would let the DL model generate a high prediction score for a class label of interest [14]. Considering an application scenario shown in Fig. 2.1, suppose we build a CDS tool to assist inexperienced physicians in diagnosis based on electrocardiogram (ECG). Physicians may question the decision made by a CDS, with no evidence provided to support the classification result. We could incorporate FS methods to generate importance scores for individual input features and to visualize them so that physicians would know what segment of ECG contributes most to a prediction of atrial fibrillation (AF). In addition, we could use DS methods to synthesize an ECG segment most representative of AF for a given model. With both FS and DS, we could help physicians to gain trust in the CDS or detect potential bias in the model or dataset.

The evaluation of the two categories of interpretation methods is relatively immature. We could evaluate FS and DS both qualitatively and quantitatively, with the former rely on clinical collaborators and the latter rely on automatic computational evaluation. The most ideal and persuasive evaluation should be qualitative. However, qualitative evaluation is quite expensive because physicians do not always have time to evaluate interpretable DL methods. For evaluating the importance scores generated by FS, we could follow a procedure called pixel flipping [15], in which we disrupt raw input features following the order of correspondent importance scores and observe whether the drop in accuracy is more dramatic than randomly disrupting input features. For quantitatively evaluating DS samples, we could apply a technique called Train on real, test on synthetic (TRTS), in which we train the DL model with original input and test the model on synthetic samples [16].

In summary, applying DL models for CDS is a field with huge challenges and opportunities. Although DL models are complex, we can interpret DL models by approximating feature importance and by synthesizing class-representative samples. We hope more researchers will join this exciting field and ultimately improve the quality of health care while reducing the cost of it.

References

[1] Hripcsak G, Albers DJ, Perotte A. Parameterizing time in electronic health record studies. J Am Med Inform Assoc 2015;22(4):794−804.

[2] Krizhevsky A, Sutskever I, Hinton GE, editors. Imagenet classification with deep convolutional neural networks. *Advances in Neural Information Processing Systems* 25 (2), 2012.

[3] Szegedy C, Vanhoucke V, Ioffe S, Shlens J, Wojna Z, editors. Rethinking the inception architecture for computer vision. In: Proceedings of the IEEE conference on computer vision and pattern recognition; 2016.

[4] Redmon J, Divvala S, Girshick R, Farhadi A, editors. You only look once: unified, real-time object detection. In: Proceedings of the IEEE conference on computer vision and pattern recognition; 2016.

[5] Mask r-cnn. computer vision (ICCV). In: He K, Gkioxari G, Dollár P, Girshick R, editors. 2017 IEEE international conference on. IEEE; 2017.

[6] Xu K, Ba J, Kiros R, Cho K, Courville A, Salakhudinov R, et al., editors. Show, attend and tell: neural image caption generation with visual attention. In: International conference on machine learning; 2015.

[7] Show and tell: a neural image caption generator. Computer vision and pattern recognition (CVPR). In: Vinyals O, Toshev A, Bengio S, Erhan D, editors. 2015 IEEE conference on. IEEE; 2015.

[8] Bahdanau D, Cho K, Bengio Y. Neural machine translation by jointly learning to align and translate. May 19, 2016. arXiv:1409.0473v7.

[9] Wu B, Fu Q, Liang J, Qu P, Li X, Wang L, et al. Hierarchical macro strategy model for MOBA game AI. December 19, 2018. arXiv:1812.07887.

[10] Graves A, Schmidhuber J. Framewise phoneme classification with bidirectional LSTM and other neural network architectures. Neural Netw 2005;18(5-6):602−10.

[11] Simonyan K, Vedaldi A, Zisserman A. Deep inside convolutional networks: visualising image classification models and saliency maps. April 19, 2014. arXiv:1312.6034.

[12] Visualizing and understanding convolutional networks. In: Zeiler MD, Fergus R, editors. European conference on computer vision. Springer; 2014.

[13] Bach S, Binder A, Montavon G, Klauschen F, Müller K-R, Samek W. On pixel-wise explanations for non-linear classifier decisions by layer-wise relevance propagation. PLoS One 2015;10(7):e0130140.

[14] Erhan D, Bengio Y, Courville A, Vincent P. Visualizing higher-layer features of a deep network. *Technical Report*, 1341.

[15] Samek W, Binder A, Montavon G, Lapuschkin S, Müller K-R. Evaluating the visualization of what a deep neural network has learned. IEEE Trans Neural Netw Learn Syst 2017;28(11):2660−73.

[16] Esteban C, Hyland SL, Rätsch G., Real-valued (medical) time series generation with recurrent conditional GANs, 2017. *arXiv*:1706.02633.

Bias and Variance Trade-off

Prediction error can be grouped into two main types. Bias is the difference between the expected prediction of the model and the correct value that the model is attempting to predict. Bias is also the inability of a ML methodology to capture the true relationship (such as linear regression not reflecting the relationship of data that would be better fitted in a curve). Variance, on the other hand, is the difference in the fits between datasets (high complexity models will therefore be more likely to have higher variance). Overall, bias is decreased and variance is increased with the increase in model complexity.

The ideal model will have both low bias as well as low variance, but it is usually a trade-off between these two parameters. Bias and variance are manifested in models that demonstrate underfitting and overfitting: the former is a result of high bias and the latter

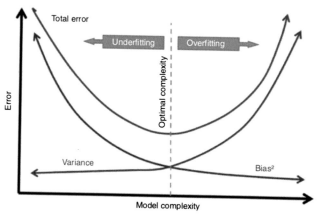

FIGURE 5.30 Bias and variance. In the figure, the y axis is degree of error and the x axis, complexity of the prediction model. As the complexity of the model increases, bias (line in red) decreases while variance (line in green) increases since errors decrease with complexity of the model in bias and the opposite is observed with variance (see the text for explanation). The optimal region is a compromise between these two forces and is where bias and variance are both jointly low (not lowest for each) (see optimal complexity). In underfitting the bias is high but the variance is low and in overfitting, the variance is high but the bias is low.

is a result of high variance; therefore the best balance, fitting, is achieved when there is low bias and low variance (see Fig. 5.30). Three strategies to have a good bias-variance balance include: regularization, bagging, and boosting.

Fitting

Underfitting occurs when the model is too simplistic: it is poorly trained on sample data (such as a linear model) or when the feature engineering is suboptimal and/or inadequate. The solution will involve a more complex model and a better feature engineering strategy. Overfitting occurs when the model is too complex: the results are too tailored to the training data (excessive training or adaptation) so that the model is overly complex for the data (opposite of the previous situation with underfitting) (see Fig. 5.31). In other words, the model will not be able to analyze new test data well (does not generalize to other data). This can be the result of a situation in which there is an excessive number of features but not enough samples. Overfitting can also occur with small studies that have concomitantly small number of attributes so the choice of attributes is critical for ideal ML. To overcome overfitting, data science strategies include using more training data or reduce the number and complexity of features, pruning, perform cross-validation sampling, and use regularization. Regularization is a technique used in machine and DL to modulate the model complexity via penalty measures so that overfitting is minimized and that the model retains its capability to predict.

Perhaps, a way of thinking about under- and overfitting is to draw an analogy with clothes: in underfitting, one has a comfortable and loosely fitted T-shirt (high bias as it is not a perfect fit but low variance as most people can fit into this T-shirt), and in

Plausible fitting   Underfitting   Overfitting

FIGURE 5.31 Underfitting and overfitting. The illustrations show plausible fitting first on the left. Underfitting—the separation is too simple and too many samples are on the other (wrong) side so therefore "misclassified." Overfitting—the classifier correctly identifies all the samples but is overly complex and therefore has a high variance.

overfitting, one has a neatly tailored shirt that is tight-fitting (high variance as it is not a good fit for most people but low bias in that it is a good fit for a few).

Curse of Dimensionality

ML often deals with large dimensional space. The curse of dimensionality occurs when there is such a large number of features so that the features space is excessively large so that there are not enough samples to fill this space. In general, as the features increase, the amount of data needed to generalize accurately grows exponentially. As the number of features (called dimensionality) increases, the classifier's performance increases until the optimal number of features is reached; this is called the Hughes phenomenon. Overall, the ideal number of features depends on the classifier involved and the amount of training data available. Beyond this optimal number of features, an additional number of features may not have a positive impact on the performance of the model (even though the sparsity of the observations may make it easier for these to be classified). Overfitting will become an issue with more added features.

There are a few strategies to reduce dimensionality (and thus reducing the curse). One popular methodology involves the PCA that was described in detail under unsupervised learning as a method for reducing dimensionality. Other possible methodologies include locally linear embeddings and LDA. In addition, another strategy involved using domain expertise as a means for more optimal feature engineering. Lastly, there is the obvious strategy of increasing the data available but this last strategy is very limited.

Correlation versus Causation

A correlation measures the relationship between any two variables (see linear regression). A common misconception is that correlation implies that there is causation; events that are correlated, therefore, do not signify that one caused the other. The common dictum is "correlation does not imply causation." This understanding, however, may lead to another misconception: you cannot infer causality from data science.

Causation states that any change in the value of one variable will cause or lead to a change in the value of another variable (cause and effect: a known, observable chain of events). Reichenbach's common cause principle states that a correlation between events A and B indicates that (1) A causes B, or (2) B causes A, or (3) A and B have a common (Reichenbachian) cause that induced the correlation. Since causes always occur before their effects, it is therefore assumed that common causes always occur before the correlated events. Causal networks represent interdependencies of events.

Machine versus Deep Learning

There are significant differences between machine and DL (see Table 5.11). Whereas traditional ML flow has manual feature extraction or engineering followed by ML algorithm (with a relatively shallow structure) that leads to output, DL flow involves an ANN that can combine feature extraction with the classification as one step in its algorithm to achieve an end-to-end learning process (see Fig. 5.32). DL, therefore, requires less domain knowledge to solve the problem assigned, but DL is more difficult to comprehend as the algorithms are largely self-directed (so-called black box).

Traditional ML, compared to DL, is relatively easy to train and test but its performance is dependent upon its features and is limited with the increasing volume of data (see Fig. 5.33). These relatively shallow models are also relatively inefficient as they need a large number of computations and high maintenance in that they require much human work for labeling.

TABLE 5.11   Machine versus deep learning.

	Machine learning	Deep learning
Era	1980s	2000s
Examples	SVM, Random Forest	CNN, RNN, GANs
Data needed	+ +	+ + +
Accuracy	+ +	+ + +
Data preprocessing	Yes	No
Training time	+ +	+ + +
Plateau in performance	Yes	No
Hardware requirement	CPU	GPU
Human involvement	Feature extraction	None needed
Correlation	Linear	Nonlinear

CNN, Convolutional neural network; *CPU*, central processing unit; *GAN*, generative adversarial network; *GPU*, graphical processing unit; *RNN*, recurrent neural network; *SVM*, support vector machine.

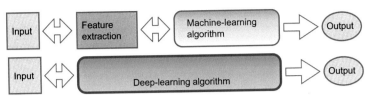

FIGURE 5.32   Machine versus deep learning. The workflow for both machine and deep learning are depicted. In machine learning, feature extraction is manually done followed by the use of the algorithm with a shallow structure. In deep learning, there is no separate feature extraction as it is inclusive of the deep learning end-to-end algorithm (which creates a "black box" perception).

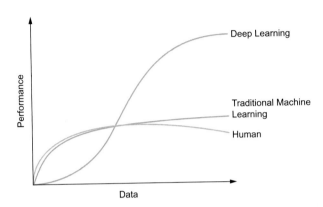

FIGURE 5.33   Performance. The data versus performance is plotted for various technologies, including the human brain. Increasing depth of the neural network will increase the level of performance but traditional machine learning will plateau earlier than even shallow neural networks. Deep neural networks will continue to increase in its performance with larger amounts of data. Humans start at a higher level of performance but will fatigue and perform less well with the increasing amount of data.

TABLE 5.12   Summary of types of learning and use in medicine and health care.

Type of learning	Use in medicine and health care	Advantages/disadvantages
Supervised learning		Advantages: Relatively easy to apply Disadvantages: Overly simplistic for some areas of biomedicine
- Classification	Medical images Phenotyping Cohort identification	
- Regression	Outcome prediction Survival prediction Risk prediction	
Unsupervised learning		Advantages: Relatively easy to apply Disadvantages: Difficult to measure performance
- Clustering	New patient and therapies Novel phenotype identification Biological hypothesis generation	

(Continued)

II. Data science and artificial intelligence in the current era

TABLE 5.12   (Continued)

Type of learning	Use in medicine and health care	Advantages/disadvantages
- Generalization	Data visualization Variable selection Data compression	
Reinforcement learning	Process optimization Decision sequence optimization	Advantages: Human-like learning Disadvantages: Needs high volume of data
Transfer learning	Models from nonpediatric data	Advantages: Faster than deep learning Disadvantages: Needs high volume of data
Deep learning	Image classification Text note classification Sequential prediction	Advantages: High level of performance Can model complex relationships Disadvantages: Needs high volume of data and long time to train (high maintenance) Prone to overfitting Not able to perform logical inferences Lack means to represent causal relationships

Modified from Bennett T.D., Callahan T.J., Feinstein J.A. et al. Data science for child health. J Pediatr 208 (2018), 12−22.

On the other hand, deep-learning performance can continue to incrementally improve with increasing data (or increasing capacity of the network). While DL can learn high-level features representation, it does require large amounts of data for training and can be expensive from a computation usage perspective. Note that with the increasing volume of data, human performance remains the same (or even deteriorates from fatigue).

In summary, the myriad of machine and deep-learning tools that are described previously have specific applications with advantages and disadvantages. These elements are organized in Table 5.12 (please see individual sections for more details):

References

[1] < tacumasolomon.com> Blog on Aug 12, 2017.
[2] Sun X, Liu L, Xu K, et al. Prediction of ISUP grading of clear cell renal cell carcinoma using support vector machine model based on CT images. Medicine 2019;98:1−6.
[3] Onisko A, Druzdzel MJ, Austin RM. Application of Bayesian network modeling to pathology informatics. Diagn Cytopathol 2019;47:41−7.
[4] Triguero I, Garcia-Gil D, Maillo J, et al. Transforming Big Data into smart data: an insight on the use of the *k*-nearest neighbor's algorithm to obtain quality Data. WIREs Data Mining Knowl Discov 2019;9: e1289.

[5] Sarkar RP, Maiti A. Investigation of dataset from diabetic retinopathy through discernibility-based *k*-NN algorithm. Contemporary advances in innovative and applicable information technology. New York: Springer; 2019. p. 93−100.

[6] Goodman KE, Lessler J, Harris AD, et al. A methodological comparison of risk scores vs decision trees for predicting drug-resistant infections: a case study using extended-spectrum beta-lactamase (ESBL) bacteremia. Infect Control Hosp Epidemiol 2019;40:400−7.

[7] Du S, Hu Y, Bai Y, et al. Emotional distress correlates among patients with chronic nonspecific low back pain: a hierarchical linear regression analysis. Pain Practice 2019;19 [ePub].

[8] Panesar SS, D'Souza RN, Yeh FC, et al. Machine learning vs logistic regression methods for 2-year mortality prognostication in a small, heterogenous glioma database. World Neurosurg 2019;2:100012.

[9] de la Fuente-Tomas L, Arranz B, Safont G et al. Classification of patients with bipolar disorder using *k*-means clustering. PLoS One 2019;14(1):e0210314.

[10] Parmar HS, Nutter B, Long R et al. Automated signal drift and global fluctuation removal from 4D fMRI data based on principal component analysis as a major preprocessing step for fMRI data analysis. In: *Proc SPIE 10953, Medical imaging 2019: Biomedical applications in molecular, structural, and functional imaging*, 2019, 109531E.

[11] dos Santos HDP, Ulbrich AH, Woloszyn V, et al. DDC-outlier: preventing medication errors using unsupervised learning. IEEE J Biomed Health Inf 2019;23(2):874−81.

[12] Yang Y, Nan F, Yang P, et al. GAN-based semi-supervised learning approach for clinical decision support in health IoT platform. IEEE Access 2019;7:8048−57.

[13] Gorczyca MT, Toscano NC, Cheng JD. The trauma severity model: an ensemble machine learning approach to risk prediction. Comput Biol Med 2019;108:9−19.

[14] Silver D, Huang A, Maddison CJ, et al. Mastering the game of go with deep neural networks and tree search. Nature 2016;529:484−9.

[15] Mnih V, Kavukcuoglu K, Silver D, et al. Human-level control through deep reinforcement learning. Nature 2015;518:529−33.

[16] Jonsson A. Deep reinforcement learning in medicine. Kidney Dis 2019;5:18−22.

[17] Lopez-Martinez D, Eschenfeldt P, Ostvar S et al. Deep reinforcement learning for optimal critical care pain management with morphine using dueling double deep Q networks. *Con Proc IEEE Eng Med Biol Soc* 2019, 3960−3963. *arXiv*:1904.11115.

[18] Yun J, Park JE, Lee H, et al. Radiomic features and multilayer perceptron network classifier for a more robust MRI classification strategy for distinguishing glioblastoma from primary central nervous system lymphoma. Sci Rep 2019;9:5746.

[19] Krizhevsky A, Sututskever I, Hinton GE. ImageNet classification with deep convolutional neural networks, Vol. 1. La Jolla, CA: Neural Information Processing Systems Foundation Inc; 2012. p. 4.

[20] LeCun Y, Bengio Y, Hinton G. Deep learning. Nature 2015;521:436−44.

[21] Porter J, editor. Deep learning: fundamentals, methods, and applications. New York: Nova Science Publishers; 2016.

[22] Arel I, Rose DC, Kanowski TP. Deep machine learning—a new Frontier in artificial intelligence research. IEEE Comput Intell Mag 2010;5 1556-603X (13−18).

[23] Esteva A, Robicquet A, Ramsundar B, et al. A guide to deep learning in health care. Nat Med 2009;25:24−9.

[24] Peng J, Guan J, Shang X. Predicting Parkinson's disease genes based on node2vec and autoencoder. Front Genet 2019;10:226.

[25] Goodfellow IJ, Pouget-Abadie J, Mirza M et al. Generative adversarial networks. *arXiv*:1406.2661.

[26] Radford A, Metz L, Chintala S. Unsupervised representation learning with deep convolutional generative adversarial networks. *arXiv*.1511.06434.

[27] Guan S, Loew M. Using generative adversarial networks and transfer learning for breast cancer detection by convolutional neural networks. In: Proceedings medical imaging 2019: imaging informatics for health care, research, and applications; 2019, p. 109541C.

[28] Cox DD, Dean T. Neural networks and neuroscience-inspired computer vision. Curr Biol 2014;24:R921−9.

[29] Yamashita R, Nishio M, Do RKG, et al. Convolutional neural networks: an overview and application in radiology. Insights Imaging 2018;9:611−29.

[30] Hannun AY, Rajpurkar P, Haghpanahi M, et al. Cardiologist-level arrhythmia detection and classification in ambulatory electrocardiograms using a deep neural network. Nature Medicine 2019;25:65−9.

[31] Bao W, Yue J, Rao Y. A deep learning framework for financial time series using stacked encoders and long-short term memory. PLoS One 2017. <https://doi.org/10.1371/journal.pone.0180944>.

[32] Yao X, Cheng Q, Zhang GQ. A novel independent RNN approach to classification of seizures against non-seizures. *arXiv*:1903.09326v1.

[33] Rajkomar A, Oren E, Chen K, et al. Scalable and accurate deep learning for electronic health records. NPJ Digital Med 2018;1 Article number: 18.

[34] Turan M, Almalioglu Y, Araujo H, et al. Deep EndoVO: a recurrent convolutional neural network (RCNN) based visual odometry approach for endoscopic capsule robots. Neurocomputing 2018;275:1861–70.

[35] Saito T, Rehmsmeier M. The precision-recall plot is more informative than the ROC plot when evaluating binary classifiers on imbalanced datasets. PLoS One 2015;10(3):e0118432.

[36] Ekelund S. Precision-recall curves: what are they and how are they used?; 2017. <www.acutecaretesting.org>.

[37] Gunning D. Talk at DARPA; 2016.

Other Key Concepts in Artificial Intelligence

Cognitive Computing

Cognitive computing is defined loosely as a science that teaches computers to think like a human mind and is, therefore, an attempt to reengineer the human brain. The cognitive computing framework leverages a portfolio of methodologies such as machine learning, pattern recognition, and natural language processing (NLP) as well as other artificial intelligence (AI) tools to mimic the human brain and its self-learning capability. Cognitive computing is a symbiotic convergence of humans (user and expert) and smart technology. The degrees of cognitive computing include first degree that entails understanding natural language and human interactions; second degree that involves the generation and evaluation of evidence-based hypotheses; and third degree that possesses adaptations and learning from interactions with users.

In 2011 the iconic supercomputer Watson was able to defeat the human champions on the game show *Jeopardy!*. The IBM supercomputer Watson was the first open cognitive platform and heralded the era of cognitive computing with its robust NLP as well as knowledge representation, information retrieval, automated reasoning capabilities along with machine learning; this supercomputer can also scan an astounding 200 million pages in 3 seconds (see Fig. 6.1) [1].

The essential part of Watson was the DeepQA technology, a massively parallel probabilistic evidence-based architecture with more than 100 different techniques for analyzing natural language, identifying sources, finding and generating hypotheses, finding and scoring evidence, and merging and ranking hypotheses [2]. The knowledge for Watson is self-contained as there is no access to the Internet. There are four essential steps in answering the questions posed on *Jeopardy!*:

1. question analysis: parsing the question into words and analyzing these parts of the question;
2. hypothesis generation: searching large volume of possible answers and narrowing these down to the more likely possibilities;

Intelligence-Based Medicine
DOI: https://doi.org/10.1016/B978-0-12-816462-4.00006-6

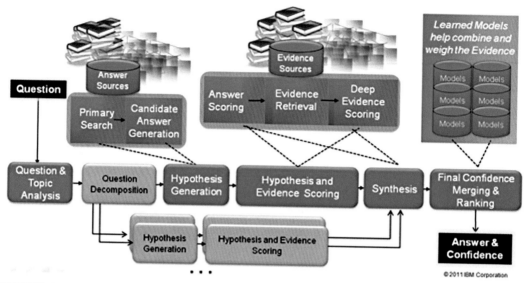

FIGURE 6.1   Cognitive computing (Watson). The DeepQA architecture of Watson involved deconstructing the question and topic analysis, followed by integrating answer sources with evidence sources to lead to a hypothesis and evidence scoring to result in the final answer with confidence scoring.

3. hypothesis and evidence scoring: collecting positive and negative evidence for passages of word associations and using algorithms to score these possibilities; and
4. final merging and ranking: weighing the evidence and deciding the final ranking of answers.

The DeepQA principles involve massive parallelism to consider multiple interpretations and hypotheses; many experts to facilitate a wide range of loosely coupled probabilistic question and content analytics, pervasive confidence estimation to find a final score and answer, and integration of shallow and deep knowledge. The massively parallel architecture coupled with a dedicated high-performance computing infrastructure of the DeepQA architecture and methodology demonstrated the systems-level approach to research in not only cognitive computing but also AI.

Just as big data has promulgated machine and deep learning into higher dimensions of performance and scope, big data has also increased the expectations of cognitive computing.

The cognitive era (with the two previous computing eras being tabulating and programmable systems) with its cognitive systems has the capabilities of learning, reasoning, perception, and language processing. The characteristics of cognitive computing include information gathering by integrating data from disparate sources, dynamic training and adaptive by learning and changing as new information is gathered, probabilistic by discovering patterns based on context, meaning-based on performing language processing, and highly interactive by providing advanced communications [3].

TABLE 6.1 Cognitive computing and artificial intelligence.

	Cognitive computing	**Artificial intelligence**
Definition	Systems that are designed to solve problems by simulating human cognitive abilities	Science of making computers do things that require intelligence by humans
Methodologies	Machine learning Deep learning NLP Rules-based systems Speech recognition Robotics Sentiment analysis	Machine learning Deep learning NLP
Purpose	Augments human intelligence	Augments human intelligence
Capabilities	Simulating human cognition and advises	Finding patterns in data and performs prediction
Applications	IBM Watson	Google DeepMind

NLP, Natural language processing.

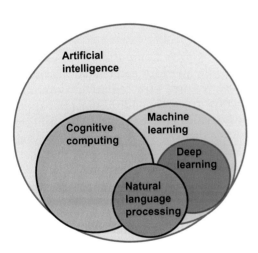

FIGURE 6.2 AI and cognitive computing. The relationship between cognitive computing and AI is delineated in this schematic diagram. Cognitive computing is usually considered within the AI realm and encompasses machine and deep learning as well as natural language processing. *AI*, Artificial intelligence.

There is sometimes understandable confusion between AI and cognitive computing, and in some ways, the latter is even more difficult to define (see Table 6.1) (see Fig. 6.2). While AI does not intentionally mimic human thought processes, cognitive computing with its origin in cognitive science does attempt to simulate the human problem-solving process in a computerized model (via AI tools such as machine learning, neural networks, and NLP, as well as sentiment analysis and contextual awareness). In the near future, cognitive computing will sense via the networks of smart devices, learn from historical data, infer by generating evidence-based hypotheses, and interact with systems and people with its natural language capabilities. There is great potential for a cognitive computing, along with machine and deep learning, to be essential parts of AI in the near future. Cognitive computing is considered under the realm of AI, and conversely, the future direction of AI is in cognitive architecture.

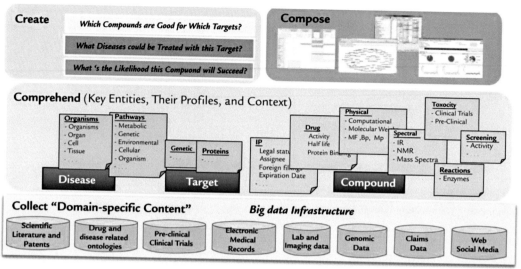

FIGURE 6.3 Watson Discovery Advisor for Life Sciences. This illustrates shows the Watson architecture for life sciences to include: data aggregation, concept recognition, and domain specific content collection.

An example in the recent biomedical literature of cognitive computing is the review article on IBM Watson with a five-part discussion of (1) the need for accelerated discovery, (2) the data hurdles that impede discovery, (3) 4-core features of a cognitive computing system, (4) pilot projects applying Watson to life sciences research, and (5) potential applications of cognitive technologies to other life science activities [4] (see Fig. 6.3). In addition, there is a Chinese experience with Watson for Oncology that showed a relatively low concordance between Watson for Oncology and oncologists in clinical practice and that there is a need to enhance Watson as a cognitive decision support tool by incorporating regional guidelines [5].

IBM Watson and the new era of health with artificial intelligence and cognitive computing

Kyu Rhee

Kyu Rhee, a medical executive with IBM Watson Health, authored this commentary to provide an overview as well as the values and principles of IBM Watson: purpose, transparency, and skills for its future journey in health care.

> The best way to predict the future is to invent it. *Alan Kay, Computer Scientist*

The history of Watson in health

IBM has played a critical role in healthcare for many years prior to making health its "moonshot." In 1962 IBM developed and implemented an early electronic medical record (EMR) with Akron Children's Hospital in Ohio. IBM has played a pioneering role in AI. At the center of

IBM's contributions to AI is Watson, a system that responds to open-ended questions posed in natural language and uses advanced NLP, machine learning, information retrieval, knowledge representation, and automated reasoning techniques to generate answers [1].

In 2008 IBM completed its first successful test case using Watson and in 2011, the company made history when Watson beat *Jeopardy!* champion Ken Jennings. Leveraging AI in health was a natural next step as IBM was hearing from doctors interested in pursuing use cases for AI in healthcare. In 2012 IBM began working with the Cleveland Clinic and Memorial Sloan Kettering Cancer Center (MSKCC) to use Watson in real-world hospital settings. The partnership with MSKCC led to the building of Watson for Oncology.

In 2015 IBM officially launched Watson Health as a stand-alone business by acquiring a group of companies with different expertise, ranging from imaging to data analytics to value-based care. This allowed IBM to access valuable health data to help further train the system and would allow for AI to be used in other important areas of health. Recognizing that 70% of our health determinants go beyond clinical, for example, where we live, how we eat, and levels of stress are huge factors in our overall health, which is why the business unit focuses on health versus just healthcare.

Scientific evidence and concordance studies

There is a growing body of evidence to support Watson technologies in health, particularly in Oncology, Genomics, and Clinical Trial Matching. (SEE GRAPHIC)

Watson for Genomics uses AI to identify and support the interpretation of genomic data and provides insights on actionable mutations for the treatment of cancer patients and links it to drug options for cancer care. In a study published in November 2017 in The Oncologist [2], UNC Lineberger researchers used IBM Watson for Genomics to assess whether cognitive computing was as effective as a panel of cancer experts in identifying therapeutic options for tumors with specific genetic abnormalities. In a retrospective analysis of 1018 cancer cases, the molecular tumor board identified actionable genetic alterations in 99% of 703 cases, which Watson also confirmed. Using the curated Watson for Genomics gene list, researchers identified additional potentially actionable genomic information in 324 patients or 32% of patients, 96 of whom were not previously identified as having an actionable mutation. The Watson for Genomics analysis took less than 3 minutes per case.

Watson for Oncology ranks the treatment options, linking to peer-reviewed studies that have been curated by MSKCC. The *Annals of Oncology* published a full study led by oncology leaders at Manipal Hospitals in India [3]. Their tumor board found Watson for Oncology was concordant with their own tumor board's treatment decisions in 93% of breast cancer cases. When the blinded study initially compared Watson's recommendations against retrospective treatment decisions from 2 to 3 years prior, the concordance was only 73%. Manipal's tumor board rereviewed the same patient cases manually, and the concordance rose to 93%. This is an indication that the corpus of data on which Watson for Oncology relies is staying up-to-date with the latest science.

Physicians from the Affiliated Hospital of Academy of Military Medical Sciences in China reported a study ($n = 1997$), which found that disclosure of Watson for Oncology recommendations resulted in treatment decision changes in 5% of cases. Attending physicians were less likely to alter their decisions (3%) than chief physicians (6%; $P < .001$) and fellows (7%; $P < .001$). Importantly, in those 106 cases where clinicians changed their treatment decision based on Watson for Oncology, adherence to professional treatment guidelines improved from 89% to 97% ($P < .01$) [4].

Watson for Clinical Trial Matching eliminates the need to manually compare enrollment criteria with patient medical data, making it possible to efficiently identify an individual's potential trial options in a list ranked by relevance and eligibility.

Data published with Highlands Oncology Group and Novartis showed that Watson for Clinical Trial Matching successfully demonstrated the ability to expedite patient screening for clinical trial eligibility, reducing processing time from 1 hour and 50 minutes to 24 minutes. It also omitted 94% of nonmatching patients automatically—reducing screening workload dramatically [5].

The Mayo Clinic reported that Watson for Clinical Trial Matching boosted enrollment in breast cancer trials by 80% following implementation (to 6.3 patients/month, up from 3.5 patients/month in the immediate 18 months prior). Enrollment was increased to 8.1 patients/month, or 140%, when including accruals to phase I trials with breast cancer cohorts in the experimental therapeutics program [6].

Values and principles

AI in medicine inherently involves the processing of sensitive health data and rendering of decisions that might affect human lives. IBM's approach to the development and deployment of AI in health reflects these values:

- *Purpose*: AI is intended to enhance and empower health professionals rather than replace them. It is the effective combination of the professional, augmented by AI that can deliver better care and transform health.
- *Transparency*: For AI to fulfill its potential trust is essential. It is vital that health professionals have confidence in their recommendations, judgments, and uses and understand how the system was trained.
- *Skills*: Health professionals must learn new skills and will require new expertise, for example, data scientists will become key members of the clinical team.

While today IBM Watson technologies are being applied to a wide variety of challenging tasks in the health domain, IBM and the broader AI industry are only at the beginning of this journey and realizing the potential in healthcare for this technology in healthcare will evolve over time.

References

[1] Kelly JE III. Computing, cognition and the future of knowing: How humans and machines are forging a new age of understanding. IBM white paper, 2015 (Computing, cognition and the future of knowing: How humans and machines are forging a new age of understanding).
[2] <http://theoncologist.alphamedpress.org/content/early/2017/11/20/theoncologist.2017-0170.abstract>.
[3] <https://www.ncbi.nlm.nih.gov/pubmed/29324970>.
[4] Jiang Z, et al. ASCO Annual Meeting, 2018.
[5] <https://ascopubs.org/doi/abs/10.1200/JCO.2017.35.15_suppl.6501>.
[6] <https://meetinglibrary.asco.org/record/158490/abstract>.

Expansive artificial intelligence (AI): convergence of deep learning and cognitive knowledge computing

Spyro Mousses

Spyro Mousses, a genomics expert with an AI background, authored this commentary to elucidate the concept of cognitive knowledge representation coupled with deep learning.

Systems Oncology, Scottsdale, AZ, United States

Genomics and big data have the promise to make a major impact against human disease, but the reality is that the data we now have access too is too big and too complex for conventional computational approaches. Over the last decade we have worked to build the next generation of intelligent machines that can augment human intelligence and greatly accelerate our ability to generate new scientific knowledge. The dream is that this new paradigm of "man + machine" will lead us to rapid progress against the diseases that plague our society. The good news is we are closer than ever to realizing that dream.

The field of machine learning has been making slow and steady progress for decades. It was not until 2012 that the field reached a major turning point when Geoff Hinton PhD, a computer scientist at the University of Toronto, published some extraordinary results [1]. Briefly, Dr. Hinton was able to show that the use of deep convolutional neural networks can leverage massive datasets of annotated images to train models which were surprisingly effective at learning how to classify objects in images. Since this seminal discovery, this new field of *deep learning* has achieved rapid evolution, with diverse applications ranging from the game "Go" to diagnosing diabetic retinopathy, demonstrating unprecedented accuracy that approximates and often surpasses the abilities of human experts. This activity has led to a true renaissance in AI, and many are now wondering where the field is going next.

In parallel, our team has been making progress in advancing the field of cognitive computing and knowledge representation. Specifically, our goal was to develop a computational framework capable of representing the complex biomedical models [2]. By leveraging the flexibility and scalability of nested hypergraphs, we were able to represent complex multidimensional relationships across multiple scales of biomedical knowledge. This new approach has completely transformed our ability to not only mine large complex datasets to recover relevant knowledge, but more sophisticated representation strategies also enabled us to explore new approaches to machine learning. This is leading us to the creation of machines that are capable of autonomously generating new knowledge. This idea, that computer systems can "imagine" novel beliefs and then test and justify those computer-generated beliefs through the use of machine learning and big-data, represents a new paradigm for how scientists can use machines to massively amplify their ability to generate new knowledge. This new ability to imagine truly novel beliefs is leading to systems that generate new knowledge, an ability has been the exclusive domain of the human mind. The difference between how humans and computer-generated knowledge is of course scale. *Scientific knowledge computing* can now generate new scientific knowledge many orders of magnitude faster than what is possible by humans.

Recently, the parallel paths of deep learning and scientific knowledge computing converged. This convergence started when we began to apply the rapidly emerging deep-learning tools to mine lined data stored in a hypergraph knowledge base. In one example, we used the features from the hypergraph knowledge base as multidimensional inputs to an artificial neural network model. Specifically, we were able to use known synthetic lethal pairs of genes (representing potential cancer drug targets) to train the model. We also fused in multiple other functional classifiers and dimensions of relationships across the entire cancer genome, which are naturally captured in the structures of the hypergraph knowledge network. The results of the experiment were impressive both from the output and in terms of the computational efficiency. The model was not only able to classify known synthetic lethal pairs but also identified completely novel gene pairs that we are now investigating as new drug targets. Unexpectedly, we also learned that combining these two converging technologies produced unexpected computational synergy. Namely, the hypergraph framework allowed us to achieve unexpected computational efficiency and flexibility in extracting high dimensional features on an unprecedented scale, so models can be trained and optimized much more efficiently.

From an epistemological perspective, this convergence of technologies is enabling a new type of consilience and fundamentally changing the science of knowledge itself. In other words, as we begin to generate knowledge on scale, we began to realize two things. The first is that imagination is at the heart of new knowledge generation. So, the rate limiting step to knowledge generation is not just data, but the number of beliefs or hypotheses that a system can generate and test. This is why we need computer imagination—to explore a massively more expansive universe of hypotheses. The second is that knowledge is not a single static entity to be discovered and archived. Instead all scientific knowledge evolves rapidly. Think of evolving knowledge like a dynamic network of beliefs that are constantly converging with evidence from multiple sources. This type of continuous consilience will disrupt our conventional notions of science and evidence-based medicine, but if done right, it will also transform our ability to make an impact.

Taken together, we now see the combination of deep learning and cognitive knowledge computing as a very powerful new platform capable of revolutionizing science; one that will enable us to explore a massively more expansive universe of knowledge—hence the name, *ExpansiveAI*. Systems Imagination will continue to advance the ExpansiveAI platform technology, and Systems Oncology has begun to apply this approach in the field of cancer therapeutics to achieve scientific breakthroughs beyond human imagination.

References

[1] Krizhevsky A, Sutskever I, Hinton GE. ImageNet classification with deep convolutional neural networks. Curran Associates Inc.; 2012. p. 1097—105.
[2] Farley T, Kiefer J, Lee P, Von Hoff D, Trent JM, Colbourn C, et al. The BioIntelligence Framework: a new computational platform for biomedical knowledge computing. J Am Med Inform Assoc 2013;20(1):128—33 Epub 2012 Aug 2..

Natural Language Processing

NLP is defined as the computer to understand spoken as well as written human language through specific set of techniques. In short, NLP is the intersection of AI, computer science, and linguistics, and is a good example of human–computer interaction. The applications of NLP include machine translation, information retrieval, document indexing, sentiment analysis, information extraction, chatbots with natural speech, virtual assistants, blocking spam, and question answering. Of note, NLP is an essential part of the supercomputer Watson.

The two components of NLP are natural language understanding (NLU) and natural language generation (NLG); NLU is usually considered the more difficult component (see Fig. 6.4). NLU involves mapping of input into a useful representation while NLG involves text planning, sentence planning, and text realization.

The connection of words or phrases to concepts can also be described in terms of tokenization, lemmatization, and mapping. Tokenization is a deconstruction of a sentence into words and phrases (or tokens) based mainly but not solely on spaces. For example, *myocardial* and *myocardial infarctions* are both tokens. Lemmatization is a standardization that maps a token onto a lemma, which is the base form of a word that is found in a dictionary. For example, *MI* can be mapped onto *myocardial infarction*. Finally, the mapping of a lemma to a concept is challenging as any word can have different meanings (just as many words can have the same meaning).

The NLP process is as follows: speech is deconstructed with phonetic analysis that consists of breaking down the speech into phonemes, while text is broken down with

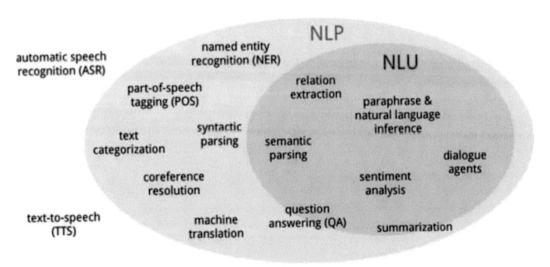

FIGURE 6.4 NLP and NLU. NLU with its parts is a subset of NLP, which has a more complete portfolio. NLU has the difficult task to interpret unstructured inputs and convert these words into structured ones for the computer to understand. *NLP*, Natural language processing; *NLU*, natural language understanding.

combined optical character recognition and tokenization. The lexical analysis is the analysis of the structure of words. Then, syntactic analysis, or dependency parsing, is the analysis of words in the sentence for grammar followed by a process of arranging the words that shows the proper relationship between these words. While syntax is the structural role of words, semantics is the meaning of words and semantic interpretation describes a meaning for these words. Next, there is discourse integration, the meaning of sentences in sequence is studied. Lastly, pragmatic analysis occurs when data are interpreted on what it actually means. An even more detailed step-by-step description will be listed (but not delineated in full detail) from the input document to the output: sentence segmentation, tokenization, parts-of-speech tagging, lemmatization, stop words, dependency parsing, noun phrases, named entity recognition, and finally coreference resolution.

Three strategies can be used in the application of NLP to biomedical data [6]:

First, a pattern-matching strategy is perhaps the simplest approach in which a sequence of characters is used for matching. Tokenization and regular expression pattern matching are part of this pattern-matching methodology.

Second, a linguistic approach can be used for more complex sentences as it treats words as symbols that are constructed together with grammatical rules. Both syntactic (rules for word arrangement in a sentence) and semantic (meaning of word in the context of the sentence) knowledge are used in this strategy to delineate the concepts.

Third, a more recent machine learning approach infers rules and patterns directly from the data by the use of elements such as features, training data, and models (discussed previously).

A chatbot or automated intelligent agent is the intelligent digital assistant that embodies NLP and used for a myriad of purposes including customer communication and information gathering. The parts of a chatbot include the following: knowledge base contains the information needed to answer the queries; data store is where the interaction history of chatbots with users is stored; an NLP layer translates user query into a usable communication; and finally an application layer where the application interface is located (see Fig. 6.5). These components is reminiscent of the early expert systems during the early AI era.

An example in the recent biomedical literature of using NLP is the recent report of NLP in extracting clinically useful information from Chinese EMRs with the development of rule-based and hybrid methods (with the former showing better results) [7].

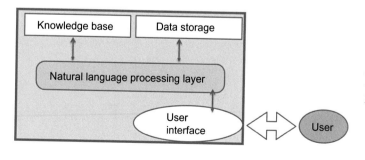

FIGURE 6.5 Chatbot architecture. The chatbot architecture is shown in this schematic diagram. The main components are knowledge base, data storage, and the NLP layer that then connects with the user interface. *NLP*, Natural language processing.

Potential of conversational artificial intelligence (AI) in medicine

Jai Nahar

Jai Nahar, a pediatric cardiologist with a keen interest in AI tools, authored this commentary on the possible utilization of conversational AI in venues such as outpatient clinic, home, and hospital for patients and families.

Introduction

Fourth Industrial Revolution has introduced many important technologies that are becoming integral part of our lives. NLP and Conversational AI are two such technologies which are increasingly being used.

NLP: This AI technology allows the computer to understand spoken as well as written human language; the two components of NLP are NLU and NLG [1].

Conversational AI: This AI technology involves human—computer interaction through the use of conversation, utilizing voice user interface (UI) and machine intelligence. This is made possible by synergistic convergence of voice technology, and AI technology (NLP, machine, and deep learning).

Natural language processing and conversational artificial intelligence: technology integration

NLP has two important applications in the field of conversational AI. These are speech recognition and conversational agents, which use voice UI.

Speech recognition systems transcribe spoken language into text. An example is speech-to-text systems that can serve as digital scribes for use by physicians to enable data entry into the electronic health record (EHR).

Two common types of Conversational Agents are virtual assistants and chatbots—these are also known as virtual agents. A virtual assistant (e.g., Apple's Siri, Google Assistant, and Amazon Alexa) is an AI-inspired software agent that is capable of performing certain tasks or services via text or voice. A chat bot is a service that is capable of conducting a conversation with a human as result of using rules governed by AI [1].

Conversational AI architecture includes speech recognition software that converts speech-to-text, and a NLP system which helps in understanding the natural language, intent of the user, and integrates this with the context, dialogue management, and knowledge base, generating response in the text form. This text response is converted to a speech response that is then delivered to the user to answer his/her input query/command.

Conversational artificial intelligence touchpoints in health-care delivery

Voice technology powered conversational AI applications are gaining increasing use in health-care delivery, in the form of provider and patient facing solutions. They help in execution of human commands through voice-enabled devices and also offer valuable insights derived from advanced analytics.

As illustrated in Fig. 1, some important use cases for conversational AI are as mentioned next.

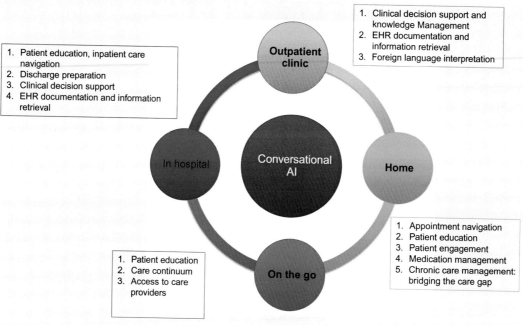

FIGURE 1   Conversational AI touchpoints in healthcare delivery

Conversational artificial intelligence in outpatient/ambulatory clinics

EHR documentation and retrieval: Conversational AI-powered voice assistants, which integrate with EHR, utilize NLP and can be used at point of care by physicians to promptly document information regarding patient encounter into EHR. This enables frictionless, prompt documentation, and compliance with coding, billing and regulatory requirements. The other important function of these assistants is to retrieve information pertaining to patient's prior records, hospitalizations, ER visits, labs, etc., from the EHR. By decreasing the nonproductive time that physicians are forced to devote for the current EHR systems, voice-assistant integrated smart EHR systems will help to decrease the incidence of physician burnout.

Clinical decision support: Integration of conversational interface in clinical decision support system can be used at point of care by the physicians for diagnosis and treatment planning.

Foreign language interpretation: Access to foreign language interpretation service can be constrained by cost and geographical location. Smart voice assistants utilizing NLP (machine translation) and machine learning can facilitate foreign language interpretation in these situations.

Conversational artificial intelligence at home

Voice user interface utilized by conversational AI has an important advantage. It can bypass the traditional mode of human interaction with computers and digital devices, which is through typing and touch screens. People (specially the elderly or visually impaired) who have difficulty interacting with computers through these traditional modes can easily do so now using the currently available intelligent voice technology.

Conversational AI-enabled virtual assistants and chatbots can help the patients in promoting wellness and disease management at home. These virtual assistants can help patients in scheduling and preparing for the upcoming appointments, serve as personal health coach facilitating disease education, chronic care management, treatment adherence, and active engagement in their health-care journey. They can offer ongoing personalized care with touch of empathy, anytime, as needed within the comfort of patient's own home. By serving as an intelligent digital interface between patients and providers they can promote two way conversation and health information exchange between them as needed.

Conversational artificial intelligence in the hospital

During inpatient hospitalization, conversational agents can be used by patients to access information regarding their care team, seek answers to simple questions, search educational content relevant to their hospitalization, and prepare for post discharge care at home.

During surgery/interventional procedure the surgeon/operator in the operating room or interventional procedure suite needs to remain in the sterile field. Smart virtual assistants powered by voice technology can be used by them in hands free mode to retrieve medical information of the patient from the EHR and timely document new information in patient's EHR [2].

Conversational artificial intelligence on the go

For patients who are traveling/or are away from home, mobile-integrated smart voice assistants and chatbots powered by machine learning can be used to access educational health content, their personal care management plan, and reach out to their health-care provider as needed.

Challenges in adoption of conversational artificial intelligence

As we look forward to widespread adoption of conversational AI in healthcare, following challenges need to be considered:

- Privacy and security: Maintenance of Privacy and Security of Health Information is integral to ensure against unauthorized access and misuse of personal health information.
- Accuracy: To obtain proper results from the use of conversational AI system, content and context of Human conversation has to be accurately comprehended by the voice-enabled devices. Failure to do so will result in inefficient use, end-user frustration, errors, and medicolegal implications.
- Optimal design/user experience: The design of these conversational agents should facilitate their smooth and painless integration in the workflow of end user.
- Reliability and trust: The content that is generated by the conversational AI system and delivered to the end user should be from a reliable source, well curated, and trustworthy.
- Ethics: Data from conversational AI system should not be unethically used to the disadvantage of end user.

Future directions

As we evolve with increase in functionality of conversational AI applications, consideration should be given to humanizing them with context specific empathy for the end user, and cognitive capabilities so they can respond proactively in an autonomous manner, appropriate to the circumstances and communicate on a human level with their end user.

Conclusion

In spite of multiple challenges, voice technology—enabled conversational AI systems have great promise in health-care sector. Optimal implementation requires identification of proper use cases where this technology can be appropriately leveraged, design of good user experience around use of voice technology, maintenance of privacy, security of the data and promotion of ethical use for the benefit of end user. With ongoing advancements in NLP and Machine Learning, Conversational AI is an emerging frontier in healthcare with great potential to transform the World of Medicine.

References

[1] Chang A. Analytics and algorithms, big data, cognitive computing, and deep learning in medicine and health care. AIMed 2017;.
[2] Small CE, Nigrin D, Churchwell K, Brownstein J. What will health care look like once smart speakers are everywhere? Harv Bus *Publishing Education* H0472H-PDF-ENG (digital article), March 6, 2018.

Robotics

The word robot (Czech for worker or servant) initially came from the Czech novelist Karel Capek as a name for a play. A robot is defined (by Robot Institute of America in 1979) as a reprogrammable and multifunctional manipulator designed to move material, parts, or specialized devices through variable programmed motions for the performance of a variety of tasks. Of course the more recent conceptualization of a robot is much more inclusive, and there has been a surge of interest and research concomitant with the more recent era of AI. This discipline involves the utilization of AI and engineering to involve the conceptualization and design as well as operation of robots; the interdisciplinary science of robotics includes electrical and mechanical engineering with computer science as well as mathematics, physics, biology, and of course AI. The recent trend in robotics is to yield robots that are more humanlike and less mechanical with living materials. As the field of robotics is vast and its topics heterogenous, a thorough discussion of robotics will not be a focus of discussion in this book, but robotics particularly as it pertains to medicine and healthcare will be covered later.

A robot share several basic elements: a mechanical frame for a task; a power source and actuation to mobilize; sensing mechanism(s) for both human and robot senses (the latter including force, tilt, proximity sensors), and computer-driven controller with a UI, and computational engine. The types of robots include manipulator (usually fixed), legged and wheeled robots, and autonomous underwater vehicle and unmanned aerial vehicle. There are many applications for robots, but the two main areas are industrial and service with the latter showing more autonomy: industrial, military, aerospace, agricultural, education, and medical along with now even nanorobot, swarm robots, and drones. The field of robotics has converged with others to even include avatars and virtual assistants. There are many taxonomies for robots based on control (preprogrammed, remote controlled, supervised autonomous, and autonomous), operational medium (location), or function (military, industrial, etc.).

Asimov's three laws of robotics have fascinating implications for AI in general and are as follows: (1) a robot may not injure a human being, or, through inaction, allow a human being to come to harm; (2) a robot must obey the orders given to it by human beings, except where such orders would conflict with the First Law; and (3) a robot must protect its own existence, as long as such protection does not conflict with the First or Second Law. One can extend these philosophical premises to AI and its implications for humans.

An example in the recent biomedical literature of robotics use is the review by Nwosu et al. for the use of robotics in palliative and supportive care [8]. These uses include supporting surgical procedures as well as assistive uses in dementia and elderly care.

Autonomous Systems

The convergence of robotics, mechatronics, and AI has led to the advent of autonomous systems. Current examples of these systems include autonomous driving cars, drones, robots in various scenarios, weapons systems, software agents, and even medical diagnostic tools. These advances are engendering many discussions around ethical and legal issues. In addition, future advances in this domain will include digital twins, computer-brain interfaces, cyborgs, and more; most if not all of these advances will have medical or health-care applications. Finally, a recent statement from the EU has set up ethical principles and democratic prerequisites for autonomous systems to achieve moral ideals and socioeconomic goals as well as legal governance and regulatory framework [9].

An example in the recent biomedical literature of the use of drones includes a comprehensive review with uses including delivery of health aid materials such as vaccines and medicines as well as test kits for diagnostic testing, and even defibrillators for cardiac arrest [10]. In addition, drones can be part of a global telemedicine network for delivery of basic healthcare.

Robotic Process Automation

This is a computer-coded program that can perform repetitive rule-based tasks (such as filling in forms, reading and writing to databases, and making calculations). Its benefits include increased throughput, reduced costs, reduced workload, and lessened errors. Although robotic process automation (RPA) is more proves-driven, machine learning and AI are more data-driven and cognitive. In a way, RPA can function as a gateway to intelligent automation, which is part of any industry's AI-enabled digital transformation with advanced robotics and IoT/IoE. RPA is often coupled with AI as it automates the prework for AI.

An emerging digital workforce for health care: powered by robotic process automation (RPA)

Sean Lane
Sean Lane, an inventor and entrepreneur with a strong background in technology and security, authored this commentary to introduce the RPA process as a digital work force of the future to mitigate the burden of administrative work in health care.

Enterprises across the globe have been digitizing their data and their processes furiously over the past couple decades. The posterity of digital storage has become commonplace and companies have started to leverage their data as assets. The ability to search and analyze data has created more intelligence about long standing industries than has ever existed in the past. The digitization efforts have also unlocked a pantheon of capabilities to offer customers and partners new products and better experiences. Technology, through enterprise software, is changing virtually every business around us.

Healthcare is no exception. Partially driven by government mandates and subsidies, healthcare systematically brought large electronic medical record (EMR) systems and other enterprise level systems of record to bring them into the digital era. This tidal wave of adoption of EMRs and other digitizing technology while on the net extraordinarily valuable had some negative side effects. These side effects created silos. Database fortresses were built at every organization. They were not built to share, nor were they built to interoperate. Not between software systems and most certainly not between organizations. No connection to insurers. No connection other providers and until recently, almost no connection to patients. The symptoms that prove the side effects are filling out the "clipboard" every time you go to get healthcare, mystical billing processes, shrouded prices for health-care services, undetectable fraud, razor thin margins at providers because of overweight administrative processes, and maybe most importantly—stunted research across populations to find cures to diseases and intelligence for drug development.

Healthcare has not ignored this problem. Thought leaders and domain experts have lamented over the lack of interoperability and there are many efforts in flight to solve interoperability. However, as we wait for a nearly magical solution to this growing problem, healthcare created a stop-gap solution. They built a router. A very sophisticated router. That router sits in a swivel chair and logs into many systems to extract and exchange data to accomplish myriad administrative tasks. The router communicates through vintage methods such as phones and fax machines to coordinate and create an analog connection across the software fortresses that power the health-care enterprise. Those routers are humans. Healthcare has literally hired millions of humans to sit between systems and organizations that do not interoperate. The phenomenon of the human router has skyrocketed the cost of doing business in healthcare. Roughly 33 cents of every dollar in the US healthcare is administrative cost. The human stopgap created and widely adopted by healthcare has bloated costs and levied a near crippling blow the entire industry. Healthcare cannot continue to operate like this. There must be a solution to rescue nearly a trillion dollars of cost and reallocate these precious resources to the delivery of care, the creation of new drugs and therapies, and the research to eradicate pernicious diseases.

Fortunately, there is a better way. Technology advances has unlocked extraordinary capabilities to power a digital workforce. A digital workforce comprises software robots purpose built to take on virtually all the administrative tasks once done by the human routers. The digital workforce logs into enterprise systems like EMRs through the UI just like a human. They are trained to understand workflows and can click, type, route, ingest, and extract data just like a human. They can be trained to think and make decisions. They never miss a day of work. Never make unprogrammed mistakes. And they learn collectively, like a network, so that you never have to solve the same problem twice. This new digital workforce is powered by RPA and machine intelligence such as machine learning and deep learning. RPA is one of the fastest growing technologies globally across every enterprise. The biggest breakthrough of RPA in the last few years was computer vision (CV). Software robots can now be trained to understand a UI through the training of advanced convolutional neural nets. This mitigates the fragility of automation and allows the digital worker to keep working even if the UI changes. Together with cloud-based platforms for orchestration and robust maintenance and learning systems, the digital workforce can now deployed reliably at scale across the enterprise.

Health-care organizations across the United States are adopting this technology and hiring their digital workforce at rapid speeds analogous with the adoption of EMRs themselves. This adoption will lead to massive increases in productivity and reductions of cost. Healthcare will become more efficient and resources will be freed up to tackle some of the most challenging clinical problems that face humanity. Most health-care providers are starting their adoption of a digital workforce inside their revenue cycle operation. The digital workforce, powered by RPA, is processing claims, dealing with benefit verification, and handling prior-authorizations at scale. After revenue cycle the digital workforce most commonly spreads to finance and accounting, supply chain, human resources, and IT.

To accommodate the rapid growth of their digital workforce, many health systems are creating a centralized digital workforce operations center that serves as the organization's center of gravity and governance focal-point for enterprise wide transformation using a digital workforce. Several models exist for buyers (employers) of the digital workforce. Organizations can purchase software tools to build their own software robots. Much like the early days of software development, the development environments for software robots are becoming easier to use and widely available. Another model is to hire integrators and services companies to build and maintain your software robots. Much like hiring a firm to build bespoke software for your enterprise, this option can not only be costly but also creates a wide range of customization opportunity. The third option is more akin to Software-as-a-Service (SaaS) where organizations can simply subscribe to a digital workforce service. This option is known as AI-as-a-Service (AIaaS). Learning from history, AIaaS leapfrogs the other models to provide an end-to-end capability without creating mini-software companies inside health systems—a lesson we learned from the early days of software adoption with other enterprise tools. The major advantage to AIaaS is that all the digital employees are connected. In an AIaaS model if one digital employee learns, all digital employees learn.

The most important thing for health systems to keep in mind during their journey toward a digital workforce is the massive difference between a simple automation and a true digital workforce. Creating a software robot to automate a task is simple and can have some initial impact on productivity. However, the real key to success is to create a flexible, learning digital employee that can learn, adapt, and change their work based on new intelligence. A true digital employee interacts with their human coworkers and managers to provide business intelligence and recommend new ways to perform tasks to continue to generate value. An automation provides fixed value on day 1 on the job. A digital employee provides increasing value from day 2 on the job and beyond. Creation of a software robot for automation is focused on building the automation and then shifting it to purely maintenance after it goes live. A digital employee, once built, will change, evolve, and get smarter over time.

Other Key Technologies Related to Artificial Intelligence

Augmented and virtual reality

Augmented reality (AR) is a technology that superimposes a computer-generated digital image on the user's perspective of the real world as to create a composite view (example is the Snapchat lenses). Virtual reality (VR) is an immersive experience of an imaginary physical environment that is completely different than your present physical environment (Oculus Rift is one choice). Finally, mixed reality (MR) combines the elements of both AR and VR so that real world and digital images interact (as observed with Microsoft HoloLens). All of these altered realities can be combined into "XR." AI with its CV dimension is essential to allow these advanced XR technologies to interact with the physical milieu with its object recognition and tracking capabilities and is at the heart of these reality altering technologies.

Artificial intelligence (AI) and applications in augmented and virtual reality: a promising path to clinical excellence

Fran Ayalasomayajula

Fran Ayalasomayajula, an advocate for population health, authored this commentary with her industry perspective on how to bring the advanced technology of XR into a coupling with AI with resulting applications in health care.

HP, Inc., San Diego, CA, United States

Greg, a 57-year-old Afghanistan war veteran suffers from severe chronic back pain. He lives an hour and 15 minutes from the nearest pain management clinic. His wife would normally drive him to the clinic, because the medication that he takes makes his "head foggy," and it is unsafe for him to drive. Six (6) months ago the clinic introduced a new form of therapy, VR pain management therapy. Now when Greg is seen in the office, he begins his visit with 20 minutes of VR therapy prior to his session with the clinicians. Greg has responded positively to the treatment, requiring less drug therapy, more self-directed care, and sustained relief.

For pain specialists, VR therapy is now a standard of practice in a growing number of clinics. Patients are introduced to the treatment and are closely monitored to determine responsiveness. For many the therapy has proven beneficial.

This is the path to increased adoption of XR technology, referring to VR, AR, and MR, in the medical field. Physicians are witnessing the value of the technology in helping them solve problems in progressive manners. The practice of medicine is resourceful. It is not all about drug therapy. With efficacy and efficiency the door is open to apply new tools and methods to address clinical challenges.

Where the technology is being deployed, XR is having a significantly positive impact. Medical education and training, surgical planning and diagnosis, rehabilitation, psychology, and telesurgery are some of the areas in which the industry has witnessed the most profound gains. Today, universities are making substantial investments in the implementation of XR powered facilities, such as is the case at the Human Simulation at the University of Colorado School of Medicine, the University of Illinois School of Medicine, the University of Gernoble, and the University of Tübingen to name a few. Around the world institutions are designing entire curriculums that incorporate VR and AR tools.

This of course leads to high-performing well-trained students with expectations that the residency programs to which they are admitted will also have integrated applications of the technology. And similar to their up-and-coming junior colleagues, an increasing number of well-established practitioners are also eager to partake of these new tools. Nearly 20% of the US physicians have used VR, and just under 70% are open to the application of VR for continuing medical education and use for therapeutic purposes [1].

Ease of integration into the existing workflow and simple protocols for administration are what will make for the vast proliferation of the technology. As the technology matures, we are also beginning to witness the convergence of XR with AI. The use of predictive modeling and learning algorithms will refine experiences for trainees, support practitioners in the surgical theater, and guide personalized therapy for rehabilitating patients.

One of the first areas in which this has proven to be most immediately within reach is in the field of medical imaging. Since voluminous amounts of data already exists to feed the intelligence of decision support tools, the combination of supportive algorithms and VR-enabled visualization tools equips surgeons to quickly examine, scope, and map surgical plans in record time with greater confidence and precision. These conditions are ripe for intuit. As Ron Schilling, PhD, former CEO of Echopixel and medical imaging innovation pioneer once described, "the perfect combination of knowledge, experience, and information."

Advances in the development of XR are rapidly taking shape and being adopted in pragmatic ways. The possibilities are endless. For example, the integration of multisensory and weight testing will create truly immersive experiences, enabling a better understanding of neurological responses to different environments and stimuli using eye tracking, heatmaps, and EEG analysis [2]. Over the course of the next 5 years, we will witness more institutional adoption, and a generation of XR hardware and software that will further empower clinicians to bring forth the best of science and talent to the practice of medicine.

References

[1] Cobos S. AR/VR innovations in surgery and healthcare. Premo Grupo. 14 August 2017. <https://3dcoil. grupopremo.com/blog/arvr-innovations-surgery-healthcare/>.
[2] Garnham, R. (March 2019). Virtual Reality: Hype of the Future? Ipsos Views. <https://www.ipsos.com/sites/ default/files/virtual-reality-hype-or-future2019_web.pdf>

Blockchain

Blockchain is a specific type of distributed or decentralized ledger or a database of information with the information stored in many servers in the network. The steps for blockchain to achieve its purpose are (1) a transaction is requested by a user; (2) the transaction is created on a "block," (3) the block is then broadcasted to the nodes of the network, (4) the nodes validate the block that was broadcasted, (5) the block is then added to the chain, and (6) the transaction is verified. While the identities of the owners are private, the transactions are not. In addition, the transactions are secured by a unique cryptographic key so the information is immutable. While AI (centralized and lacking in transparency) and blockchain (decentralized and transparent) seem like an odd pairing, the combination of AI with its machine learning tools can enable both to be more efficient. The convergence of these disruptive technologies can create an ideal situation: a distributed computing substrate for AI with a universal anonymous blockchain data for AI work with more equity [11]. In addition, geospatially enabled blockchain solutions that use a crypto-spatial coordinate system to add an immutable spatial context can be even more useful in healthcare [12].

Health data and blockchain

E. Kevin Hall

Kevin Hall, a pediatric cardiologist with an ardent passion for blockchain technology, authored this commentary on how blockchain will be an impactful technology in bring security and access to health care data.

Pediatric Heart Failure and Cardiomyopathy, Yale University School of Medicine, New Haven, CT, United States

(Short version) health data and blockchains

The patient has long been touted as the owner of his or her medical data but this has never yet been effectively true. While hospitals are required to give a patient's records upon request, the process typically means that the institution will provide copies printed on paper and occasionally disks of imaging data when specifically sought. What happens to the patient's copy of the data when the hospital adds new data to his or her record? Nothing. The records duplicated for patients are copies at that particular point-in-time only. As soon as new data is added to a patient's record, the copy that had been provided to the patient is no longer an accurate representation. This problem becomes more complex when one considers medical data from multiple institutions such as two separate hospitals, or a hospital, and a private practice. Each of these institution's records contain a partial and incomplete representation of "truth" for any given patient. The truth would instead be a sum of the data at both institutions. Multiply this issue by many organizations over a patient's life and the problems are only more compounded. One of the benefits of digital data is that it can be copied with no expense. But in health data, these "copies" are typically one way copies only; it remains the rarest of exceptions when multiple sources, acting as equal peers, cooperatively synchronize data bidirectionally.

We can certainly do better, but what would a better patient-centered, patient-controlled repository for medical data require? As compared to today's EHRs, true solution must be architected differently from its conception. The most forward-looking answer to this question must fulfill several key features. First, the system must be decentralized and fault-tolerant, available and robust. A decentralized system is one where any part or group of parts can become unresponsive but the system itself remains functional. The data remains available and robust or correct even if a provider's computers go offline.

Data contributed to the system must be stored indefinitely and it should be immutable. When data are permanent, it cannot be destroyed, erased, or corrupted, a factor of great importance in building a longitudinal health record of individuals and populations. Prior to the move to digital, data immutability was not a foreign concept to us. It was only with the advent of early computer storage and its astronomical costs that we designed into the storage media the ability to alter data. This is not to say that a fact cannot be amended later, only that we do not lose the record of what any value once was even after it is modified or updated.

Last, data must be safely and securely sharable when required. A data's owner (e.g., a patient) should be able to share his or her data selectively with other trusted individuals or groups via cryptographic signatures. Cryptography is widely used today in securing communications and financial transactions, although the complexities are hidden from the end users. While the technical details are beyond the scope of this discussion, all parties typically possess both public and private cryptography keys. A message for a recipient can be encrypted with the recipient's public key, but readable by the recipient only after it is decrypted by his or her private key. Such an approach would allow the secure exchange of data between two parties.

Blockchain

By taking into account the requirements stated previously, we have recreated a data structure called a "blockchain." The idea of a blockchain was first introduced in a 2008 paper describing the virtual distributed currency Bitcoin. A blockchain is a growing list of records (so-called blocks) where each record is cryptographically stamped by a mathematical hash of the preceding record. In this way a "chain" of records is built with each subsequent record marked by the one preceding it. As the chain of records grows, any single record can only be modified by altering the mathematical signatures of all the blocks that follow, therefore ensuring that the blocks within the chain remain unchanged.

Blockchains are typically managed by distributed networks of independent computers called nodes. Blocks to add are agreed upon by a distributive mathematical process termed "consensus" that is automatically run according to an agreed upon protocol. In traditional implementations a single block is added at fixed intervals and each one contains a payload of data to be stored.

Data stored within a blockchain answers the requirements described previously: data are immutable, robust and distributable, and cryptographically signed and obscured by the owner. A key question for healthcare is that if data are distributed and encrypted on a blockchain and can only be shared with the permission of an owner (the patient), what would happen in the case of an emergency where the patient is consciously unable to grant access to caregivers? Modern implementations of blockchains provide a feature called "smart contracts." Smart contracts are simply programmatic functions publicly stored on the blockchain which enable more advanced functionality. In cases of emergencies a smart contract might enable health-care workers who are first verified to be legitimate and working nearby to access key health information from an incapacitated patient.

Due to factors including their distributed architecture and consensus mechanisms, blockchain transaction speeds are currently one to two orders of magnitude slower than many centralized systems; this should improve. When this does improve, perhaps to tens or hundreds of transactions per millisecond, it changes the data's medium from a batch format to one that is streaming. We have seen how streaming has changed how we access music and movies: we no longer need to possess a full copy of a song or movie before enjoying it, instead streaming it in real-time across the network as we continue to receive subsequent portions that data. We can begin to imagine a world where health data is streamed in real time. This would likely enable better real-time monitoring of patient data; during high risk procedures such as surgeries, one could imagine insurance companies risk adjusting policy costs moment-to-moment based on various intraoperative vital signs and cues.

This type of data sharing fundamentally changes how we imagine healthcare. Many have spoken about the coming age of truly personalized medicine. The arrival of this age has been hampered by our collective and current limitations in sharing, accessing, and acting on actionable health data from patients. Current limitations drastically limit our ability to use a patient's data to directly benefit him or her across the health-care arena. The patient is ultimately the true and best arbiter of his or her own data, and this data can only be optimally employed when the data is properly empowered by its owner, the patient. The blockchain data structure is so far the best potential answer to the current limitations in safe and broadly accessible, permissioned health data exchange.

Cloud

Cloud types include public, private, and hybrid types. The private cloud (such as hospitals) (see Fig. 6.6.) has potential advantages that include security and autonomy, while the public cloud (such as Google) has potential advantages that include scalability and cost-effectiveness. There is also the hybrid cloud that can potentially offer both security as well as scalability with an attractive cost structure as well. With the advent of machine learning capabilities in the cloud, there is a higher attractiveness about storing and managing biomedical data in the hybrid or public cloud. This is especially true now with increased security capabilities of the public and hybrid clouds.

Cloud computing in healthcare is at present in the form of singular, individual features such as elasticity, pay-per-use, and broad network access rather than as cloud paradigm on its own [13]. Cloud computing is, therefore, often in the "OMICS-context" with computing in genomics, proteomics, and molecular medicine with little use in other domains. Most of biomedicine data is presently stored in local storage or in private clouds due to concern for HIPPA compliance and privacy [14,15]. A recent survey of cloud computing adoption in the health-care sector revealed that 83% of IT executive in healthcare reported use of cloud services today with majority utilizing SaaS-based applications (67%) [16]. The cloud infrastructure with convergence to mobile computing, wireless networks, and sensor technology can enable a health-care service to be delivered as delineated by Kaur in his Cloud Based Intelligent Health Care Service (CBIHCS) for monitoring chronic illnesses such as diabetes [17]. A community cloud, serving a common interest or purpose, can also be acceptable for a biomedical group or system (such as a specific subspecialty or hospital system) [18].

Although the storage of genomic data into the public cloud raises issues such as form of security and privacy, initial efforts to secure computation techniques that can enable the comparative analysis of human genomes have been productive. One such effort is the NIH-funded National Center for Biomedical Computing iDASH (integrating Data for Analysis, anonymization, and Sharing) and its Critical Assessment of Data Privacy and

Protection competition to evaluate the capacity of the cryptographic technologies for protecting computation over human genomes in the cloud while promoting cross-institutional collaboration [19]. In addition, high-performance computing platforms such as clusters, grids, and clouds can be used by neurologists, radiologists, and researchers for imaging such as neuroimaging to increase both storage and/or computational power [20]. Finally, IBM Watson debuted SleepHealth, an app and ResearchKit (Apple) study designed to investigate the possible connection between sleep habits and health. This app requires the support of HIPPA-compliant Watson Health Cloud and will gather the crowd-sourcing data for an unprecedented study of sleep and health.

Health care cloud computing

Timothy Chou

Timothy Chou, one of the progenitors of cloud computing, authored this commentary from is vast industry experience on how the cloud can be an enabling platform for all the relevant medical devices and equipment to converge to become a medical internet of everything.

For medical intelligence textbook

You cannot go thru any airport today or watch any sporting event without hearing about "the cloud". So what is cloud computing? Quite simply compute and storage cloud computing is the management of the security, availability, and performance of compute and storage. Managing security insures the right patch levels are applied. Managing performance insures that a predictable instance of computing is available and by managing availability a cloud computing service insures there are three copies of the data stored. By centralizing and standardizing the management of compute and storage the management can be automated, thereby not only decreasing costs but also increasing quality.

Compute and storage cloud services have enabled new business models. Rather than purchasing hardware up front, locating a place to house it and hiring people to manage it you can buy managed compute for an hour. These economics open up totally new use cases, including making high performance computing available to small teams with minimum investments. At the beginning of cloud computing I launched a class at Tsinghua University, one of the most famous universities in China. The Amazon team gave me $3000 worth of cloud computing time. I told the class at the time $3000 would buy you a computer in Northern CA, Ireland, or Northern Virginia for 3.5 years OR 10,000 computers for 30 minutes. It was an example to inspire them to think not about how we do computing today, but what might be economically possible in the future. Of course, today it would be even more computing for even less money. So what might be possible? How could cloud computing reshape healthcare?

The dramatic changes occurring in the application of AI technologies, in particular deep learning, are based on the confluence of three powerful forces. The first is the advent of large amounts of computing, for little cost (mentioned previously). The second is the arrival is software for building many kinds of neural networks. Neural network technology has been in the labs for a lot of years, but it is only recently that software such as Tensor Flow or Pytorch has become available. And finally, with the arrival of the Internet we have large amounts of data.

Facebook facial recognition can work so well precisely because they have lots and lots of photos of people's faces.

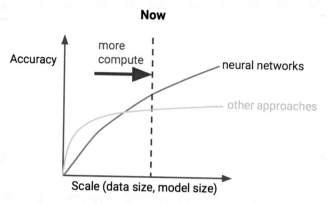

The confluence of these three powerful forces shown above is propelling a dramatic ability for machines to be more accurate. A simple case study is found in the ImageNet competition. Begun in 2011, the ImageNet contest pitted man against machine to determine who could recognize images more accurately. In the first contest the machines could only be 75% accurate versus humans who were 95% accurate. In just 4 years machines hit 97% accuracy. Advances are starting in healthcare. Researchers at Stanford University [1] had four radiologists annotate 420 chest X-rays for possible indications of pneumonia. Within a month the team had developed CheXnet, which outperformed the four radiologists in both sensitivity (identifying positives correctly) and specificity (identifying negatives correctly). While this is a good beginning, it is just a beginning. There are at least four challenges and opportunities in front of us.

Connecting health-care machines to the cloud

The Stanford university project used 420 chest X-rays. You can easily guess there are 100,000 or more chest X-rays being taken every day. Unfortunately these are locked away in isolated X-ray machines. What if they were all connected and what if we could bring years' worth of chest X-rays from around the world? Can you imagine the accuracy of a pneumonia digital assistant? And what if every MRI scanner, gene sequencer, ultrasound, blood analyzer, cell plate reader, etc. were connected? While we have the compute and storage cloud services, we need the data to take the next big step.

Digital service revenue

Health-care machine makers have the opportunity to create digital service products. Today most executives think service is break-fix, a guy with a wrench to fix the machine. But service is information—information about how to maintain or optimize the performance, availability, or security of the machine. In the connected state the machine manufacturer would have access to all of the information from the experiences of all of the machines. If you charged just 1% of the price of the machine per month for a personalized (or should I say thing-ized) digital service, most health-care machine manufacturers could double their revenues and quadruple their margins and furthermore build a significant recurring revenue business.

Analytic applications

Cloud computing is not only bringing low cost compute and storage, but it is also allowing us to build more powerful applications quicker. Up until now most enterprise software companies have focused on building workflow applications and improving back office operations like purchase-to-pay, order-to-cash, or hire-to-fire. In the world of analytics, we have supplied people tools. Building modern analytic applications require at least 16 classes of software tools. In the old days much of this would have to be functionality you'd have to build. Instead today with the vast amount of open source (e.g., TensorFlow, Hadoop, Kubernetes, Kafka, Airflow, Databricks, Cassandra, Superset, D3, Django, Puppet) as well as variants also available as managed cloud service. It is now possible to stand on the shoulders of giants and build health-care analytic applications. Makers and users of health-care machines (blood analyzers, MRI scanners, ultrasounds, gene sequencers) do not build their own financial, HR, or purchasing software instead they purchase packaged enterprise application cloud services. Perhaps in the future rather than buying tools and hiring data engineers, DevOps people and ML experts will be able to purchase packaged health-care analytic application cloud services.

Global applications for global health

Once an application is available as a cloud service, whether that is the ability to purchase a book or providing a pneumonia digital assistant then it has the potential to be available to everyone around the globe instantaneously. By the year 2040 25% of the globes population will be in Africa. There is no way to be able to build a first world health-care system. There is not enough time or resource to build the medical schools, train the people, and construct the traditional hospitals. Instead we have the opportunity to build new applications that can be deployed around the world to analyze, diagnose, and prevent diseases. Software technology has near zero costs. Its only requirement is the energy and creativity of people. Many students today do not want to work on the next dating website or social network, imagine if we could bring together their creativity along with the economics of modern compute and storage cloud services.

Maybe cloud computing can change the world and all those ads in the airport will really have predicted the future.

References

[1] <https://spectrum.ieee.org/the-human-os/biomedical/diagnostics/stanford-algorithm-can-diagnose-pneumonia-better-than-radiologists>.

The hybrid cloud has recently emerged as a type of cloud infrastructure that will take advantage of both the public and the private clouds [21] (see Fig. 6.6). The hybrid cloud combines the customization and efficiency as well as security and privacy of the private cloud with the capital preservation and standardization of the public cloud that could be essential for the biomedical milieu [22]. The usual arrangement is for the private cloud to store data and the public cloud to render services with both communicating via a secured

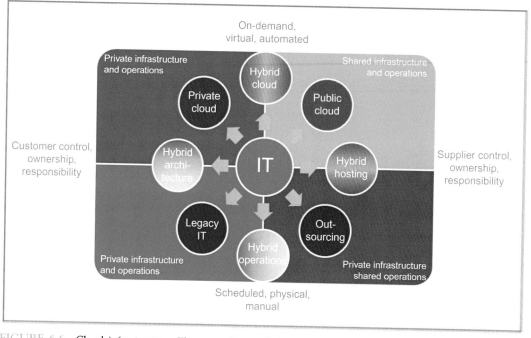

FIGURE 6.6   Cloud infrastructure. The many forms of cloud infrastructure can be based on supplier versus customer control as well as on-demand, virtual and automated operations (vs scheduled, physical, and manual). There are far more types of cloud than simply public, private, and hybrid (see text for more details).

connection. This strategy can also be dynamic with the private cloud utilizing its public cloud partner(s) on an as-needed basis (the so-called dynamic hybrid cloud). Logistical challenges of the hybrid cloud include connectivity as well as management of the arrangement on a continuous basis. The best alternative cloud infrastructure for medical care, however, may be an even more sophisticated cloud infrastructure system that is customized for each need in health-care data storage and security. For instance, if supplier control with ownership and responsibility are needed but with shared and private infrastructure, a "hybrid hosting" is desired.

Cybersecurity

Cybersecurity in healthcare is very vulnerable to attacks and is considered to lag behind other sectors [23]. An issue related to the aforementioned blockchain is AI security, which can have different levels of consequences depending on narrow vs general AI failures. In other words a single isolated failure of a super intelligent system can be utterly catastrophic and fulfill the dire warnings of some of our leaders. There is the additional concern of an algorithm being hacked and misinterpret a medical image or misdirect a decision support system.

Internet of Things

The recent explosion of physical devices and embedded sensors from equipment, buildings, vehicles, and appliances, as well as wearable devices has led to network connectivity which will enable all these devices to collect and exchange data. An IoT platform enables connectivity for these devices that are capable of being part of the network. The method of communication is radio-frequency identification but can also be wireless, quick response codes, or other sensor technologies. The connected devices together can form ambient type of intelligence but will need to be AI-enabled. For embedded AI, AI software can be embedded in the applications and devices so that AI is pushed peripherally without the necessity of having all the data pushed into a central repository for machine- or deep-learning processes.

Artificial intelligence (AI) in medicine: the need for early education and continuous awareness campaign

George Murickan

George Murickan, an expert in the domain of IoT, authored this commentary on his personal experience with a medical device and how important AI will be in the future of medical devices that will need to be monitored.

While AI in Medicine has been around us for a while now, the awareness of it is still not to the level it needs to be. AI in Medicine especially seems like a "taboo" subject for the general population and understandably with the proliferation of news stories about cyber hacks; many grow even more weary. So how do we manage this in a sensible and practical manner, so we can make meaningful progress in Medicine through AI? The answer is education and is provide early and often to increase the awareness.

This educational effort needs to start from the high school level to make our children aware of what is available for them to monitor and improve their wellness, with less medication and more importantly, make them aware of how one can prepare for careers in such a field. It may seem like a daunting effort, but it is much easier with the younger generation. If you introduce the concept of AI, at a younger age, you allow the natural curiosities and interests of the young minds to explore courses that better prepare them for AI. The collegiate level infusion of preparatory courses is a more immediate need and requires the medical professionals and collegiate leaders to work together to advance these efforts. Departments such as advanced mathematics, computer science, and natural sciences all need to work collaboratively to knead courses that provide the proper cross training for aspiring AI-assisted medical professionals. The more daunting effort is for Medical School programs that must quickly arm currently graduating medical students with enough information and familiarity to embrace AI in Medicine, so that they do not become the "lost generation" in the brave new world of AI assisted Medicine. This too can be averted by working with experienced physicians in organizations such as AI Med and other similar organizations that can provide online Seminar courses to give them enough of an overview to better prepare themselves after entering the medical profession.

The awareness campaign to make AI in Medicine a household concept should also take precedence. Why? At the core of any AI technology is its data, and lots of it, to have as complete a spectrum for better predictions and breakthroughs, and more importantly, reduce misapplication of the technology. So, the more people that adopt the use of AI assisted Medicine and become willing and consenting data providers, the faster we can democratize the data sharing and accelerate true innovation for the benefit of all. Thus it is a symbiotic relationship between the medical professionals and the masses. There will be a lot of public advisories and even advertising campaigns that need to be deployed for this. But this can only happen if we can provide practical use cases of how it can immediately and affectively help a person today and ultimately, how it can help us go from the Healthcare (or "Sick-care," with intended sarcasm) that we have today to true Wellness-care. The task then is to not only take advantage of the early adopters but also to reach the more skeptical population and have them use and show actual benefit from the use.

This is a true-life account of how such a skeptical person became an adopter. My tax advisor has suffered from cardiac issues for many years that stemmed from early onset of Type 2 Diabetes. With a recent medical complication from inflammation of internal tissues, he was forced to use prednisone (steroids) daily to control the inflammation in the body. The addition of the steroids to the dozen carefully balanced medicines he was already taking completely threw off the sugar levels in his body to the point that his blood sugar levels were running up to 300s and down to low 50s, daily. In order to help him, I did some research on medical devices/ IoE (Internet of Everything—sensor-based smart solutions) devices that can track, collect data, analyze it, and provide meaningful predictions to help him and his physicians get this under control. I recommended him to purchase an IoE medical device that was FDA approved in March of 2018. The product was the 6th version from the same medical device company, and it consists of a sensor to be attached in the abdominal area of the user that samples blood automatically, several times a minute and then uses the colocated transmitter to send the data to the accompanying app on the user's phone and simultaneously to a secure cloud server site. Based on these readings, the device and its App is smart enough to warn the user of their current levels and predict whether it is headed up or down or will hold steady and how fast or slow the next change will take place. Based on set points one chooses (high and low), it can warn the user and five other people (on their mobile devices), so they can intervene and help the user. In addition, the app also has the feature for the user to enter daily meal and drink intake, medicine intake and exercise and activities, so it can be saved for further analysis. The cloud portion of the application saves historical information that can be sent to the physician or nutritionist to help them further analyze and provide recommendations. The physician can then recommend adjustments of medication or timings of medication or determining interactions of medications. The nutritionist can also better assess what the user should or should not eat and when to eat and what portion sizes to help the user make better lifestyle choices and changes.

Once he started using the device, the family was woken up several times during the late-night hours to warn that his sugar had hit a low of 50s and many times during the day, when he spiked as high as 300s. Slowly, they adjusted his diet and activities to balance this out. After a few months, they sent the data to their physician and nutritionist, who analyzed and made some appropriate recommendations and adjustments to meal patterns and medication. Consequently, the levels became steadier with less spikes. I am glad to report that after 6 months he is a much healthier and happier individual with a much more peaceful and happier family. They were not techies, and truly AI doubters and all involved members of the family were above 40 and up to 75 years old. The online video instruction and app were simple enough for them to navigate, adapt to, gain knowledge, and then embraced it. And that is a small way in which the awareness of AI and how its helpfulness can lead to better adoption and wellness-care. If we can get more people to try, insurance companies to start covering the expenses of these devices and the acquired data to be analyzed and utilized to help others, then we are not only increasing the awareness and education, but the adoption of AI in Medicine to positively help society.

Key Issues Related to Artificial Intelligence

There is a lengthy list of issues that pertain to AI, but only three main issues will be briefly covered here but will be covered in more detail in the health-care context later in the book (these topic are also well covered in the books listed in the compendium section of this book).

Bias

As data can contain racial, gender, political, religious biases, the AI algorithm can inherit these biases, and this bias is then perpetuated into other algorithms. Bias can be introduced at any stage of the learning process: framing the problem, data collection (from past or present biases), data preparation, and any part of the deep-learning algorithm [24]. There are five types of bias in AI: dataset bias, association bias, automation bias, interaction bias, and confirmation bias [25]. The possible solutions include constructing algorithms that can detect bias in the data and having higher sense of awareness for bias by creating an ethical culture with a diverse team as well as being transparent. In addition, the recent Algorithmic Accountability Act, which requires companies to study and fix flawed computer algorithms that result in inaccurate or unfair decisions, can help mitigate bias in AI. Closely tied to bias is the issue of inequality. This issue of inequality caused by AI is not only economic but also racial and gender inequity as well. In her insightful book *Automating Inequality: How High Tech Tools Profile, Police, and Punish the Poor*, Virginia Eubanks delineated how algorithmic technologies are upending bureaucracies and disadvantage the poor [26].

Ethics

There is a myriad of AI-related ethical issues but there are very few guidelines for ethics in AI. The most obvious ethical issue is regarding inducing harm with AI, such as use of military drones or cyberattacks. According to the World Economic Forum, additional issues include unemployment, inequality, humanity, mistakes, bias, security, unintended consequences, singularity, and, finally, robot rights [27]. One of the most highly regarded efforts in forming an AI ethical framework is "The 23 Asilomar Principles" from the Future of Life Institute in 2017 with over 100 scholars, scientists, philosophers, and industry leaders [28]. These principles mostly center around AI doing good and include safety, judicial transparency, research goal to do good, responsibility, value alignment with humans, and shared benefit and prosperity. Solutions include a dual philosophy of engineers being aware of ethical challenges and the autonomous systems being self-aware of ethical issues [29].

Safety

There is incessant concern that the AI can become self-empowered or be manipulated by evil forces to be antagonistic or lethal to humans. Concepts about sentience and consciousness for AI render this discussion even more relevant. Traditional regulatory processes are quickly becoming woefully inadequate (and perhaps inappropriate) to be used as a strategy to have oversight over ultrafast-evolving AI software that can change in real-time in a matter of seconds. Alan Turing, the father of AI, would also remind us that machines should be created to engage machines, and, therefore, perhaps we eventually will need to develop "regulatory" algorithms. Stuart Russell, one of the progenitors of AI in this modern era, proposed three principles of AI safety (human-compatible AI): (1) the robot's only objective is to maximize the realization of human values, (2) the robot is initial uncertain about what those values are, and (3) human behavior provides information about human values [30]. In addition, there is growing concern over the rise of autonomous weapons and their potential danger. This concern culminated in the employees of Google objecting to Project Maven (autonomous weapons at the Pentagon) and influencing its company to promulgate its AI Principles to prohibit AI use in weapons and human rights abuses. Another tangible action from the Asilomar conference and this august group's gathering was the Lethal Autonomous Weapons Systems pledge.

Legal

There is a myriad of legal issues that range from data privacy to transparency as well as regulatory issues and intellectual property rights. Other pertinent issues include duties of competence and diligence, supervision, and client confidentiality and privilege; all of these duties mandate the legal community to maintain an adequate knowledge base of AI. Several of these issues have been or will be covered in various parts of the book. Just as in ethical and political discussions of AI in general, the legal realm has been caught off guard by the steep exponential rise in the capability and proliferation of AI in recent years.

Legal aspects of artificial intelligence (AI) in medicine and health care

Dale C. Van Demark

Dale C. Van Demark, a legal counsel with a special interest in AI in medicine, authored this commentary to elucidate several key issues in the legal realm, in particular transparency and standards, and how these general principles will need to be defined for a legal framework.

McDermott Will & Emery, Chicago, IL, United States

The application of AI systems to the delivery of health-care services falls within a vast and complicated legal and regulatory environment that is not designed to be nimble or welcoming of "the new," There are strong policy reasons for this. New health-care services need to be safe effective and, increasingly cost-effective. In addition, this legal and regulatory environment, generally, was not created at a time when information technology was as ubiquitous as it is today. Accordingly, it is generally poorly designed for the existing and increasing ubiquity of information technology.

Current AI systems that employ machine learning to health-care settings are proving their effectiveness, but few have been deployed. The full legal and regulatory implications of this technology in the health-care delivery environment, then, are not yet known and attempting to address them all in this article would be counterproductive. Accordingly, this article will address two issues that are fundamental to modern AI systems and the legal and regulatory health-care environment: transparency and applicable standards in the health-care delivery context.

Transparency

The "AI black box" is a reference to the inability to understand why modern AI systems produce the output they produce. The black box issue is more than an inconvenience; however, it creates thorny legal issues.

For example, a fundamental notion to US jurisprudence is that of due process, a concept enshrined in the Constitution.[1] In healthcare the clearest and most direct application of the notion of transparency to AI comes through the 21st Century Cures Act, passed at the end of 2016. In the Act the Federal government excluded from the definition of "device," and thus from the requirement to obtain FDA clearance or approval for marketing purposes certain clinical decision support tools, including those that might make recommendations to a practitioner.[2] The software that is excluded from the definition must satisfy certain criteria including allowing a practitioner to independently review the basis for the recommendation provided by the software.[3] In draft guidance the Food and Drug Administration explains what this might require:

> …. the intended user should be able to reach the same recommendation on his or her own without relying primarily on the software function. The sources supporting the recommendation or underlying the rationale for the recommendation should be identified and easily accessible to the intended user, understandable by the intended user …, and publicly available ….[4]

While this guidance may not be formally adopted, clearly the thinking behind the draft guidance includes a concept of transparency that is missing from some modern AI systems.

---

[1] In health care, due process may be implicated in the context of the administration of benefits. Transparency in this context, as well as the need to use appropriate and accurate data, is critical. See K. W. v. Armstrong, Case No. 1:12-cv-00022-BLW, Memorandum Decision and Order, March 28, 2016, pp. 17–25, available at <https://www.acluidaho.org/sites/default/files/field_documents/summary_judgment_decision.pdf>.

[2] 21 USC §360j(o)(1)(E).

[3] 21 USC §360j(o)(1)(E)(iii).

[4] Clinical and Patient Decision Support: Draft Guidance for Industry and Food and Drug Administration Staff, December 8, 2017, p. 8, Available at <https://www.fda.gov/downloads/medicaldevices/deviceregulationandguidance/guidancedocuments/ucm587819.pdf>.

Standards

The professional judgment of health-care providers is exercised in the context of past and ongoing learning, training, experience, and the utilization of tools that assist the professional in exercising that judgment. The professional takes the facts and circumstances and weighs various conclusions regarding a course of action. In this context, AI tools can be extremely beneficial as they can speed analysis, expand the knowledge base of the provider, and speed the review of critical or vast amounts of data.

The general standard of care applicable to physicians requires the application of a reasonable degree of skill, knowledge, and care ordinarily possessed and exercised by members of the medical profession under similar circumstances. Generally, product liability emanates from negligence, strict liability, or breach of warranty standards. Putting aside breach of warranty (which has a contractual nature), negligence and strict liability are vastly different standards—one requires there to be a duty and a breach of the duty which is the proximate cause of an injury (negligence), while the other requires proof of an unreasonably dangerous defect that causes harm (strict liability). With this in mind, it is useful to employ a focused thought experiment to see how, in the not so distant future, AI systems may start to challenge these standards.

At some point, under some circumstances, AI systems will start to look more like a practitioner than a device—they will be capable of, and we will expect them to, render judgments in a health-care context. The AI system will analyze symptoms, data, and medical histories and determine treatment plans. In such a circumstance, questions related to the appropriate allocation of responsibility among device, practitioner, and patient (particularly if the AI system is consumer oriented) become difficult. Should practitioners be required to rely on high functioning AI systems, or should the output of AI systems be only a factor in a practitioner's ultimate decision-making? What should be the consequences if the AI system is right and the practitioner wrong? What standard should an AI system be held too—particularly if it is a system that continues to evolve through it experience? This latter question is relevant given the "black box" nature of advanced AI systems.

Developers of AI systems for health-care delivery should take very seriously both of these general principles, even if there are no existing or apparent applications to the system under development. The law is developed, legislatively (frequently), and through case law in a reactive context. Bad facts, the saying goes, make bad law. Good facts and responsible behavior on the part of developers will go a long way to mitigate the development of bad laws.

In conclusion the fast ascent of deep-learning performance has promulgated the entire field of data science and AI to new heights. As deep learning usually mandates large training sets and the medical field often lack such large-sized databases, it behooves the medical field to be better organized and more collaborative in such endeavors and concomitantly for the data science stakeholders to accommodate this limitation with innovative approaches in deep learning. The data conundrum in healthcare is a particularly important issue for AI applications but with efforts such as data sharing and innovations such as graph databases, this can be vastly improved. The assessment as well as the transparency of AI tools will need to be mature for wider AI adoption amongst clinicians.

Key Concepts

- The recent advent of an AI "triad" (conveniently "ABC") consisted of (1) the emergence of sophisticated algorithms, particularly, machine and deep learning with all its variants, (2) the increasingly large volumes of available data that requires new computational methodologies (or simply "Big Data"), and (3) the escalating capability of computational power (with faster, cheaper, and more powerful parallel processing that defied Moore's Law) with coupling to the widely available cloud computing (with nearly infinite storage). These elements have converged to engender this new resurgence of AI.
- Much of the future of AI in medicine and its success will be rooted in the quality and integrity of biomedical data and databases.
- It is estimated that about 80% of health-care data is unstructured.
- Data that have escalated in a myriad of ways to the point that traditional data processing applications are no longer adequate is termed "Big Data."
- Despite the large volume, variety, velocity, and veracity of big data in biomedicine, there is little dividend in the form of information from this health-care big data.
- An ETL (extract, transform, and load) process is employed in order to extract data out of the system and configure the data for the data warehouse that is favored by business professionals as the data are usually structured (but storage usually more costly). A data lake is a lower cost data storage repository preferred by data scientists and can hold large amounts of raw data, including unstructured data, for later analytic use.
- Interoperability, according to HIMSS, is "the ability of different information systems, devices, or applications to connect, in a coordinated manner, within and across organizational boundaries to access, exchange, and cooperatively use data amongst stakeholders with the goal of optimizing the health of individuals and populations."
- Most of the present health-care data remain embedded in flat files or at best, in relatively simplistic hierarchical or relational DBMS with most of the data centralized and locked into local operating systems that reside in hospitals or offices.
- There are limitations to relational DBMS for health-care data: these lack sufficient infrastructural support for the larger health-care data (such as time-series data, large text documents, and image/videos). In addition, queries are difficult due to the structure of relational DBMS.
- A graph DBMS can store data in the form of graph elements (nodes, edges, and properties) in order to facilitate relationship definitions for data elements. This type of database is more three-dimensional and has advantages over the traditional relational database.
- The graph DBMS with these search algorithms is especially well designed for complex queries in healthcare such as chronic disease management, acute epidemiological crises, and health-care resource allocation. The location of a similar patient to an index patient can also be performed using this strategy.
- Machine learning, a term initially coined by Arthur Samuel in 1959, is an increasingly popular subdiscipline of AI and is the art of computer programming that enables the computer to learn and improve its performance without an external program instructing it to do so.

- In traditional programming (and statistical analysis), a top-down approach provides rules for the input data and output is derived. In machine learning, both the input data as well as output data (labeled by humans) are entered into the computer and the rules are derived from the data. The new rules are then applied to the new set of data.
- The steps of collecting and processing the data can easily make up the majority of the effort and time needed to do a project for the data scientist, especially in a clinical setting.
- The current paradigm of AI in medicine for biomedical data science is adding another domain of knowledge to computer science and mathematics: the domain knowledge of biomedicine (bioinformatics and clinical informatics as well as biology, genetics and genomics, medicine, and health sciences).
- Machine learning, or, more accurately, classical machine learning, is better suited for smaller and less complicated datasets and clinical scenarios with less features. Classical machine learning is categorized into two types of learning: (1) supervised learning and (2) unsupervised learning.
- Supervised learning develops a predictive model from both input and output data (the later labeled by humans) and this model is then used to make predictions on a new set of data.
- These supervised learning methodologies lead to classification (dichotomous or categorical) or regression (to a continuous variable).
- For classification the popular methodologies are support vector machines, naive Bayes classifier, k-nearest neighbor, and decision trees (with boosting or bagging); logistic regression is a misnomer and is in fact a classification methodology. For regression, linear and polynomial regression methods are most commonly used, but other types (such as ridge and lasso regression) may become more popular in the future.
- Unsupervised learning takes unlabeled data and uses algorithms to predict patterns or groupings in the data set without any human intervention. These unsupervised learning methodologies lead to clustering, generalization, association, or anomaly detection.
- A hybrid technique of supervised and unsupervised learning is semi-supervised learning, which uses a small amount of labeled data and then a relatively large amount of unlabeled examples.
- This ensemble learning strategy (bagging, boosting, and stacking) involves training a large number of models that together will surpass the performance of a single model; in short, it is the creation of a meta-model that has better prediction and more stability. This ensemble of models reduces noise, bias, and variance.
- In addition to the aforementioned supervised (task-driven with classification or regression) and unsupervised (data-driven with clustering) learning, another type of learning is reinforcement learning. Although reinforcement learning is often described as a third or additional type of machine learning along with supervised and unsupervised learning, it is distinctly different than the former two types of machine learning.
- In reinforcement learning the model is not relating itself to data but rather finding the optimal method via exploration to achieve the most desirable outcome while receiving input data in a dynamic environment.

- Reinforcement learning and its AI congener deep reinforcement learning are particularly valuable assets for biomedicine as these methodologies are well designed to make sequential decisions in an uncertain environment toward a long-term goal that can be to minimize error (leading to morbidity and/or mortality).
- Compared to the machine-learning techniques just discussed, the more sophisticated neural networks and deep-learning techniques (with sometimes hundreds of layers of neurons) are particularly well suited for nonlinear and complex relationships, which are not uncommon in biomedicine and healthcare.
- Current applications of deep learning include speech recognition and NLP, CV with visual object recognition and detection, speech recognition, and autonomous vehicle driving.
- The building blocks of the CNN architecture consist of convolution layers, pooling layers, and fully connected layers as well as rectified linear unit (ReLU). These layers are constructed to enable CNN to learn spatial hierarchies of features.
- CNN is different from conventional machine learning in that CNN requires large amounts of data for model training; CNN, on the other hand, neither requires manual (human-derived) feature extraction nor image segmentation.
- In short, CNN is good for spatial data while RNN is designed for sequential or temporal data. There is, however, a hybrid "CNN-RNN" model (also called recurrent CNN, or RCNN but not to be confused with regional CNN, R-CNN) that has some potential in biomedical data, such as multilabel image classification and serial complex biomedical data.
- The evaluation of a model with a test set of data (that the model has not seen before) can follow two methods: cross-validation or holdout method.
- For a binary classification model the performance is measured by the confusion matrix, the area under the curve (AUC) in a receiver operating characteristic curve, and AUC in a precision−recall curve (PRC).
- The F1-score, also called F-measure or balanced F-score, is probably the least familiar to most readers; it is the harmonic mean between precision and recall and can be used to assess binary or multiclass classification models for accuracy.
- In short, accuracy and error rate are not good indicators of performance especially when the incidence of disease is very low (like for cancer) because the true negatives, a majority of the cases usually in low incidence disease states, is a relatively high number to make accuracy and error rate look more favorable. This is a good example of how precision and the F_1 score will be more realistic reflection of the classification model prediction performance, especially when there is an imbalance in the classes as in the case of true negatives being a very large number. When the true negative is a large number, one can also consider the PRC.
- While there is a difference between interpretability and explainability, AI methods need at least interpretability for it to be widely adopted by clinicians. Interpretability is the capability to observe a cause-and-effect, while explainability is the understanding of the inner workings of a system or technology.
- Some of the higher prediction accuracy machine-learning methodologies (deep learning, random forest, support vector machines, etc.) have the least explainability, whereas

others (Bayesian belief nets, decision tress) have more explainability (but relatively lower prediction accuracy).

- The ideal model will have both low bias as well as low variance, but it is usually a trade-off between these two parameters. Bias and variance are manifested in models that demonstrate underfitting and overfitting: the former is a result of high bias and the latter is a result of high variance; therefore the best balance, fitting, is achieved when there is low bias and low variance.
- Underfitting occurs when the model is too simplistic: it is poorly trained on sample data (such as a linear model) or when the feature engineering is suboptimal and/or inadequate. The solution will involve a more complex model and a better feature engineering strategy. Overfitting occurs when the model is too complex: the results are too tailored to the training data (excessive training or adaptation) so that the model is overly complex for the data (opposite of above situation with underfitting).
- The curse of dimensionality occurs when there is such a large number of features so that the features space is excessively large so that there are not enough samples to fill this space. In general, as the features increase, the amount of data needed to generalize accurately grows exponentially. As the number of features (called dimensionality) increases, the classifier's performance increases until the optimal number of features is reached; this is called the Hughes phenomenon.
- A common misconception is that correlation implies that there is causation; events that are correlated, therefore, do not signify that one caused the other.
- Traditional machine learning, compared to deep learning, is relatively easy to train and test, but its performance is dependent upon its features and is limited with increasing volume of data.
- On the other hand, deep learning performance can continue to incrementally improve with increasing data. While deep learning can learn high-level features representation, it does require large amounts of data for training and can be expensive from a computation usage perspective.
- The cognitive computing framework leverages a portfolio of methodologies such as machine learning, pattern recognition, and NLP, as well as other AI tools to mimic the human brain and its self-learning capability. Cognitive computing is a symbiotic convergence of humans (user and expert) and smart technology.
- NLP is defined as the computer to understand spoken as well as written human language through specific set of techniques. In short, NLP is the intersection of AI, computer science, and linguistics and is a good example of human—computer interaction.
- A robot is defined as a reprogrammable and multifunctional manipulator designed to move material, parts, or specialized devices through variable programmed motions for the performance of a variety of tasks.
- Traditional regulatory processes are quickly becoming woefully inadequate (and perhaps inappropriate) to be used as a strategy to have oversight over ultrafast-evolving AI software that can change in real-time in a matter of seconds.

Ten Questions to Assess your Data Science and Artificial Intelligence Knowledge

These are 10 short questions to test your overall data, data science, and AI general knowledge in medicine (answers in compendium section):

	TOPIC	Score	Notes
		Two points per correct answer (10 questions $\times$ 5 answers for total of 100 points)	
AI	Timing	Name reasons why AI is so relevant now for medicine and healthcare: — — — — —	
Data	Big Data	List the "V"s of big data: — — — —	
AI	Components	What are components of AI? — — — — —	
AI	Types	Name different types of AI:—————	
Data	Conundrum	List five issues with health-care data: — — — —	
Data science	Supervised learning	List five common supervised learning methodologies: — — — —	
AI	Neural networks	Determine if the following are correct in methodology and use: CNN–MRI interpretation RNN–ICU data analytics Machine learning–patient clustering Cognitive computing–EKG analysis Natural language processing–HER	
AI	Subspecialties	Name five subspecialties that are most active in the area of AI applications: — —	

(Continued)

II. Data science and artificial intelligence in the current era

(Continued)

	TOPIC	Score	Notes
		−	
		−	
		−	
AI	Obstacles	What are the five main obstacles for AI adoption?	
		−	
		−	
		−	
		−	
AI	History	Name five significant events in history of AI:	
		−	
		−	
		−	
		−	
AI	History	(BONUS) Name the AI methodology associated with the event: MYCIN: Deep Blue and Chess: Jeopardy: Go Match: ImageNet:	

AI, Artificial intelligence.

Ten Steps to Become More Knowledgeable in Artificial Intelligence in Medicine

These are 10 ways one can start to learn and know data science and AI in medicine and healthcare (not in any particular order so one can pursue these in parallel or in series):

Review data/database and statistics. Much of the data science in medicine is biostatistics propriate to be used as a strategy to have oversight over ultrafast-evolving AI software that can change in real-t. particularly as it relates to healthcare and statistics (regression, confusion matrix, sensitivity, and specificity, etc.) is essential for a better appreciation and understanding of data science and AI.

Become familiar with health informatics. Another part of the foundational layers (data-information-knowledge) necessary to have for data science and AI in healthcare and medicine is information (informatics). Informatics, like the aforementioned data, is a key element in the full understanding of data science and AI in medicine. Much of the time spent on AI projects is spent on data and informatics so the ability to navigate in these two domains is helpful and productive.

Identify data science and AI educational resources. There is a comprehensive educational compendium that consists of books and textbooks, journals, articles, and websites at the end of this book. There are many helpful video clips on the Internet that focus on the many topics that were discussed in this book. There are also ongoing publications and blogs on this topic that can be helpful.

Attend a meeting on data science and AI in medicine or healthcare. It is important to start a personal educational strategy for yourself (as well as the members of the division or department who may be interested as well). There are several meetings that focus on AI in healthcare and medicine; for those of you who are not yet fully educated in data science or AI, one caveat is that some of these meetings are very heavy on the data science and you may find some or most of the talks at these meetings too esoteric. Many of the ML/DL and AI papers on subspecialties are not necessarily published in the medical journals but an aggressive search strategy can enable one to find good papers (see list in back of the book).

Meet and get to know a data scientist. It is good to meet a data scientist from the community or at a meeting to simply understand what they do and how they do their projects. If there is good interpersonal dynamic, one can invite each other to their respective domains for a visit. It is usually a valuable experience for a clinician to visit a data science department just as it is usually a meaningful experience for a data scientist to spend time in a clinic or hospital setting with a clinician.

Seek hands-on data science experiences. The next step would be to spend some significant time with a data scientist and see their programming and analyzing skills at work. One can consider taking an online course (MOOCs such as Coursera, edX, Udemy, and Khan Academy) on data science and/or programming (such as R or Python). Another option is to invite a data scientist to do a workshop on computer programming at a frequency that is workable for both you and the data scientist. If you are willing and have the resources and time, a degree in data science at the Masters or PhD level is especially helpful; even more important than the education and experience is the network you form during these years in the program.

Recruit data science support. The initial effort to collaborate with a data scientist does not have to be hiring a full time data scientist but rather some data science resource either part-time or virtual. Often the local colleges and universities have good talent in data science and usually some of these students are very enthusiastic to be involved in healthcare. Occasionally there is even a premed student who is savvy with programming and computer science.

Start with a small data science project. With the support of a data scientist, one can start a very small project with a small patient population with more straightforward machine learning for analytics to learn about data mining and analytics. One good strategy is to work with available data sources such as MIMIC-III, which is a publicly available ICU database that one can do analytics with. An additional resource is the NIH Big Data to Knowledge (BD2K) initiative which was launched to support research and development of tools for integrating data science into biomedical research.

Select a clinical project. Once you have gained sufficient experience with a small project, one can go bigger with a clinical project with a bigger scope and higher complexity. It is also a good idea to collaborate with a center that already has an active data science program with ongoing projects; these centers usually can use more data that you may have access to so it is a "win—win." The experience and insight one can gain from this arrangement can be more than you expect.

Build a coalition of AI enthusiasts. It is important to start gathering all those interested in the AI domain in your geographical vicinity and have a diversity of leaders to include administration as well a physician and nursing leadership. This network of networks effect can be very productive and very often meaningful relationships and even projects can promulgate from these monthly or periodic gatherings.

II. Data science and artificial intelligence in the current era

References

[1] Chen Y, Argentinis E, Weber G. IBM Watson: how cognitive computing can be applied to big data challenges in life and science research. Clin Ther 2016;38(4):688–701.

[2] Ferrucci D, Brown E, Chu-Carroll J, et al. Building Watson: an overview of the DeepQA Project. AI Mag 2010;31(3):59–79.

[3] Noor AK. Potential of cognitive computing and cognitive systems. Open Eng 2015;5:75–88.

[4] Chen Y, Argentinis E, Weber G. IBM Watson: how cognitive computing can be applied to big data challenges in life sciences research. Clin Ther 2016;38(4):688–701.

[5] Pan H, Tao J, Qian M, et al. Concordance assessment of Watson for oncology in breast cancer chemotherapy: first China experience. Transl Cancer Res 2019;8(2):389–401.

[6] Cai T, Giannopoulos AA, Yu S, et al. Natural language processing technologies in radiology research and clinical applications. RadioGraphics 2016;36:176–91.

[7] Chen L, Song L, Shao Y, et al. Using natural language processing to extract clinically useful information from Chinese electronic medical records. Int J Med Inform 2019;124:6–12.

[8] Nwosu AC, Sturgeon B, McGlinchey T, et al. Robotic technology for palliative and supportive care: strengths, weaknesses, opportunities, and threats. Palliat Med 2019; [Epub ahead of print].

[9] European Group on Ethics in Science and Technologies. Statement on artificial intelligence, robotics, and autonomous systems. March, 2018.

[10] Balasingam M. Drones in medicine—the rise of the machines. Int J Clin Pract 2017;71(9) [Epub].

[11] Mamoshina P, Ojomoko L, Yanovich Y, et al. Converging blockchain and next-generation artificial intelligence technologies to decentralize and accelerate biomedical research and health care. Oncotarget 2018;9 (5):5665–90.

[12] Boulos MNK, Wilson JT, Clauson KA. Geospatial blockchain: promises, challenges, and scenarios in health and health care. Int J Health Geogr 2018;17:25.

[13] Griebel L, Prokosch HU, Kopcke F, et al. A scoping review of cloud computing in health care. BMC Med Inf Decis Mak 2015;15:17.

[14] Regota N, et al. Storing and using health data in a virtual private cloud. J Med Internet Res 2013;15(3):e63.

[15] Kaur PD, Chana I. Cloud-based intelligent system for delivering health care as a service. Comput Methods Prog Biomed 2014;113(1):346–59.

[16] Columbus L. 83% of health care organizations are using cloud-based apps today. Technology 7/17/2014.

[17] Kaur PD, et al. Cloud-based intelligent system for delivering health care as a service. Comput Methods Prog Biomed 2014;113(2014):346–59.

[18] Yao Q, et al. Cloud-based hospital information system as a service for grassroots health care institutions. J Med Syst 2014;38(9):104–12.

[19] Tang H, Jiang X, Wang X, et al. Protecting genomic data analytics in the cloud: state of the art and opportunities. BMC Med Genomics 2016;9(1):63.

[20] Shatil AS, Younas S, Pourreza H, et al. Heads in the cloud: a primer on neuroimaging applications of performance computing. Magn Reson Insights 2016;8(Suppl. 1):69–80.

[21] Your cloud in health care by VMware. <http://www.vmware.com/files/pdf/VMware-Your-Cloud-in-Healthcare-Industry-Brief.pdf>.

[22] Nagaty KA. Mobile health care on a secured hybrid cloud. J Sel Areas Health Inform 2014;4(2):1–6.

[23] Kruse CS, Frederick B, Jacobson T, et al. Cybersecurity in health care: a systematic review of modern threats and trends. Technol Health Care 2017;25:1–10.

[24] Hao K. This is how AI bias really happens- and why it's so hard to fix. MIT Rev February 4, 2019.

[25] Chou Jm Murillo O, Ibars R. How to recognize exclusion in AI. Medium September 26, 2017.

[26] Eubanks V. Automating inequality: how high tech tools profile, police, and punish the poor. New York: St. Martin's Press; 2017.

[27] Bossmann J. Top 9 ethical issues in artificial intelligence. World Economic Forum October 21, 2016.

[28] <futureoflife.org>.

[29] Torresen J. A review of future and ethical perspectives of robotics and AI. Front Robot AI January 15, 2018.

[30] Stuart Russell, TED Talk, May 15, 2017.

The Current Era of Artificial Intelligence in Medicine

The future is already here. It is just not evenly distributed. *William Gibson, American Canadian author*

There is currently an escalating interest and enthusiasm for artificial intelligence (AI) in medicine and health care, particularly in a few areas such as medical imaging and decision support; the origin of this surge in accommodation of AI in medicine has been mostly the rapid proliferation as well as adoption of machine and deep learning in many areas. In the past 5–7 years, there have been more available resources for AI in medicine and health care than ever before. In addition to the proliferation of AI tools and availability of educated personnel, venture capital for AI in medicine has also escalated into high levels and is now more than 10% of all AI-related deals. Of the various areas of AI applications in health care, the most active sectors in terms of number of deals made in AI in medicine are imaging and diagnostics, followed by insights and risk analytics. Other areas that are active include lifestyle management and monitoring, emergency room and hospital management, wearables, and mental health projects. In addition, based on institutional capability maturity as well as outcomes, certain health-care enterprises are particularly strong in their AI strategies and execution: Aetna, Anthem, Cigna, Humana, Mayo Clinic, and Sutter Health [1].

Health care and medical applications of deep learning is particularly robust but still in the relatively nascent stages. Deep learning and its full capabilities will be increasingly applied to the escalating data available in health care and medicine as this methodology can be applied to a wide range of medical data, from medical images to genomic data (such as DNA sequence and regulatory features) [2] as well as phenotypic expressions of disease (such as clinical measurements, biomarkers, imaging data, and disease subtypes) [3]. A prime example of the use of deep learning and clinician–data scientist collaboration is the recent National Health Services (NHS) and Google DeepMind collaboration for the detection of eye disease in over a million patients (see next section on Applications). Another notable example is the skin cancer work by the Stanford group that was published in *Nature* (see below). Finally, Dudley's group elegantly described the "deep patient" concept in which a patient's electronic medical records are used to predict the patient's future disease profile [4]. Overall, prediction from deep learning still may not meet expectations [5].

The power of collaboration—building infrastructure toward artificial intelligence (AI)—ready healthcare

David Cox[1] and Dean Mohamedally[2,3]

[1]*Neonatal Grid Trainee, London, United Kingdom; Clinical Fellow for Artificial Intelligence work stream of the Topol Review*
[2]*NHS, London, United Kingdom*
[3]*Department of Computer Science, University College London, London, United Kingdom*

David Cox, an advocate for AI in health care and his colleague Dean Mohamedally, authored this commentary on the Industry Exchange Network, a successful UK-wide initiative to promote collaboration between academia and industry as well as with health care providers. Cox was the clinical fellow for the AI and Robotics workstream of the Topol Review.

During discussions that contributed to "The Topol Review" [1], the UK National Health Service's recent review of how to prepare a health-care workforce for the digital future, Professor Neil Lawrence (Director, IPC Machine Learning at Amazon; Founder of Amazon Research Cambridge) commented, "The history of automation has been that the humans have had to adapt to the machines, the promise of AI is that the machines should adapt to us."

This promise is exciting but is likely to require fundamental examination of, and changes to, relationships between health-care providers, academic institutions, and industry [2]. For too long, health-care systems worldwide have suffered from a system-wide lack of education and expertise regarding health-care technologies, an IT infrastructure often not fit for purpose, and procurement of technological solutions from industry that have not addressed local patient or clinician needs.

To suggest that there is a simple, universal solution—a "magic bullet"—is naïve. However, some of the more progressive health systems are reviewing and redefining their collaborations, both with local and global universities and technology companies in an attempt to realize long-term relationships of mutual benefit. To exhaustively document the range of initiatives occurring globally would double the length of this book and would probably be outdated at the point of publication. Instead, this chapter will document a blueprint being explored in the United Kingdom's NHS; it is a framework that is replicable in most health systems looking to harness the potential of AI, and one that in my opinion holds more promise than other models being explored.

The premise of the "Industry Exchange Network (IXN) for the NHS," which was established as a UK-wide initiative in early 2019 [3], is that the development of user-centered, patient- and clinician-driven technology should be paired with the evolution of a diverse, modern workforce with complimentary skillsets; this workforce should be strategically designed for a health system to get the best out of academic and industry collaborations.

Machine learning and AI have a core focus in this generations' computer science education syllabus. The concept of the IXN was pioneered by the Department of Computer Science at University College London (UCL) where it has been in development and operation for 8 years [4]. The UCL IXN is one of the world's largest taught problem-based learning initiatives for computer science engagement. Each year over 500 undergraduate and Master's level students work on an array of real-world computer science problems supported by over 300 industry partners, including Microsoft, IBM, NTT Data, and ARM. The students' outputs are proof-of-concept technology deliverables, which can be progressed if considered likely to be of true benefit.

Since the creation of the UK Government-backed Apperta Foundation CIC (which supports the NHS with open source technologies), UCL Computer Science students in the IXN model have performed increasing numbers of healthcare—related projects that address local clinical or health system problems. Last year over one-third of the IXN projects were healthcare related. NHS mentors and industry partner clients work collaboratively to generate suitable, and relevant, projects; each project (undertaken by a group of students) is provided with a named technical mentor from industry, an academic supervisor, and a clinician mentor. All incoming projects are designed and selected based on their interoperability, efficiency, or innovation merits. The UCL IXN initiative is now the largest FHIR (Fast Healthcare Interoperability Resources) and OpenEHR (Open Electronic Health Record) training environment in the world. An ethos of the IXN is that students must publish their projects to allow external visibility.

The UCL IXN is complemented by the state-of-the-art unit at Great Ormond Street Hospital (GOSH), called DRIVE—Digital Research, Informatics and Virtual Environments—which is the first of its kind in the world. Built through partnership between GOSH, UCL, and leading industry experts in technology, AI, and digital innovation, DRIVE is a unique informatics hub created to harness the power of the latest technologies to revolutionize clinical practice and enhance the patient experience [5]. DRIVE is both a physical and conceptual unit; such units are essential to support the IXN model—providing dedicated physical space and a common culture and governance that brings clinicians, academics, and industry together to develop and test their new technologies in a safe, nonclinical arena.

Student projects developed through the UCL IXN have been significant contributors to NHS technologies, ranging from mental health chatbots, to a web-based treatment and management system for hepatitis C, and machine learning readiness tasks for clinicians. All software developed is open sourced and held at Apperta Foundation's government-backed GITHUB repository. The default license for open source is AGPLv3, meaning that any further work on the code base by anyone must be shared back to everyone; the distribution clause requires that if the open source code is embedded into proprietary code, the proprietary code must then be released as open source. The underlying intention is to prevent people and organizations "taking but not giving"; however, this does not preclude an AGPLv3 application running on a proprietary platform, for example, Azure; many of the IXN projects are based upon AGPLv3 software running on proprietary platforms.

Following the success of the UCL Computer Science IXN model in rapidly producing and testing new health-care technologies, it has been proposed that many other UK universities can also contribute with their Maths, Physics, Engineering, Computer Science, and IT classes working on projects for the NHS. The scaling of operations to manage the pipeline, and to determine the priorities of projects at a UK level, is reviewed by an independent panel of doctors.

UCL's open methodology has also been demonstrated internationally. In the United States, several universities have begun developing Capstone Projects classes for their final year computer science students. While the Capstone Projects mirror the core collaborative components for any individual IXN project, the UCL IXN methodology adds additional value by applying real-world requirements to all year groups of study, identifying small to large projects.

The potential benefits of an IXN model reach far beyond the creation of new AI and digital technologies for health care. These models encourage open collaboration between health systems/providers, academia, and industry, which strengthens relationships. Clinicians gain better understanding of the potential for technology in health care, learn the rudiments of different health-care technologies, and are encouraged to codesign and develop technologies to improve their local or national health system and the care they provide. The IXN model also allows individuals with complementary skills—computer scientists, engineers, mathematicians—to experience working with clinicians and health systems at an early stage of their career, with the intention of engaging them to work with health-care providers in a more symbiotic way in the future.

While scaling of the IXN model holds great potential, its greatest impact will be realized when allied with the curation of modernized career pathways for clinicians and scientists, creating approved, and valued, flexible careers that allow individuals to share experience and knowledge between health-care providers, academia, and industry. The IXN model will start to build relationships upon which future career pathways can develop.

The future of health care is undoubtedly one augmented by technology, but one which will need to be powered by a diverse, modern workforce. Delivering wide-scale change through health-care technologies is a wicked problem, but the principles and infrastructure upon which the IXN concept is based are replicable, and could hold the key to successful long-term collaborations between academia, industry, and health-care providers. The result will be patient, clinician, and system benefit on local, national, and global scales.

References

[1] Topol E. The Topol review: preparing the healthcare workforce to deliver the digital future, <https://topol.hee.nhs.uk/>; 2019.
[2] Rimmer A. Technology will improve doctors' relationships with patients, says Topol review. BMJ 2019;364:l661.
[3] Cox DJ, Mohamedally D. A UK-wide IXN for the NHS—a new solution for a digital NHS. BMJ Opinion 2019. <https://blogs.bmj.com/bmj/2019/03/20/a-uk-wide-ixn-for-the-nhs-a-new-solution-for-a-digital-nhs/> [accessed 21.03.19].
[4] Microsoft and Industry Exchange Network Projects. A white paper from Microsoft in association with UCL Computer Science, <www.cs.ucl.ac.uk/staff/D.Mohamedally/msixn.pdf>; 2017 [accessed 21.03.19].
[5] Crouch H. GOSH looks to DRIVE healthcare technology forward. *digitalhealth.net* October 11, 2018. <https://www.digitalhealth.net/2018/10/gosh-drive-healthcare-technology/> [accessed 21.03.19].

Overall, the fastest growing trends in deep learning and AI use in medicine and health care are observed in the following areas: (1) medical imaging: the use of GPU-accelerated deep learning and computer vision with classification, detection, and segmentation can automate analysis of a myriad of medical images such as electrocardiograms, CT, MRI, X-rays, and even moving images such as echocardiograms and angiograms [6]; (2) decision support: deep learning techniques can be used to collate and analyze the heterogeneous electronic health record data pools such as doctors notes, laboratory data, drug information, and medical images in order to facilitate diagnosis and therapy as well as complex event prediction; and (3) precision medicine: the genomic sequencing data and phenotypic expression information can be coupled to devise an individualized diagnosis and therapy for patients as well as for the overall population. In addition, natural language processing in medicine and health care is becoming increasingly important with an escalating number of publications as we delve into getting unstructured data better organized and complete [7,8]. Lastly, AI can now also be applied to other domains such as drug discovery, blockchain security, altered reality (augmented, virtual, and mixed reality), and even administrative tasks and workflow with robotic process automation.

AI in medicine and health care at the international level is rapidly proliferating as well. Other than the United States, this interest in AI for medical applications is also at a high level in certain countries and regions around the world such as Canada, China, and parts of Asia, United Kingdom and the European Union, Israel, and Australia. The AI in medicine and health-care domain is particularly interesting domain as it is the intersection of each country's health delivery system and AI ecosystem. Each country has each own strength(s): the United States and Canada for AI talent and academic productivity as well as resources for startups, the United Kingdom for also its academic reputation, Israel for medical image work, and China for its work on facial recognition as well as access to its vast amount of health-care data and growing talent pool of AI in medicine.

Artificial intelligence (AI) in medicine in China, a quick overview

Linhua Tan[1] and Yu Uny Cao[2]
[1]The Children's Hospital of Zhejiang University School of Medicine, Hangzhou, China
[2]Zhejiang Intellectual Property Exchange Center, Hangzhou, China

Linhua Tan and Yu Uny Cao, program chairs for AIMed China this past year, authored this commentary on the current state of AI in medicine in China with its exponentially increasing AI capabilities as well as its huge need for innovative health care solutions.

Background

With one-fifth of the world's population and a fast expanding economy, "the principal contradiction facing Chinese society in the new era is that between unbalanced and inadequate development and the people's ever-growing needs for a better life." [1]. Among the needs is better medical and health care.

In 2016 the State Council issued the "Healthy China 2030" Planning Outline, which is both a guiding document and China's commitment the United Nations 2030 Agenda for Sustainable Development. And in 2018 legislations were proposed for the Healthy China Initiative. In addition, 2018 saw the establishment of the International Development Cooperation Agency (IDCA), which manages China's global health engagement when China shifts from being a recipient country to a donor one.

Key difficulties

As observed by many, a key difficulty in China's health-care system is a demand and supply problem with a Chinese tint. On the supply side, it is a shortage of doctors, low salaries for doctors, a lack of local clinics, a historical neglect of family doctors; and on the demand side, it is patients often crowding into large, specialist hospitals in major cities for ailments both severe and minor [2,3].

The size of the problem is enormous. By 2020 China's health-care spending is expected to reach 1 trillion US dollars equivalent, and double that by 2030 as stated in the "Healthy China 2030" Planning Outline, and the high growth occurs in the backdrop of the health sector accounting to less than 7% of the GDP, far below figures of developed nations.

The innovative China

To solve the contradiction of unbalanced and inadequate development and the need for better medical and health care among other needs, the country has been implementing the innovation-driven development strategy, and the emerging innovative China is for all to see: in the WIPO/Cornell Global Innovation Index 2018, the big news was that China joined the world's top 20 most innovative economies for the first time and provided an example for other middle-income economies. And the rise is a result of China's sustained investment in science and technology in the recent decades; as reported by the United Nations, in 2017 China's R&D investments were 452 billion PPP dollars, close to the US number of 511 billion PPP dollars, and China has 1.69 million R&D personnel (researchers, technicians, and support staff engaged in R&D), which is the most in the world, followed by the United States at 1.38 million [4].

Therefore it is no surprise that innovation plays an important role in delivering better medical and health.

Among the over 100 awards of the State Science and Technology Prizes announced in early 2018, the health sector scored more wins than any other discipline in the Science and Technology Progress section.

And university technology transfer has been encouraged by legislators, regulators, and university administrators. For example, West China Hospital of Sichuan University, a top-ranked hospital in China, has instituted aggressive measures in commercializing medical research and in the last 5 years has licensed more than 60 new drugs attracting more than 10 billion RMB (1.5 billion US dollars equivalent) investment. The Hospital's measures have been taken up by the National Health Commission with the intention of promulgating them nationwide.

Internet and medicine in China

A quarter century ago China entered the Internet era, and since then China has developed a vibrant e-commerce sector that is a pride of the nation, and in 2015 the "Internet Plus" model was elevated to a national strategy.

In medicine the Internet Plus model first helped to solve the "registration" step, with which patients take a number before seeing doctors, at hospitals. The pioneering company, We Doctor, grew into a large conglomerate in a short few years, and points the way for the Internet + medicine model in solving problems in shortage of doctors, lack of family doctors, among others.

Another Internet + medicine company, DXY, focuses on doctors and has created one of China's biggest online networks for the professional growth of millions of medical doctors. DXY has since branched into online consultation for patients with chronic diseases.

Artificial intelligence and medicine in China

After the wave of the Internet + model came AI. Lee, a world-famous AI pioneer and investor wrote in his 2018 book [5], *AI Superpowers*, that today the United States and China are the forerunners in AI, and he predicts that in 5 years China will either catch up or overtake the United States in a few domains of AI applications, due to China's large population, vibrant startup culture, numerous engineers, large amounts of data, national policies, and other favorable factors.

In 2016 the National Health Commission selected four cities as pilot sites for big datasets in health and medicine, including regional health data, administrative data, public health services data, birth and death registries, and electronic medical records [6]. The initiative has trickled down to large technology companies who are tasked with new data center technologies, computer servers and storages, and data analytics with AI.

In recent years, there has emerged dozens of well-known AI technology companies, some of which have achieved the unicorn status (a unicorn company is a private company with valuation exceeding 1 billion US dollars). Many of these AI companies have worked closely with leading hospitals around country. There have been enough progress that the China Food and Drug Administration has incorporated AI diagnostic tools into its list of permitted medical devices.

In early 2019 Ping An Good Doctor, China's leading one-stop health-care platform, released its 1-minute Clinics across eight provinces and cities in China and signed service contracts for nearly 1000 units, providing health-care services to more than 3 million users, who can use the vending-machine-like One-minute Clinics for online consultation backed by Ping An's "AI Doctor" technology.

Alibaba has been testing AI ambulances in the city of Hangzhou, where the City Brain project, also by Alibaba, has laid foundation for smart traffic control.

And last but not least, Traditional Chinese Medicine (TCM) has received increasing attention from technologists and entrepreneurs. 2017 say the first AI-based TCM clinic opened, where a medical doctor works with a system can deliver a diagnosis write a prescription its huge database of knowledge from TCM masters from both ancient and modern times.

References

[1] China Daily. The 19th CPC National Congress [bi-lingual guide]. China Daily 2017 October 19. Available from <http://language.chinadaily.com.cn/19thcpcnationalcongress/2017-10/19/content_33442254.html>.

[2] Reuters. AI ambulances and robot doctors: China seeks digital salve to ease hospital strain. 2018. Available from <https://www..com/article/us-china-healthcare-tech/ai-ambulances-and-robot-doctors-china-seeks-digital-salve-to-ease-hospital-strain-idUSKBN1JO1VB>.

[3] The New York Times. China's Health Care Crisis: Lines before Dawn, Violence and 'No Trust'. The New York Times 2018 September 30. Available from <https://www.nytimes.com/2018/09/30/business/china-health-care-doctors.html>.

[4] The UNESCO Institute for Statistics. R&D data release, <http://uis.unesco.org/en/news/rd-data-release>; 2018 [accessed 31.01.19].

[5] Lee KF. AI superpowers: China, Silicon Valley, and the New World Order. Houghton Mifflin; 2018.

[6] Zhang L, Wang H, Li Q et al. Big Data and Medical Research in China. BMJ 2018; 360:j5910.

The National Health Service (NHS) and its Adoption of Artificial Intelligence (AI)

Christopher Kelly[1,2] and Tony Young[1,3,4]
[1]NHS England, London, United Kingdom
[2]DeepMind, London, United Kingdom
[3]Southend University Hospital, Southendn-ea, United Kingdom
[4]Anglia Ruskin University, Cambridge, United Kingdom

Christopher Kelly and Tony Young, both physicians in the clinical entrepreneur program of the NHS with Tony being the clinical lead, authored this commentary on the current state of AI adoption at the NHS and its future role in the international coalition of AI in health care leaders.

The United Kingdom's NHS is the world's largest and longest established universal healthcare system. Throughout its 70 year history it has been at the forefront of many medical advances ranging from intraocular lens implants and total hip replacements to the development of CT and MRI technology. The NHS is uniquely positioned to be a world leader in the development, evaluation, and successful deployment of a range of new, emerging technologies, including AI.

In this article, we summarize the potential for AI in the NHS, explore its current state and challenges, review recent UK initiatives to support AI research and development, and summarize some steps that can translate the use of this potentially transformative technology into safe clinical practice.

Can artificial intelligence help the National Health Service address the challenges it faces?

Since the NHS was founded in 1948, health care and society have changed dramatically. We now have an aging population, often with multiple complex long-term health conditions and a range of lifestyle-related diseases contributed to by excess alcohol consumption, smoking, poor diet, and lack of exercise [1]. UK health spending has increased from 3.5% of GDP in 1948 to 9.8% in 2016 [2].

The NHS is not unique in facing these challenges. Many countries are grappling with the same issues and struggling to deliver the quadruple aim [3] of improving clinical outcomes, enhancing patient experience and the work life of health-care providers, while at the same time reducing costs. AI has the potential to enhance our ability to deliver on the quadruple aim and unlock the benefits of personalized prediction/prevention, earlier detection, and treatment optimization. By utilizing the vast datasets now available—including our genome, exposome, health-care records, and other digital information collected from our daily lives—AI has the potential to make this vision a reality.

With early detection, AI offers the opportunity to start treatment sooner, which in turn could reduce morbidity, mortality, and complications. The ability to provide specialist-level diagnostics in the community through AI systems could also allow health care to be delivered closer to the patient, reducing costs and improving patient experience. Improved operational performance and productivity through better patient triage, speed to diagnosis, and improved scheduling could free up clinicians giving them back the time for direct patient care. New AI technologies that enable the monitoring of disease via self-care, wearables, and identification of new disease signals also have the potential to enable new paradigms of care that will be transformational for the NHS.

The National Health Service is positioned as a world leader in artificial intelligence research, development and deployment

The NHS is the single largest health-care organization in the world, employing 1.3 million staff and seeing over 1 million patients every 24 hours [4], while generating vast amounts of data that are potentially suitable to develop world-leading algorithms for health care.

Furthermore, the United Kingdom is home to a number of internationally recognized academic institutions with access to significant government and charity research funding, a successful life sciences industry, an advanced biopharmaceutical infrastructure, and a rich network of venture funding institutions that focus on health-care technologies. The Alan Turing Institute, for example, was founded as a national institute for data science and AI in 2015 to support this thriving ecosystem.

When the NHS partners with research institutions and technology companies, there is great potential to accelerate progress in health care. For example, Microsoft has partnered with GOSH and Addenbrooke's Hospital to collaborate on the development of AI tools. DeepMind has built partnerships with Moorfields Eye Hospital, Imperial College London, and University College London Hospital to develop novel algorithms in health care. Amazon has partnered with the NHS to enable Alexa to offer health advice. Startups are also embracing the NHS, including Sensyne Health, which has developed exclusive partnerships with a number of UK hospital trusts to develop clinical AI technology, and Kheiron Medical, which is trialing a system to read mammograms in Leeds and the East Midlands, supported by NHS England's Test Bed Programme.

Government strategy to accelerate the United Kingdom artificial intelligence industry

The UK Government has recognized the importance of AI to the NHS and has outlined a number of high profile strategies to drive the industry forward. The Industrial Strategy

Challenge Fund comprises a £4.7 billion investment in R&D over 4 years "to meet the major industrial and societal challenges of our time" by bringing together the NHS, academia, the charitable sector, and industry [5]. One of the "Grand Challenge" missions is to use data, AI and innovation to transform the prevention, early diagnosis and treatment of chronic diseases, with £210 million available to industry and research. The first funded centers in Leeds, Oxford, Coventry, Glasgow, and London were announced in November 2018 and will focus on imaging, pathology, and diagnostics.

The NHS has also recognized the need to support the introduction of AI algorithms into the health-care system and has put together an initial "code of conduct for data-driven health and care technology" [6]. It aims to deepen trust between patients, clinicians, researchers, and innovators, while simplifying the regulatory and funding landscape, reviewing contracting and procurement arrangements, and encouraging the system to adopt clinically effective innovations at scale.

In January 2019 the NHS Long Term Plan sets out key ambitions for the next 10 years. Digital technology underpins many of the plan's most ambitious patient-facing targets, including using decision support and AI to "help clinicians in applying best practice, eliminate unwarranted variation across the whole pathway of care, and support patients in managing their health and condition."

The National Health Service can and will do more to realize the benefits of artificial intelligence

In order to optimize the NHS for AI health research that ultimately benefits patients, key areas for focus must include improving the NHS's data infrastructure, developing regulatory frameworks and clinical evaluation strategies, and supporting workforce education.

Although the singular name "NHS" might imply one entity, it is actually made up of hundreds of different organizations. Differing systems between organizations with variable levels of interoperability means that there is no single approach to bring the NHS dataset together. However, despite this challenge there are incredible examples in disease-specific areas where disparate datasets have been united to reveal new insights, for example, the national cancer registry at Public Health England [7]. A key challenge to applying machine learning to real-world datasets is curating high-quality labels at scale, but this can be overcome as exemplified by the team at Moorfields Eye Hospital and DeepMind with their work on optical coherence tomography [8].

If the NHS is to lead the realization of the potential benefits that AI can bring, procedures for accessing data for research purposes need to be streamlined. Initiatives, including the Local Health and Care Record Exemplars and Digital Innovation Hubs, are already working to improve this [9].

To give confidence to health-care providers to adopt innovations, work is already in progress to launch a trusted approval scheme for digital health products, allowing suppliers to demonstrate that their product complies with the code of conduct, giving assurance to health-care providers and commissioners [10].

Another important area where the NHS could facilitate AI research is the development of standardized test sets to benchmark future AI algorithms in a consistent and comparable manner. The NHS could curate these labeled datasets for key strategic disease areas (i.e., mammography screening, diabetic retinopathy screening, and lung nodule detection) in order to assure patients that all algorithms that are deployed meet certain quality standards and have been independently evaluated in a representative population.

Education and training of the health-care workforce in this rapidly evolving field is crucial. The NHS has led the way with the recent publication of the Topol Review [11] that sets out the vision and strategy for how we deliver the benefits of advanced digital technology, including AI, to patients. The NHS needs to consider further how we educate patients and the wider public about the benefits of embracing data science in health care.

Conclusion

The NHS has the potential to be an ideal environment to develop, evaluate, and deploy AI systems at scale to benefit patients. In order to fully realize this potential, it needs to further modernize its data infrastructure, interoperability of current and future NHS systems, foster an appropriate regulatory environment, and develop effective mechanisms to support the onward commercialization of these technologies.

References

[1] Khaw K-T, Wareham N, Bingham S, Welch A, Luben R, Day N. Combined impact of health behaviours and mortality in men and women: the EPIC-Norfolk prospective population study. PLoS Med 2008;5(1):e12.

[2] Office of National Statistics. UK Health Accounts: 2016 [Internet]. 2018 April. Available from: <https://www.ons.gov.uk/peoplepopulationandcommunity/healthandsocialcare/healthcaresystem/bulletins/ukhealthaccounts/2016>.

[3] Bodenheimer T, Sinsky C. From triple to quadruple aim: care of the patient requires care of the provider. Ann Fam Med 2014;12(6):573−6.

[4] NHS England. The NHS long term plan [Internet]. 2019 January. Available from: <https://www.longtermplan.nhs.uk/wp-content/uploads/2019/01/nhs-long-term-plan.pdf>.

[5] UK Research and Innovation. Industrial Strategy Challenge Fund: for research and innovation [Internet]. 2017 May. Available from: <https://www.gov.uk/government/collections/industrial-strategy-challenge-fund-joint-research-and-innovation#from-data-to-early-diagnosis-and-precision-medicine>.

[6] Department of Health & Social Care. Initial code of conduct for data-driven health and care technology [Internet]. 2018 Sep. Available from: <https://www.gov.uk/government/publications/code-of-conduct-for-data-driven-health-and-care-technology/initial-code-of-conduct-for-data-driven-health-and-care-technology>.

[7] National Cancer Registration and Analysis Service (NCRAS) [Internet]. Available from: <https://www.gov.uk/guidance/national-cancer-registration-and-analysis-service-ncras>.

[8] De Fauw J, Ledsam JR, Romera-Paredes B, Nikolov S, Tomasev N, Blackwell S, et al. Clinically applicable deep learning for diagnosis and referral in retinal disease. Nat Med 2018;24(9):1342−50.

[9] NHS England. Local health and care record exemplars [Internet]. Available from: <https://www.england.nhs.uk/publication/local-health-and-care-record-exemplars/>.

[10] Department of Health and Social Care. Code of conduct for data-driven health and care technology [Internet]. 2019 February. Available from: <https://www.gov.uk/government/publications/code-of-conduct-for-data-driven-health-and-care-technology>.

[11] NHS Health Education England. The Topol Review [Internet]. 2019 February. Available from: <https://topol.hee.nhs.uk/>.

References

[1] Data from the Everest Group, 2019. www2.everestgrp.com.

[2] Rusk N. Deep learning. Nat Methods 2016;13(1):35.

[3] Deo RC. Machine learning in medicine. Circulation 2015;132:1920–30.

[4] Miotto R, Li L, Kidd BA, et al. Deep patient: an unsupervised representation to predict the future of patients from the electronic health records. Nat Sci Rep 2016;6(26094):1–10.

[5] Chen J, Asch SM. Machine learning and prediction in medicine- beyond peak of inflated expectations. N Engl J Med 2017;376(26):2507–9.

[6] Greenspan H, van Ginneken B, Summers RM. Guest editorial/Deep learning in medical imaging: overview and future promise of an exciting new technique. IEEE Trans Med Imaging 2016;35(5):1153–9.

[7] Kreimeyer K, Foster M, Pandey A, et al. Natural language processing systems for capturing and standardizing unstructured clinical information: a systematic review. J Biomed Inform 2017;73:14–29.

[8] Friedman C, Rindflesch TC, Corn M. Natural language processing: state of the art and prospects for significant progress, a workshop sponsored by the National Library of Medicine. J Biomed Inform 2013;46:765–73.

Clinician Cognition and Artificial Intelligence in Medicine

The Rationale for Intelligence-based Medicine

The physicians in this era are facing the perfect storm: exponentially increasing medical information with a relatively flat trajectory for personal knowledge acquisition due to time constraints; more patients with higher degree of complexity of chronic diseases with increasingly more data volume of many diverse types in many separate locations; mounting pressures to produce work units with diminishing reimbursements and constant denials for procedures and tests; and high level of stress and burnout from the mounting burdens of electronic health record (EHR) and workload.

Informatics in the next decade: can artificial intelligence (AI) redefine a physician's interaction with the EMR?

William W. Feaster[1]
William W. Feaster, an anesthesiologist and informatician, authored this commentary on how AI can render the EHR much more valuable by extracting important knowledge from this resource and utilizing AI toward automated processes.
[1]*Chief Health Information Office, CHOC Children's Hospital, Orange, CA, United States*

In his 1995 book, *The Road Ahead*, Bill Gates [1] observed that "we always overestimate the change that will occur in the next 2 years and underestimate the change that will occur in the next 10." That seems particularly prescient when trying to define how artificial intelligence (AI) will reshape the clinician's interaction with the electronic medical record (EMR). Physicians are laboring under the current burden of clicks, data entry and documentation for meaningful use and compliant billing. They are dissatisfied with this administrative overhead and the time it takes from their patients. They are burning out in high percentages [2]. They are demanding relief from these burdens. Some have turned to scribe as a temporary solution and other organizations are looking to virtualize scribes as a more cost-effective solution using a tool such as Google Glass in the exam room.

Intelligence-Based Medicine
DOI: https://doi.org/10.1016/B978-0-12-816462-4.00007-8

Many companies including Google, Microsoft, and EMR vendors such as Cerner are outfitting exam rooms with cameras and microphones, recording multiplex conversations and interactions, and utilizing voice recognition and natural language processing (NLP) to both record the provider visit, populate data elements in the EMR, create a note, and control the EMR (place orders, etc.). These companies are demonstrating future products now, but to think that the widespread use of these types of technologies will occur in the next 2 years would be highly optimistic. They will be in lengthy pilot phases at organizations doing this cutting-edge work: cutting-edge at least by today's standards of keyboard and mouse computer interactions augmented by voice recognition software. But it's still the same old EMR. Kind of like putting lipstick on a pig, to use a current vernacular.

The EMR won't be replaced any time soon as it is a useful repository of patient data (though much is less accessible to analytics in text format) and helps organizations run their clinical services. Most will agree that the EMR improves the quality and safety of care. Gone are the scribbled and illegible notes and hand-written prescriptions and the medication errors that accompanied them. It makes no sense, however, that many of these benefits fall on the backs of the highly paid and scarce physician. Many of the routine functions that providers perform in the EMR are candidates for robotic process automation (RPA), and some of these tools are already available, just not widely distributed. For example, billing for a patient visit is often a manual process where a provider manually chooses a CPT (current procedural terminology) code based on the level of service provided or procedures performed. There are already advanced tools that can scan the provider's documentation via NLP and assign the proper code, but these tools are expensive. No individual should need to further intervene until the bill is analyzed by computerized claim scrubbers and submitted for payment. Only those claims failing the automated process will need to be touched by a human. This is available today in sophisticated systems.

There are many other aspects of the EMR and patient care on which providers and vendors are focusing AI-related technologies. These include security systems analyzing external threats, imaging systems that prioritize worklists based on predetermined image analysis or interpret the studies themselves (CXR for pneumonia), diagnostic systems that can interpret an eye exam and diagnose cancerous skin lesions, among others. A lot of these systems are available now, but adoption will be slow due to technology costs.

The most exciting area of applying AI technologies to the care of patients is through provider intelligence augmentation. By this, I refer to pairing the machine with the provider and patient, utilizing the power of machine intelligence to synthesize the full scope of medical knowledge and the full-spectrum of a patient's data from all sources from genomics to wearables and deliver to the provider and patient the best diagnosis and treatment recommendations for a given illness. Our first, more straightforward experience with this in the EMR is rule-based alerting. Challen et al. [3] describe the current rule-based decision support systems based on predefined rules that have predictable behavior and have been shown to improve quality. But predefined rules struggle when faced with complex data and diagnostic uncertainty, opening the way for machine-learning (ML) algorithms.

The true power of ML on big data will result from not only the analysis of data from one patient or even one institution but instead the sharing of data from multiple patients with a similar condition across multiple organizations. This is the ultimate big data challenge, and it will likely replace much of today's traditional medical research. The biggest block to doing this is the privacy protections under Health Insurance Portability and Accountability Act (HIPAA) as it is difficult to deidentify data for sharing. One emerging technology that may facilitate this data exchange is Blockchain. With Blockchain, data privacy can be maintained, and the data can be retrieved when needed while still permanently residing in a primary database. Once retrieved, these data can be mined for determining the best treatment for the best patient outcomes. The computer then becomes an even more useful intelligence assistant as in addition to current knowledge in the medical literature, the computer is now learning from thousands or even millions of patients with a similar condition. These technologies may seem further off but are mostly limited by the inability to overcome the privacy regulations surrounding sharing patient data.

AI initiatives won't be confined to care providers...they will also be patient facing. Another current government initiative is to give ownership and custody of data to the patient. If this in fact evolves, patient-facing AI applications will guide a patient through their care, present appropriate educational resources, and will them to when they need to see a doctor or other professional. Genomic testing is available to all via companies such as 23andme. These companies provide a detailed interpretation of the patient's genetic predisposition for health risks, though still a bit limited in scope. If given access to the patient's own genomic data, a pharmacogenomic analysis of the drugs they are taken could determine which drugs are appropriate for the patient and which are not, subsequently informing the clinician to prescribe more appropriate therapies. A potential benefit of patients "owning" their own data also means that the patient then controls the use of their medical data and can choose to share it without the restrictions of HIPAA. Perhaps that is the way we'll overcome these data-sharing obstacles.

All this sounds quite futuristic, but it is starting to happen now. We'll see more and more AI tools in our EMR workflow, and well within the next 10 years. The only problem for many is that it won't be evenly distributed!

References

[1] Gates W. The road ahead. Viking Penguin; 1995.
[2] Verghese A, Shah N, Harrington R. What this computer needs is a physician: humanism and artificial intelligence. JAMA 2018;319(1):19−20.
[3] Challen R, Denny J, Pitt M, et al. Artificial intelligence, bias and clinical safety. BMJ Qual Saf 2019;28:231−7.

There is a myriad of reasons that physicians in any subspecialty could benefit from the incorporation of AI into their practices.

First, AI with the clinician can become a powerful dyad (akin to Doctor Watson and Sherlock Holmes or Captain Kirk and Mr. Spock from *Star Trek*) with AI being a very capable second pair of eyes and an additional brain in a myriad of activities ranging from medical image interpretation to decision support. This partner adds perhaps additional dimensions as well as capabilities (such as endurance and objectivity) that can neutralize natural human frailties.

Physician acceptance of artificial intelligence (AI)

Kevin Maher[1,2,3]

Kevin Maher, a pediatric cardiologist with an ardent passion for data science, authored this commentary on the augmented intelligence for physician acceptance of AI as well as the future physicians who understand AI having a potential advantage over colleagues who do not.

[1]*Pediatrics, Emory University School of Medicine, Atlanta, GA, United States*

[2]*Cardiac Intensive Care, Children's Healthcare of Atlanta, Atlanta, GA, United States*

[3]*Pediatric Technology Center, Georgia Institute of Technology, Atlanta, GA, United States*

It has become increasingly clear that AI will have a major impact on the future of medicine and how healthcare is delivered. As the various applications of AI are introduced to the medical field, there will be variability in how quickly they will be accepted and adopted by physicians. The rate of adoption of AI will likely vary by the type of medicine practiced, geographic location, age and experiences of the individual practitioner, as well as legal, regulatory, and reimbursement aspects.

Change tends to be difficult, slow, and resisted by many, especially when the change is poorly understood or even feared. This is where we sit with AI, in part because the final integration of AI in medicine is unknown, even by the experts in the field. AI is rapidly advancing and novel applications are developed at a dizzying pace. If one thinks back 15 years, the electronic health record (EHR) was quite limited and paper charting still existed at many institutions across the country. When the US Government pushed the adoption of the EHR, dramatic increases in the use and application of the EHR occurred. Institutions without an EHR today would be well outside the accepted clinical practice norm. Will this occur with AI? I suspect that the answer is "yes," and change will occur.

Providing medical care is more complex and challenging today than ever before. Physicians feel limited in the amount of time they can spend with patients, are overwhelmed with information, and are frustrated with time spent on the EHR. With the "OMIC" revolution, wearable data, and continued addition of knowledge from medical research, tomorrow's healthcare providers will not come close to keeping up with the patient information presented to them. This is where AI meets clinical care and perhaps saves the clinician from the tidal wave of data that is useful, but not manageable.

For healthcare providers to accept and feel comfortable with AI, a certain level of understanding of AI needs to exist. Knowledge of how AI can support medical care is needed for the healthcare system, getting away from the "black box" mentality of something that is poorly understood. For physicians, AI education should begin in medical school, with all trainees having a thorough understanding of AI, informatics, and the future and current applications of this burgeoning field. The opportunity for advanced training such as a master's degree incorporated into the curriculum should be considered at medical schools, as future educators and leaders in the field will be in very high demand.

In the near future (and present is some cases), AI can be used in clinical medicine to augment decision-making, improve safety and quality, read radiographic and pathologic studies, provide early warning for changes in clinical status, and the list goes on [1–4]. What is lacking at the present time is a "gold standard" for AI algorithms. What AI programs are the best for each patient and disease? Is there an ideal algorithm for population x or y and how will this be determined? If a clinician makes a treatment plan using AI support and the plan is wrong and the patient suffers, who pays? The federal government in conjunction with AI physician experts will need to provide some level of oversight to regulate this field to enhance widespread implementation. A detailed assessment of many components of AI still needs to be reviewed, studied and presented for discussion. This entails the legal, regulatory and fiduciary aspects of AI and how it applies to medicine. Finally, what may drive acceptance and adoption of AI will be the recognition that AI is not expected to replace physicians, but rather physicians that use AI will replace physicians that don't [5].

References

[1] Poplin R, Varadarajan AV, Blumer K, et al. Prediction of cardiovascular risk factors from retinal fundus photographs via deep learning. Nat Biomed Eng 2018;2(3):158–64.

[2] Gulshan V, Peng L, Coram M, et al. Development and validation of a deep learning algorithm for detection of diabetic retinopathy in retinal fundus photographs. JAMA 2016;316(22):2402–10.

[3] Esteva A, Kuprel B, Novoa RA, et al. Dermatologist-level classification of skin cancer with deep neural networks. Nature 2017;542(7639):115–18.

[4] US Food and Drug Administration. FDA permits marketing of artificial intelligence-based device to detect certain diabetes-related eye problems. FDA news release, <https://www.fda.gov/newsevents/newsroom/pressannouncements/ucm604357.htm>; 2018.

[5] Obermeyer Z, Lee TH. Lost in thought—the limits of the human mind and the future of medicine. N Engl J Med 2017;377:1209–11.

Second, the amount of medical knowledge is exponentially increasing and doubling at a rate of a few months, and yet physicians do not have enough time to read and maintain their knowledge capacity. AI can be a useful up-to-date knowledge resource that can be even part of the EHR.

Third, AI can help to reduce the repetitive tasks and therefore inevitable burden that physicians do not enjoy performing (clinical fatigue), and this phenomenon is observed especially more senior clinicians as they become very seasoned in their clinical work. This burden includes interpreting normal studies, refilling medications, checking laboratory data, communicating with various stakeholders' routine information, etc. An element that can contribute to the weight of this burden is avoiding errors in these tasks.

Medical error and the promise of artificial intelligence (AI)

John Lee[1]
John Lee, a CMIO and long-time informatics expert, expounds on the large number of deaths from preventable medical errors and how AI, in its appropriate utilization, can decrease this mortality and improve the quality of care.
[1]*Edward-Elmhurst Healthcare, Naperville, IL, United States*

Much of modern medical technology is focused on high profile applications. All of these are understandably exciting and garner many headlines, but as shiny as these objects are, the potentials of these technologies are limited to their specific use cases. However, an oft ignored but far more ubiquitous use case is medical error and its myriad, very boring contributors such as medication reconciliation, communication breakdown, power distance, and their ilk. They are mundane and not sexy but are exceedingly common features of the sewage of medical care [1].

IOM's seminal treatise, "To Err is Human" estimated that 44,000—98,000 deaths per year could be attributed to *preventable* errors, exceeding deaths due to motor vehicle crashes, breast cancer, or AIDS [2]. More recently, a review of four studies estimated that the rate of iatrogenic harm may be 10 times higher than IOM's estimate [3]. In any case the numbers are staggering.

We often have some sense of the cause of these errors, but these insights are uniformly borne from exhaustive retrospective manual abstraction. IOM's report relied heavily on a Harvard study that was a retrospective review of 30,000 random discharges from 51 randomly selected New York hospitals in 1984 [4]. The more recent analysis by James also leans on manual, retrospective review of hundreds to thousands of medical charts [3].

To truly make an impact, we must identify and catalog these errors. We cannot rely on a retrospective review by humans long after they occur to identify them. Just like we know that a myocardial infarction is caused by a ruptured plaque or that a cancer is caused by activation of a previously dormant gene, we need to know the pathophysiology of the "disease" of systemic medical error.

Much as we had to study the anatomy and physiology of disease states using biochemical, histologic, anatomic and radiographic techniques, we must use techniques that expose the unfiltered causes of these errors. These data will be the equivalent of airline "black boxes" that collect all sorts of data *as the errors occur* without requiring some sort of retrospective manual abstraction.

Systemic errors occur because of our human tendencies to make mistakes when our human faculties are overwhelmed or distracted. Instead of using biochemistry, histology, radiographic, and other familiar techniques with physiologic disease, we need to use techniques that allow us to analyze these human tendencies.

If we can detect and record these errors and associated metadata, the data can be joined with the data produced to our clinical data warehouses. Then, we can use the magic of data to connect the systematic errors with outcomes.

But how do we do this? The mainstay of safety data is the safety report. However, most such safety reporting systems require reporters to fill out several fields and a barrage of required metadata. The process is onerous and creates significant friction to data collection. As a result, at best, 10% of true errors and systematic safety problems are reported or detected [5]. This is where tools that fall under the rubric of "AI" and other advanced data tools can help provide clarity and shine light on errors and safety events that otherwise might be hidden.

We must reduce the effort it takes for an observer to abstract and enter such safety reports. The actual description of the safety event is only a small part of these reports. The laborious part of the process is adding metatags and other data that gives structure and statistical substance to these reports. This activity is steeped in tedium and is a significant deterrent to submission. If we can alleviate or eliminate the tedious parts of the data entry process, we should be able to facilitate the creation of safety reports and get a clearer picture of the >90% of errors that are hidden from our view.

The data that exists in our EMRs that record clinical transactions can often be manifestations of underlying systematic safety. Fortunately, we now have a broad adoption of digital documentation in our healthcare system resulting in large quantities of data. Simultaneously, we have access to advanced data tools to consume and analyze these data. Organizations have started to combine these tools to more rapidly identify errors and safety events [6]. This presents better opportunities to address systematic problems before they become more ingrained or widespread.

III. The current era of artificial intelligence in medicine

Because errors that cause harm are borne from physical workflows, computer vision is a particularly promising technology to detect and log these events. There is technology that can identify and highlight shoplifting behavior [7]. Similar technology could identify providers who don't wash their hands, patients at risk for falling or any other physical event that is a precursor to harm.

AI can create a path to identify the pathophysiology of error much in the same way we have been able to identify the pathophysiology of other disease processes. This will allow us to study mistakes at scale and eventually make the same strides against medical error that we have against heart disease and cancer.

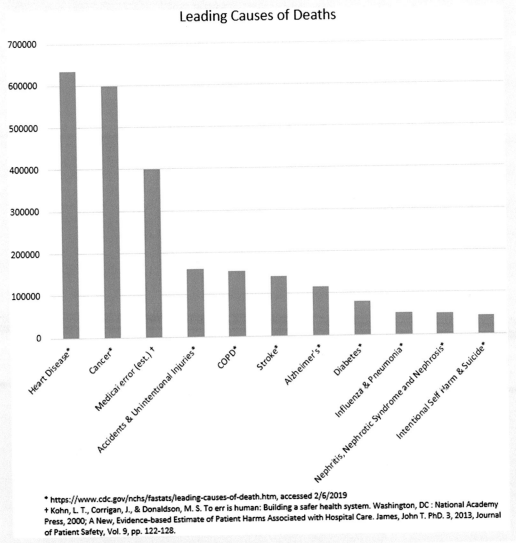

Leading Causes of Deaths

* https://www.cdc.gov/nchs/fastats/leading-causes-of-death.htm, accessed 2/6/2019
+ Kohn, L. T., Corrigan, J., & Donaldson, M. S. To err is human: Building a safer health system. Washington, DC : National Academy Press, 2000; A New, Evidence-based Estimate of Patient Harms Associated with Hospital Care. James, John T. PhD. 3, 2013, Journal of Patient Safety, Vol. 9, pp. 122-128.

III. The current era of artificial intelligence in medicine

References

[1] Bush J. How AI is taking the scut work out of health care. Harv Bus Rev [Online] March 5, 2018. <https://hbr.org/2018/03/how-ai-is-taking-the-scut-work-out-of-health-care>.

[2] Kohn LT, Corrigan J, Donaldson MS. To err is human: building a safer health system. Washington, DC: National Academy Press; 2000.

[3] James JT. A new, evidence-based estimate of patient harms associated with hospital care. J Patient Saf 2013;9:122–8.

[4] Brennan TA, Leape LL, Laird NM, et al. Incidence of adverse events and negligence in hospitalized patients: results of the Harvard Medical Practice study. N Engl J Med 1991;324:377–84.

[5] Griffin FA, Resar RK. IHI innovation series white paper IHI global trigger tool for measuring adverse events. 2nd ed. Cambridge, MA: IHI; 2009.

[6] Classen D, et al. An electronic health record–based real-time analytics program for patient safety surveillance and improvement. Health Affairs 2018;37:1805–12.

[7] Vincent J. This Japanese AI security camera shows the future of surveillance will be automated. The Verge [Online] June 26, 2018. <https://www.theverge.com/2018/6/26/17479068/ai-guardman-security-camera-shoplifter-japan-automated-surveillance> [cited 17.01.19].

Fourth, AI can help organize and facilitate the care coordination of chronic and complex diseases in many of the patients especially as they have more relevant data from disparate sources such as genomic sequencing, medical imaging, and wearable technology. This also mandates a central repository of data and information that AI can help gather and organize.

Fifth, physicians have currently a high rate of stress and many are facing or having had burnout from their careers. The use of AI can mitigate particularly the EHR burden that is often a prime source of frustration and thereby automate and simplify their workload. AI can even help monitor physician burnout with metrics that reflect burnout and dissatisfaction.

Artificially intelligent medicine can combat burnout and improve patient outcomes

Addison Gearhart

Addison Gearhart, a pediatric resident who has been interested in innovative solutions in health care, authored this commentary on the use of artificial-intelligence tools such as natural language processing and robotic process automation in mitigating the high levels of stress and burden for the young and senior clinicians.

Climbing rates of young physician burnout are at an alarming high, an ominous harbinger to potentially dangerous consequences unless addressed. While the epidemic of physician burnout typically elicits an image of a seasoned physician downtrodden by years of long working hours, it is now clear that burnout spans all levels of training. A study by the Mayo Clinic found 45% of resident physicians report symptoms of burnout [1]. The statistic suggests our current medical education system is inefficient and suboptimal. We devote significant resources to a system that produces a growing number of detached and emotionally exhausted physicians at high risk of dropping out when predicted physician shortages are reaching an unprecedented high. The implications of these discouraging statistics threaten the quality of healthcare delivery in the United States. Piling evidence associates physician burnout with major medical errors [2].

While no single event precipitated the epidemic, the rise in burnout parallels an increased utilization of the electronic medical record (EMR) [3]. The advent of the EMR shifted practice to spend more time facing the computer screen than the patient. It is not surprising that physician trainees now report symptoms of burnout earlier: computer-centric tasks often fall disproportionately on the shoulders of resident trainees and despite duty hour regulations, residents tend to work extended hours. While the cited benefits of the EMR are long, the implementation comes at a steep cost. Younger physicians are losing precious time at their patient's bedside and patients are noticing. This depersonalization dismantles the sacred physician—patient relationship known to be protective against burnout. For young physicians, this occurs at a valuable and irreplaceable time, a limited training period intended to cultivate and refine this relationship—the averred heart of medicine.

A recent report issued by a collaborative of Massachusetts healthcare giants deemed *burnout* a public health crisis releasing a "call to action" to address the escalating numbers [4]. Hiring medical scribes exploded in popularity obvious solution with cited improvements in physician job satisfaction rates. The solution is likely a Band-Aid approach to a big problem that warrants an even bigger solution: scribes turn over rapidly, receive minimal training, and produce high rates of error. In "Why Doctors Hate Their Computers," by Dr. Atul Gawande in *The New Yorker*, he details many of the health IT challenges that plague doctors but also recognizes a role for computerization and emerging technologies such as AI to allow clinicians to help patients in ways previously unattainable and to prevent physician burnout [5].

The breadth of tools that fall under the umbrella term AI may be just the medicine healthcare needs to modernize how we address current challenges to improve patient satisfaction and outcomes as well as physician attrition rates. Undeniably patient-generated data and medical information continue to rise at rates faster than our current technologies can make manageable for humans to cognitively process and interpret. The situation overwhelms physicians, especially those in training already at a knowledge disadvantage. While the EMR contributed to burnout, it also has a notable wealth of complex medical, scientific, and patient-generated data. Applying components of AI to the EMR could serve to interpret and translate the massive amounts of stored data into clinically relevant insights for better counseling and intervention based on patient-specific screening, prediction, and prognosis analytics. Furthermore, AI could reduce the redundancy of the notes, the volume of unnecessary alerts, and the packed inbox coined the "4000-keystroke-a-day" thought to accelerate rates of burnout.

Other branches of AI, such as NLP, promise to create more reasonable and time-efficient systems for the future practice of medicine by relieving the time burdens of documentation. NLP helps computers understand, interpret, and manipulate language to fill the gap between human communication and computer comprehension. Emerging technologies employing this system are sophisticated enough to detect the physicians' voice in the clinical encounter and communicate with the computer to utilize the speech in a clinically effective manner. Imagine dictating to the patient your physical exam findings, decision-making process, and future orders for working up the patient and then leaving the clinical encounter with all the orders imputed and a note typed ready for your review. In such a scenario, the patient feels more involved in their care plan. Reciprocally, the physician can place more time into forming a differential and evaluating different etiologies over documenting and placing orders. We are closer to this model than one may think.

III. The current era of artificial intelligence in medicine

Other systems such as RPA promise to relieve physicians of repetitive tasks to improve efficiency, reduce costs, and combat burnout. RPA comprises smart software that automates repetitive, standardized tasks that humans otherwise perform while delivering a wealth of data in a continuous feedback loop that can be used for performance improvement and optimization. RPA collects data on how the process is working and analyzes that information to allow the program to improve itself to become more accurate and helpful for the provider's workload. It is now used in patient scheduling, coding, clinical documentation, Medicare billing and compliance, and documentation. By streamlining processes, RPA offloads time-heavy task that requires less brainpower returning employees to cognitively higher functioning roles known to increase job satisfaction.

AI will fundamentally change the way we practice medicine and its impact extends well beyond that of combating physician burnout. Successful implementation will require us to anticipate and acknowledge the potential challenges that lie ahead. The evidence for AI is building and now is the time to transition to an AI model, reinvigorating younger physicians by allowing them to refocus on the most important aspect of medicine, the physician–patient relationship.

References

[1] Dyrbye LN, Burke SE, Hardeman RR, et al. Association of clinical specialty with symptoms of burnout and career choice regret among US resident physicians. JAMA—J Am Med Assoc 2018;. Available from: https://doi.org/10.1001/jama.2018.12615.

[2] Tawfik DS, Profit J, Morgenthaler TI, et al. Physician burnout, well-being, and work unit safety grades in relationship to reported medical errors. Mayo Clin Proc 2018. Available from: https://doi.org/10.1016/j.mayocp.2018.05.014.

[3] Downing NL, Bates DW, Longhurst CA. Physician burnout in the electronic health record era: are we ignoring the real cause? Ann Intern Med 2018. Available from: https://doi.org/10.7326/M18-0139.

[4] Jha AK, Iliff AR, Chaoui AA, Defossez S, Bombaugh MC, Miller YR. A crisis in health care: a call to action on physician burnout. Boston, MA: Massachusetts Health and Hospital Association; 2019. <https://cdn1.sph.harvard.edu/wp-content/uploads/sites/21/2019/01/PhysicianBurnoutReport2018FINAL.pdf>.

[5] Gawande A. Why doctors hate their computers. New Yorker 2018. Available from: https://doi.org/10.1162/POSC_a_00184.

Physician burnout and artificial intelligence (AI) in medicine

Alan S. Young

Alan S. Young, a surgeon and physician entrepreneur, authored this commentary to explore using AI to mitigate the physician's burden as a means to reduce the high incidence of physician burnout in this current era.

Years before the recent release of the article titled *A Crisis in Healthcare: A Call to Action on Physician Burnout* by A.K. Jha et al.,[1] physicians around the United States began exhibiting signs and symptoms consistent with the spectrum of a serious clinical condition. Medscape's 2019

---

[1] https://cdn1.sph.harvard.edu/wp-content/uploads/sites/21/2019/01/PhysicianBurnoutReport2018FINAL.pdf

National Physician Burnout, Depression and Suicide report[2] of 15,000 physicians indicated 44% had experienced burnout and 59% cited bureaucratic burdens such as charting and paperwork as contributing factors. The prevalence of certain "diseases of despair"[3] such as suicide, opiate abuse and overdose, and alcohol-related liver cirrhosis is undeniably underreported among practicing physicians. This translates into substance abuse patterns, mental illness, attempted suicides, relationship challenges, and other common struggles across all physician groups starting as early as medical school and residency. Stories of suicide committed by a medical student and a resident in New York[4] taking place only a few days apart shocked the medical community and put a spotlight on the underlying issues, but only momentarily. The world continues to demand doctors to care for the sick, poor and elderly. The question being asked recently is who will take care of the doctors?

My own personal story of physician burnout started shortly after I began my internship year of residency in a General Surgery PGY-1 program before transitioning to an Orthopedic Surgery PGY-2 and beyond. Back then, the work restriction hours had recently been implemented to "protect" residents from exceeding 80 hours of training per week. After a few intense years of scut work, surgical training under an apprentice model and numerous tests, the switch to working as an independent physician seemed like a blessing at first. During residency, the idea of completing all medical documentation in an electronic health record started to take root and fairly soon a large number of hospitals and health systems were announcing their decisions to move away from paper charts to electronic medical records (EMR). My first job at Kaiser Permanente required 2 full days of Epic EMR training before I was left to my own in clinic. Considering myself computer literate after having grown up with a computer engineer for a father and playing a wide range of computer games, the heavy reliance on typing and word processing skills didn't faze me as I welcomed the ability to rapidly create legible documents using a mouse and keyboard as opposed to pen and paper. While the mandate to move to EMR systems caused some attrition among older physicians who couldn't adapt quickly enough, the majority of the physician workforce survived the dawning of the EMR era only to face growing challenges for the years to come.

The topic of EMR and documentation being key factors in the level of physician burnout is well described across various journal articles and thought pieces. One other influence on the incidence of physician burnout is the growing opportunities for physicians outside of the traditional clinical role. In business school, a common fallacy is the idea that an individual makes forward-looking decisions biased by his or her sunk costs when the decision should be based on opportunity cost instead. The time, money, and energy invested by physicians during college, medical school, residency, fellowship, and more are incredibly burdensome and usually several years go by before physicians experience the fruits of their labors. The average medical student debt at the time of graduation in 2018 was $196,520 according to the Association of

---

[2] https://www.medscape.com/slideshow/2019-lifestyle-burnout-depression-6011056

[3] Dr. Sanjay Gupta.

[4] https://www.medscape.com/viewarticle/896460

III. The current era of artificial intelligence in medicine

American Medical Colleges (AAMC).[5] This financial pressure motivates physicians to begin their careers in order to climb out of the debt and take on roles that can perhaps pay better in the near term. The choices made my younger physicians because of enormous debt can lead them to unpleasant work environments or locations where they are forced to live. The more recent generations of college graduates are more tech-savvy and can quickly identify new jobs, new roles and new adventures that match their interests more closely. Technology-enabled platforms can share job-postings along with unique travel or career experiences that serve to inspire or influence the next cohort of job seekers. The value of having an experienced physician for safety or clinical acumen has diminished gradually with the massive proliferation of medical science and other hot topics. The speed at which new medical knowledge is being created is unsustainable for an individual to keep up with his or her field of practice without automated resources. These factors contribute to the ease and accessibility to alternate career paths by physicians who historically had very few options outside of clinical practice. A remark given by a health system Chief Medical Information Officer (CMIO) was lamenting the fact that only 80% of medical students graduating from his prestigious institution were matching into one of their top three picks for residency. Other similar organizations across the country boast high 90% statistics for this key metric. However, the revelation that his medical school resides in close proximity to California's Silicon Valley explains how some medical students forgo a residency spot to pursue a career in consulting, biotechnology or entrepreneurship with comparable or more lucrative compensation packages.

Is there a mass migration of physicians into alternate careers such as health tech, consulting or venture capital? The answer isn't clear, but the rise of AI in medicine has brought new attention to the workforce challenges facing providers taking care of an aging and more demanding patient population. The threat of replacing physicians with AI "bots" is not realistic as human physicians will be necessary for higher level complex decision-making. However, there is a role in applying AI technology to help reduce burnout among various physician groups. If a clinical decision support (CDS) solution can help identify and triage patients to allow providers to function at the top of their license capabilities, they can provide higher value care to the patients who need it the most and reduce the overall number of patients that need to be seen. Medical training can be augmented by the use of AI to enhance the learning experience and provide greater flexibility to give students a more customized and personal teaching journey. The documentation requirements for physicians can be alleviated with well-designed systems that help generate the necessary structured data to meet financial, regulatory, and clinical imperatives. Physicians need additional support to navigate a career path that is long, arduous and no longer gives complete job satisfaction. AI will not cure the problem of physician burnout but has the potential to offer some relief for this growing public health crisis.

[5] https://store.aamc.org/medical-student-education-debt-costs-and-loan-repayment-fact-card-2018-pdf.html

III. The current era of artificial intelligence in medicine

Lastly, AI tools can provide an important resource for medical education as well as clinical training at all levels with smart tools coupled to the EHR as well as the implementation of augmented reality (AR) and virtual reality (VR) technologies.

Naylor also identified seven factors driving AI adoption in clinical medicine and health care: (1) the strengths of digital imaging over human interpretation; (2) the digitization of health-related records and data sharing; (3) the adaptability of deep learning to analysis of heterogeneous data sets; (4) the capacity of deep learning for hypothesis generation in research; (5) the promise of deep learning to streamline clinical workflows and empower patients; (6) the rapid-diffusion open-source and proprietary deep-learning programs; and (7) of the adequacy of today's basic deep-learning technology to deliver improved performance as data sets get larger [1]. Along with these factors, the escalating volume of healthcare data as well as exponential increase in medical knowledge are additional forces as well.

Data sharing for a learning system

Allana Cummings

Allana Cummings, a CIO of a large children's hospital, authored this commentary about the vision of data sharing leading to an innovative artificial intelligence—enabled learning system so that healthcare providers can perhaps have a different perspective on data and databases in a healthcare organization.

Visionary healthcare leaders have used AI innovations to create a new generation of advances, resulting in the development of interventions, new standards of care, and population health strategies that otherwise would not have been possible. What may be underappreciated with these accomplishments is that the most significant advances, and likely the only reasonable path forward, depend on a framework of large-scale data sharing by multiorganizational learning healthcare systems. For many clinical and business issues, no single organization can produce the large data sets needed to enable reliable, nonlinear models with ML methods. Combined experience is needed to supply an adequate number of instances of events. We are no longer in the era of computers helping us develop "rule-based" predictions. With ML, "[m]ore data beats a clever algorithm." [1].

Trailblazers in AI have recognized that data sharing is fundamental to the success of AI and exploit the added benefit that these same data-sharing strategies are critical tools for their Leaning Healthcare System efforts. By finding creative approaches to mitigate the risks and challenges of substantial data exchanges, they have overcome barriers to achieve the levels of data sharing needed to support both a learning healthcare system and sophisticated AI development.

The reluctance to share data has become perhaps the most significant barrier to an organization's success as a learning health system and, as a consequence, significant innovator with AI. Health systems that successfully deploy AI solutions do so by establishing a deliberate culture of data sharing as described by the Institute of Medicine and others. The opportunities arising from learning and sharing expertise create a win-win for participating organizations, as rich, diverse datasets are created that accelerate innovation and improvements. Conversely, health systems that adopt a posture of data secrecy foster siloed expertise, isolating their organizations from advanced learning and limiting them to ineffective and inefficient healthcare delivery practices.

While the potential value of a learning health system is high, incentives are often not aligned with requirements to make this happen, according to Yale School of Medicine cardiologist Harlan Krumholz [2]. In an opinion paper for the National Academy of Medicine, Krumholz, Terry, and Waldstreicher observed that organizations providing stewardship of data generally bear the costs associated with data sharing [3]. Financial incentives to reward data sharing and interoperability, coupled with penalties for not doing so, are needed to help organizations mitigate financial barriers. We must also advocate for the creation of regulatory pressure for data sharing, including requirements for studies that use public funds.

The healthcare IT (HIT) industry has made efforts to standardize application programing interfaces (APIs), which has improved data-sharing capabilities for the masses. Healthcare organizations that are committed to building learning systems must also contribute to group expertise by sharing APIs developed in-house with other health systems. As API connections and other tools supporting broad aggregation of data across entities increase, a more cost-effective model of data exchange will be developed. It is imperative that these efforts be encouraged, if we are to unleash the power of AI to all.

In order to encourage our patients to participate in broad data-sharing agreements, we must do a better job of sharing data with patients. This may require new tools and infrastructure as well as regulatory changes. Patients are exposed to the benefits of AI in their everyday lives and expect that healthcare will also leverage it for improvements in their care experience and ability to actively participate in their care decisions. As an example, with the growing popularity and acceptance of smart devices, AI can help patients better manage their health by allowing them to access timely, personalized information to make better decisions in their health management.

Even when organizations commit to data and expertise sharing, they must address the ever-present risk of violating HIPAA requirements and the costly penalties associated with infractions. Organizations and researchers often struggle to develop effective data-sharing practices that start with the creation of sound data-sharing agreements. Stewardship of patient and organizational data custody must be ensured through appropriate compliance monitors established to mitigate any undue risk.

The public health imperative to find ways to share deidentified data sets has never been more acute. One only needs to consider if Flint residents' electronic medical record data were not aggregated, residents would still be at risk for additional lead poisoning from their water. However, it is naïve to believe that HIPAA-based "deidentify and release" methods can protect patients' interest in the era of predictive analytics, persistent databases, and unregulated data brokers. The scale of available sets of data has allowed for possible data reidentification. This has exposed the population to the potential threat of health data discrimination, thereby creating additional risks for broader data sharing. Such barriers stifle the idea that amongst our collective health data lies the cure for diseases and illness waiting to be harvested with AI. To maintain the public trust, we must continue to improve ways to protect the use of deidentified data sets for research. We must find ways to reduce administrative and legal burdens to arrive at appropriate data-sharing agreements while ensuring the protection of patients' privacy and from discrimination.

We must remain optimistic that these data-sharing challenges can be addressed. Other industries, such as banking and e-retail, have overcome data exchange and security challenges to emerge with new services that have transformed customer experience and reduced costs. At the same time, we need to assure our patients that we are their caring healthcare providers, and distinct from the almost daily instances of escalating public concerns that led Apple's Tim Cook in a recent op-ed piece in *Time* to call for a Federal Registry of all data brokers [4]. Healthcare can find solutions to allow for more secure and seamless data exchange. Providers and researchers can help HIT and the government leads the way, and patients can benefit by improving their own, and everyone's, healthcare with a trusted partner.

The deep learning from AI through aggregated data sets from large-scale data-sharing initiatives would allow for expedited advances in medicine and improvements in care quality and delivery, whose opportunities might otherwise never be identified if data remains siloed. In time, advances in broader data exchange will lead to greater sharing of expertise between healthcare entities, helping all who participate to become effective learning organizations with continuously improving AI innovations—and continuously improving care for their patients.

References

[1] Domingos P. A few useful things to know about machine learning. Department of Computer Science and Engineering, University of Washington. Accessible at: <https://homes.cs.washington.edu/~pedrod/papers/cacm12.pdf>.
[2] Krumholz HM. Big data and new knowledge in medicine: The thinking, training, and tools needed for a learning health system. *Health Aff* 9millwood **33** (7) (2014) 1163–1170. Section of Cardiovascular Medicine, Department of Internal Medicine, Yale School of Medicine, New Haven, CT.
[3] Krumholz HM, Terry SF, Waldstreicher J. Data acquisition, curation, and use for a continuously learning health system JAMA 2016;316(16):1669–70.Vital directions from the National Academy of Medicine, October 25, Available from: https://doi.org/10.1001/jama.2016.12537.
[4] Cook T. You deserve privacy online. Here's how you could actually get it. Time Jan 16, 2019.

Adoption of Artificial Intelligence in Medicine: The Challenges Ahead

There are realistic challenges for wide AI adoption in biomedicine amongst clinicians (see Table 7.1).

TABLE 7.1 Challenges of artificial intelligence (AI) in adoption in biomedicine.

Data	Relational database	Data inaccurate	Data incomplete
	Data sharing	Data security	Data standardization
	Data storage and transfer	Data ownership	Data volume
Technology	Black box	Difficult to interpret	Relevance
	Cost	Regulation	Workflow
People	Lack of AI education	Trust	Cultural differences
	Shortage of data scientists	Lack of clinician champions	Hubris
Other	Legal	Bias	Ethics
	Inequity	Data privacy	Business model

III. The current era of artificial intelligence in medicine

Why artificial intelligence (AI) and machine learning (ML) are moving so slowly in health care?

Bill Vorhies

Bill Vorhies, an engineer and a keen observer of the recent surge in AI, authored this commentary on why there seems to be a schism between the medical world and the data science domain and offered a few suggestions for both groups.

As a practicing data scientist for the last 18 years and as Editorial Director of DataScienceCentral.com for the last 4 years I've been tracking the adoption of AI and ML across all industries. Like everyone else I have high hopes for its application in healthcare.

The popular press would have us believe that advances are being implemented in healthcare like gangbusters so I was delighted last December when I was invited to attend Anthony Chang's 3-day AIMed conference. This event devoted to advances in AI/ML in healthcare counts 80% of its attendees as clinicians or hospital CIO/Administrators. I was looking forward to the first-hand accounts of how the tools of my profession had benefited patients, doctors, and hospitals. I was caught up short by two facts revealed during the conference.

First only about 1% of US hospitals have active data science programs. That compares to between 20% and 33% of larger corporations who are sponsoring full-on programs to remake themselves with AI/ML tools, and the better than half of large and midsize companies that have some project underway.

Second, although the clinicians and administrators attending the conference were largely from the 1% early adopters, the message in almost every presentation given by clinicians was that AI/ML may be coming but it's not ready for prime time in healthcare. Indeed, you had to come away with the feeling that 99% of clinicians welcome AI/ML-driven change as much as a fish needs a bicycle.

It seems that all those glowing articles I'd been reading were all written from a data science point of view about what's possible but not from the perspective of the doctors and clinicians who must adopt these breakthroughs.

Nowhere else in the public or private sector is there such an imbalance between adoption and expectation.

Causes of slow adoption

What I learned is that the root causes of this slow adoption is partially financial, but most important for data scientists, particularly those in the hundreds of AI/ML healthcare startups and big-tech innovation labs looking to follow the path of disruption to financial success and glory is to listen a lot more closely to medical professionals on whom you are relying to adopt your breakthroughs.

Let me be clear. It remains true that perhaps the greatest good that AI/ML will achieve over the next 10 or 20 years lies in improving the health and wellness of every human being.

But to get there we who are creating these tools and techniques have many lessons to learn about the unique nature of healthcare.

Where it's working—where it's not

Today, the areas in which AI/ML are making the most inroads depends largely on who's paying.

Of all the AI/ML opportunities, drug discovery and innovation are actually furthest along largely because it's big pharma that pays, not the insurers.

Second most readily adopted seems to be the business of healthcare. The operational world of the clinician may be unique, but at a business level, hospitals and healthcare organizations share some marked similarities with the commercial world.

Where AI/ML is decidedly lagging, but where great promise awaits is in clinical applications—what happens between clinician and patient aka the AI/ML augmented physician. If you want AI/ML to succeed in improving healthcare it needs to get into the space between the doctor and the patient and prove its value there.

Here, in brief, are the major challenge areas for data scientists as described by clinicians.

Too many false positives

Data scientists don't need to think twice about the fact that all of our techniques are probabilistic and contain both false-positive and false-negative errors.

In healthcare however, false negatives, that is failing to detect a disease state, is the ultimate failure to be avoided at all cost. As a result, applications designed, for example, to automatically detect cancer or other diseases in medical images are tuned to minimize these type 2 errors.

This necessarily increases false positives that can only be reduced by increasing overall model accuracy. And that can only happen where a large amount of training data is available.

Radiologists and pathologists complain that false positives slow them down too much as they are forced to examine all the portions of the image flagged by the model. And in fact spend even more time on the false-positive indications, not wanting to miss something important.

Similarly, with the many IoT type applications to monitor in-patients for critical events, clinicians reported "alarm fatigue" from too many false positives, reducing the likelihood that they would respond with urgency.

Turn down the hype

On the topic of automated image evaluation, radiologists and pathologists would like the press to turn down the hype on these "breakthroughs" often described as new levels of accuracy in the detection of this or that cancer.

They remind us that the job of radiologists and pathologists is not to tell the treating doctor that they have discovered a cancer in the image, but rather to say that a specific area looks suspicious and requires the doctor to examine it more closely

Don't disrupt my workflow bro

It's in the nature of running a hospital not to have too few or too many of any particular clinical specialty for reasons of cost. That means that healthcare professionals are quite possibly the most overworked or at least critically scheduled category of workers anywhere. There is seldom a moment when they are underutilized.

The result is the adoption of natural workflow patterns that allow attending physicians to see as many patients as possible (without causing harm) or for radiologists and pathologists to examine as many images or slides as possible in the shortest amount of time.

It's the heart of a unique culture that is the opposite of the equally unique culture of the healthcare data science startup to disrupt the status quo with their innovative breakthrough de jour.

Fear of small startups versus fear of missing out

FOSS (fear of small startups) has more weight than FOMO (fear of missing out). Even where AI/ML solutions like automated image classification were shown to be promising, hospital administrators showed the same reticence to contract with new, small startups that any competent commercial enterprise would.

The electronic healthcare record (EHR)—a deal with the devil

A survey from the AMA this year continues to show that the EHR and related clinical systems are the chief reason for physician burnout.

And yet to get to the benefits of AI/ML in healthcare requires the data that starts here in the EHR.

There are many structural and procedural problems with health data but key among them is extracting that data from these EHRs.

This practically screams out NLP to any data scientist and some applications are making inroads. However, some still haven't learned lesson 2 above about integrating into existing workflows.

A major challenge to NLP solutions and indeed all types of data capture in healthcare is interoperability among different data sets. The consistency and standardization aren't there today restricting most data sets to relatively small size and making the blending of data sets chancy at best.

Data is too thin and won't generalize

The problems with the healthcare data necessary to train AI/ML solutions don't end with extraction. The first major problem is that the data is simply too thin and won't generalize.

There are a few large public databases in the 100,000 record range but the effort to rollup patient data into data science worthy DBs is early in the process.

Continuous learning is broken

AI/ML modeling is based on the presumption of continuously improving our models as new data becomes available. One barrier to getting good feedback information lies in the hospital itself. Some of this is an organizational problem. Some results from the fact that data from similar machines made by different manufacturers, or simply with different settings isn't comparable.

An even more formidable barrier is raised by the Food and Drug Administration (FDA) in approving imaging-based solutions. On the one hand, the FDA has taken a very permissive approach to approve AI/ML image classifying solutions based on training with as few as 100−300 images.

However, those approvals are then frozen and require reapplication before an improved solution can be released.

III. The current era of artificial intelligence in medicine

Not so fast with those rollouts

This problem lies squarely at the feet of those data science healthcare vendors still in the "move fast and break things" mindset.

The widely touted rollout just a few months ago of a chatbot called Babylon implemented in the United Kingdom to provide diagnostic advice on common ailments without human interaction is a good example. A group of auditing physicians found that around 10%–15% of the chatbot's 100 most frequently suggested outcomes were just flat-out wrong.

The problem was simply shortcuts taken during training and too great an emphasis on rolling out fast before being audited.

Finally

For data scientists hoping to capitalize in this market, there are a few important lessons.

- Slow down a little and make sure you understand how your disruptive application can actually be integrated into this world of specialized workflows.
- Make sure you understand both the current restrictions on data size and accuracy, and how long it may take before that gets better.
- Don't rush the rollout. People's health is at stake.

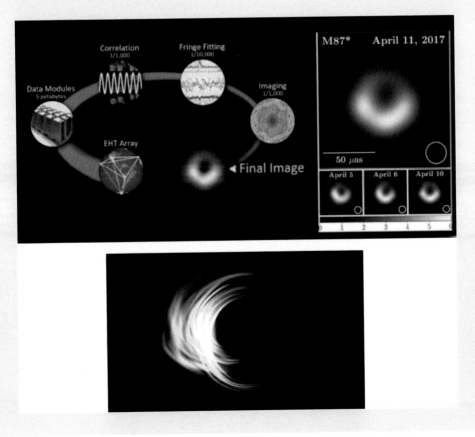

III. The current era of artificial intelligence in medicine

Free innovation and matchmaking in healthcare in the artificial intelligence (AI) age

Chris Yoo

Chris Yoo, a seasoned healthcare IT executive, authored this commentary to elaborate on his perspective of the innovations in AI in health care and how this has created a large middle space that will require strategic navigation for all stakeholders.

The most impactful technological changes throughout human history have two common characteristics. First, innovative technologies enable us to accomplish basic human needs much more efficiently than before. For example, a bicycle takes us from point A to point B. Second, these are technologies that alter the way we perceive time and space both individually and as a group. For example, an airplane takes us from one side of the planet to the other. Healthcare is an industry where technologies that possess these two characteristics result in immediate impact on the human condition and are therefore much easier to recognize as innovative. When a surgeon can "see" disease using an imaging device before taking action to heal a patient, it can be said that technology has made surgery faster, more accurate, and results in improved outcomes. The change introduced by AI in Healthcare, however, is without precedent and likely far greater than what we can immediately perceive. AI is introducing a third characteristic—friction-free diffusion of meaningful, actionable knowledge across purposely connected groups—because AI can enable solutions that autonomously evolve with us without the need for us to direct it to do so. Our role as the "human" part in a "human + machine" [1] future for medicine is likely to focus our attention on socializing best practices and tailoring care for groups of future patients before disease can progress.

Improvements in radiology and pathology are the first examples of AI detecting patterns in image data as accurately as—or in some cases better than—humans [2,3]. Deep neural networks interpreting images of the retina have also been developed that are as accurate as the decades-old gold standard of predicting risk based on physical and metabolic measurements [4]. In these early examples, AI is enabling us to explore a modality for diagnosing disease that is not the result of a new physical technology (i.e., a better microscope or a better imaging device). The digitization of large amounts of image data collected from several sources provided AI with the raw materials needed to produce accurate results. AI's requirement for "big data" will accelerate collaborative sharing across providers of care as more applications will clearly show that very large amounts of data equal the highest accuracy in prediction. The current healthcare business ecosystem will have to evolve into a multisided platform, and matchmakers who provide value efficiently across the platform for participants will benefit the most [5].

As AI's fuel of big data collected across multisided platform networks becomes established, the role of the physician—to heal patients with knowledge of medicine—becomes even more important as the matchmaker in the system. Hospitals and other institutions providing advanced care are responding to tremendous cost pressures by consolidating, and more specialty procedures are increasingly being performed in remote centers. "Medical tourism" is a byproduct of this slow change [6]. Physicians, however, are the one participant group with the expertise to take the output from AI and connect the right providers with the right consumers. Their current core capability of ingesting large amounts of very complicated, disparate data about a patient and cerebrally synthesizing solutions will be turbocharged by AI. This change is occurring rapidly and will require the matchmakers to become experts not just in medicine but also in the multisided platform model.

Looking further ahead, AI's predictive accuracy will enable patients to become "prepatients" living healthier, longer lives at the edge of the network. The household market demand for knowledge of beneficial behaviors and more accessible healthcare solutions that can be deployed in the home setting will grow as quickly as AI's need for big data as fuel. With an efficiently operating multisided platform healthcare ecosystem, a free innovation model for enabling producer collaborations that result in more valuable healthcare solutions for consumers will result in an exponential network effect on productivity and resulting value creation [7]. A virtuous cycle, where physicians use AI to treat patients who use AI to find matchmakers with the right solutions, will further create a need for AI to determine the most efficient distribution of the right knowledge to the correct participant in the network. Rather than replacing workers in healthcare, AI will help create a new business model with new producers of solutions, and more opportunities for addressing prepatients' and physicians' needs at earlier and more durable interaction points. Business models will be disrupted, as they always are in the creative destruction process, but those who embrace the multisided platform change will benefit from the advent of AI.

References

[1] Daugherty PH, and Wilson, HJ., Human + Machine: *Reimagining* Work in the Age of AI, Harvard Business Review Press, Boston, MA, 2018.

[2] Bejnordi, et al. JAMA 2017. Available from: https://doi.org/10.1001/jama.2017.14585.

[3] Esteva A, et al. Nature 2017;1−4. Available from: https://doi.org/10.1038/nature21056.

[4] Poplin, et al. Nat Biomed Eng 2018. Available from: https://doi.org/10.1038/s41551-018-0195-0.

[5] Evans DS, Schmalensee R., Matchmakers: The New Economics of Multisided Platforms, Harvard Business Review Press, Boston, MA, 2016.

[6] Christensen CM., The Innovator's Prescription: A Disruptive Solution for Health Care, McGraw-Hill, New York, 2009.

[7] von Hippel E., Free Innovation, MIT Press, Cambridge, MA, 2016.

Education. First and foremost, there is a lack of focus on data science and AI in the medical school and training program as well as later in continuing medical education. This knowledge gap, if it persists, will increase even more as more sophisticated AI technologies are made available for application in medicine and health care. It is important to emphasize that given the present milieu of work burden and escalating knowledge across many other domains, clinicians may simply not have the time nor the desire to learn a difficult domain no matter how interesting or relevant. Perhaps, a taxonomy of AI in medicine and health care will be useful for an easier understanding of AI as it pertains to medicine and health care; clinicians have a biology mindset and such a classification system can help. To further exaggerate the former issue of undereducation and lack of exposure of data science for clinicians, AI (particularly deep learning but to some degree some of the other tools as well) has an inherent "black box" nature in that it is not easily explainable nor transparent even to the data scientists themselves (see explainable AI previously). These two aforementioned not only issues combined render AI adoption for clinicians a daunting (but solvable) challenge but also presents a great opportunity to learn an entirely new domain for the willing and even enthusiastic clinician cohort. While regression analyses have yielded some insights into patient risk scores, more sophisticated data science in the form of random forest, convolutional neural network (CNN), and even generative adversarial networks will need to analyze these cohorts in the future (see Fig. 7.1). Overall, some if not most clinicians feel that

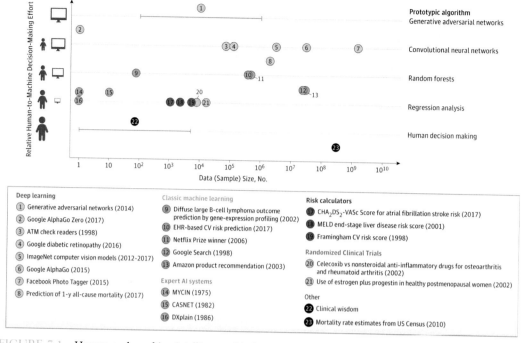

FIGURE 7.1   Human and machine intelligence. Medicine has progressed from methodologies that are mainly reliant on human intelligence to technologies that are predominantly machine intelligence. Currently, most of medicine is still at the level of logistic regression and not yet in the domain of machine and deep learning.

these AI methodologies should be tested in a clinical setting, either as a randomized controlled trial or as cluster randomization of time periods [2].

A new day has dawned: the converged scientist

Roderic Ivan Pettigrew[1,2]

Roderick Ivan Pettigrew, a physician and an engineer, authored this commentary on the concept of training and educating a cohort of physicianeers, physicians with an educational background in data science and engineering to have that very valuable dual perspective.

[1]*Texas A&M University, Houston, TX, United States*

[2]*Houston Methodist Hospital, Houston, TX, United States*

Today's medicine needs a new practitioner: not a physician who has subsequently mastered the myriad engineered tools of the profession and not an engineer who has subsequently learned diagnosis and treatment through a medical education. Today's medical challenges require a mind with a blended understanding of biological science, engineering, and data science coupled with skills to build effective new solutions to improve health outcomes. The convergence of life sciences, quantitative sciences, and engineering is the most promising approach to solving our greatest health challenges. What is needed is a practitioner of this convergence [1].

Like raising a bilingual child where neither language overpowers or shapes perception there are inherent advantages to immersion in a blended education in engineering and medicine.

Students are not saddled with prejudices as a legacy of prior training, and they learn the natural convergence of disciplines that give rise to life. New and innovative approaches to practicing medicine will arise most rapidly from this greenfield approach, not from incremental advances in prior practice.

The fundamental value of convergence and the dual clinician data scientist

Data is the currency of modern medicine. The tools of the modern physician generate reams of data, enough to overwhelm even the most devoted clinician. Adding data generated in the research sphere furthers the challenge. However, the physician who can harness data for the benefit of patients is the opportunity and the promise. Moreover, data science merged with the natural sciences can give new insights.

The fundamental power of such convergence was dramatically demonstrated when the Event Horizon Telescope 200-person team imaged a black hole. The unseeable was made visible through collaboration across disciplines. Telescopes positioned around the globe contributed multiple perspectives yielding petabytes of data, algorithms reconstructed probable images of the black hole's associated emissions which generated the data and converged experts determined which image or set of images looked most reasonable [2]. It was a triumph for collaborative science and for convergence.

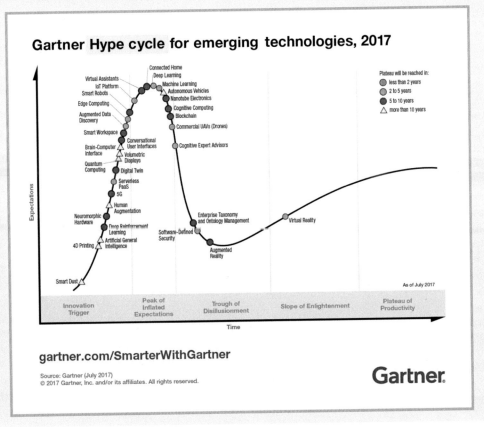

III. The current era of artificial intelligence in medicine

Emerging opportunities for the converged clinician data scientist

This new age of data science, afforded by the power of digital recording, transmission, and sophisticated analysis, is further enhanced by AI and gives rise to another type of natural convergence poised to pay huge dividends. The next generation of medical scientists must be trained as bilingual clinician-data scientists to utilize this data. AI in medicine is not about replacing physicians, it's about giving them superpowers. AI can lead to earlier disease detection, more precise diagnoses, new physiological observations, and a future of more personalized healthcare. As with the black hole, the greatest impact may be seeing the unseen: new discoveries that transform through the integration of correlated information and purposeful analysis.

As is becoming apparent, the clinician data scientist will be well positioned to leverage the ubiquity of data generated from wearable devices, external sensors, and medical information—even pharmacy compliance using smart pills—and integrate this with environmental and population-based data to streamline, prioritize and sharpen physician decision-making and treatment.

Already, AI is quickly becoming integrated into clinical practice. The US FDA has approved AI-backed devices for diagnosis of diabetic retinopathy, stroke, breast cancer, and automated analysis of 4D cardiovascular images. However, these early applications rely on "locked algorithms" and don't yet take advantage of the full potential of ML [3]. A key task at hand for the converged clinician data scientist is to capture the highly variable real-world experience to build AI that can adapt and improve its performance.

Enter the physicianeer

Texas A&M University and Houston Methodist Hospital recently instituted a joint Engineering Medicine (EnMed) blended initiative that leads to simultaneous MD and Masters in Engineering Innovation degrees in 4 years. This starts with students having an undergraduate degree in engineering or a comparable background in a 4-year curriculum blending engineering and medicine from the start. At its core, EnMed trains students to be intellectually and conceptually fluent in multiple languages across the scientific landscape. Graduates will have an integrated engineering and medicine mindset with the tools necessary to solve big problems. Students will emerge from this blended training as *Physicaneers*. Given the goal of developing invention-minded medical professionals, EnMed students are required to invent a practical solution to a medical or healthcare problem, with this requirement serving as an anchor in training problem-solvers.

The 4-year curriculum consists of three 6-month semesters of preclinical studies followed by comprehensive clinical electives with immersed ideation, invention, and translation exercises. Training begins with a prematriculation course on engineering innovation in medicine, and in each of the first three semesters, the weekly course learning objectives, study assignments, and formative learning assessment tools are laid out. These are inclusive of correlated medicine and engineering concepts for the topics under study. EnMed employs the flipped classroom approach with team-based learning; online lectures and reading material mapped to the week's learning objectives are made available. Labs, faculty discussions, and facilitated sessions include both clinical and engineering experts. During clinical rotations, students will identify a problem and then design a solution.

The promise

AI will deliver new insights, develop efficient pathways, and produce better healthcare. The next generation of healthcare professionals must understand medicine through the convergence of the life sciences, quantitative sciences, and engineering. With data science increasingly guiding patient care the role of the clinician data scientist will be important in realizing the yield from this convergence.

Medical imaging, where data science is central, may be an indicator of what lies ahead. Dramatic advances in the speed of image data acquisition and new methods for data handling, processing efficiency, and information extraction are already enhancing diagnostic content and efficiency. Targeted protocols now allow scans and analysis to be much shorter. Full studies can be acquired obviating the need for further imaging. Future interpretations will leverage quantitative data and assessments in which analytical tools integrate information across biological and physical datasets. The integration of genetic coding, protein expression, metabolic action, and tissue and organ physiology with detected image features will characterize tomorrow's medical imaging science. As we look toward the goal of health throughout our lives, those that are scientifically multilingual such as physicianeers and the clinician data scientist will be central in the transformation of the medical ecosystem from one centered on disease treatment to one focused on sustained health.

References

[1] Chen S, Bashir R, Nerem R, Pettigrew R. Engineering as a new frontier for translational medicine. Sci Transl Med 2015;7(281):281fs13.

[2] Event Horizon Telescope. [Internet]. Cambridge, MA: Harvard University; 2019. Available from <https://eventhorizontelescope.org/science> [cited 29.04.19].

[3] Food and Drug Administration. [Internet]. Silver Spring, MD: FDA; 2019 Available from: <https://www.fda.gov/MedicalDevices/DigitalHealth/SoftwareasaMedicalDevice/ucm634612.htm> [cited 25.04.19].

Building an artificial intelligence (AI) in medicine ecosystem

Piyush Mathur[1] and Francis Papay[2,3,4]

Piyush Mathur and Francis Papay, both physicians who are AI-focused, authored this commentary on the concept of building an AI ecosystem with a pyramid approach, starting with problem, and need assessment first and ending at the top with implementation, scaling, and generalizability.

[1]*Anesthesiology Institute, Cleveland Clinic, Cleveland, OH, United States*

[2]*Dermatology and Plastic Surgery Institute, Cleveland Clinic, Cleveland, OH, United States*

[3]*Department of Surgery, Cleveland Clinic, Cleveland, OH, United States*

[4]*Lerner College of Medicine, Case Western Reserve University, Cleveland, OH, United States*

Background

Healthcare can be described as a complex adaptive system (CAS) with multiple nodes of care delivery, various data inputs and interventions. At the center of any healthcare delivery system is the patient who is ultimately impacted by decision-making processes from a variety of patient care levels. Patient diagnoses systems which links healthcare decisions, the healthcare providers (decision makers) and an environment such as an electronic health record (EHR), supporting both needs to be designed in an effort to increase healthcare efficiencies, lower cost, increase patient access and deliver better quality outcomes. AI offers a solution that can be leveraged in this complex environment to collect continuous mixed quality data from multiple sources which can be processed through a variety of strategies into an actionable and impactful decision-making information. The starting point of any AI model development is recognition and clarification of a problem, clinical or administrative, and a need to develop a solution that requires AI strategies [1].

Model considerations

As described, the center of any clinical AI solution or even nonclinical administrative tool is the "patient−diagnosis" couplet. This couplet is important since all the decision-making and management of the patient's disease process is linked to his/her diagnosis. Data source supporting such a system has to be timely (near real time) and verified for accuracy and quality. Direct device or monitor derived variables are better solutions than human entered ones although large data loads, artifacts, and validation of the quality of data in former are still a challenge. Interaction of monitored or laboratory derived variables with various other data elements is essentially through EHR including medications, clinical notes, and imaging reports. Multimodal ML solutions can be leveraged not just for predictive outcomes but also for cross-validation of data, imputation (interpolation) of missing variables and the development of trends, alert systems and other patient−healthcare provider interactive features. Once processed, this information has to enable the patient and the healthcare provider through various interactive mediums, such as EHR and others through electronic monitoring devices or smart devices such as mobile phones, smartwatches and a variety of wearable electronic devices. Actionable information is tantamount to the ML model strategy and patent−healthcare provider interaction since display of nonactionable data has shown to decrease process efficiencies through data overload and has been proven to decrease quality of care and safety. Going from retrospective batched data to continuous data input from a variety of strategic data sources is a key initial step before enabling clinical validation and deployment.

While varying AI techniques have been used and success claimed in validation, most of the recent FDA approved ML systems are enabling niche focused care environment solutions. A system-wide approach for AI healthcare design has been described (Fig. 1) for ML platform developed and deployment [2]. Development and deployment of such medical AI platforms will take significant work as integrated solutions in healthcare such as electronic health records are disparate, noncommunicative and still in evolution.

Business plan

At the core, key components of an AI project development are not too different from any other project management solutions. Identification of a customer and development of a project plan is critical to its success. Building a team with expertise from a variety of healthcare providers, various levels of healthcare management, project managers and ML experts is important. These are sometimes best leveraged and accelerated through contracted collaboration with various industry or academic partners rather than the creation and the employment of a de novo AI healthcare project team. Identification of data sources can be of varying types such as monitor derived waveform data, clinical imaging, unstructured clinical notes, numerical lab variables,

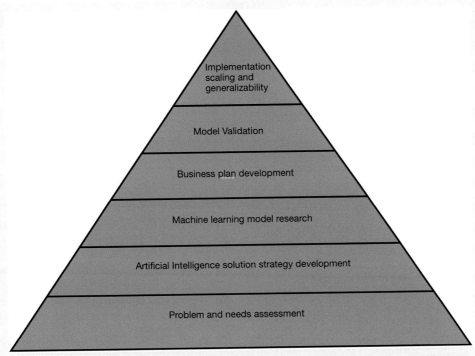

FIGURE 1   Pyramid approach to building and implementation of artificial intelligence for healthcare systems.

and patient-reported texts most often historically documented within the EHR at variable periods of time [3]. Having a clear initial AI design strategy and development timeline is critical as it's well recognized that 80% of total AI model development time is spent in data preprocessing, rather than actual model development. Significant financial resources are required in research and development phase which can be derived through grants, investor funds or other sources. Key performance indicators and their timeline milestones are essential for measurement of project development, resource efficiency, and new venture legitimacy and reputation.

Research and validation

Building an ML model is similar to any other after data preprocessing build which includes dividing the general data pool into training and test sets for model development, output features, and the assessment of accuracy parameters such as area under the curve (AUC) or F_1 score amongst others. Final selection of the best ML model and refining the model are cycled steps for continued ML strategy improvement and verification of process. Rigorous clinical validation is different from model validation using various research methods and has to follow various levels of trials and ultimately publication into peer-reviewed journals [4]. FDA has supported approval of AI solutions through its Digital Health Innovation Action Plan and its innovative Software Precertification (Pre-Cert) Pilot Program.

Success

Metrics of successful AI deployment:

1. Patient safety
2. Clinical impact (quality improvement)
3. Administrative impact (process improvement)
4. Financial (lowering of healthcare delivery costs)
5. Adaptability (user-ability interface and ease-of-use)
6. Sustainability and scalability.

Future considerations

Ultimately, many rigorous validation studies are in development and needed to address the complex questions of impact of AI in healthcare. Are we just developing better data representation tools or are these impactful healthcare solutions? Can these systems be generalized into various different healthcare systems or will they require individualized deployment approaches that might be highly inefficient and inconsistent? Can these AI solutions be scaled or interact with other AI solutions being deployed in the same healthcare environment? These questions and more are still being researched and require further innovation. The ultimate question is whether AI can save lives? The answer lies in a successful collaboration between healthcare providers and ML scientists to work together in development, validation and deployment of complex AI solutions to benefit us all.

References

[1] Topol EJ. High-performance medicine: the convergence of human and artificial intelligence. Nat Med 2019;25 (1):44–56.
[2] Maheshwari K, Cywinski J, Mathur P, Cummings III KC, Avitsian R, Crone T, et al. Identify and monitor clinical variation using machine intelligence: a pilot in colorectal surgery. J Clin Monit Comput 2018.
[3] Rajkomar A, Oren E, Chen K, Dai AM, Hajaj N, Hardt M, et al. Scalable and accurate deep learning with electronic health records. NPJ Dig Med 2018;1(1):18.
[4] Mathur P, Burns ML. Artificial intelligence in critical care. Int Anesthesiol Clin 2019;57(2):89–102.

A personal journey in artificial intelligence: in artificial intelligence hub

Hamilton Baker

Hamilton Baker, a pediatric cardiologist with a passion to learn about AI, is an inspiring role model of someone who started with only natural curiosity about AI just a few years ago and has now started an AI "hub" that is making impact at his hospital and beyond.

I attended my first AIMed meeting in Dana Point, California, early December 2017. I was honored to be selected as a winner of the abstract competition. I was enthralled at that conference and began moving my career toward applied AI in medicine. Since that conference, I have learned something new every single day about AI and its relationship with medicine. I have pivoted my research focus and career as a pediatric cardiologist in this direction. Specifically,

I have made it my mission to raise awareness about the incredible progress at the intersection of AI and medicine.

At Medical University of South Carolina (MUSC), we have recently created the MUSC AI Hub. Its purpose is to bring together those working, researching, and developing an interest in the field. The Hub will be a nexus for the AI talent at MUSC. Furthermore, it will serve as a connection point between MUSC AI and the wider community, a portal for philanthropy, investors, industry, and other university partners.

Ultimately, the goal is to move toward the development of AI tools to improve population health, empower providers, enhance value and reduce cost in a manner that keeps ethics, diversity, and inclusion at the forefront. We believe the best way to accomplish this is to provide the necessary leadership, resources, and project management to enable early-stage investigators to succeed. All projects will have the necessary domain experts represented (ML expert, informatics, clinician, ethicist, etc.) throughout the entire process from ideation to clinical implementation and adoption. Impactful clinical pain-points will guide selection of project support. Specifically, our future plan is to develop the Secure Artificial Fast Testing Environment Zone. This will provide access to a secure, deep data environment utilizing a clinical data warehouse within the infrastructure of the Hub.

I am impressed each day with the growing interest within our community, and we have recently integrated student leaders at the MUSC College of Health Professionals as well as Clemson University. It has become clear that much of the best future work will come from the incredible enthusiasm of these students.

Another important initiative is to support the creation of a formalized vision and strategy for the institution. Healthcare institutions will soon be required to make significant investments in AI tools and technologies. The development of a clear institutional vision and strategy brings together disparate AI efforts to reduce overlap and improve efficiency. I am indebted to the coalition of people that have come together over recent years through conferences such as AIMed. They have both inspired and supported me throughout this journey.

In addition, there are additional challenges that relate more to AI applications in biomedicine that render AI even more complicated than AI in other sectors (see the previous section for a more detailed discussion on these topics). The following are major issues from a societal perspective [3]:

Bias. A recurrent theme throughout this book as well as in meetings and discussions on AI in general is the issue of bias (see the previous section for a broader discussion). Clinicians are wary of bias as a result of algorithms designed for certain patient populations or single institutions that do not necessarily reflect a more heterogeneous population or additional institutions. There is also the bias that is derived from differences in samples (such as differences in quality of images as well as interpretation of these images by certain groups of specialists).

Equity. There is also an additional issue of inequity (see also the previous section). Many aspects of ML in studies may not be fair and equal to all members of the population; these elements can exacerbate healthcare disparities that already exist.

Distributive justice and its principles can potentially be incorporated into the model design and deployment of the models [4].

Ethics. An issue related to the aforementioned bias and inequity is ethics (again, see the previous section). Among the more often debated areas under this topic for medicine and health care are: (1) Who/what is liable if there is an adverse medical outcome? (2) What if there is an inequity as a result of this new technology not being accessible to everyone, especially those in the third and fourth world? (3) Should/could clinicians and hospitals financially benefit from using patient data to start an AI in medicine startup and company? (4) Who owns and have rights to medical data? There is even discussion of an autonomy algorithm, one which takes data about the patient as input and consequently yields a confidence estimate for a particular patient's predicted health care—related decision as an output with its ethical nuances [5].

Artificial intelligence (AI) in medicine and ethical considerations

Danton Char

Danton Char, a pediatric anesthesiologist with a special interest in ethics and AI, authored this commentary on the ethical issues of AI in medicine, such as biases, intended as well as unintended uses of AI tools, and the fiduciary physician—patient relationship.

Though AI is still early in its clinical adoption, it is likely to have both enormous impact on future care and to raise unique ethical challenges as it has done with implementations in nonhealthcare contexts [1]. However, a true understanding of the specific ethical issues emerging with AI in healthcare has been limited by the lack of examinable clinical AI implementations. The FDA only approved the first autonomous diagnostic system in 2018 (the IDx system for diagnosis of diabetic retinopathy) [2]. Consequently, the potential ethical issues that can be currently identified for AI implementation to healthcare are still vague and largely extrapolated from ethical challenges that have emerged with AI implementations in nonmedical contexts. Broadly, these potential ethical issues with AI can be grouped into four areas: (1) bias in the underlying training data; (2) the design agenda or intent; (3) unintended uses of the AI; and (4) the impact of the AI system on the fiduciary physician—patient relationship.

Algorithms introduced outside of healthcare have already demonstrated mirroring of *biases* inherent in the algorithm's training data, leading to problematically biased recommendations. For example, programs designed to aid judges in sentencing by predicting an offender's risk for recidivism have shown a disturbing propensity for racial discrimination [3]. It is certainly possible that racial biases such as these can occur in AI for healthcare uses as well. An algorithm applied to data will only reflect what is already present in the data. If there have been few (or no) genetic studies for certain populations, an AI system designed to predict outcomes from genetic findings will be biased. When used to predict cardiovascular event risk in nonCaucasian populations, Framingham study data has shown bias both over- and underestimating risk for different populations [4].

Bias from other subtle discriminations inherent to healthcare delivery may be harder to anticipate and prevent from being incorporated into an AI system's training and consequently, its decisions. Such biases may lead to self-fulfilling prophesies. If clinicians always withdraw intensive care in patients with findings like extreme prematurity or traumatic brain injury, AI systems may learn such findings to be always fatal, recommend withdrawal of care and the opportunity to improve outcomes for such conditions, missed.

The *agenda or intent* of the designers behind the AI creation also needs to be considered. Algorithms have already been built to prioritize a business' interests over social responsibilities. Recent high profile examples are Uber's Greyball and Volkswagen's nitrogen oxide emissions algorithm, both designed to circumvent regulations. American healthcare exists in a tension between the often-conflicting goals of health and profit. This tension needs to be considered with AI implementation to healthcare, since the builders and purchasers of AI are unlikely to be the individuals delivering bedside care. With the growing use of quality metrics to guide reimbursement, AI might be deployed to "game the system" either through guidance toward clinical actions that would improve quality metrics or, more egregiously, might simply skew data provided for public evaluation and regulators. CDS systems might also embed recommendations that enhanced the profit of its designers or purchasers (for drugs, tests, devices they hold stake in or within-insurance-plan referrals) without the clinical users being aware.

Many AI systems present a "black box" problem: their workings are not transparent to users (clinicians and patients). In the case of certain AI approaches (such as neural networks) the learning methods of the system are opaque even to the system's designers. The agenda or design intent embedded in AI systems becomes even more important against a background of opaque AI functioning.

Unintended uses of an AI system may emerge after it has been deployed. For example, studies have shown that the majority of Americans would like to spend their final days at home, but few do. An AI system designed to predict within-1-year mortality could help, bringing patients who are predicted to die to the attention of the Palliative Care team. Unfortunately, since the last year of life is often the most healthcare expensive, such an algorithm could also be coopted to help triage away potentially expensive patients from a healthcare system.

In addition, AI and AI output may *impact the fiduciary physician—patient relationship*. There has been an ongoing shift to systems-based approaches in clinical medicine, with the emergence of the electronic record and a move toward shift-work in residencies and newer clinical specialties such as emergency medicine. Though AI is only the latest addition to this trend, as clinical medicine moves progressively more toward this shift based, systems-based model, the number of clinicians who have followed diseases from presentation through resolution is also diminishing. This gives the electronic record and AI approaches both a necessary role in healthcare but also an unintended power and authority. As opposed to an individual clinician's memory and experience, the medical memory responsible for patient care is becoming the data captured in healthcare systems and the AI design decisions about how that information is managed.

Because of their role in managing clinical information and providing CDS, these AI tools may become present actors in the therapeutic relationship and will need to be similarly guided by a fiduciary responsibility to the patient. The nature of the fiduciary relationship between the patient and the AI system, impact on clinician and patient autonomy [5], and

III. The current era of artificial intelligence in medicine

what, if any, agency and responsibilities AI systems and designers should have, are now urgent ethical problems.

Ethical guidelines still need to catch up with the emerging use of AI in healthcare. Clinicians who use AI systems will need to become more educated about their construction and consequently, their limitations. Ignorance, or complicit permission of allowing AI systems to be black boxes, can lead to ethically problematic outcomes.

References

[1] Char DS, Shah NH, Magnus D. Implementing machine learning in health care—addressing ethical challenges. N Engl J Med 2018;378(11):981−3.
[2] Abràmoff MD, Lavin PT, Birch M, Shah N, Folk JC. Pivotal trial of an autonomous AI-based diagnostic system for detection of diabetic retinopathy in primary care offices. NPJ Dig Med 2018;1:39.
[3] Angwin J, Larson J, Mattu S, Kirchner L. Machine bias. ProPublica; 2016. <www.propublica.org> [accessed online at 06.10.17].
[4] Gijsberts CM, Groenewegen KA, Hoefer IE, et al. Race/ethnic differences in the associations of the Framingham risk factors with carotid IMT and cardiovascular events. PLoS One 2015;10(7):e0132321.
[5] Cohen IG, Amarasingham R, Shah A, Xie B, Lo B. The legal and ethical concerns that arise from using complex predictive analytics in health care. Health Aff (Millwood) 2014;33(7):1139−11347.

Regulation. Intertwined with bias and ethics is the area of regulation of this new technology in medicine. Even if we treat AI and its panoply of tools as "software-as-a-device," how can we effectively and expediently approve all these upcoming AI tools as these emerge and converge with other advanced technologies (such as AI and AR, see next). Perhaps, we need to match this paradigm shift in technology with a parallel philosophical shift in how we regulate. Both the FDA and the AMA deserve much praise for starting just such a shift in the way we conceptualize regulation and oversight in this new technological era [6]. The FDA has proposed new submission type and data requirements based on risk in the form of 510(k) notification, De Novo, or premarket approval (PMA) application pathway, with its Center for Devices and Radiological Health (CDRH) as an important resource. The paper reflects a more innovative strategy (including an algorithm change protocol) to the total product lifecycle regulatory approach that will be a more appropriate regulatory process for these new software-as-a-medical-devices (SaMDs) (see Fig. 7.2). Overall, the FDA and its good ML practices is a much more congruent regulatory strategy with the exponential increase in AI technologies in clinical medicine and health care. The AMA, on the other hand, has just recently passed its first policy recommendations on augmented intelligence. The recommendations include oversight and regulation of healthcare AI systems based on the risk of harm and benefit accounting for a host of factors as well as payment and coverage for all healthcare AI systems conditioned on complying with all appropriate federal and state laws and regulations [7]. In addition, there are also differences in this regulatory domain between regions of the world [8]. For the future, one potential solution is to not regulate device but rather teams or individuals working on the AI tools (akin to clinician licensure). Another possible answer lies in the Turing philosophy of "machines to deal with machines" and devise regulatory algorithms that will overlook algorithms.

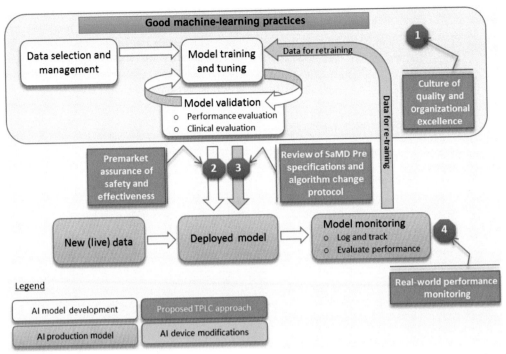

FIGURE 7.2   FDA and TPLC regulatory approach on AI/ML workflow. The general four main principles elucidated in this diagram include (1) establishing the clear expectations on quality systems and GMLP; (2) conducting a premarket assurance of safety and effectiveness of those SaMD that require premarket submission; (3) expecting these manufacturers to monitor the AI/ML devices and incorporate a risk management approach; and (4) enabling a real-world performance monitoring with transparency. *AI*, Artificial intelligence; *FDA*, Food and Drug Administration; *GMLP*, good ML practice; *ML*, machine-learning; *SaMD*, software-as-a-medical-device; *TPLC*, total product lifecycle. Source: *Adapted from FDA white paper.*

Artificial intelligence (AI) in medicine and the Food and Drug Administration (FDA)

Kevin Seals
Kevin Seals, a radiologist and an innovator, authored this commentary to discuss his experience with the new FDA precertification process with its categories as the new era of regulatory process of AI in health care begins.

ML represents one of the hottest and most potentially transformative technologies in healthcare. Although technical ML ability is critical to building a revolutionary product, it is also fundamentally important to understand the relevant pathways of regulatory approval, a major gateway between powerful technology and practically useful clinical tool. This requires interfacing with the FDA, the governmental agency regulating medical technologies in the United States.

In this text, we will provide an overview of key FDA approval concepts, with a focus on areas that are relevant to the development and release of AI technologies in healthcare.

Central to FDA approval is the classification of a particular device by the degree of associated risk. ML software has previously been classified by the FDA as a medical device, and medical devices are categorized into one of three categories based on the associated level of risk to the patient [1]:

Class I: Low risk of illness or injury. Example: a toothbrush.
Class II: Moderate risk of illness or injury. Example: computed tomography scanner.
Class III: Severe risk of illness or injury. Example: artificial heart valve.

There are a variety of pathways for medical device approval, and the pathway that is appropriate for a particular device is largely related to its associated risk level [2]. These pathways can be summarized as follows:

- Exempt from premarket review: Particularly in low-risk devices (class I), such as a simple medical bandage, an exemption can be granted such that the approval process is avoided.
- Premarket notification, also known as a 510(k) application: This is considered "fast-track" approval and is currently the most common approach for ML technologies in healthcare. Qualifying for this category has two primary requirements: (1) the risk of the device is Class II (moderate) and (2) a predicate device exists that is similar to the technology in question, with comparable safety and effectiveness. Technology in this category is said to be "evolutionary" rather than "revolutionary," as revolutionary technology lacks a predicate. This approval requires approximately 4–8 months of time, 20 FDA work hours, and $5000 in monetary costs.
- PMA: The strictest and most rigorous level of FDA approval is required when there is no preexisting predicate technology and/or for class III devices. This pathway typically requires 3–7 years of rigorous clinical trials and approximately 1200 FDA work hours.
- De novo: This is a special category for devices that have no similar prior technology but are low to moderate in risk (Class I or II). In this case the approval process reduces to the less rigorous 510(k) fast-track system, simplifying time, and cost burdens for both developers and the FDA.
- Humanitarian device exception: A special approval case in which a medical device is designed to benefit patients with a rare disease, defined as fewer than 8000 cases per year. Given the combination of medical need and difficulty of collecting robust datasets in these small populations, the FDA requirements are less stringent.

ML technologies and other digital healthcare applications have created unique challenges for the FDA. Whereas traditional medical technology required years of development, this timescale is shifting to months for digital products, where code can be rapidly written and deployed at a low cost. This has resulted in a dramatic increase in the number of FDA approval applications and significant workload burdens for the FDA. Furthermore, the regulatory system was created well before the popularization of software applications, and there can be significant ambiguity regarding which software requires FDA approval.

In the past several years the FDA and Congress have taken steps toward providing guidance and reform on these topics. Congress enacted the 21st Century Cures Act in 2016, clarifying the role of the FDA in regulating digital health products. Specifically, Section 3060 of the Cures Act and a subsequent FDA draft guidance exclude medical software from regulatory evaluation when certain criteria are met [3]. Various categories are identified as not requiring approval, including "wellness" software encouraging a healthy lifestyle, certain types of electronic medical record (EMR), and particular types of CDS technology. This clarity regarding regulatory need is critically valuable to developers of medical technology. The Cures Act further reduced regulatory burden in a number of other ways, including streamlining the approval of "breakthrough devices" and making the requirements for Humanitarian Device Exception less stringent.

A streamlined oversight process is of particular importance for ML technologies, which offer immense potential patient benefit and are emerging rapidly and in high volume. In a key approach to this problem, the FDA established a Digital Health Software Precertification Pilot Program as a part of its Digital Health Innovation Action Plan. In this pilot selected technology companies (such as Apple, Fitbit, Google's Verily, and Pear Therapeutics) are being evaluated using an experimental precertification approach. The basic logic of precertification is, should a company work with the FDA to establish a baseline of excellence in safe, high-quality medical software development, they should face less onerous regulatory oversight in the future. This parallels the precertification approach of the Transportation Security Administration, where prevetted travelers enjoy a streamlined travel process.

Although the FDA precertification program is currently at the pilot stage, it represents one of the most promising long-term solutions for the strained FDA-approval process. Of note, it may have particular value and relevance for ML technologies; in that, their dynamic nature has associated regulatory ambiguity. That is, although an ML technology may attain FDA approval, it is currently unclear if that approval remains when an additional training data is gathered and used to update and theoretically enhance the algorithm. There is a risk that the new data will be problematic in some way and break a previously functional algorithm. Precertification offers a system for avoiding repeat FDA evaluation with each training data update, as precertified companies can perhaps freely update their technology while undergoing periodic FDA audits to verify continued quality and safety.

ML technologies in healthcare are undergoing a period of unprecedented growth, offering the potential to fundamentally transform the practice of medicine. While new software was previously considered the regulatory equivalent of a new artificial heart, regulatory bodies are developing a modern, proactive approach to the evaluation of new digital technologies. Through the 21st Century Cures Act of US Congress, the Digital Health Innovation Action Plan of the FDA, and other novel and progressive efforts, digital products are beginning to be considered in a category of their own with new protocols for evaluation and approval. The willingness of governing bodies to consider digital products outside of preexisting medical device paradigms is

III. The current era of artificial intelligence in medicine

encouraging for the many developers working to build digital technologies with the potential to revolutionize healthcare [4].

References

[1] The device development process—step 3: pathway to approval [Internet]. Available from: <https://www.fda. gov/ForPatients/Approvals/Devices/ucm405381.htm>, 2018 [cited 03.12.18].

[2] Van Norman GA. Drugs, devices, and the FDA: part 2. JACC Basic Transl Sci 2016;1(4):277—87.

[3] Bennett JD-M, Heisey CM, Pearson IM. FDA's evolving regulation of artificial intelligence in digital health products. Lexology [Internet]. Available from: <https://www.lexology.com/library/detail.aspx?g = cf6351a3-a944-4468-9214-d141a689955e> [cited 03.12.18].

[4] Press announcements—FDA selects participants for new digital health software precertification pilot program [Internet]. Available from: <https://www.fda.gov/NewsEvents/Newsroom/PressAnnouncements/ucm577480.htm> [cited 03.12.18].

Economics. Last, and certainly not the least important, is the issue of economics of AI in medicine and health care as it is not clear at present just how the payment models would be designed for AI implementation in health care. One potential solution is to have AI be embedded in a clinical program (and thus overall hospital budget) in the form of an AI-centered team and service. This strategy will enable this team to have a budget and manpower to take on AI-centric projects. Other AI experts have rendered an "AI-as-a-service" model where it is treated as a resource that is utilized when needed (akin to electricity). This concept is especially fitting since Andrew Ng has stated, "AI is the new electricity" (perhaps we can add that health care is a primitive hut with a single lightbulb as the data infrastructure has ample room for improvement). In addition, if AI is deployed in countries in which there is insufficient tertiary and quaternary care, how are procedures, medications, and interventions going to be financed after more new cases are diagnosed?

Barriers to artificial intelligence (AI) dissemination and implementation from an entrepreneurial perspective

Arlen Meyer

Arlen Meyer, a surgeon with vast entrepreneurial experience, authored this commentary on barriers of AI in terms of four categories: technical, human factors, business models barriers to entry, or environmental factors such as regulatory, societal, ethical, political, or legal.

AI, ML, neural networks and deep learning are the new, new thing. Applications in medicine are potentially vast and, as most things on the upside of the hype cycle, there is a proliferation of papers, conferences, webinars, and organizations trying to stay ahead of the curve. Doing so, however, means you are on the leading edge to the trough of disillusionment.

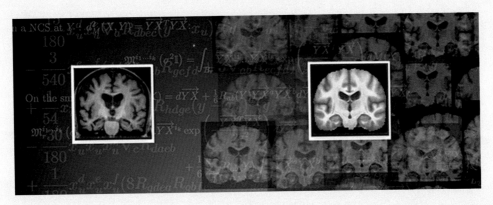

Despite advances in computer technology and other parts of the fourth industrial revolution, there are many barriers to overcome before ML *crosses the chasm.* Here are some things you should know about *dissemination and implementation, and innovation diffusion basics.*

There are four basic categories of barriers: (1) technical; (2) human factors; (3) environmental, including legal, regulatory, ethical, political, societal, and economic determinants; and (4) business model barriers to entry.

Technical

A recent Deloitte report *highlighted the technical barriers.* Here are the vectors of progress (Fig. 1):

Explainability, for example, is a barrier. Say one can predict the onset of Type 2 diabetes. It's one thing to say that we think there's a propensity but, typically the next question is "Why?" Most algorithms don't, but we make sure that if we provide a prediction we can also answer those types of questions.

Human factors

Human factors, such as how and whether doctors will use AI technologies, can be reduced to the *ABCDEs of technology adoption. Research suggests the reasons* more ideas from open innovation aren't being adopted are political and cultural, not technical. Multiple gatekeepers, skepticism regarding anything *"not invented here,"* and turf wars all hold back adoption.

Attitudes: While the evidence may point one way, there is an attitude about whether the evidence pertains to a particular patient or is a reflection of a general bias against "cookbook medicine."

Biased behavior: We're all creatures of habit and they are hard to change. Particularly for surgeons, the switching costs of adopting a new technology and running the risk of exposure to complications, lawsuits, and hassles simply isn't worth the effort.

Cognition: Doctors may be unaware of a changing standard, guideline or recommendation, given the enormous amount of information produced on a daily basis, or might have an incomplete understanding of the literature. Some may simply feel the guidelines are wrong or do not apply to a particular patient or clinical situation and just reject them outright.

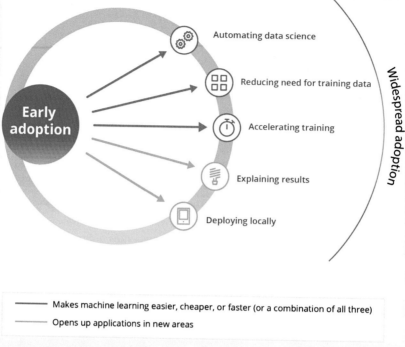

FIGURE 1   The five vectors of progress.

Denial: Doctors sometimes deny that their results are suboptimal and in need of improvement, based on "the last case." More commonly, they are unwilling or unable to track short term and long term outcomes to see if their results conform to standards.

Emotions: Emotion is perhaps the strongest motivator, and includes: fear of reprisals or malpractice suits, greed driving the use of inappropriate technologies that drive revenue, need for peer acceptance to "do what everyone else is doing" or ego driving the opposite need to be on the cutting edge and winning the medical technology arms race or create a perceived marketing competitive advantage.

Ethics

The UK House of Lords Select Committee on Artificial Intelligence has asked the Law Commission to investigate whether UK law is "sufficient" when systems malfunction or cause harm to users.

The recommendation comes as part of a report by the 13-member committee on the "economic, ethical and social implications of advances in AI."

One of the recommendations of the report is for a cross-sector AI Code to be established, which can be adopted nationally, and internationally. *The Committee's suggested five principles for such a code are:*

1. AI should be developed for the common good and benefit of humanity.

2. AI should operate on principles of intelligibility and fairness.
3. AI should not be used to diminish the data rights or privacy of individuals, families, or communities.
4. All citizens should have the right to be educated to enable them to flourish mentally, emotionally, and economically alongside AI.
5. The autonomous power to hurt, destroy or deceive human beings should never be vested in AI.

The Nuffield Council on Bioethics identified the ethical and societal issues as

1. reliability and safety;
2. transparency and accountability;
3. data bias, fairness, and equity;
4. effects of patients;
5. trust;
6. effects on healthcare professionals;
7. data privacy and security;
8. malicious use of AI.

Finally, the parts of the environmental *SWOT analysis are more wild cards in the game.*

Environmental and business model barriers to entry

Startup developers of commercial AI applications operate in a competitive market. They compete with the data available to them and meet a market need for AI applications for midsize companies, which, in turn, enables those companies to compete with larger companies that often develop AI applications internally.

Here are some regulatory and reimbursement issues.

AI in medicine is advancing rapidly. However, for it to grow at scale and provide the promised value will depend on how quickly the barriers fall.

Arlen Meyers, MD, MBA is the President and CEO of the *Society of Physician Entrepreneurs.*

Overdiagnosis. With all the AI tools that will be available, there can be potential issues of overdiagnosis of diseases that are subclinical [9]. This can lead to the risk-to-benefit ratio not in favor of the patient if therapy has more risk than the subclinical diagnosis of certain conditions. An example would be a subclinical cardiac arrhythmia detected on a wearable device that a primary care or cardiologist decided to treat with an antiarrythmic agent.

Complexity. The nature of biomedicine is intrinsically very complex as it is a biological system with constantly evolving milieu. In addition, many variables in clinical medicine do not have a dichotomous or categorical result and can have a continuous (or fuzzy) nature to the data. Lastly, there is a system's complexity to the person or population with conditions or diseases that may be challenging for any prediction model.

III. The current era of artificial intelligence in medicine

Chaos and medicine

Peter R. Holbrook[1,2]

Peter R. Holbrook, a renowned pediatric critical care physician who is known for his insights, authored this commentary in an effort to elaborate chaos theory and complex adaptive system in biomedicine to set a framework for any possible congruence with an artificial-intelligence solution.

[1]*George Washington University, Washington, DC, United States*

[2]*Children's National Health System, Washington, DC, United States*

Medical science is dominated by reductionism. Dissections, microscopes, gene isolation, knock-out animals, and DNA sequencing allow deeper and more detailed looks at smaller and smaller units until the function of fundamental pieces is understood. Then, the thinking goes, the units are reassembled to recreate the now understood whole. To paraphrase the brilliant Simon-Pierre LaPlace, give a man the positions and velocities of all the particles in the universe, and he can predict the future [1].

But in the 20th century, Relativity, Quantum Theory and the study of control systems which showed great variance in results despite minor differences in initial conditions caused a rethinking. Examination of the "noise" led to the discovery of order in apparent chaos, and this has become codified as Chaos Theory. Fortified with mathematics and Chaos Theory, control systems in many disciplines (e.g., anthropology, sociology, biology, and evolution) have been intensively investigated [2]. Many now are described as Complex Adaptive Systems (CASs).

A *CAS* is a system in which understanding of the individual subsystems does *not* convey a perfect understanding of the whole system's behavior. Key characteristics are summarized in the table.

Attribute	Comment
Deterministic	Caused by stimuli and responses of system; not random
Self-organizing	Resulting system comes together spontaneously, not directed from another force
Nonlinear	One or more exponential amplifying or suppressing functions
Nonrepetitive	May be very similar but never identical responses because initiating conditions never identical
Spatial limitations	System behavior is contained within a particular space of possible options
Organize around attractors	System results coalesce around geometric structures
Emergent behavior	Whole system behavior not a sum of parts, e.g., walking or creativity is not obvious from parts of human body
Sensitivity to perturbations	System brings internal and external variances into control
Adaptive Resistant to change	Robust and able to process variances

	System evolved in face of multiple variables; able to manage similar stimuli
Subject to phase transitions	If confronted with major change may oscillate or shift to new phase
Fractal characteristics	Statistical order in chaos (see text)

Characteristics of complex adaptive systems.

A critical observation is that CAS data sets have *fractal* components [3,4]. A geometric fractal is an object that has self-similar organization, that is, details of the structure at smaller scales have a similar form to that of the whole structure. Data can also be fractal over temporal scales, that is, data fluctuations over long periods have similar structures to data from a shorter time-scale. Furthermore, when examined at greater levels of magnification, progressively greater details of the structure are observed (scaling) and there exists no specific scale to describe the structure (scale-invariance) (Fig. 1).

Fractal geometric forms are abundant in nature (e.g., snowflakes, mountain ranges, and coastlines) and medicine (e.g., bronchial tree, arterial system, and bone structure). An efficient development of maximal surface area in a constrained space is fractal (leaf exposure to the sun, blood exposure to air, calyceal collecting system in the kidney). Not surprisingly, filling of small spaces is fractal (DNA and protein folding, and cancer metastases) as is the transmission of signal from origin to periphery (e.g., the cardiac conduction system).

In the typical reductionist scenario when a physician encounters a patient with a disorder, the disorder is the insult and treatment is focused on eliminating it, for example, a bug deserves a drug. Scant attention is paid to the efficacy of the patient's intrinsic response system, despite the fact that for billions of years organisms have been responding to threats.

Consider the body's control system. Intracellular chemical reactions occur simultaneously and in harmony with the function of tissues, organs, and organ systems, and whole organism level processes such as fight or flight. Diurnal cycles, growth, development, maturation, and senescence overlay daily functions. Biomes make demands on the organism. Traumatic, infectious, and neoplastic threats are continual.

Is it possible that the control systems which evolved at micro scales have adapted into self-similar control systems at macro scales, and that the accumulated signal is fractal? Human data gathered over the last three decades support this hypothesis. Variations in heart rate, breathing rate, and gait in normal humans—each a result of multiple interacting control systems—exhibit fractal nature [3,4]. These fractal patterns are noted to mature then deteriorate with aging. Also, in several conditions a narrowing range of fractal patterns is associated with disease, deterioration, and death [4]. A special case is sepsis where the body's overreaction to an infectious stimulus reaction (multiple organ failure syndrome) leads itself to deterioration and death [5].

The observations are consistent with a large, multifaceted, self-organized control system, sampling data from many different scales, and poised to deal with a perturbation coming from any direction. Reduction in the breadth of the control system leads to less resilience and greater risk of deterioration or demise. Thus to the physician trained in Chaos theory, disorder may be due

III. The current era of artificial intelligence in medicine

Spatial self-similarity   Temporal self-similarity

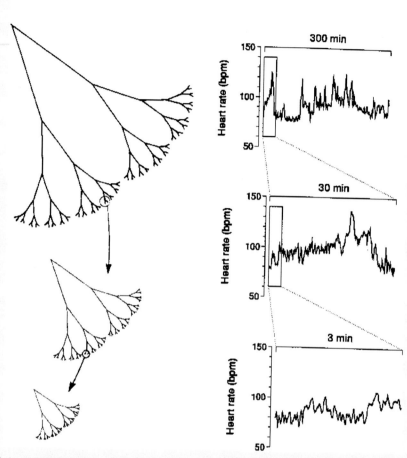

FIGURE 1   Schematic representations of self-similar structures and self-similar fluctuations. The tree-like, spatial fractal (*Left*) has self-similar branchings, such that the small-scale structure resembles the large-scale form. A fractal temporal process, such as healthy heart rate regulation (*Right*), generates fluctuations on different time scales (temporal magnifications) that are statistically self-similar. Source: *From Goldberger AL. Non-linear dynamics for clinicians: chaos theory, fractals and complexity at the bedside. Lancet 1996;347:1312–4 with permission.*

to the direct effect of an insult, a failure of the inherent control systems, or an uncoupling of the components of the control system leading to an exaggerated response which is itself deleterious.

This Chaos Theory–based CAS construct provides a new framework for addressing disease and disorder. Specifically, an emphasis on restoring the control system to "health" may allow it to better deal with threats. Current findings from a CAS analysis are diagnostic and prognostic, but manipulation of the control system to improve therapeutic results is surely coming [5].

III. The current era of artificial intelligence in medicine

References

[1] Hoefer C. Causal determinism. In: Zalta EN, editor. The Stanford encyclopedia of philosophy, Metaphysics Research Lab, Stanford University, Palo Alto, CA, 2016.
[2] Coffey DS. Self-organization, complexity and chaos: the new biology for medicine. Nat Med 1998;4(6):882−5.
[3] Goldberger AL. Non-linear dynamics for clinicians: chaos theory, fractals and complexity at the bedside. Lancet 1996;347:1312−14.
[4] Varela M, Ruiz-Esteban R, De Juan M, Chaos MJ. Fractals and our concept of disease. Perspect Biol Med 2010;53 (4):584−95.
[5] Buchman TG. The community of the self. Nature 2002;420:246−51.

Clinician Cognition and Artificial Intelligence in Medicine

System 1 and System 2 thinking. Daniel Kahneman, the Nobel Prize−winning psychologist noted for his work on decision-making, described System 1 versus System 2 thinking (fast and experiential vs slow and analytical, respectively) (see Table 7.2) [10]. This dichotomy conveniently delineates some of the key differences between clinicians (more prone to System 1 thinking) and data scientists (with their affinity for System 2 thinking). Physicians, especially those in the acute care clinical setting (such as emergency room, ICUs, and operating and procedure rooms), often rely on a fast intuition-based system I thinking that is based on past experiences and judgments. Data scientists, on the other hand, more frequently approach problems with slower and more logical progressive thinking that is rationality-based System 2 thinking. Both of these types of thinking are

TABLE 7.2   System 1 versus System 2 thinking.

	System 1	System 2
Brain location	Limbic system	Neocortex
Velocity	Fast	Slow
Thinking type	Intuitive	Rational
	Qualitative	Scientific
	Pattern	Deliberate
Decision type	Simple	Complex
Conscious state	Unconscious	Conscious
Effort level	Lower	Higher
Error rate	Higher	Lower
Characteristic	Associative	Analytical
Advantage	Fast	Accurate
Disadvantage	Biased	Slow

III. The current era of artificial intelligence in medicine

necessary in medicine, but physicians often lack the time nor the discipline to utilize proportionally more System 2 thinking.

Utilizing different allocations of System 1 and System 2 thinking depending on the situation may be the best strategy; this is a potential future application of AI in aiding this allocation in various case scenarios. For instance, rather than relying on mostly System 1 thinking in urgent or emergent situations, AI can provide the complementary and relatively fast (compared to humans) System 2 thinking to support the former. This combined System 1 and 2 thinking would be ideal in the ICU where urgent and emergent decisions are made often but could always use more System 2 thinking especially in an expeditious manner. On the other hand, excessive or redundant System 2 thinking can be balanced with System 1 thinking to render the entire thinking process more efficient and expedient. For instance, a nonurgent diagnostic dilemma is better off with mostly System 2 thinking than System 1 cognition but perhaps cognitive shortcuts can be made with System 1 input. In short, the human clinician and AI can decide how much System 1 and 2 thinking are necessary to be allocated on a case-by-case basis; this balance or symmetry between the two systems has been termed dual process thinking [11].

Artificial intelligence in medicine and collective wisdom

Geoffrey W. Rutledge

Geoffrey W. Rutledge, an internist with a strong background in medical informatics, authored this commentary on the value of a combined knowledge-based and data-driven diagnostic prediction model and the special collective wisdom of many doctors.

The popularity of direct-to-consumer health information services is driving a resurgence of interest in developing tools that help people understand the possible explanations for their symptoms. New AI-based methods are being developed that we expect will perform as well as (or better than) a good doctor at this task: predicting which of about 1000 possible conditions could best explain a set of symptoms based only on their medical history, including prior conditions, medications, and risk factors.

Knowledge-based diagnostic prediction models

The field of AI in medicine began with attempts to create diagnostic models. The first rule-based "expert systems" [1] and other heuristic symbolic methods [2] performed well at narrowly constrained diagnostic tasks [3]. Early "Naive Bayes" models [4] led to more powerful Belief Networks (causal probabilistic models) [5] and to methods that improved the scope and accuracy of these models while controlling their complexity [6].

However, knowledge-based expert systems did not achieve wide adoption or use [7] for a number of reasons, including

1. limitations imposed by model assumptions (e.g., the common assumption in Bayesian models that symptoms are conditionally independent);
2. challenges in knowledge management, given that a general-purpose diagnostic system should include the collective wisdom of doctors in every specialty; and

3. difficulties in measuring the performance of the models, and of predicting how new knowledge added affects the performance of an established and validated model.

Data-driven diagnostic prediction models

Recent advances in AI in medicine demonstrate that data-driven deep neural network (DNN) models can provide accurate and reliable "better than expert" performance in a variety of medical applications [8], so it is natural to ask how DNNs could create better prediction models for clinical diagnosis. DNN methods apply massive computing power to discover the complex relationships among inputs in large training sets to create a diagnostic model that predicts the most likely explanations for any given set of inputs (e.g., features of a patient scenario).

DNN models require training data sets that include a large sample of possible variations for each diagnosis. For the problem of clinical diagnosis in ambulatory primary care, there are at least 1000 possible diagnoses/explanations, and conservatively, about 3000 symptoms and other features that are relevant. If we limit the number of relevant features of each diagnosis and each patient scenario to no more than 20, then there are more than 10^9 possible patient scenarios, out of the larger set of 10^{51} possible combinations of all symptoms.[6]

The challenge for the creation of a data-driven diagnostic model is chiefly the need for massive number of patient scenarios, including all the conditions that are less common (and therefore observed less frequently).

Models that combine expert knowledge and clinical observations

We can solve the problem (1) of model assumptions and simplifications by building DNNs that are trained with sufficient data. We solve the problem (2) of finding the consensus knowledge across multiple specialties by tapping the collective wisdom of a large network of contributing experts (doctors), which is the basis for the knowledge model that is incorporated in the first version of the Dr.AI service on HealthTap [9]. We use the same network of experts to solve the problem (3) of measuring predictive accuracy, by eliciting the consensus for the diagnosis or set of diagnoses for a large set of benchmark cases. The benchmark cases enable an objective measure of diagnostic performance.

Finally, we solve the problem (4) of the need for massive training data by combining the available observed clinical case data with simulated data derived from a knowledge model built from the collective wisdom of a large crowd of expert doctors.

Collecting the wisdom of doctors

How is it possible to collect the wisdom of a sufficient number of doctors across the range of medical specialties, and at the scale needed to create and fine-tune a knowledge model of all of medicine? The RateRx project shows how a large network of contributing experts makes this task possible.

In the RateRx project, more than 7000 doctors contributed their knowledge of the clinical effectiveness of medication treatments; they provided 215,979 individual reviews of 6577 medication—indication pairs to create a knowledgebase of the collective wisdom of doctors on the

---

[6] The number of combinations of symptoms (n) taken m at a time is $n!/[m!(n-m)!]$. For all possible combinations of 20 symptoms, $n = 3000$ and $m = 20$, this is 1.3×10^{51}. The sum of possible patient scenarios for 1000 diagnoses, with $n = 20$ and m from 5 to 20, is $1000 \times 1,042,380 = 1.04 \times 10^9$.

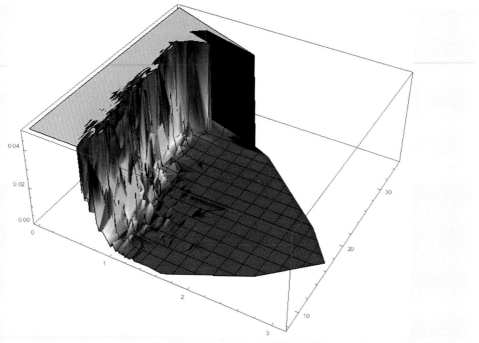

FIGURE 1   *P*-value (*y*-axis, .0−.05) for the accuracy of star-ratings of medication effectiveness (*x* axis, 0−3) as a function of the number of ratings (*z*-axis, 10−30). For average ratings with at least 30 assessments, differences of 0.3 stars were significant at the $P = .05$ level.

clinical effectiveness of medication-treatments [10]. The accuracy of assessment of doctors' ratings of clinical effectiveness was ± 0.3 stars on a rating scale of 1−5 stars (see Fig. 1).

The network of contributing doctors is now working on the ShareKnowledge project, which assesses the knowledge of which risk factors contribute to each condition, and how often each symptom is likely to occur in each condition. We combine that information with the prior probability of each condition and the gender and age distributions for each condition to refine a comprehensive knowledge model. This model is then used to generate simulated patient data that complement the observed patient data, to create the training data at a sufficient scale to develop a more effective DNN model for predicting clinical diagnoses.

An improved clinical diagnostic model

Thanks to the participation and collective wisdom of thousands of doctors, it will soon be possible to deploy improved clinical diagnostic models to help people understand what may be the explanation for their symptoms and to support doctors' ability to make accurate diagnoses for their patients.

References

[1] Van Melle W. MYCIN: a knowledge-based consultation program for infectious disease diagnosis. Int J Man Mach Stud 1978;10:313−22.

[2] Szolovits P, Patil RS, Schwartz WB. Artificial intelligence in medical diagnosis. Ann Intern Med 1988;108 (1):80−7.

[3] Arene I, Ahmed W, Fox M, Barr CE, Fisher K MD. Evaluation of quick medical reference (QMR) as a teaching tool. Comput: Comput Med Pract 1998;15(5):323−6.

[4] Betaque NE, Gorry GA. Automated judgmental decision making for a serious medical problem. Manage Sci 1971;17(B):421−34.

[5] Andreassen S, Jensen FV, Olesen KG. Medical expert systems based on causal probabilistic networks. Int J Bio-Med Comput 1991;28(1−2):1−30.

[6] Heckerman DE, Nathwani BN. An evaluation of the diagnostic accuracy of Pathfinder. Comput Biomed Res 1992;25(1):56−74.

[7] Duchessi P, O'Keefe RM. Understanding expert systems success and failure. Expert Syst Appl 1995;9(2):123−33.

[8] Esteva A, Kuprel B, Novoa RA, Ko J, Swetter SM, Blau HM, et al. Dermatologist-level classification of skin cancer with deep neural networks. Nature 2017;542:115−18.

[9] Mukherjee S. You can now download an artificial intelligence doctor. Fortune January 10, 2017.

[10] Lapowsky I. Doctors on this site rate drugs to give patients more power. Wired Magazine March 11, 2015.

Uncertainty in biomedicine. One of the major issues with applying data science to biomedicine is the necessary bundling of the dichotomous or categorical element of the former with the highly empirical and "fuzzy" nature of the latter. There is a relatively high degree of error in human interpretation of medical data in general and this degree of error may even be exaggerated in the future as a result of computer-aided diagnosis (automation bias). In addition, there is a high degree of interobserver variability in interpreting medical images or data so any data science applied to this data will need to consider this variability. There is also intraobserver variability in which one single clinician may have varying degrees of interpretation for the exact same data. This variability can occur simply from being influenced by medical history attached to the data or innate human mental and perceptual capabilities at different times. Lastly, uncertainty can also occur as a result of a time continuity to the patient's health outcome. For example, if an echocardiogram is performed in a teenager with a family history of hypertrophic cardiomyopathy and is interpreted as "normal," should this study be reinterpreted (to be "not normal") if he turned out to have hypertrophic cardiomyopathy at age 27 years?

Clinician cognitive biases and heuristics. Physicians are often vulnerable to cognitive biases and heuristics with their decision-making processes, especially when they are overburdened with their workload and stressed with their expectations to know everything as well as the expediency that they are expected to make decisions today. While cognitive biases distort pure cognition as a result of circumstances based on cognitive factors, heuristics are mental shortcuts to reduce the effort in cognition and can therefore lead to biases. In Jerome Groopman's How Doctors Think [12] as well as the review on clinicians' biases and heuristics by Klein [13], both described several cognitive deficiencies in the way physicians think. Table 7.3 delineates some of the common cognitive biases as well as heuristics that may distract the clinician from making the best decision in diagnosis and/or therapy; a few of these biases and heuristics are described in more detail later.

One such cognitive deficiency is confirmation bias, which is the tendency for physicians to search for information that confirms one's preexisting hypothesis. In Sherlock Holme's parlance: "It is a capital mistake to theorize before one has data. Insensibly one begins to twist facts to suit theories, instead of theories to suit facts." A clinical example would be a

TABLE 7.3   Clinician cognitive biases and heuristics.

Bias or heuristic	Description
Confirmation bias	Proactive and selective seeking of information that confirms our prior and/or existing beliefs
Availability heuristic	Tendency to rely on events or examples more available in one's memory or experience
Illusory correlation	Tendency to overestimate a relationship between variables when the evidence is not supportive
Representativeness heuristic	Rendering a decision based on representativeness of a group and not evidence
Dichotomous thinking	Situation viewed only in two categories or outcomes instead as a continuum of possibilities
Automation bias	Tendency to rely or favor automated decision support systems more than human cognition
Anchoring and adjustment	Tendency to rely too much on one piece of information for decision-making
Selection bias	Bias from selection of data or individuals that renders a randomization not possible
Affect heuristic	Decisions made with strong influence by emotions that are present
Groupthink	Group dynamic deficiency when the group desires conformity and makes a suboptimal decision together
Belief bias	Tendency to judge the strength of the argument based on the plausibility of the conclusion only
Dunning–Kruger effect	Cognitive bias in which someone mistakenly assesses his/her cognitive ability greater than it really is
Framing effect	Cognitive bias in which decisions are made depending on situation's positive/negative semantics
Semmelweis effect	A reflex-like tendency to reject new evidence because it directly contradicts the established paradigm

cardiologist thinking that the emergency room patient with chest pain has a myocardial infarction by mistakingly thinking that the borderline EKG is abnormal rather than a normal variant (even if the cardiac enzymes are normal).

Heuristics, on the other hand, are mental shortcuts that anyone who thinks use to decrease the cognitive load. One example such a shortcut that can lead to an error is the availability heuristic, or an intellectual shortcut that relies on immediate recall when evaluating a situation. A third cognitive error is illusory correlation, or the tendency to perceive two events as causally related when these events are not related and occurred by chance. Klein also further describes the representativeness heuristic: this heuristic makes the assumption of a certain conclusion based on available information but without attention to base rates.

In face of all these biases and heuristics, Klein summarized rules for good decision-making as follows: (1) be aware of base rates; (2) consider whether data are truly relevant, rather than just salient; (3) seek reasons why your decisions may be wrong and entertain alternative hypotheses; (4) ask questions that would disprove, rather than confirm, your current hypothesis; and (5) remember that you are wrong more often than you think.

Overall, perhaps the myriad of human biases and heuristics can potentially be neutralized with an objective AI-supported strategy in the decision-making process to reduce the proportion of human-related biases and heuristics that often lead to errors.

Levels of evidence in medicine. Physicians often discuss and perhaps revere evidence-based medicine (EBM), which is centered on using evidence to help make sound clinical decisions. The cornerstone of EBM is a hierarchical system of evidence called the level of evidence and clinicians aim to practice with higher/highest levels of evidence. This system started with the Canadian Task Force on Periodic Health Examination in 1979 [levels I, II.1 and II.2, and III with I being at least one randomized controlled trial (RCT) and III being solely expert opinions]. Following this, the United States Preventive Services Task Force (USPSTF) published its three-level system with modifications. A metaanalysis strategy with the Cochrane Collaboration takes systematic reviews to a higher level of rigor.

The entire continuum or pyramid for EBM is shown below with three broad categories starting from the bottom: observational studies, experimental studies, and at the highest echelon, critical appraisals (see Table 7.4). The bottom level is information from anecdotes, publicly available information, and even opinions and recommendations from experts. One level higher is case reports and case series that delineate the conditions and relevant information without a study design of any kind. A case-controlled study is a retrospective examination of the condition with diagnosis or therapy implications. A cohort study (also called longitudinal study) is a prospective study of patients of a particular condition,

TABLE 7.4   Evidence-based medicine levels.

Level of evidence	Level
Critical appraisal	*Highest*
Metaanalyses	
Systematic review	
Critically appraised topics	
Critically appraised individual articles	
Experimental studies	
Randomized controlled trials	
Nonrandomized controlled trials	
Observational studies	
Cohort studies	
Case-controlled studies	
Case series and case reports	
Expert opinion and information	*Lowest*

III. The current era of artificial intelligence in medicine

disease or risk factor and can also involve an intervention or observation of some kind with before and after study parameters. An RCT is a study with the design involving treated and untreated (or having different interventions) patients in a randomized assignment format and seeing the results of these different groups. RCT, especially in the double-blinded form, is considered the most reliable clinical test of a study. The critically appraised articles and reviews are highly regarded articles from usually the leaders and experts in that particular realm. Finally, the metaanalyses are sometimes performed by the Cochrane Collaboration, which is an international virtual effort to study available information in an unbiased way; this final level of evidence is considered the highest level of evidence in biomedicine.

A technology solution for the mining of medical literature

Todd Feinman

Todd Feinman, a physician with decades of dedicated work in the realm of medical knowledge, authored this commentary on utilizing artificial-intelligence technology to perform automated search, extraction, and analysis of evidence in the medical literature to help clinicians make the best decisions.

Every day millions of consumers go online to find real evidence before they purchase cars, mutual funds, books, and appliances. From Carfax to E-trade to Amazon reviews to Consumer Reports, these online resources leverage ML and natural language processes to enable consumers to use evidence/data to identify the best products that match their personal preferences. But why is there is no comparable user-friendly mining resource online for consumers who want to also use real evidence to identify the best healthcare therapies for their unique medical condition?

The volume of medical literature is rapidly increasing every year; Over 1 million new articles are published each year in the existing 25,000 journals globally. Policymakers, scientists, healthcare providers, and patients are struggling to stay current with the evidence that matters to them when using these traditional search platforms. Online healthcare information sites, such as WebMD or Up To Date, provide healthcare "content" (i.e., opinions, recommendations, information) but only provide a fraction of the relevant evidence found in clinical studies regarding the safety and efficacy for medical therapies. The available online evidence resources, such as Google Scholar or PubMed, are too hard to use. Users, who do not have expertise in search strategies, will get search results that have either too much noise (irrelevant articles) or too many misses (relevant articles not retrieved).

To solve the above search and retrieval problem requires the creation of a software program that facilitates the efficient mining and analysis of all types of healthcare literature. This search platform would be able to quickly retrieve a wide range of relevant medical articles from multiple sources, such as PubMed, ClinicalTrials.gov, newsfeeds, guidelines, and some proprietary databases. Then, the platform would rank the search results based on how relevant the article is to the search query.

An innovative search engine will use state of the art semantic web, NLP, and ML technologies in order to provide a superior alternative to existing search platforms. This AI platform will include features that will make it easy for consumers, doctors, and others to use real evidence to make informed decisions. These features include, but not limited to, the following:

- Digital reference lists of the available literature including links to most relevant abstracts that answer any medical question(s)
- Literature monitoring with automated signal detection and notifications when new articles found
- Mapping and identification of influential authors and/or healthcare providers
- Interactive high-level analytics that include stimulating data visualizations (i.e., distribution graphs of timelines and geography, patient population mapping, categorization of studies)
- Ontology system: Users enter natural language search terms and then rely on the system to automatically include all synonymous variations of the search term
  - Enables users to easily iteratively expand or contract search parameters.

Some software companies in the information space report that they have already built the above AI-enabled search engines. But most of the platforms have major problems with accuracy and/or ease of use. To overcome these limitations, computer scientists with deep expertise in this domain realize that the following architectural features must be included in a search engine program:

- Reliance on entities and embeddings that overcome limitations of existing tags
- Structured for flexibility to incorporate and harmonize massive and diverse evidence sources
- Powerful ontology systems that can deal with millions of concepts
- Indexing of entities + the full pass of NLP on all documents
- State-of-the-art programing methodologies
- Focus on interactivity and enjoyability of experience
- And the ability to eventually deal with "word sense disambiguation" (i.e., know when "stroke" is referring to the brain and not golf) and "iterative search" (more refined search results based on ML of users' profiles).

Many pharmaceutical companies and medical societies are already adopting premium search platforms that leverage ML and AI to explore the medical literature. Hundreds of patients and doctors all over the country are pressure testing and using the same premium search engines to make informed healthcare decisions. This growing cohort of patients has proven that these AI technologies in the evidence space improve the quality of care and reduce costs. Patients are using the search results and associated features to reverse insurance denials, identify surgeons with best outcomes, and avoid unnecessary surgeries. The next big leap in this evidence technology space will be the creation of an AI software solution that will enable the automated extraction of relevant data from clinical studies. Automated search, extraction, and analysis of the evidence found in the medical literature will lead to a major paradigm shift in the American healthcare system.

Major criticisms of EBM include a publication bias, which refers to the clinician and investigator tendencies to publish only studies with a positive diagnostic or therapeutic result. In addition, the terms of the criteria for the levels of evidence are not always agreed

III. The current era of artificial intelligence in medicine

upon and these imprecise definitions are often confusing and misleading. These guidelines or recommendations are very often out of date and do not accommodate more recent ideas and/or study results; in short, the information is not timely and therefore lacks real-time relevance. Even with well-designed RCTs, often there is no clear actionable recommendation and the phrase "further studies or investigations are warranted" is frequently the conclusion. Finally, even with clinical guidelines that are published and discussed in conferences, clinicians often do not have incentives to attain adherence to these recommendations [14]. It is entirely possible that EBM, for a myriad of reasons including some not listed above, is simply becoming more obsolete and needs to be modified to reflect the current era of AI in medicine (at least as a strategic complement to each other).

Artificial intelligence (AI) as game changer for a deeply flawed medical knowledge system

A commentary byLouis Agha-Mir-Salim[1] and Leo Anthony Celi[2,3]
Leo Celi, an active intensivist and one of the progenitors of MIMIC and his colleague Louis Agha-Mir-Salim, authored this commentary on the new paradigm of using AI in achieving real-time relevance in patients to engender a much more effective and accurate knowledge system.
[1]*Faculty of Medicine, University of Southampton, Southampton, United Kingdom*
[2]*Institute for Medical Engineering and Science, Massachusetts Institute of Technology, Cambridge, MA, United States*
[3]*Division of Pulmonary, Critical Care and Sleep Medicine, Beth Israel Deaconess Medical Center, Boston, MA, United States*

Rapid developments in the field of AI over the past decade have sparked unprecedented levels of excitement within the healthcare sector: the expectation is that AI can be leveraged to tackle some of healthcare's greatest challenges. Beyond the hopes and the hype, we need to define and determine which issues constitute the primary challenges to better care. This list includes but is not limited to eliminating preventable medical errors and adverse events due to medical care; reducing inefficiencies and rectifying faulty pathways in clinical workflows; identifying and addressing disparities in care delivery; ending the waste of resources caused by overtesting, overdiagnosis, and overtreatment; and ameliorating the current high level of dissatisfaction evident in the healthcare workforce. All of these examples are to some extent fueled by the large information gaps that arise from a flawed medical knowledge system.

Knowledge is the output derived from drawing conclusions and creating models from observations and experiments. For medicine, it is the basis of clinical practice, the armamentarium of tests and interventions with which doctors diagnose and treat. It would appear to consist of the collection of "ground truths": from the pathophysiology to the epidemiology of diseases, from information gained from blood tests and medical images, to the effectiveness of medications and procedures. But, in fact, there are no ground truths in medicine because truths in medicine are impermanent, having half-lives whose lengths vary depending on the advent of new information. For example, an intervention that was found to be effective in 1978 might well be beneficial for the patient population of that time. But as patient demographics change, and new concepts, tests, and treatments are incorporated into practice, it becomes less and less likely that the "ground truths" of 1978 will stand the test of time. Indeed, the 1978 edition of Harrison's Principles of Internal Medicine lists the following for the management of acute myocardial infarction: bed rest for 6 weeks, avoid beta-blockade, lidocaine to suppress arrhythmia and no cardiac catheterization given that the atheromatous plaque is unstable. Current practice involves essentially *the opposite* of all these recommendations!

Current clinical practice is dictated by the existing evidence, whether it is based on randomized prospective trials, observational studies, or expert opinion, and without undue consideration of the time- (and space-) limited validity of medical truths. After all, clinicians must act to care for patients even if at some philosophical level, they realize that their current practices are likely to be disproved as optimal ones in the future. At present, large, randomized, controlled, prospective trials are heralded as the final verdict on whether an intervention provides some benefit (on average) to some patient population. In the intensive care unit, the TRICC trial [1] from 20 years ago is still referenced to inform decisions about blood transfusion and NICE-SUGAR [2] from a decade ago is still cited in protocols for blood sugar control. If we are to revamp the medical knowledge system, the implication would be to repeat every clinical trial (and for every patient subpopulation) on a regular basis. Given that the median cost of a clinical trial is around $19M [3], such a knowledge system is not feasible (and cost is not the only limiting issue here).

This is where the excitement around AI comes into play. Its focus on (and requirement for) real-world, real-time data represents the first clinical approach that can begin to address medical truths in a near-continuous, data-driven manner. Hence, investments toward a health data infrastructure that would support a more dynamic medical knowledge system are needed. We need to divert some of the $28B per year that is wasted on irreproducible biomedical research [4] to develop and maintain a data health infrastructure that will not only fuel AI research but will form the backbone of a true learning health system.

AI will be a game changer—not from the opportunities of AI itself but rather because of how it shifts the focus to continuous learning from the data. This emphasis on how we build, maintain, and validate the medical knowledge system is the real value of AI. The past two decades have addressed many of the data infrastructure issues that required resolution before AI could be fairly instituted in clinical practice. Now, to a large extent, we have much of the necessary data to begin the design of a system that can most productively utilize the data. The volume, complexity, and dynamic nature of the data do not lend itself to analyze by even the most brilliant of bedside clinicians. The answer lies in formulating a hybrid system that combines the best of data analytics, including AI, with the best of human clinical endeavors.

References

[1] Hébert PC, Wells G, Blajchman MA, Marshall J, Martin C, Pagliarello G, et al. A multicenter, randomized, controlled clinical trial of transfusion requirements in critical care. N Engl J Med 1999;340(6):409—17.

[2] NICE-SUGAR Study Investigators, Finfer S, Chittock DR, Su SY-S, Blair D, Foster D, et al. Intensive versus conventional glucose control in critically ill patients. N Engl J Med 2009;360(13):1283—97.

[3] Moore TJ, Zhang H, Anderson G, Alexander GC. Estimated costs of pivotal trials for novel therapeutic agents approved by the US Food and Drug Administration, 2015-2016. JAMA Internal Med 2018;178(11):1451.

[4] Baker M. Irreproducible biology research costs put at $28 billion per year. Nature [Internet] 2015. Available from: <http://www.nature.com/doifinder/10.1038/nature.2015.17711> [cited 13.05.19].

Clinician perception/cognition. Most, if not all physicians in their subspecialties, can be configured by three basic areas of thinking that involves different areas of the brain in which they perform their tasks (how much time they spend in these areas rather than whether they are capable of performing these functions) (see Fig. 7.3): perception, cognition, and operation.

1. Perception—These are tasks clinicians do that machines can currently do with reasonable accuracy and speed. One such task is medical image interpretation as

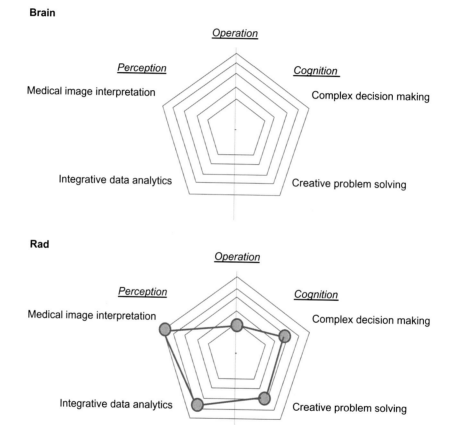

FIGURE 7.3   The clinician's brain. A spider diagram can be drawn for each subspecialty to see how much perception, cognition, versus operation that subspecialist does on a day-to-day basis. The concentric pentagons can be levels 1–5 going outward (with "1" the innermost pentagon being no or little activity in that area to "5" being the highest level of activity in that area). For instance, a radiologist would rank "medical image interpretation" to be a "5" as a radiologist spends much of the time performing that task. Next to the original diagram, one is filled in for a radiologist who does not perform interventional work (so no "operation" task). One can see that the brain is performing more perception work than cognition work (about 60–40 ratio based on the areas designated for perception vs cognition). Note. This spider diagram denotes what the clinician *does* and not what the clinician is *capable of doing* or *likes to do best*.

III. The current era of artificial intelligence in medicine

computer vision and CNN are able to interpret medical images as good if not better than single or groups of a certain image-focused subspecialty. In addition, integrative data analytics with machine and deep learning have capabilities that surpass what humans can do. With deep learning and its variants, some of the traditional image interpretation and integrative data analytics can be performed by machines but humans can still be able to provide oversight and design.

2. Cognition—These are tasks that humans are more able to do compared to machines but machines may be able to take on these tasks in the near future. One such task is complex decision-making, especially in real-time, as biomedicine in most subspecialties is full of this type of task. In addition, creative problem solving remains uniquely human and an essential part of many areas in medicine and health care. While deep learning and its variants (such as deep reinforcement learning) are making progress in both of these areas, these methodologies are not yet mature nor sophisticated enough for full deployment in medicine.

3. Operation—These are manual tasks, especially advanced tasks, that computers are not able to do at this point. As much as robotics and related technologies have advanced in the past decade, much of complicated procedures and operations will not be able to be performed entirely by robotic methodologies in the near future. This area can be explored for a hybrid model in which humans benefit from the relative strengths of the robots (such as better in situ vision, elimination of physiological tremors, and better ergonomics for surgeons) while providing creativity and oversight.

Of course the important aspect of empathy and human touch is also under exploration by AI experts to mimic but not replace humans.

If the subspecialty has predominantly perception-focused tasks, such as radiology or pathology, it behooves the physician to utilize AI to both relieve the burden (as image volume and complexity will escalate the coming decades) and to incorporate its use for advanced interpretations and for quality improvement. This will also enable the subspecialist to explore the cognition as well as the operation aspects of his/her job. On the other hand, if the subspecialty has more cognition-focused tasks, such as psychiatry, the physician can either explore new perception tasks that the computer can now do with relative ease that was not available before, or find ways of incorporating AI into the cognition aspects for the future. In short, perception tasks will be augmented by AI and cognition tasks can also be explored and also augmented by AI in the near future.

Operation and manual tasks will also be augmented in the near future with more advances in robotics and its panoply of capabilities. In addition, for these procedure-oriented subspecialties (such as cardiology or surgery), preoperative assessment [such as using AI and mixed reality (VAMR) modalities such as AR or VR for preoperative planning] and postoperative planning (such as deep learning for risk stratification and precision postoperative care) can also be explored.

In short, perception-focused tasks can be mitigated by AI for the subspecialist so he/she can either have oversight over these tasks or take on tasks in the cognition areas with cognition-centric tasks that can be explored in the near future. Perhaps, an AI strategy with more symmetric or individualized task allocation will ensure less burnout for the overburdened clinician.

III. The current era of artificial intelligence in medicine

Current Artificial Intelligence in Medicine Applications

The present state of AI in medicine includes a myriad of applications, and these application areas will be generalized into 10 main categories and separately described in more detail below. These medical applications in AI orientation will be useful to set the framework for the section to follow (Specific Strategies and Applications of AI for Selected Subspecialties).

Medical imaging. There is much promise in the utilization of AI methodologies such as deep learning (in particular CNN) (see above for details) for automated medical image interpretation and/or augmented medical imaging [15]. The image interpretation tasks include classification, regression, localization, and segmentation. The images that this AI capability can be made available include not only static images [such as a chest X-rays, pathology slides, fundus or skin photographs, and MRI (magnetic resonance imaging) images] but also starting to become available for moving images (such as ultrasound images, pictures during procedures such as endoscopic examinations, and echocardiograms) [16]. This AI computer vision capability, already as accurate if not more so than single or groups of clinicians, is especially needed in the image-focused subspecialties such as radiology, pathology, dermatology, ophthalmology, and cardiology in the near future as both the volume as well as the complexity of imaging continue to escalate beyond what human specialists (even excellent ones) are able to accommodate and perform at a high level.

Medical imaging and mathematical modeling geometric statistics for computational anatomy

Nina Miolane

Nina Miolane, a statistician with a special interest in geometric statistics, authored this commentary on a method to mathematically model anatomical structures by using geometric statistics and principal component analysis.

AI in medicine has fostered the development of computational tools to assist clinicians in their daily practice. As an example, more and more ML pipelines perform automated diagnosis from medical images. In this commentary, we present the motivations and mathematical foundations of statistics on organ shapes extracted from medical images.

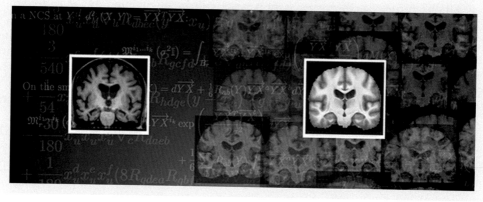

From image intensities to organ shapes diagnosis

ML pipelines take the medical images as inputs: each input is thus a function from an image voxel to an intensity value and the dimension of the inputs is the number of voxels. The higher the dimension of the inputs, the more data are usually needed to train an ML algorithm. The ImageNet database [1] has around 14 millions of 2D images that are on average of size 469×387, thus of dimension 181,503. In contrast, medical images databases are often smaller in size, with 10,000 images at maximum, while having a higher dimension, for example, typical brain structural MRIs are of size $256 \times 256 \times 192$, thus of dimension around 12 million. In order to deal with smaller databases, we are interested in a methodology to reduce the dimension of the inputs, by first extracting meaningful features from the images, for example, organ shape features.

Computational anatomy

Medical images show organ shapes with healthy and pathological variations. Extracting organ shape features from images is interesting as many diseases are correlated with organ shape changes. For example, Alzheimer's disease can be seen on brain's MRIs as shown in the previous figure: on the left, a brain with Alzheimer's diseases; on a right, an average healthy brain. Computational anatomy is a field of research that mathematically models and analyzes organ shapes variations in the population [2]. What is a mathematical representation of a shape? And how do we perform statistics on organ shapes?

Mathematical modeling of organ shapes

There are two main ways in which we can represent an organ shape from a mathematical perspective: either indirectly or directly. With the indirect method, the shape is encoded as the deformation of a "template." Let us consider brain shapes for example. We fix a template brain shape that represents the average healthy brain shape in the population [3]. Then, the subject's brain is represented as the deformation needed to map the template brain to the subject brain. With the direct method, there is no use of template shape: organ shapes are encoded for themselves. One can first put landmarks on salient features of the organ: the set of 3D coordinates of the landmarks encodes the organ shape. Alternatively, one can mesh the surface of the organ: the 3D coordinates of the mesh's nodes encode the organ shape. In each case, the mathematical representation chosen provides a way to represent an organ shape on the computer. This representation is less complex than taking the full medical image and thus easier to handle with small datasets.

Statistical analysis of the variability of organ shapes

We now have a representation of shape. How can we extract the most important features of an organ shape dataset? Consider a dataset of heart shapes, represented as deformations from a template healthy heart shape. How many parameters do we need to encode the principal modes of shape variations? In statistics, a procedure called Principal Component Analysis (PCA) allows answering this question. Yet, usual statistical methods, including PCA, were developed for statistics on numbers or vectors. How can we do statistics on shapes, such as heart deformations in our example? Heart deformations, as mathematical objects, are not vectors but elements of what is called a "Lie group." Therefore the very theory of statistics needs to be extended to analyze these shapes. We call this new theory "Geometric Statistics" [4]. Interestingly enough, Geometric

Statistics relies on the mathematics of "Riemannian Geometry," the framework used by Einstein to formulate the geometry of space-time in General Relativity. In the field of computational anatomy, geometric statistics generalizes PCA to "tangent PCA" to be able to compute the principal directions of variations of heart shapes. In Ref. [5], we see that 10 parameters are enough to represent 90% of the shape variability in the right ventricle of 18 patients with repaired Tetralogy of Fallot, a congenital heart disease. Thus we have reduced a medical image of originally around 12 millions of parameters down to 10 shape parameters. These 10 parameters, albeit automatically learned, make intuitive and even clinical sense, for example, the first one represents the size of the right ventricle.

Machine learning on shape data

This low-dimensional representation opens doors for ML procedures on small datasets. For example, Ref. [5] correlates the body surface index of a patient to a linear combination of the 10 parameters describing its right ventricle's shape. In the clinics, this enables to predict the right ventricle shape just from the patient's body surface area. In the case of repaired Tetralogy of Fallot patients, this allows predicting the optimal time for the often required reintervention.

With computational anatomy, geometric statistics and associated ML paradigms, and innovative mathematical models are feeding AI to advance the computational field of medicine.

References

[1] Deng J, Dong W, Socher R, et al. ImageNet: a large-scale hierarchical image database. CVPR'09; 2009 IEEE conference on computer vision and pattern recognition, 2009, Miami, FL.
[2] Grenander U, Miller M. Computational anatomy: an emerging discipline. Q Appl Math 1998;56.
[3] Guimond A, Meunier J, Thirion J-P. Automatic computation of average brain models. in Wells WM, Colchester A, and Delp S (eds) *Medical image computing and computer assisted intervention- MICCAI'98*. MICCAI1998. *Lecture notes in computer science*, vol 1496. Springer, 1998, Berlin, Heidelberg.
[4] Miolane N. Geometric statistics for computational anatomy [Ph.D. thesis]. 2016.
[5] Mansi T, Voigt I, Leonardi B, Pennec X, Durrleman S, Sermesant M, et al. A statistical model of right ventricle in Tetralogy of Fallot for prediction of remodelling and therapy planning. IEEE Trans Med Imaging 2011. Available from: https://doi.org/10.1007/978-3-642-04268-3_27.

Altered reality. The area of AR, VR, and VAMR will be able to leverage AI technology and use this resource to the fullest for a variety of purposes; these include education and training as well as simulation and immersive scenarios for all stakeholders (including patients and families) and preoperative and intraoperative imaging and planning for certain medical and surgical subspecialists. In a recent review, the application of VR and AI is especially useful in surgical training (especially laparoscopic and orthopedic surgery), pain management, and therapeutic treatment of mental illnesses [17]. The three altered reality technologies, combined with AI, will enable any clinician to wear a different "lens" in the practice of his/her specialty.

Augmented reality/virtual reality (AR/VR) in medicine and the role of AI

David M. Axelrod, MD[1]

David M. Axelrod, a pediatric cardiologist with a penchant for using VR in congenital heart disease, authored this commentary about the future role of AI in AR/VR in the context of congenital heart disease and medicine.

[1]*Stanford University, Stanford, CA, United States*

Upon initial contemplation by the technologically oriented clinician, the link between AI and VR/AR may appear superficial. What really links these burgeoning technologies as they blaze a trail through their respective hype cycles, other than opportunity and enthusiasm? Most medical VR and AR applications currently focus on clinical treatment and patient and provider education, and AI rarely intersects with existing applications on a meaningful level. However, upon further inspection, the incorporation of AI and deep learning into extended reality (XR) has revealed remarkable avenues for future pursuit. A few developments in the artificial and virtual worlds are worth highlighting, as examples of key entry points to a collaborative path forward.

Leveraging the power of completely immersive VRs, mental health providers have demonstrated impressive treatments for depression, anxiety, and posttraumatic stress disorder among others. Philosophically oriented psychologists have even crossed the threshold into the afterlife, with an out-of-body VR death experience [1]. Observation of behaviors in VR along with the collection of physiologic data such as pupillometry, electroencephalogram, and other biometric data can provide astounding insight into the cognitive status of a patient (or any VR user) [2]. Theoretically, a VR mental health experience can provide an AI-enhanced physician a wealth of information about their patients. For example, as more users interact with the environment, AI and ML may eventually diagnose a patient's depression based on their biometrics, then present a new therapeutic virtual world tailored specifically to that patient's treatment.

Similarly, VR simulation programs have provided insight into the behavior and physiologic responses of medical providers in high-impact clinical scenarios [3]. VR simulation in medicine—as in military, education, and other sectors—can utilize AI to present the student with nearly unlimited clinical scenarios that occur randomly within the training platform. The artificially intelligent VR simulation will then learn from users' previous actions and responses, creating an increasingly challenging scenario tailored to each student's needs. Biologic monitoring within VR simulations may allow AI to predict suboptimal performance—in a virtual "Code Blue" situation, for example.

Data analysis, representation, and advanced modeling may represent the most promising intersection between AI and VR. Sophisticated VR data acquisition tools already offer an immersive and three-dimensional interface for complex data analysis in various industries [4]. In medicine, "big data" approaches to diagnosis and management have attracted significant appeal; however, the ability to represent these unwieldy data sets will benefit from AI integration. In pharmacology, if biologic response to anticoagulants were assessed via a data set of every patient ever treated with anticoagulants for a cerebrovascular accident, one would need AI to comprehend the analysis. How, then, will physician scientists interact with the AI-enhanced output? VR/AR (and their natural extension—XR) will likely provide the multidimensional, interactive, and completely immersive interface to view not only the analysis we might expect but also to open avenues for data interpretation we never thought possible.

Sophisticated VR models of human biologic systems will utilize AI and ML, as has been demonstrated in initial models created by our group at Stanford [5] (Fig. 1), among others. In future AI-enhanced VR experiences, ML will process medical data, provide an immersive environment for clinicians to interact with physiologic models, and predict patient outcomes. In complex anatomic and physiologic scenarios such as congenital heart disease, an AI-enhanced ventricular and vascular modeling system will create a VR environment for the surgeon to evaluate the borderline left ventricle of a specific patient. ML will produce increasingly accurate models based on medical imaging, and the surgeon will plan a procedure in VR, manipulating intracardiac patches and suture lines based on prior experience. AI can provide a predictive model, even allowing the surgeon to observe results of a given surgical approach as the heart ages 5 or 10 years in VR: "This intracardiac baffle appears adequate in today's VR model, but will it become obstructed in 5 years (or about 250 million heartbeats)?" If the virtual surgical results are suboptimal, the system can test alternative treatments, compare future outcomes, and even provide AI-empowered suggestions for a surgical technique.

These tools, among many others still to be developed, outline potential directions for AI and VR/AR in medicine. While these ideas may represent science fiction today, they also offer great potential for today's physician technologist. And, as always, with vast potential comes significant responsibility—to maintain technology's orientation toward improving local and global health, improving equity in future medical treatments, and preventing undue harm that might arise from unintended consequences.

FIGURE 1 A user interacts with a virtual–reality visualization of the blood flow in a con genital heart defect, imported from a simulation generated on the open-source software SimVascular.com. This is the first VR representation of vascular simulations, performed at Stanford University. *VR*, Virtual reality. Source: *Photo courtesy Alison Marsden, PhD*.

References

[1] Barberia I, Oliva R, Bourdin P, Slater M. Virtual mortality and near-death experience after a prolonged exposure in a shared virtual reality may lead to positive life-attitude changes. PLoS One 2018;13(11):e0203358.

[2] Juvrud J, Gredebäck G, Åhs F, Lerin N, Nyström P, Kastrati G, et al. The immersive virtual reality lab: possibilities for remote experimental manipulations of autonomic activity on a large scale. Front Neurosci [Internet] 2018; [cited 07.05.19]; 12. Available from: <https://www.ncbi.nlm.nih.gov/pmc/articles/PMC5951925/>.

[3] Chang TP, Beshay Y, Hollinger T, Sherman JM. Comparisons of stress physiology of providers in real-life resuscitations and virtual reality-simulated resuscitations. Simul Healthc J Soc Simul Healthc 2019;14(2):104–12.

[4] Donalek C, Djorgovski SG, Cioc A, Wang A, Zhang J, Lawler E, et al. Immersive and collaborative data visualization using virtual reality platforms. In: 2014 IEEE international conference on big data (big data); 2014. p. 609–14.

[5] Lighthaus Inc. & Stanford Children's Health. Axelrod DM (collaboration with Alison Marsden PhD). The stanford virtual heart. Stanford, CA: Lucile Packard Children's Hospital [Internet]. Available from: <https://www.youtube.com/watch?v=xW1EMBVmAW4> [cited 08.05.19].

Decision support. CDS, especially with an urgent timeline, is one of the most difficult challenges in medicine; just how machine intelligence can help a clinician to solve a difficult and complex clinical situation will be one of the holy grails of AI in medicine. The number of publications and companies now involved in this domain of CDS using AI reflects both the interest as well as the potential in this area, albeit with its nuances and challenges with data and EHR in health care [18]. A recent review of the history of CDS states the dramatic improvement in this sector due to the advent of cognitive aids and AI tools to support diagnosis, treatment, care-coordination, surveillance and prevention, and health maintenance or wellness [19]. While it is laudable that AI was able to defeat the human champion in the game *Go*, the practice of medicine, especially in the chaotic domains of the emergency room, intensive care unit, and operating rooms are more akin to the real-time strategy games like *StarCraft*.

Outcome drift and implications for decision support

Sybil Klaus

Sybil Klaus, a hospitalist and researcher with a focus on patient outcomes, authored this commentary on the principle of outcome drift in decision support, or nonstationarity, and its influence on value-based care and outcome assessment.

The United States is moving toward a value-based healthcare delivery model in which providers and hospitals are rewarded for helping patients improve their health through evolving reimbursement models linked with patient health outcomes [1]. The quality of clinical care delivery is compared across the system using benchmarks for quality measures of these patient health outcomes [2]. Many hospitals are building algorithms that identify high-risk patients on which to focus interventions which may lead to greater improvements in patient outcomes and quality measures. Recent research has demonstrated predictive algorithm can reduce hospital readmissions and improve patient outcomes [3]. However, one study found the performance of these algorithms can be impacted by cohort selection, data source and the definition of prediction target [4]. These factors influence the discrimination performance of the algorithms because the processes and data within the healthcare system are nonstationary. Nonstationarity is broadly defined as occurring when the data generating process being modeled changes over time (e.g., clinical processes and patient characteristics) [4]. Further, the output of the predictive models influences interventions and thus has the potential to alter the prevalence of the outcomes.

A variation in prevalence and definition of patient outcomes is relevant for a value-based healthcare system and particularly important when benchmarking quality measures across different healthcare providers and hospitals. This is especially true for outcomes with definitions that depend on clinical processes, interventions, or cohort selection (e.g., sending a blood culture for sepsis). One study demonstrated a prediction performance difference for a sepsis algorithm when applied to different patient cohorts (e.g., patients with sepsis present on admission vs those that developed sepsis later) [5]. Many current predictive models are developed at the institution level and cannot easily be transferred to other institutions because of differences in the data structure, clinical processes, and prevalence of disease. Together these factors may challenge the current system's approach to benchmarking quality measures if there are significant differences or disparities in the use of predictive analytics and large variance in the algorithm performance and outcomes [6]. Consider the following scenario to illustrate these factors: two similar urban academic hospitals develop their own algorithm to predict patients at high-risk of sepsis. In Hospital A the algorithm successfully predicts sepsis in patients 8 hours prior to deterioration with an AUC (0.91). In Hospital A, patients with sepsis are treated earlier and thus do not require intensive care. The hospital reimbursement model changes as the prevalence of the sepsis continually decreases. In Hospital B the algorithm predicts patients with sepsis 2 hours prior to deterioration with an AUC (0.65). In Hospital B, patients require intensive care and the overall prevalence of the sepsis is marginally decreased. In Hospital B the staff and patients require more clinical care and have worse outcomes as compared to Hospital A. Is it possible to compare the quality of care and offer similar reimbursement models for these hospitals if there is significant variation in the number of patients, amount of clinical care provided and prevalence of sepsis? Further, as the performance of the predictive algorithm continually improves, the prevalence of disease may continue to decrease, and the target prediction outcome definition may drift (outcome drift). This is particularly important for outcomes and quality measures that inherently rely on clinical processes.

References

[1] What is value-based healthcare? NEJM Catalyst January 1, 2017.
[2] <https://www.cms.gov/Medicare/Medicare-Fee-for-Service Payment/sharedsavingsprogram/Quality_Measures_Standards.html> [accessed 09.06.19].
[3] Arkaitz A, Andoni B, Manuel G. Predictive models for hospital readmission risk: a systematic review of methods. Comput Methods Prog Biomed 2018;164:49–64.
[4] Jung K, Shah N. Implications of non-stationarity on predictive modeling using EHRs. J Biomed Inform 2015;58:168–74.
[5] Rothman M, Levy M, Dellinger P, Jones S, Fogerty R, Voelker K, et al. Sepsis as 2 problems: Identifying sepsis at admission and predicting onset in the hospital using an electronic medical record–based acuity score. J Crit Care 2017;38:237–44.
[6] Bates D, Heitmueller A, Kakad M, Saria S. Why policymakers should care about "big data" in healthcare. Health Policy Technol 2018;7(2):211–16.

Revolutionizing medical decision-making with an informatics consult service

Alison Callahan[1] and Nigam H. Shah[1]

Allison Callahan and Nigam H. Shah, both involved in utilizing AI and data science in EHR, authored this commentary on the concept of an informatics consult that can provide reassurance and/or confirmation for clinicians in various situations.

[1]*Stanford Center for Biomedical Informatics Research, School of Medicine, Stanford University, Stanford, CA, United States*

Most medical decisions are not based on "level A" evidence produced by randomized clinical trials, due to the cost and complexity of carrying out such studies. Clinicians must operate in an inferential gap between the evidence needed to make a care decision informed by risks and benefits relevant to their patient and the limited trial-based results available, often relying solely on their personal experience and the collective experience of their colleagues. The advent of the electronic health record (EHR) creates the potential to fill this inferential gap by analyzing the records of similar patients and comparing their outcomes. We now have a tremendous opportunity to learn from the experiences of millions of patients by searching EHRs for "patients like mine."

It is often assumed that the increasing availability of EHR and commercial claims data, combined with advanced computing and data science methods, will enable clinicians to learn from data via a "self-serve" paradigm. Given the limitations of EHRs in terms the patient population they capture, incompleteness in those patients' records, and systematic biases resulting from certain types of patients getting certain treatments, producing reliable evidence from this rich observational data is not trivial—especially in time to inform the course of clinical care that unfolds on the scale of days or even hours.

Depending on the evidence needed to inform a given care decision, different methods are required [1]. In many cases the methods themselves are being actively researched and questions remain about their implementation. In light of this rapidly evolving intersection of statistical methods and medical decision-making, we believe it is best to offer an on-demand service that uses available data to produce the most up-to-date evidence possible with a human-in-the-loop design to contextualize the findings for a clinician to incorporate in their decision-making.

Therefore we have operationalized the opportunity to produce evidence from EHRs in the form of an informatics consult, which clinicians solicit the same way they would solicit other specialist consults. Obtaining a consult is a familiar process to clinicians and eliminates the friction between researchers and practitioners. Instead of just producing standalone "reports," we create a dialog between the clinician and data scientists on our consult team, with the goal of making use of all the evidence on hand to arrive at the best decision for patient care.

To offer the consult service at Stanford, we first developed a search engine to find similar patients from our clinical data warehouse. Given a specific case from a requesting clinician, we use this search engine to create, and then analyze a set of similar patients, and provide a report summarizing the common treatment choices made for those patients, and the observed outcomes after specific choices [2].

In an IRB-approved pilot study of our consult model, we have received 120 consult requests and completed 100 reports (Fig. 1). Of the 20 unfulfilled requests, 12 were due to lack of patients

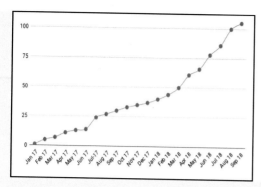

FIGURE 1   The cumulative number of consults received over time (left) and the top sentiments expressed by clinicians after receiving a consult report and debriefing (right).

similar to the motivating case, and 8 were due to lack of the relevant outcome in the data. Of the completed consults, approximately 55% were descriptive analyses, 30% required causal inference methods, and 15% were survival analyses. For example, one request asked whether patients who have mildly elevated kappa or lambda free light chains are at increased risk of developing malignancy. Based on data from 1012 patients, we found that those with mildly elevated light chains had significantly lower malignancy-free survival. In another request, for an 18-year-old man with no past medical history found to have an interatrial septal aneurysm, the requesting clinician wanted to know whether the risk of thrombus differed with anticoagulation or antiplatelet medications. Using a large health insurance claims data set, we found that in 3688 patients treated with either antiplatelet or anticoagulant medication, all three outcomes were rare (<10 patients) or not observed, and therefore the choice of treatment did not seem to matter. In another instance, a treating team wanted to know the incidence of erythema nodosum in patients on BRAF or MEK inhibitors. We reported that 7 of 7940 patients (0.09%) prescribed only a BRAF inhibitor who were diagnosed with erythema nodosum within 2 years, while only 58 patients were prescribed only a MEK inhibitor, and none of those patients were diagnosed with erythema nodosum within 2 years. Overall, 100% of the clinicians who have used our service would recommend it to a colleague, with about half of the clinicians using the service more than once with one clinician making seven consult requests in 1 year.

In most cases, our reports led to either *assurance* or *confirmation*, preventing second guessing in the clinicians' final decision. In many cases, we provided an accurate estimate of the risks of specific adverse outcomes that clinicians were worried about before initiating treatment. Several reports led to subsequent additional analyses, resulting in peer-reviewed publications.

Over the past year, our informatics consult service has explored a unique opportunity to generate actionable insights from health data that are routinely generated as a byproduct of clinical processes. If scaled to multiple sites, we have the potential to transform the way existing EHR data are used to inform a learning healthcare system (Fig. 2).

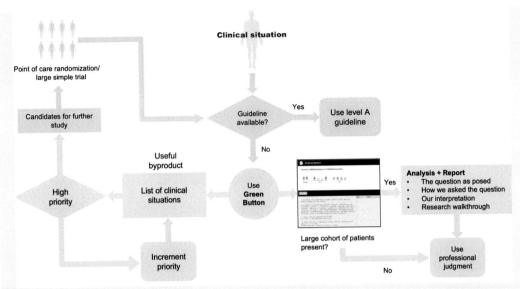

FIGURE 2 Overview of the Green Button consult service as part of a learning health system. Source: Adapted with permission from Longhurst CA, Harrington RA, Shah NH. A 'green button' for using aggregate patient data at the point of care. Health Affairs (Millwood) 2014;33(7):1229−35, appendix exhibit no. 2. Available from: <www.healthaffairs.org/>. Reused with permission from Project HOPE/Health Affairs.

References

[1] Schuler A, Callahan A, Jung K, Shah NH. Performing an informatics consult: methods and challenges. J Am Coll Radiol 2018;15:563−8.
[2] Callahan A, Gombar S, Jung K, Polony V, Shah N. Clinical informatics consult at Stanford [Internet]. Available from: <http://greenbutton.stanford.edu>; 2018 [cited 29.01.19].

Biomedical diagnostics. Bedside biomedical monitoring has been unidirectional: displaying data such as vital signs in a continuous fashion but not analyzing and understanding data internally so therefore not at all "intelligent." AI has the potential to change this paradigm by deploying machine and deep learning to this rich data milieu (with RNN described above) and deriving knowledge and intelligence in a real-time fashion. The hope with monitoring in the hospital setting is for AI to provide real-time analytics coupled with EHR to provide an even more robust decision support tool than either aspect alone. This change in mindset is the concept of AI systems for complex decision-making in acute care medicine, with more focus on time pattern-based analysis and decisions by AI and communication to the human team with a feedback loop [20].

Precision medicine. The paradigm of precision medicine with its complexity of decisions that can be made and enormity of data to be analyzed is particularly well suited for the portfolio of AI methodologies such as deep learning (especially its AI congener deep reinforcement learning) as similar patients can be identified and assessed [21]. Precision

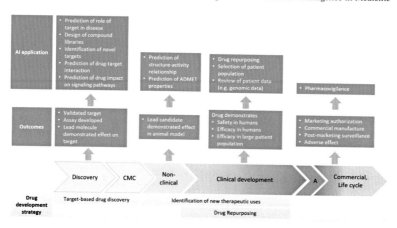

FIGURE 7.4 AI and drug development. Application of AI (in blue) at each stage of the entire drug development strategies (in gray) and process (in green) is shown. The outcome (in orange) is the result of such a strategic coupling of AI and drug development. *AI*, Artificial intelligence.

medicine at its highest level will need a disruptive graph or even hypergraph database configuration and a computational platform for new biomedical knowledge discovery. An essential part of the precision medicine paradigm is individualized therapy based on genotype–phenotype coupling and pharmacogenomic profiles that will provide a health "GPS" for every person. A recent review paper on AI and precision medicine emphasizes the key to success of AI for precision medicine is data quality and relevance and also ML application in functional genomic studies with physiological genomic readouts in disease-relevant tissues with advanced AI [22]. In short, AI can be a very useful tool for individualized risk prediction and medicine can change toward prevention, personalization, and precision.

Drug discovery. Diverse disciplines such as language, neurophysiology, chemistry, toxicology, biostatistics, and medicine can converge to leverage AI and ML/DL to design novel drug candidates [23]. The cognitive solutions are designed to fully integrate and analyze relatively large data sets such as in life sciences for drug discovery. Such strategies include collecting domain-specific content in the form of scientific literature and patents, drug- and disease-related ontologies, preclinical clinical trials, electronic medical records, labs and imaging data, genomic data, and even claims data and social media data. In addition, there are many potential applications of deep learning for large datasets in pharmaceutical research (such as physicochemical property prediction, formulation prediction, and properties such as absorption, distribution, metabolism, excretion, toxicity, and even target prediction) [24] as well as molecular informatics and computer-assisted drug discovery [25]. Mak [26] delineated the entire process of AI utilization in the drug development process with applications of AI at each stage below (see Fig. 7.4).

Digital health. The current digital health portfolio entails software as well as hardware, so it includes telemedicine, cellular phone communications, web-based tools, and wearable technology; this will create the futuristic internet of everything in medical and healthier care delivery (see more detailed discussion later). Of particular significance will be the ability to provide data in multiple formats (even videos of patients) via cell phone. An essential part of digital health is the use of information and communication technologies as well as AI in the form of data mining of the incoming data as well as machine and deep learning for anomaly detection, prediction, and diagnosis/decision-making in a continuous manner. AI in digital health, if applied strategically and efficiently, also has the potential to revive human-to-human bonding with the application of emotional intelligence (EQ) in health care [27].

Wearable technology. The coupling and synergy of AI with wearable technology are essential for both technologies to thrive in the next decades [28]. The data mining process for wearable technology data can be daunting and includes a feature extraction/selection process for modeling/learning to yield detection, prediction, and decision-making for the clinician. Expert knowledge and metadata can influence modeling and learning for this incoming data "tsunami." In addition, wearable technology has also taken a different perspective in this time of AI, especially with the possibility of wide adoption of simple AI tools embedded in medical devices (such as the recent FDA-cleared, AI-enabled wearable device that measures multiple vital signs from Current Health that utilizes ML algorithms to detect problematic changes in data). Monitoring can even include smart pills with a sensor that can track compliance (Abilify MyCite).

Embedded artificial intelligence (AI) transforms wearable technology into a diagnostic powerhouse

Sharib Gaffar

Sharib Gaffar, a soon-to-be pediatric cardiology fellow with a strong interest in AI applications, authored this commentary on the coupling of AI with wearable technology to enable these devices to become very valuable monitoring assets in disease management.

The two current approaches to ML are embedded and remote learning, both with their respective advantages and disadvantages. Embedded deep learning is a form of ML composed of processors that mimic neuronal activity in the human brain while thinking and remains the long-term goal for medicine [1–3]. However, deep learning currently requires large-scale onboard processing power to interpret data and run appropriate algorithms. The large onboard processing requirement hinders portability, and limits the current reach of embedded AI. Our current workaround thus far is to rely on remote learning, with both medical- and consumer-grade wearable technology harboring simple processors that mainly collect and transmit data, rather than understand it. Current processors found in both consumer- and medical-grade wearable technology are able to collect health data, but lack the processing power to mimic neuronal activity levels of interpretation (Fig. 1). Instead, wearable data are analyzed at large-scale remote data centers and an interpretation is given to the user once data processing is complete [2,4] (Fig. 1).

Cardiology wearables are particularly demanding: they must remain precise, dependable, responsive, mobile, sensitive to small clinical changes, and continually connected. Fortunately, Moore's law is reaching its final stages, and processor companies have shifted toward deep learning functionality in smaller, less power-intensive mobile chips [1–3]. It seems apparent that future cardiology wearables will be small, low power, unobtrusive devices with embedded AI capable of profound deep learning.

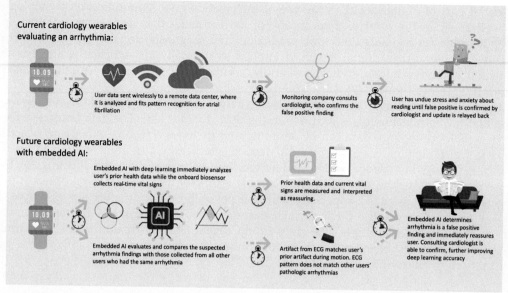

FIGURE 1   Modern cardiology wearables are limited to remote learning, causing significant delays between detection and diagnosis, especially with false-positive diagnoses. Future embedded AI with deep learning overcomes these obstacles with improved onboard thinking and continual connectivity. *AI,* Artificial intelligence. Source: *Vectors in the figure designed by rawpixel.com, freepik.com, Yurlick, Alekksall, Makyzz, Photoroyalty, and macrovector.*

A wearable with onboard deep learning can deliver a far greater impact on patient care than one that predominantly monitors and collects data but relies on remote learning. Embedded deep learning could differentiate false-positive pathologic findings in a less time-intensive process. While current wearables can assess ECG patterns and maybe transmit a suspected diagnosis based on these patterns, they are unable to accurately clarify the difference between a false signal and an actual pathologic arrhythmia (Fig. 1). A "thinking" ECG wearable with embedded AI can avoid misdiagnosis by combining the historical information collected by its sensors since the start of device use with advanced pattern recognition (Fig. 1). Deep-learning ECG wearables could understand if the detected atrial fibrillation is indeed a patient with early asymptomatic signs, or simply confounding artifact from patient motion (Fig. 1). Deep learning is therefore the most powerful method of transforming vast amounts of raw health data picked up by a wearable into an intelligible trend of the user's health.

Deep learning embedded into a consumer wearable will eventually be able to harness a large amount of health data on the user's "normal" health baseline. It can then use this collected information to determine in real-time if an abnormal finding is indeed pathologic or a false positive, thereby avoiding patient anxiety and stress. A large-scale version of this is the Apple Heart Study, a prospective study evaluating the accuracy of the Apple Watch's photoplethysmography in detecting and diagnosing atrial fibrillation or atrial flutter [5]. The Apple Heart Study and similar investigations directly address embedded deep learning AI's largest obstacle, namely, the onus of collecting a large amount of data on a large number of diverse subjects [2,5]. Wearables independently collect a large amount of data on a single subject, and any current embedded AI only draws conclusions based on this information. With a large population of users employing wearables, future embedded deep-learning processors with continual connectivity could compare a single user's abnormal rhythm to both prior health information trends of the same user and prior health information trends from all other wearable users. The accuracy of the wearable will greatly increase even without any actual change in the biosensor capabilities.

Wearables with embedded AI also improve the human aspect of medicine by allaying user's anxiety with false-positive results. While the diagnostic capability of medical AI improves in its current state, we would be remiss to ignore the toll a false-positive diagnosis can take on a patient's mental well-being. Embedded AI processors will be able to dismiss false positives in real time, preventing unnecessary hospital visits and keeping patients at ease (Fig. 1). Embedded AI will be able to provide a level of patient care that cannot be met with devices utilizing remote learning.

The connectivity of wearables is constantly improving, allowing embedded deep learning to refine its judgment every minute a user keeps his or her device attached. Embedded deep learning also improves with every minute of collected data from all other wearable users, to the point that it may eventually be able to predict the development of pathologic outcomes and notify physicians in the early stages of a disease for further preventative management. Deep-learning embedded wearables can shape the future of medicine toward a preventative or early intervention pathway, rather than a limited repair of an end-stage disease.

References

[1] Hof R. Deep learning: with massive amounts of computational power, machines can now recognize objects and translate speech in real time. Artificial intelligence is finally getting smart. Available from: <https://www.technologyreview.com/s/513696/deep-learning/>; 2013 [accessed 24.02.19].

[2] Johnson KW, Torres Soto J, Glicksberg BS, Shameer K, Miotto R, Ali M, et al. Artificial intelligence in cardiology. J Am Coll Cardiol 2018;71(23):2668–79.

[3] Simonite T. Moore's law is dead. now what?. Available from: <https://www.technologyreview.com/s/601441/moores-law-is-dead-now-what/#comments>; 2016 [accessed 14.02.19].

[4] Lau E, Watson KE, Ping P. Connecting the dots: from big data to healthy heart. Circulation 2016;134(5):362–4.

[5] Turakhia MP, Desai M, Hedlin H, Rajmane A, Talati N, Ferris T, et al. Rationale and design of a large-scale, app-based study to identify cardiac arrhythmias using a smartwatch: the Apple Heart Study. Am Heart J 2019;207:66–75.

Robotic technology. Surgical robotics such as the da Vinci system has penetrated even community hospitals and has recently advanced to include 3D visualization and data analytics. Meanwhile, other uses of robotic technology in health care include delivery and sterilization of healthcare equipment and devices, management of pharmaceutical products, and physical therapy in various venues. Human–robot interaction and relationship are being evaluated in a variety of clinical scenarios such as physical or psychiatric rehabilitation and education and training [29]. There is an ongoing debate about robotics and its ethical implications in the future of society including an exacerbation of healthcare disparities and creation of new disparities [30]. Lastly, some of the ethics that involve the use of robotics are extended for use of AI (see the "Ethics" section).

Virtual assistance. AI is very involved in the evolution of virtual assistants as these are dividends of NLP (including natural language understanding and generation).

There are also AI-inspired software agents that are capable of performing certain tasks or services via text or voice for health care (examples are Apple's Siri, Google Assistant, and Amazon's Alexa). A chatbot (or bot) is a service that is capable of conducting a conversation with a human as a result of using rules governed by AI (as discussed previously). Sophisticated chatbots can even use ML and therefore can get "smarter" as it converses with people; other names for this entity include virtual or conversational agent. While it is still relatively early for these virtual assistants to have an impact in medicine and health care, it is certainly a robust area ready for future applications.

Emotional intelligence centered design

Leila Entezam

Leila Entezam, with her strong background in neuroscience and psychology, authored this commentary on the importance of AI intaking into account emotional intelligence, especially as the natural language processing and virtual assistance domain becomes more relevant in health care.

EQ is defined by Google as, "the capacity to be aware of, control, and express one's emotions, and to handle interpersonal relationships judiciously and empathetically." In essence, it's the ability to be well-versed in feelings, which are the body's way of communicating with you. To create a feeling in engagement design, everything matters: every image, word, word sequence, color, feedback loop...it all affects the feelings created by the product/platform the user is interacting with.

Why is it necessary to focus on emotions? Scientists have uncovered that humans feel first and think second. When confronted with sensory information, the emotional section of the brain can process the information in one-fifth of the time the cognitive section requires. Emotions also have a large impact on brand loyalty, according to the Tempkin Group. In a 2016 study, they found that when individuals have a positive emotional association with a specific brand, they are 8.4 times more likely to trust the company, 7.1 times more likely to purchase more and 6.6 times more likely to forgive a company's mistake [1].

In emotion-centered design, the goal is to identify the emotion that you want the user to feel (e.g., excitement, fear, love), then identify the design elements necessary to create that experience. Humans form emotional connections with objects on three levels: the *visceral*, *behavioral*, and *reflective* levels. A designer should address the human cognitive ability at each level—to elicit appropriate emotions so as to provide a positive experience. A positive experience may include positive emotions (e.g., pleasure, trust) or negative ones (e.g., fear, anxiety), depending on the context (e.g., a horror-themed computer game) [2].

As 54% of healthcare professionals expect widespread AI adoption in 5 years, the need for EQ centered design in order to maximize the benefits of technology to users is all the more paramount [3]. Inclusion of diverse voices in the programing and planning of technology solutions will move us closer to this aim.

One area of great potential for AI in healthcare is with voice-activated assistants such as Alexa, specifically as used in senior care. Gartner predicts global spending on virtual personal assistant wireless speakers will surpass $3.5 billion by 2021, but analyst Ranjit Atwal thinks devices customized for senior care and healthcare will roll out in the retail sector beginning in 2020 [4]. In this context, awareness of EQ will make us sensitive to design factors that will maximize impact and engagement. For example:

- More "forgiving"—allow for greater flexibility in command/response, perhaps offering suggestions to the user if they do not offer exactly the right command
- Offer to speak more slowly ("Would you like to me to repeat that more slowly?")
- Offer to speak more loudly ("Would you like me to repeat that more loudly?")
- Reminder of task completion ("I reminded you to take your medication an hour ago. Did you remember to do so?").

With consideration to human decision-making factors, the opportunities for AI in healthcare are endless.

References

[1] Jenblat O. Let's get emotional: The future of online marketing. *Forbes agency council post*, February 26, 2018.
[2] <https://www.interaction-design.org/literature/topics/emotional-design>.
[3] Bresnick J. 54% of healthcare pros expect widespread AI adoption in 5 years. *Health IT Analytics* post, July 9, 2018.<https://healthitanalytics.com/news/54-of-healthcare-pros-expect-widespread-ai-adoption-in-5-years>.
[4] Balk G. Voice activation and virtual assistants modernize health, senior care. *HealthTech* post, September 27, 2018.<https://healthtechmagazine.net/article/2018/09/voice-activation-and-virtual-assistants-modernize-health-senior-care>.

Table 7.5 summarizes each of the 10 application categories as well as examples of applications and the clinicians that would be in the best position to directly benefit from these applications.

TABLE 7.5   Artificial intelligence (AI) application category and applications with clinical relevance.

AI application category	Application examples (clinicians with most relevance)
Medical imaging	Static images (all clinicians, especially image-oriented subspecialties)
	Moving images (all clinicians, especially image-oriented subspecialties)
	Hybrid image (all clinicians, especially image-oriented subspecialties)
	Radiomics with therapy (all clinicians, especially image-oriented subspecialties)
	Facial recognition of syndromes and conditions (all clinicians)
Altered reality	Education and training for clinicians (all clinicians, including nurses)
	Education for patients and families (all clinicians, including nurses)
	Preoperative planning (surgery, procedure-oriented subspecialties)
	Operation visual augmentation (surgery, procedure-oriented subspecialties)
	Pain management (anesthesia, ICU, procedure-oriented subspecialties)
	Psychiatric treatment (psychiatry, primary care)

(Continued)

TABLE 7.5　(Continued)

AI application category	Application examples (clinicians with most relevance)
Decision support	Rehabilitation (PM&R, orthopedic surgery, primary care)
	ICU patient decisions (ICU, surgery, cardiology)
	Hospital patient decisions (all clinicians, including nurses)
	Outpatient chronic disease management (all clinicians, including nurses)
	Risk prediction and intervention (all clinicians, especially primary care)
	Calculation of scores (all clinicians)
	Patient triage (all clinicians, especially emergency medicine)
Biomedical diagnostics	Real-time data analytics (ICU, ED, anesthesiology, surgery, cardiology)
	Embedded AI in monitoring (ICU, ED, anesthesiology, surgery, cardiology)
	AI for biomedical testing (ICU, ED, anesthesiology, surgery, cardiology)
Precision medicine	Precision medicine including pharmacogenomic profile (all clinicians)
	New disease subtypes (all clinicians)
	Chronic disease management (all clinicians, especially primary care)
	Population health management (all clinicians, especially primary care)
	Clinical trial candidates (all clinicians, especially primary care)
Drug discovery	New drugs for therapy (all clinicians, pharmacists, and researchers)
Digital health	Chronic disease management (all clinicians, especially primary care)
	Population health management (all clinicians, especially primary care)
	Telehealth and telemedicine (all clinicians, including nurses)
Wearable technology	Chronic disease management (all clinicians, especially primary care)
	Population health management (all clinicians, especially primary care)
Robotic technology	Robot-assisted surgery (procedure-oriented subspecialties)
	Physical rehabilitation (neurologists, PM&R, orthopedic surgery)
	Administrative task automation (all clinicians and administrators)
Virtual assistants	Medical advice and triage (all clinicians, especially primary care)
	Health coaching and education (all clinicians, especially primary care)
	Chart review and documentation (all clinicians, including nurses)
	Psychiatric treatment (psychiatry, primary care)

Image-oriented subspecialties: cardiology, dentistry, dermatology, gastroenterology, ophthalmology, pathology, and radiology. Procedure-oriented subspecialties: cardiology, dentistry, dermatology, gastroenterology, ophthalmology, pulmonology, radiology, and surgery. *ED*, Emergency department clinicians; *ICU*, intensive care unit clinicians; *PM&R*, physical medicine and rehabilitation.

TABLE 7.6 Artificial intelligence (AI) application category and current AI availability.

Category of AI application	Current AI availability
Medical imaging	$+++$
Altered reality	$+$
Decision support	$++$
Biomedical diagnostics	$+$
Precision medicine	$+$
Drug discovery	$+$
Digital health	$+$
Wearable technology	$+$
Robotic technology	$++$
Virtual assistants	$+$

Table 7.6 summarizes each of the 10 AI application categories and their current AI availability. The table illustrates low (+), medium (+ +), or high (+ + +) AI availability currently for the category. Medical imaging with deep learning is clearly ahead of the other areas, with decision support and robotic technology in the second tier. Most other AI application categories are available but not as mature. This table will be used as a template for each of the subspecialties in a later section (as these AI application categories and level of availability related to clinical relevance) and as a convenient "gap analysis" tool for both clinicians and others in the domain of AI in medicine.

References

[1] Naylor CD. On the prospects for a (deep) learning health care system. JAMA 2018;320(11):1099−100.
[2] Park SH, Han K. Methodologic guide for evaluating clinical performance and effect of artificial intelligence technology for medical diagnosis and prediction. Radiology 2018;286(3):800−9.
[3] Vellido A. Societal issues concerning the application of artificial intelligence in medicine. Kidney Dis 2019;5:11−17.
[4] Rajkomar A, Hardt M, Howell MD, et al. Ensuring fairness in machine learning to advance health equity. Ann Intern Med 2018;169(12):866−72.
[5] Lamanna C, Byrne L. Should artificial intelligence augment medical decision making? The case for an autonomy algorithm. AMA J Ethics 2018;20(9):E902−910.
[6] Proposed regulatory framework for modifications to artificial intelligence/machine learning (AI/ML)-based software as a medical device (SaMD): discussion paper and request for feedback. <regulations.gov>.
[7] Personal communications with Sylvia Trujillo and Jesse Ehrenfeld (AMA); 2018−2019.
[8] Pesapane F, Volonte C, Codari M, et al. Artificial intelligence as a medical device in radiology: ethical and regulatory issues in Europe and the United States. Insights Imaging 2018;9:745−53.
[9] Komorowski M, Celi LA. Will artificial intelligence contribute to overuse in health care? Crit Care Med 2017;45(5):912−13.
[10] Kahneman D. Thinking, fast and slow. New York: Farrar, Straus, and Giroux; 2011.

[11] Norman GR, Monteiro SD, Sherbino J, et al. The causes of errors in clinical reasoning: cognitive biases, knowledge deficits, and dual process thinking. Acad Med 2017;92(1):23−30.

[12] Groopman J. How doctors think. Boston, MA: Houghton Mifflin; 2007.

[13] Klein JG. Five pitfalls in decisions about diagnosis and prescribing. Br Med J 2005;330:781−3.

[14] Greenhalgh T, Howick J, Maskrey N. Evidence based medicine: a movement in crisis? Br Med J 2014;348: g3725.

[15] Ranschaert ER, Morozov S, Algra PR. Artificial intelligence in medical imaging. Cham: Springer; 2019.

[16] Madani A, Arnaout R, Mofrad M, et al. Fast and accurate view classification of echocardiograms using deep learning. NPJ Dig Med 2018;1:6.

[17] Li L, Yu F, Shi D, et al. Application of virtual reality technology in clinical medicine. Am J Transl Res 2017;9 (9):3867−80.

[18] Shortliffe EH. Clinical decision support in the era of artificial intelligence. JAMA 2018;320(21):2199−200.

[19] Middleton B, Sittig DF, Wright A. Clinical decision support: a 25 year retrospective and a 25 year vision. Yearbook Med Inform 2016;(Suppl. 1):S103−116.

[20] Lynn LA. Artificial intelligence systems for complex decision making in acute care medicine: a review. Patient Saf Surg 2019;13(6).

[21] Castaneda C, Nalley K, Mannion C, et al. Clinical decision support systems for improving diagnostic accuracy and achieving precision medicine. J Clin Bioinforma 2015;5:4.

[22] Williams AM, Liu Y, Regner KR, et al. Artificial intelligence, physiological genomics, and precision medicine. Physiol Genomics 2018;50:237−43.

[23] Dana D, Gadhiya SV, Surin LG, et al. Deep learning in drug discovery and medicine; scratching the surface. Molecules 2018;23:2384.

[24] Ekins S. The next era: deep learning in pharmaceutical research. Pharm Res 2016;33(11):2594−603.

[25] Gawehn E, Hiss JA, Schneider G. Deep learning in drug discovery. Molecular Informatics 2016;35(1):3−14.

[26] Mak MM, Pichika MR. Artificial intelligence in drug development: present status and future prospects. Drug Discov Today 2019;24(3):773−80.

[27] Fogel AL, Kvedar JC. Perspective: artificial intelligence powers digital medicine. NPJ Dig Med 2018;1:5−8.

[28] Banaee H, Ahmed MU, Loutfi A. Data mining for wearable sensors in health monitoring systems: a review of recent trends and challenges. Sensors (Basel) 2013;13(12):17472−500.

[29] Sheridan TB. Human-robot interaction: status and challenges. Hum Factors 2016;58(4):525−32.

[30] Russell S, Hauert S, Altman R, et al. Robotics: ethics of artificial intelligence. Nature 2015;521(7553):415−18.

Artificial Intelligence in Subspecialties

The Present State of Artificial Intelligence in Subspecialties

In a PubMed search for published reports for "artificial intelligence" (AI) and "medicine" yielded 15,718 reports since 1950, but 10,975 (70%) and 7454 (47%) just in the last 10 and 5 years, respectively. A search using solely "artificial intelligence" under PubMed yielded more publications with 88,727 reports (with 53,729 and 29,932 reports or 61% and 34% in the last 10 and 5 years, respectively). In 2019 the number of AI-related reports are 737 or 2179 (based on the aforementioned search terms) out of more than 220,000 articles published listed with PubMed, or less than 1% of total number of biomedical publications. This relatively low percentage may be even less as some estimate that the number of biomedical reports to be actually over 1 million per annum.

AI-related reports in various subspecialties since 1950 (see Table 8.1) are separated into "high," "medium," and "low" use groups based on the current year (2019) published reports. This is a cursory analysis and not an in-depth one with every single reference accounted for (there is much overlap amongst subspecialties in AI-related publications). The subspecialties range from almost 300 reports in radiology [not surprising given its focus on deep learning (DL) and computer vision], to only a few in several fields that are not image-intensive (such as infectious disease). Interestingly, the neurosciences (neurology and neurosurgery as well as psychiatry) are on this list of medium/high user groups that have an AI presence as there is much emphasis of neuroscience currently in AI. Finally, most subspecialties, even those in low and medium use subspecialties, have seen a sizable increase in the annual number of published reports in this past decade.

The subspecialties and artificial intelligence strategy and applications

For each of the selected subspecialties given later a brief description of the scope of clinical work is first presented so that all readers, especially the nonclinicians, can comprehend the portfolio and scope of that clinician's work. Following this, the *Published Reviews and Selected Works* section reflect the current state of the AI with that particular

TABLE 8.1   Artificial intelligence (AI)-related publications.

Subspecialty	Total published reports (since 1950)	Published reports (last 10 years)	Published reports (last 5 years)	Published reports (through 6/2019)
High level (> 100 articles in 2019)				
Radiology	5778	3484	2106	286
Oncology	10,825	7606	3741	253
Surgery	14,390	9524	3632	230
Pathology	7861	5606	3098	186
Medium level (50–100 articles in 2019)				
Epidemiology	3168	2349	1373	75
Neurology	1110	885	751	64
Cardiology	662	475	370	55
Critical care medicine	1245	721	464	53
Low use (<50 articles in 2019)				
Psychiatry	1268	939	778	49
Genomic medicine	520	492	425	44
Gastroenterology	307	257	202	44
Neurosurgery	1249	788	445	39
Internal medicine/ primary care	852	562	352	32
Ophthalmology	367	284	231	31
Pediatrics	563	447	381	27
Pulmonology	331	244	204	26
Dermatology	241	190	152	25
Emergency medicine	298	209	175	21
Anesthesiology	444	303	193	20
Obstetrics and gynecology	591	473	210	18
Infectious disease	374	306	224	16
Endocrinology	193	174	130	8

For each of the subspecialties listed, the number of publications that are AI-related since 1950 as well as 10 and 5 years ago are listed. The subspecialties are categorized into high, medium, and low level of AI-based publications in first half of 2019.

III. The current era of artificial intelligence in medicine

subspecialty as published in the literature and experienced in national or international meetings. Subsequent to this section, a *Present Assessment and Future Strategy* of the AI in that subspecialty then delineates opportunities, both present (within the next few years) and future (within a decade and beyond), that are particularly relevant for that subspecialty. The length of these sections reflect, to some extent, the level of current activity (e.g., cardiology and radiology sections are longer than pulmonology and rheumatology) but does not portend the future potential of AI in that particular subspecialty.

A Clinical Relevance to Current AI Availability table (mentioned previously for a preamble) is presented for each subspecialty within the respective subspecialty section and reflects the clinical relevance and the current AI availability (or maturity) for each of the AI application categories. Each cell is categorized as +, ++, or +++ for low, medium, or high clinical relevance for that category for each subspecialty; in addition, the table also illustrates low, medium, or high AI availability currently for the AI application category. For example, if the clinical relevance is +++ but the current AI availability is +, there is an ample opportunity for AI to better accommodate the clinical relevance or need. On the other hand, if the clinical relevance is + and the current AI availability is ++ or +++, perhaps there are future applications that the clinicians can consider.

Artificial Intelligence Applications for all Subspecialties

Most reviews of AI in medicine to date are specific for a given subspecialty, but an excellent comprehensive review of AI for both biology and medicine (more the former) by Ching et al. discussed both the opportunities as well as challenges for DL [1]. For all subspecialties, however, certain general AI application strategies are listed in the following for now (or in the near future) and can be considered by almost every clinician from radiology to primary care:

1. Use of DL in medical image interpretation. This is perhaps the most impressive AI dividend in biomedicine with DL/convolutional neural network (CNN) performing at human (or above) levels of performance [measure by area under the curve (AUC)]. One of the best potential uses of this methodology is a fast and efficient screening tool for life-threatening diagnoses that need to lead to an acute intervention (pneumothorax and drainage or cerebral vascular accident and intervention) for any clinician; this AI capability is ideal for any institution with a large volume of image studies or for a health-care facility lacking specialized expertise. Another aspect of this screening is the accelerated diagnosis timeline to diagnose non–life-threatening diagnoses but nevertheless significant conditions (such as a small pneumonia that can be treated more expeditiously with antibiotics). Of course this screening strategy may be counterproductive if there is automation bias (over reliance on this mechanism) or if the AI tool is not able to pick up small or early findings of these diagnoses. Another dividend from this sophisticated DL-enabled image interpretation is novel discoveries in medical images (such as tumor–stroma interfaces) that can help with subtyping, therapy modification, and/or prognostication. Yet another aspect of this DL approach to medical image interpretation is the capability to add other information [such as electronic health record (EHR) or genomic analyses] to enable a radiomic approach to

medical image interpretation. Overall, there is an ongoing question about the absolute validity of the preliminary reports of CNN and medical image interpretation to date as most have not had clinical validation by clinicians nor have these reports been published in a peer-reviewed format; in addition, these early studies have not been clinically tested across many hospitals (outside the study institutions). The Food and Drug Administration (FDA), however, has approved the first autonomous diagnosis software (for diabetic retinopathy).

2. Use of machine or DL for decision support. The complex nature of patient care, especially with escalating amount and types of data as well as increasing number of clinical guidelines and treatment plans, is better suited for machine learning (ML)/DL given enough clinician direction and input. The main use is for risk prediction using EHR data for improved allocation of scarce health-care resources with so many health facilities and patient cohorts burdened by unplanned hospital admissions with morbidity and mortality risk. One aspect of risk prediction with utilization of DL models is whether these AI tools can be eventually applied to the holy grail of risk prediction: real-time risk prediction. Although there is recent work on deploying deep reinforcement learning in, particularly, the intensive care unit (ICU) setting for predictions of events such as septic shock, this work is preliminary and needs more real-world experience. In addition, these AI tools can also be used for workflow in venues such as the emergency department for better resource allocation.

Artificial intelligent application in disease prediction: weighing bias in data and model design

Arta Bakshandeh[1]

Arta Bakshandeh, a practicing hospitalist with an appreciation for data science in risk prediction models and puts this in practice at the population level in an impressive data analytics center, authored this commentary on bias in data and model design in this discipline.

[1]Alignment Healthcare, Orange, CA, United States

As ML models begin providing clinicians with insights into patient illness, it is important for us all to have a true understanding of how we got here and the direction we should be heading, not only with clinical partnerships with data scientists but also with an understanding of bias in our data sets and minds of the model makers. As clinicians, we yearn for the data, both tangible and intangible, to allow us to provide accurate diagnosis to our patients. AI and ML has a unique opportunity to take what we know as clinicians and augment our workflows, actively triage populations, make diagnosis and provide treatment plans with the most up-to-date medical information available.

An exciting application of ML comes in the form of disease prediction. However, we must first ask ourselves several questions before embarking on this journey. For instance: what is the problem we are trying to solve? Is the data relevant or reliable? Do we even have enough data to come to a clinical conclusion? Once these questions have been answered, the hardest still looms—can we intervene to halt any disease progression? Or simply, what can we do about it?

There seems to be an insatiable appetite for creating AI models in medicine, but often without the input of the clinical teams delivering the care. A DL model can already predict a diagnosis of dementia 6 years before a clinical diagnosis can be made [1]. Thus far, we have no medicines or treatments that can cure or effectively slow the disease, leading to years of anxiety anticipating when that day will come. In the United Kingdom a simple algorithm glitch in the QRISK Calculator deployed by the National Health Services caused thousands of patients to be wrongly given or denied statins for presumed cardiovascular risk [2]. The level of harm created remains to be seen and begs the question as to why physicians were not given an option of overriding the system. This also illustrates the importance of transparency in algorithm outputs through practices such as local interpretable model-agnostic explanations (LIME) which explains predictions in an interpretable manner and leaves the intervention to a human [3].

On the other side of the spectrum a DL model has successfully been trained to diagnose diabetic retinopathy with an algorithm that had 97.5% and 96.1% sensitivity and 93.4% and 93.9% specificity in the two validation sets, which is better than the world's leading ophthalmologists [4]. This algorithm can lead to early diagnosis and treatment of this debilitating complication of diabetes along with the primary illness itself. This outcome depends solely on how we as clinicians understand and chose to execute the data presented.

With the World Health Organization estimating a shortage of almost 4.3 million doctors, midwives, nurses, and support workers worldwide in the coming years [5], AI is set to become the frontline in guiding patients to seek medical attention and provide direct insights to the consumer. Wearables like the Apple Watch are already providing consumers with alerts notifying them to see their doctor through AI-enabled applications reading periodic electrocardiograms (ECGs) and identifying irregular heart rhythms such as atrial fibrillation [6]. As we will likely see, a surge in healthcare utilization initially, future AI models should focus on efficiencies which can negate the need for a clinician to be involved at all. This, however, will be very difficult given no known data sets contain genetic, behavioral, social, socioeconomic, and biometric data points which provide clinicians with the context of patient care needs.

We must urge partnerships with academic and practicing clinicians alongside data scientists to solve problems that have interventions. While never forgetting that machines are only as good as the data they have been trained on and the biases of their trainers. The clinician should take the output from ML as a new data point, always questioning its relevance. The moment we blindly trust an algorithm output is the moment we as clinicians turn our back on the oath we took to do no harm. Ultimately, only we can put into context the ML outputs, knowing that what is at stake is the most precious thing in this world—human lives.

References

[1] Ding Y, et al. A deep learning model to predict a diagnosis of Alzheimer disease by using 18F-FDG PET of the brain. Radiology 2018;. Available from: https://doi.org/10.1148/radiol.2018180958.

[2] Statins alert over IT glitch in heart risk tool: Thousands of patients in England may have been wrongly given or denied statins due to a computer glitch. *BBC News/Health.*<https://www.bbc.com/news/health-36274791>.

[3] Ribeiro M, Singh S, Guestrin C, et al. "Why should I trust you?" Explaining the predictions of any classifier. <https://www.kdd.org/kdd2016/papers/files/rfp0573-ribeiroA.pdf>.

[4] Gulshan V, Peng L, Coram M, et al. Development and validation of a deep learning algorithm for detection of diabetic retinopathy in retinal fundus photographs. JAMA 2016;316(22):2402–10. Available from: https://doi.org/10.1001/jama.2016.17216.

[5] World Health Organization. The world health report 2006: working together for health. Geneva; 2006.

[6] Apple Inc. <https://www.apple.com/newsroom/2018/12/ecg-app-and-irregular-heart-rhythm-notification-available-today-on-apple-watch/Cupertino>; 2018.

3. **Use of AI tools for administrative support.** AI tools can be utilized for automating processes in healthcare that are mundane and tedious as performed by humans. Relatively primitive AI tools such as robotic process automation (RPA) has already made some impact in reducing human burden in administrative tasks in other business sectors and can have the same impact with health-care administrative tasks such as prior authorization, insurance eligibility, and clinician credentialing.

4. **Use of natural language processing (NLP) for communication.** Simple tools such as virtual assistants and chatbots can be used for communication in various situations like in operating or procedure rooms with all the stakeholders. Physician-to-patient or physician-to-physician or caretaker communications can also be achieved with NLP-related tools. Overall, NLP is relatively underleveraged and not well understood in the medical and health-care arenas but can have impact if this resource is given more appreciation and utilization.

5. **Use of AI tools for mining data.** Use of ML, particularly unsupervised learning, can be effective for mining patient data for new patterns of diseases or novel diagnoses and therapies. This is a new bottom up approach to medicine (as opposed to traditional diagnosis of diseases based on clinical criteria that clinicians often have to memorize). This paradigm shift to have new subtypes of diagnoses discovered in order to risk stratify as well as to achieve better success with therapeutic interventions can be very productive as these are tenets of precision medicine. In addition, this strategy can also help to detect diseases in patients who may not have met full criteria (especially ones with one significant criterion but perhaps few or none of the other criteria) but nevertheless have the disease. Lastly, mining data can also be used for the selection of the appropriate patients for a clinical trial for intervention; much of the expenditure and labor for a clinical trial selection of patients can be obviated with intelligent use of ML/DL tools.

6. **Use of NLP for medical reports.** Many aspects of automating reports as well as organizing these reports can be accomplished with intelligent application of NLP. There is much preliminary work in this area that has already been completed in radiology. In addition, NLP can be used to bring about more consistency in how these reports or notes are generated.

7. **Use of machine or DL for risk assessment and intervention.** Another area in terms of expenditures and potential savings is the domain of risk assessment and timely interventions to reduce this risk. One strategy is to delineate patients that would warrant an alert for a number of reasons (abnormal vital signs, inappropriate communication, decreased mobility, etc.). Another strategy is that ML/DL can be used to detect anomalies is medication error. All of this requires sophisticated algorithms and also tools to execute and follow-up on action plans.

III. The current era of artificial intelligence in medicine

Challenges to deep learning (DL) in the electronic medical record

Tasha Nagamine and Mayur Saxena
Tasha Nagamine and Mayur Saxena, data scientists in health care, authored this commentary to elucidate her experience in deploying DL models in electronic health record (EHR) and to detail the nuances and challenges of using DL in deciphering the bias and noise issues of EHR.

The role that AI will play in healthcare cannot be overstated. As more and more health-care information is digitized, medicine and society as a whole are poised to benefit immensely from the technological advances that have already transformed many aspects of the world we live in. One of the most promising trends observed over the past several years has been the simultaneous rise of electronic medical record (EMR) systems coinciding with the unprecedented advances in AI using DL. As a result, many experts in both medicine and ML are exploring using EMR data in secondary ML applications.

DL has already shown promising results in clinical data, where neural networks have been able to predict a number of clinical outcomes such as disease onset and unexpected complications [1–3]. However, at present these models are limited to research use and have not been validated in real world situations. In this commentary, we will discuss a few common but important pitfalls for DL in EMRs.

Machine learning versus deep learning

ML encompasses a subset of methods in AI in which algorithms learn hidden patterns and relationships in data. Traditionally, ML algorithms work by building a mathematical model, or mapping function, based on example data that can make predictions about future or unseen data points. The advantage of using ML methods is that they automatically learn the "rules" that are needed to make a prediction. As a result, ML algorithms are more effective and scalable than other rule-based algorithms.

DL is a specialized subset of ML. The neural networks used in DL are able to learn patterns in data by using multiple layers of nonlinear functions instead of one (often linear) function as in traditional ML; as a result, DL is able to estimate extremely complex functions in cases where traditional ML models may perform poorly [4].

Deep learning in healthcare

There is no question that healthcare is complex. For example, consider the multitude of factors that can influence a measurement as seemingly simple as blood pressure (BP). Short-term factors include measurement error, physical activity, posture, time of day, medications, smoking, caffeine intake, emotional state (e.g., white coat hypertension), etc.; long-term factors include diet, BMI, chronic conditions, age, and lifestyle. Unexpectedly, the result is that analyzing BP measurements over time can be quite complicated.

Now imagine combining BP measurements with thousands of other similar measurements and clinical variables in a patient's EMR to build a model to predict if and when a patient will develop heart failure. Just like BP, most clinical variables are subject to many sources of variability, which can be quite different for different patients (see Fig. 1A). The vast complexity of

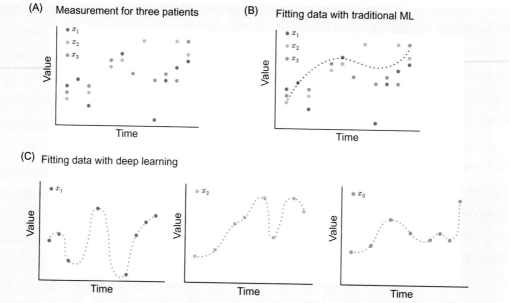

FIGURE 1 Clinical variables are subject to many sources of variability, which can be quite different for different patients (see Fig. 1A). Traditional ML models generally try to fit a single function to the data (Fig. 1B), while DL (Fig. 1C) is an intuitively attractive option for trying to model more complex situations because it can learn a variety of specific functions that can be tailored to different data points.

medical variability is astounding. This makes DL an intuitively attractive option for trying to model such phenomena in healthcare because traditional ML models generally try to fit a single function to the data (Fig. 1B), whereas a DL model can learn a variety of specific functions that can be tailored to different data points (Fig. 1C).

However, using the typical end-to-end approach in DL might make this modeling infeasible for use in a real world example. The problem is that there are certain statistical issues with EMR data that will increase the amount of data required for DL applications so much so that this approach becomes impractical. We will walk through a few toy examples to illustrate this point.

Example 1: billing codes

Billing codes such as ICD-10 reflect the diagnosis of a patient. Therefore ICD-10s are commonly used as a label or a feature in multiple health-care prediction tasks. However, ICD-10 codes suffer from heavy reporting bias due to multiple reasons. One key reason is that reimbursement influences how they are entered into the EMR. Under certain plans the hospital is paid a fixed amount for a particular medical service (fee for service), whereas under capitated plans, services are be covered based on the complexity of the patient's overall disease state or health risk (i.e., Medicare Advantage plans). In the latter, hospitals are required to document all ICD-codes applicable to get maximum reimbursement. These two scenarios can result in similar patients being assigned different ICD-10 codes for the same underlying condition.

TABLE 1 Three similar patients suffering from type 2 diabetes, diabetic nephropathy, and hypertension.

	Patient 1	Patient 2	Patient 3
Name	Anton Lewis	Thomas Schmidt	Eric Smith
Age	56	56	56
Sex	Male	Male	Male
Problems	Type 2 diabetes mellitus with diabetic nephropathy (stage 3 CKD). Hypertension	Diabetes type 2, chronic kidney disease. Diabetic nephropathy. Hypertensive heart disease	Hypertensive heart disease. Insulin-independent diabetes with diabetic nephropathy
Blood sugar (mg/dL)	126	130	130
BP	160/94	160/95	165/96
eGFR (mL/min/1.73 m^2)	45	42	43
ICD-10	E11.21: Type 2 diabetes with diabetic nephropathy	E11.21: Type 2 diabetes with diabetic nephropathy I13.10: Hypertensive heart disease and chronic kidney disease	I11.0: Hypertensive heart disease without heart failure
Treatment	Metformin, ACE inhibitor	Metformin, ACE inhibitor	Metformin, ACE inhibitor

BP, Blood pressure.

Consider a hospital that wants to use a model to assign ICD-10 codes from problems described in the clinical text as a means of improving compliance, documentation, efficiency, and billing quality. In this hospital, there are three patients (Table 1). They have the same complaints (type 2 diabetes, diabetic nephropathy, and hypertension), demographics, and lab values. However, they are covered by different insurance types and were treated by three different doctors; as a result, their complaints are written slightly differently and were given different sets of ICD-10 codes.

The data science team at the hospital decides to train a DL model to predict the ICD-10 codes by analyzing the rest of patients' EMR data. By looking at Fig. 1 one can easily tell that a DL model, therefore, can falsely infer that differences in phrasing between doctors should be predictive of using different ICD-10 codes, when in fact all the model has learned to do is replicate the reporting bias in the EMR. There are multiple other situations where these types of considerations need to be taken into account, because billing codes are just one piece of a very large and complicated collection of data.

Example 2: practice variation

For example, there are other sources of bias in EMR data. There is a common phrase in medicine that in any given patient's problems—a cardiologist sees heart problems, a nephrologist sees kidney problems, an oncologist sees cancer, and so on. Different doctors will notice different symptoms, order different tests, document differently—and in the end, they may even disagree on treatment plans. This is due to *practice variation*, and it can be challenging to create one ML or DL system that can account for this across specialties or centers [5].

Consider a more general example, where the hospital wants to create a system that can predict what will happen to patients in the future. This outcome can be disease progression, adverse events, mortality, and unplanned hospital readmissions, to name just a few. In general, these predictive tasks take on a standard structure:

$$\text{problems} * + \text{measurements} + \text{treatment} \rightarrow \text{outcome}$$

*Items in the problem list, disease signs, symptoms, diagnoses, etc.

Again, it is obvious that this is a very complex problem and may seem like a good candidate for DL. However, when trying to predict outcomes, practice variation makes it difficult for a single model to learn the patterns of individual patients without a massive amount of data.

Furthermore, it is typical for models to succeed at learning the patterns for the most common conditions and presentations where there is more data to learn from. But these same models fail to perform on rarer cases, which is arguably where the models have the most clinical value.

Conclusion

Because of the systematic noise and bias in health-care data, it is difficult to tackle large problems with same end-to-end DL models that have proven so effective in other areas. DL systems can learn to deal with noise and bias in data, but EMR data presents a bias in the source that current models and preprocessing methods cannot overcome. Therefore it is more effective to break down these predictive models into smaller pieces, which can be tackled individually with DL or other methods.

References

[1] Esteva A, Robicquet A, Ramsundar B, Kuleshov V, DePristo M, Chou K, et al. A guide to deep learning in healthcare. Nat Med 2019;25:24–9.

[2] Miotto R, Wang F, Wang S, Jiang X, Dudley JT. Deep learning for healthcare: review, opportunities, and challenges. Brief Bioinform 2017;19(6):1236–46.

[3] Shickel B, Tighe PJ, Bihorac A, Rashidi P. Deep EHR: A survey of recent advances in deep learning techniques for electronic health record (EHR) analysis. IEEE J Biomed Health Inform 2018;22:1589–604. Available from: <doi.org/10.1109/JBHI.2017.2767063>.

[4] LeCun Y, Bengio Y, Hinton G. Deep learning. Nature 2015;521:436–44.

[5] Corallo AN, Croxford R, Goodman DC, Bryan EL, Srivastava D, Stuken TA. A systematic review of medical practice variation in OECD countries. Health Policy 2014;114(1):5–14.

8. Use of robotic technology for rehabilitation or therapy. The range of robotic technology for particularly rehabilitation varies from country to country but is advancing quickly. With chronic disease burden increasing and number and availability of specialists not keeping pace, this area can grow further. Uses for robotics already in place include robots used for physical therapy and rehabilitation after strokes or injury.

9. Use of virtual assistants for patient communications. The proliferation of chatbots and virtual assistants outside of healthcare will render these tools available for healthcare as well, especially for chronic disease management. Many aspects of patient communications can be enhanced by these tools and further improved into situations such as medical triage and health advice.

10. Use of altered reality in planning and education/training. There is preliminary work on use of different modes of altered reality for rendering a visual representation of anatomy/physiology or even a procedure for multiple purposes. This strategy can be used for education, training, or preprocedure planning.

Artificial intelligence (AI)-enabled care

Uli Chettipally
Uli Chettipally, an ER clinician with a strong interest in AI applications in health care, authored this commentary on his personal perspective on how the three types of AI can be leveraged for health care and how business models can accompany these uses.

AI is here to stay, grow, and interweave into every transaction and interaction in healthcare. Health-care businesses are looking to figure out how to put AI to use to gain the benefits of using this technology. It may be to improve cost efficiency, get better clinical outcomes, decrease cognitive burden for physicians, and improve convenience for patients while increasing their satisfaction with the care. This phasing in of AI technology will be gradual as the leaders get more comfortable as they implement and learn from the use of AI systems. This process can be split into three phases based on the level of support these AI systems provide in decision-making. I will call this the Artificial Intelligence Spectrum. Although this separation is arbitrary and there is no clear demarcation where one phase ends and the other begins, it can help us understand the capabilities of AI better. The following are the three phases:

1. Assistive intelligence: Here the AI system gathers and analyzes large amounts of data into understandable information, where a human can look at it and make a decision. The technology itself provides a summary of the data, where the information is presented to the user. This stage can also be called the descriptive stage. Example: a system that curates the latest information from the literature for patients with a specific condition. Based on this information, the physician (or the patient) will decide on the next steps in evaluation and management of the condition.

2. Augmentative intelligence: In this stage the AI system will be able to work side by side with a human being. This stage can also be called the predictive stage where the machine predicts an outcome of interest. It does not recommend a solution but a human has to decide on the best solution given the prediction. Example: an AI system highlighting an area of abnormality on a mammogram, which is suspicious for a malignancy, while separating the normal scans from the list. The radiologist will look at the abnormal ones and decide what the diagnosis is and prepare a report for the treating physician.

3. Autonomous intelligence: Here the system makes decisions without the help of a human being. AI system with this kind of capability can also be called prescriptive. Humans are not essential in the loop. Example: an AI system that collects data from a patient with urinary

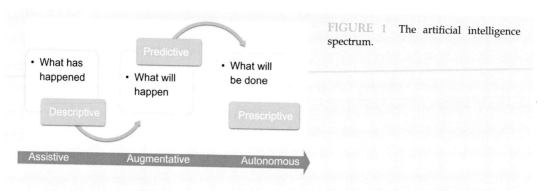

FIGURE 1   The artificial intelligence spectrum.

tract infection over the phone about symptoms and comes up with the diagnosis and a treatment plan where a prescription is sent to the pharmacy, based on the patient's profile. This is the advanced stage where AI functions autonomously (Fig. 1).

As the AI technology gets successfully tested with use cases, the possibility of being used in operations increases. With this application in real world situations the opportunities for new models opens up. These business processes will replace some of the mundane tasks that are currently being done by humans. It starts at the lower skill levels and gradually builds up to the higher skill levels. All of these tasks use large amounts of data and are analytical in nature. Connecting with patients, showing empathy and building trust, are what humans are good at. Outsourcing to machines some of the analytical tasks will provide humans a much better appreciation of quality using more time available to them. The importance of these soft skills increases as machines take over more of the mundane analytical tasks.

Here are some examples of new business processes and models that are made possible with the use of AI technology:

1. Providing user-driven appointment service and customer service to the patients using an autonomous AI engine.
2. Triage service using autonomous AI as the first line to connect to various health-care services based on the severity and acuity of the condition and availability of services.
3. Delivering real-time clinical decision support (CDS) to the physicians at the point of care as the workup is underway to reach desired clinical outcomes using augmentative AI.
4. Managing resources such as staffing, beds, and ambulances based on the capacity, capability, and volume predictions using augmentative AI.
5. Running real-time research results to decision makers and researchers on the current knowledge on a particular subject using assistive AI.
6. Sorting of imaging data based on severity, acuity, or urgency of a clinical condition, so that a physician can prioritize their work load for the day using assistive AI.

In summary, AI technologies have an opportunity to change the business processes and models which will in turn have an effect on cost, quality, efficiency, job satisfaction, patient engagement, and staffing mix of health-care businesses. These changes will free humans to allow them to do jobs at which only humans are good at—being human.

III. The current era of artificial intelligence in medicine

Anesthesiology

Anesthesiologists work almost exclusively in the operating or recovery room with the surgical (or procedural) team or a critical care setting in conjunction with the critical care team (and, therefore, very rarely in the outpatient setting). These physicians pay close attention to vital signs and monitor data throughout procedures and care for these patients before and after the procedure as well ("perioperative" care). The anesthesiologist usually has very little time to learn about the patients and their potentially lengthy and/or complicated medical history in great detail prior to the procedure. This subspecialty has been described as "99% boredom, 1% terror" not unlike perhaps airline pilots. These clinicians are amongst the best doctors in dealing with the airway as well as general resuscitations, so there is much overlap with ICU clinicians in practice and scope (although anesthesiologists usually deals with a single patient at a time). As opposed to the ICU clinicians, anesthesiologists usually do not usually know about the short or longer term outcome of these patients after they depart from the operating room or recovery area.

Published reviews and selected works

There is still relatively low publication activity currently in anesthesiology on AI applications (especially since some references are actually in the ICU literature). A recent review paper by Connor astutely emphasized that it is timely for anesthesiologists to learn and to adopt AI now because the practice of anesthesiology embodies a requirement for high reliability as well as a pressured cycle of interpretation, physical action, and response (as opposed to a single cognitive act like medical image interpretation) and AI is becoming capable of meeting this practice need of the anesthesiologist [2]. A commentary by Alexander agreed with the prior review and discussed the history of many prior failed attempts to incorporate automation into the practice of anesthesiology due to underlying complexity of anesthesia practice as well as inability of rule-based feedback loops to influence decisions [3]. Another editorial in anesthesiology commented on the landmark paper (see Lee et al. as given later) that heralded a new era of ML in clinical anesthesiology [4]. The aforementioned editorial ended with an interesting departing comment: "For those of us who consider ourselves experts in modeling drug behavior, Uber is hiring." Finally, a review that is more extensive on the methodologies in AI but in the critical care medicine literature also placed emphasis on the current AI capabilities for decision support [5].

One report suggested an interesting collaborative strategy between AI and human intelligence for periprocedural medical safety in the form of "simplexity-oriented decisional approach" where there is fusion of sufficient complexity of thought with necessary simplicity of action [6]. Another group reported their work with an automated, clinically curated surgical data pipeline and repository that can identify high-risk surgical patients from complex data with a ML strategy as a risk-prediction model [7]. The cohort of more than 66,000 patients and 194 clinical features were studied with LASSO regression as well as random forest and boosted decision trees methodologies, but LASSO regression was the highest performing risk prediction model with an AUC of 0.92. This ML strategy proved to be superior to traditional heuristics and risk calculators.

Finally, a recent landmark paper described a DL approach to link target controlled infusion of propofol and remifentanil to the bispectral index (BIS) in 231 patients, and this collection of 2 million data points trained a neural network to predict the BIS based on the infusion rates of these medications [8]. What is impressive about this DL approach is that it is totally independent of pharmacokinetic and pharmacodynamic interaction data and simply matched the input (medication infusions) to the output (BIS).

Artificial intelligence (AI)-guided automated, safe, and continuous patient monitoring—current evidence and future horizons

Ashish K. Khanna[1,2] and Kamal Maheshwari[2,3]

Ashish K. Khanna and Kamal Maheshwari, anesthesiologists with a penchant for AI and clinical outcomes, authored this commentary on how important AI will be for hospital monitoring as there are often not enough signals in the noise of physiological data.

[1]Section on Critical Care Medicine, Department of Anesthesiology, Wake Forest University School of Medicine, Wake Forest Baptist Health, Winston-Salem, NC, United States

[2]Outcomes Research Consortium, Cleveland, OH, United States

[3]Department of General Anesthesiology, Center for Perioperative Intelligence, Cleveland Clinic, Cleveland, OH, United States

Most of us have heard of or used a heart rate monitoring device. Heart rate is a measure of how fast the heart is pumping blood per second and to an extent heart health. Similarly, healthcare providers have used other physiologic parameters such as BP, oxygen saturation, respiratory rate, and temperature to understand the health state of a person. These physiologic parameters or vital signs are, mostly, checked intermittently generating numerical values that are vital to medical decision-making. Now, continuous vital sign monitoring is possible, using smart devices, providing us with complex and large amount of data. This data is a treasure trove of information, which will need careful and intelligent curation to yield best results. AI will play a key role in analyzing this data and generating useful information to guide medical decision-making.

Some form of continuous real-time monitoring is used in the ICU, the post anesthesia care unit and the operating room. However, there is continual evolution of intraoperative monitoring technology, to allow for improvement in hard outcomes. In surgical patients continuous real-time BP monitoring nearly halved the amount of intraoperative hypotension when compared to intermittent monitoring (as a usual standard practice) every 5 minutes [1]. Furthermore, the pressure waveform generated by continuous monitoring can provide more information about patients' condition, which was impossible before. For example, a ML algorithm can be trained, with large data sets of high-fidelity arterial waveforms, to predict future hypotension states [2]. The waveform analysis can also help in guiding appropriate therapy. Most patients during surgery require intravenous fluid administration, however, it is not clear how much is appropriate. Assisted Fluid Management is an algorithm that is currently being tested (ClinicalTrials.gov NCT03141411) to determine fluid responsiveness in a patient by estimating the predicted change in stroke volume which guides the anesthesiologist when fluid is required. If successful, the algorithm will help reduce cognitive burden of clinicians to choose the right amount of fluid and reduce variation in clinical care.

However, clinical harm can occur outside these traditional high intensity care areas, such as the operating room and the ICUs, and most likely at home or hospital wards. For example, nearly half of all adverse events in hospitalized patients occur on the general care ward [3]. Overall, 30-day postoperative mortality rates are at 1%–2%, a significant 1000 times more than anesthesia-related intraoperative mortality [4–6]. About three quarters of patients who experience in-hospital postsurgical mortality are never admitted to an ICU [7,8]. A large registry identified 44,551 index events across more than 300 US hospitals. More importantly these acute respiratory events on inpatient wards had an associated in-hospital mortality of approximately 40% [9–11].

Current ward monitoring protocols leave our patients dangerously under monitored [12]. There are typically large intervals of time (anywhere between 4 and 6 hours at any given instance) when vital signs can show a pattern of deterioration, with a consequent "code blue" event. Because we monitor patients in snap-shots of time, we fail to detect these patterns and our response to acute cardiorespiratory compromise via rapid response teams is largely retroactive. Patients recovering from noncardiac surgery experienced severe prolonged hypoxemia ($SpO_2 < 90\%$ for an hour or more), and this was unfortunately missed about 90% of the times by routine vital sign monitoring [13]. Nearly all patients after major abdominal cancer surgery had continuous monitoring detected hypoxemia while standard vital sign assessments by nurses (linked to an early warning score algorithm triggering rapid response teams) detected a $SpO_2 < 92\%$ in less than a fifth [14]. The heart–lung apparatus functions in close cohesion, and it would be unreal to separate ward hypotension from respiratory compromise. Postoperative hypotension (mean arterial pressure <65 mmHg) occurred for at least a continuous 15 minutes in about one fifth of patients, and for at least a continuous 30 minutes (mean arterial pressure <70 mmHg) in a quarter. These episodes detected by continuous monitoring were missed at least half of the time using traditional monitoring [15]. Most opioid-induced respiratory compromise occurs within 2 hours of the last nursing check and can be prevented by better continuous monitoring and education [16].

Wherein lies the answer to safer, smarter, continuous monitoring on the general care floor? Simply stated, this would mean adding a multiparameter vital sign monitor to every patient room, tethering the patient to this and having a bedside nurse respond to every real and false alarm (which there would be many more than real ones). This clearly is an undesirable outcome—and one which would never make a difference in patient safety. Herein, risk stratification scores such as that developed by large observational, internally validated continuous monitoring datasets such as the PRODIGY trial, will serve to help allow the bedside health-care provider triage and reallocate resources and proactive continuous monitoring in those at a higher risk [17,18].

Availability of increased amount of complex data from continuous vital sign monitoring calls for appropriate and quick analysis to guide medical decision-making. AI will play a key role in the future, ultimately, helping clinicians and patients. With increasing amounts of sensors and continuously monitored real-time data, alarms and alerts are likely to increase. We are also seeing increasing amounts of complex feature scores and derivatives being developed to improve the predictive value of the sensed data. Such high volumes of complex data requires interpretation to make it actionable, else its likely to either be ignored, cause alarm fatigue, or lead to medical errors. AI has not only been postulated but is also being applied to solve this problem [19]. Multimodal ML techniques using patient vitals, laboratory data, and clinical notes are being developed and implemented in critical care units and have been shown to have significant predictive ability [20].

In the very near future, AI will present the solution to monitoring and providing precise and actionable decision support in real time to clinicians in these complex adaptive systems to prevent imminent deterioration and improve outcomes of our patients.

References

[1] Maheshwari K, Khanna S, Bajracharya GR, Makarova N, Riter Q, Raza S, et al. A randomized trial of continuous noninvasive blood pressure monitoring during noncardiac surgery. Anesth Analg 2018;127(2):424−31.

[2] Hatib F, Jian Z, Buddi S, Lee C, Settels J, Sibert K, et al. Machine-learning algorithm to predict hypotension based on high-fidelity arterial pressure waveform analysis. Anesthesiology. 2018;129(4):663−74.

[3] de Vries EN, Ramrattan MA, Smorenburg SM, Gouma DJ, Boermeester MA. The incidence and nature of in-hospital adverse events: a systematic review. Qual Saf Health Care 2008;17(3):216−23.

[4] Fecho K, Lunney AT, Boysen PG, Rock P, Norfleet EA. Postoperative mortality after inpatient surgery: incidence and risk factors. Ther Clin Risk Manag 2008;4(4):681−8.

[5] Semel ME, Lipsitz SR, Funk LM, Bader AM, Weiser TG, Gawande AA. Rates and patterns of death after surgery in the United States, 1996 and 2006. Surgery 2012;151(2):171−82.

[6] Smilowitz NR, Gupta N, Ramakrishna H, Guo Y, Berger JS, Bangalore S. Perioperative major adverse cardiovascular and cerebrovascular events associated with noncardiac surgery. JAMA Cardiol 2017;2(2):181−7.

[7] Li G, Warner M, Lang BH, Huang L, Sun LS. Epidemiology of anesthesia-related mortality in the United States, 1999-2005. Anesthesiology 2009;110(4):759−65.

[8] Pearse RM, Moreno RP, Bauer P, Pelosi P, Metnitz P, Spies C, et al. Mortality after surgery in Europe: a 7 day cohort study. Lancet 2012;380(9847):1059−65.

[9] Perman SM, Stanton E, Soar J, Berg RA, Donnino MW, Mikkelsen ME, et al. Location of in-hospital cardiac arrest in the United States—variability in event rate and outcomes. J Am Heart Assoc 2016;5(10).

[10] Morrison LJ, Neumar RW, Zimmerman JL, Link MS, Newby LK, McMullan PW, Jr., et al. Strategies for improving survival after in-hospital cardiac arrest in the United States: 2013 consensus recommendations: a consensus statement from the American Heart Association. Circulation 2013;127(14):1538−63.

[11] Andersen LW, Berg KM, Chase M, Cocchi MN, Massaro J, Donnino MW, et al. Acute respiratory compromise on inpatient wards in the United States: incidence, outcomes, and factors associated with in-hospital mortality. Resuscitation 2016;105:123−9.

[12] Leuvan CH, Mitchell I. Missed opportunities? An observational study of vital sign measurements. Crit Care Resusc 2008;10(2):111−15.

[13] Sun Z, Sessler DI, Dalton JE, Devereaux PJ, Shahinyan A, Naylor AJ, et al. Postoperative hypoxemia is common and persistent: a prospective blinded observational study. Anesth Analg 2015;121(3):709−15.

[14] Duus CL, Aasvang EK, Olsen RM, Sorensen HBD, Jorgensen LN, Achiam MP, et al. Continuous vital sign monitoring after major abdominal surgery—quantification of micro events. Acta Anaesthesiol Scand 2018;62(9):1200−8.

[15] Turan A, Chang C, Cohen B, Saasouh W, Essber H, Yang D, et al. Incidence, severity, and detection of blood pressure perturbations after abdominal surgery: a prospective blinded observational study. Anesthesiology 2019;130(4):550−9.

[16] Lee LA, Caplan RA, Stephens LS, Posner KL, Terman GW, Voepel-Lewis T, et al. Postoperative opioid-induced respiratory depression: a closed claims analysis. Anesthesiology 2015;122(3):659−65.

[17] Khanna A, Buhre W, Saager L, Stefano PD, Weingarten T, Dahan A, et al. 36: Derivation and validation of a novel opioid-induced respiratory depression risk prediction tool. Crit Care Med 2019;47(1):18.

[18] Khanna AK, Overdyk FJ, Greening C, Di Stefano P, Buhre WF. Respiratory depression in low acuity hospital settings—seeking answers from the PRODIGY trial. J Crit Care 2018;47:80−7.

[19] Sessler DI, Saugel B. Beyond 'failure to rescue': the time has come for continuous ward monitoring. Br J Anaesth 2019;122(3):304−6.

[20] Nemati S, Holder A, Razmi F, Stanley MD, Clifford GD, Buchman TG. An interpretable machine learning model for accurate prediction of sepsis in the ICU. Crit Care Med 2018;46(4):547−53.

Present assessment and future strategy

Overall, anesthesiologists are technology-savvy and early adopters of technological advances, but interestingly there has not been wider adoption of AI to date. The main reason is probably that AI technology, while maturing in the area of medical image interpretation, still needs to be more sophisticated in the realm of decision support; real-time, complex decision support with feedback mechanism is the crux of the need for AI-enabled anesthesia practice (not medical image interpretation as with other clinicians). This special type of decision support also needs to be inclusive of all data from the patient prior to the procedure for completeness. In addition, anesthesiologists usually have a heavy and hurried workflow, and AI still has not had major impact in this domain, although AI has started to have presence in this area.

For the future, monitoring in various venues can be embedded with simple algorithms in the form of embedded AI (eAI) so there is less monitor fatigue for the anesthesiologists. ML and DL and deep reinforcement learning will be more sophisticated for both decision support and biomedical diagnostics and will be able to accommodate the challenging real-time, complex decision-making processes in anesthesia for enthusiastic support for AI by this group. There can be much more clinical application of AI (especially ML and fuzzy logic) with medication infusions so often used in anesthesia practice. In addition, there is high potential for the application of AI in altered reality for pain management, robotic, and virtual assistance in the OR for certain areas such as intubation and other procedures and for the utilization of RPA in automating much of the workflow for the OR and anesthesia team (Table 8.2).

TABLE 8.2   Anesthesiology.

Category of AI application	Clinical relevance	Current AI availability
Medical imaging	+ +	+ + +
Altered reality	+ + +	+
Decision support	+ + +	+ +
Biomedical diagnostics	+ + +	+
Precision medicine	+ +	+
Drug discovery	+ +	+
Digital health	+	+
Wearable technology	+	+
Robotic technology	+ +	+ +
Virtual assistants	+ +	+

AI, Artificial intelligence.

Cardiology (Adult and Pediatric) and Cardiac Surgery

Cardiology is a multidimensional subspecialty that deals with disorders of the heart and blood vessels: while an adult cardiologist sees adults with usually acquired cardiovascular disorders (coronary artery disease, congestive heart failure, elevated cholesterol, atrial fibrillation, or systemic hypertension), a pediatric cardiologist treats fetuses, children, and adolescents with both acquired and congenital heart disease. Typical tests that a cardiologist orders include EKG (a tracing of the heart rhythm), echocardiogram (an ultrasound study of the heart often confused with the former test), CT or MRI of the heart and chest, nuclear imaging, cardiopulmonary exercise testing, and tests that are designed to detect heart rhythm disturbances such as an external (including recent miniature portable devices such as Kardia and the Apple watch) or an internal implanted monitor. Cardiologists also perform cardiac catheterizations for diagnostic purposes as well as for interventional procedures such a balloon angioplasty, stent and/or valve implantation, biopsies, electrophysiologic study, or a pacemaker insertion. A cardiologist sees patients both in the hospital setting (including often in the ICU setting) or in the outpatient clinic.

Published reviews and selected works

For a subspecialty particularly rich with imaging and clinical data already with more sources to come in the very near future (especially with EKG apps, implantable monitors, and biosensors), cardiology remained relatively dormant in AI-related publications until recently and is in its early discovery stages of adoption of AI [9]. A short review by Bonderman discussed some of the very early AI works in cardiology in the context of reducing both bias and noise in cardiology [10]. A much more comprehensive review by Johnson discussed the many dimensions how AI can affect cardiology: from research to clinical practice and population health (see Fig. 8.1) [11]. This review also delineated the basics of ML (supervised/unsupervised learning and even reinforcement learning). Another review presented a broad and concise overview of AI and its implications in cardiology [12]. An excellent and concise review by Shameer focused on all aspects of cardiovascular medicine in the context of ML [13] and another review also showed how effective AI can be in the context of precision cardiovascular medicine [14]. A somewhat esoteric review on an overall DL approach (vs rule-based expert systems) in cardiology discussed the advantages and limitations of this new paradigm with an emphasis on the data science aspects such as CNN, recurrent neural network (RNN), and deep belief networks as well as the limitations of data such as standardization and quality [15]. A very recent review of AI in cardiovascular imaging summarizes recent promising applications of AI in cardiac imaging with innovative data visualization [16].

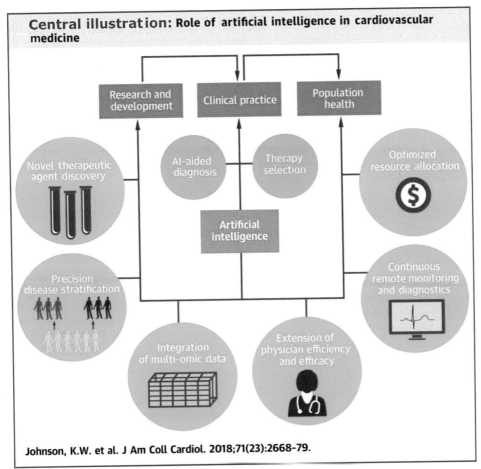

FIGURE 8.1   Role of AI in cardiovascular medicine. The role of AI in cardiovascular medicine is wide-ranging, from research and development to clinical practice as well as population health with applications such as AI-aided diagnosis, continuous remote monitoring and diagnostics, integration of multiomic data, and precision disease stratification. *AI,* Artificial intelligence.

Will artificial intelligence (AI) disrupt cardiology?

John S. Rumsfeld

John S. Rumsfeld, a cardiologist and the chief innovation officer for the American College of Cardiology, authored this commentary to give his insight into both the reasons for the relatively slow adoption in cardiology as well as the promise of AI in cardiology in the future.

A seemingly endless string of blogs and news headlines declare that AI is improving cardiac care and is poised to cure heart disease altogether. Based solely on headlines, it would seem that AI has already fundamentally changed cardiac care, even supplanting clinicians in the interpretation of cardiovascular imaging, risk prediction, and the care of patients with cardiovascular risk factors and disease. However, a visit to almost any hospital or clinic demonstrates that this is far from the case. There is a huge gap between headlines touting AI and the front lines of care.

This gap is somewhat surprising given massive investments in health-care AI. The overall health-care AI market exceeded $2 billion as of 2018 and is estimated to become a $30 + billion business sector by the mid-2020s [1]. Beyond the many start-up health-care AI companies, large technology companies continue to make major investments in health-care AI and medical device and pharmaceutical companies are increasingly focused on AI-driven solutions. Cardiovascular risk factors (e.g., diabetes and hypertension) and diseases (e.g., heart failure and atrial fibrillation) are often a major—if not primary—foci for the many hundreds of "AI solutions" being developed.

How can there be so much investment in AI, with so much focus on cardiology, and so little adoption in cardiac care? There are many reasons, but three factors worthy of emphasis include the following:

1. *Lack of evidence*: While we are still in the early phases of AI application development in relation to cardiovascular care, there is a surprisingly thin evidence base thus far. Some AI technology companies are committed to developing clinical evidence, but the overall "hype to evidence" ratio is extremely high. Evidence—or clinical validation—is essential since these tools will be used in health and clinical decision-making. The high stakes of health and healthcare warrant evidence before adoption.
2. *Uncertainty whether AI will solve clinical problems*: Particularly in relation to risk prediction and CDS, AI models may only yield nominal, or incremental, improvements over existing risk models. And it is unclear whether, or how, clinicians will utilize these models. Use cases are also critical: for example, classifying older, sicker patients as higher risk for worse outcomes and resource utilization is not particularly helpful. There are methodologic concerns as well, particularly where AI tools depend on observational data (e.g., EHR) [2].
3. *Lack of integration into care delivery*: For AI tools to be effectively utilized, they will need to be integrated in to clinical care delivery. Even as clinical care evolves (e.g., more virtual care and remote monitoring), AI solutions will need to be integrated in to workflow—and be actionable based on their interpretation—to achieve their promise.

Combining these challenges with other factors (regulatory, payment model alignment, legal/ethical considerations, etc.), it becomes much less surprising that AI has not—yet—transformed CV care.

Still, the *promise* of AI to improve CV care remains unsullied. If successfully developed and implemented, AI solutions can potentially improve access, efficiency, quality, and outcomes, across the spectrum of prevention to disease management. Ideally, AI solutions will serve as "augmented intelligence," promoting humanism in CV care, with positive engagement of both clinicians and patients [3]. Many feel that "we will get there," with AI-driven CV applications focused on image interpretation, risk prediction, diagnosis, and treatment. Certainly, there are some early promising studies that reinforce this direction, particularly with regard to imaging and risk prediction [4,5].

What is needed, then, to help close the gap between the promise of AI and the current reality of CV care? There is obviously no single answer, but a roadmap acknowledging the factors needed to successfully develop and adopt AI to advance cardiac care is a start. To that end, the American College of Cardiology (ACC) hosted a multistakeholder innovation summit to develop such a roadmap [6]. At a high level the roadmap emphasizes that just having AI analytics and related tools (e.g., digital health or imaging applications) alone is not sufficient. For example, AI tools should be created based on what clinical problem(s) are being addressed; they should be developed with clinical integration and interpretability in mind; there needs to be clinical validation and evidence, and there must be payment model alignment to support their use in evolved care models [7].

Following this roadmap, with an overarching goal of transforming CV care to improve heart health, ACC has been developing novel partnerships—bringing the clinical and technologic worlds together to codevelop AI and related digital health solutions, to support evidence development, and to promote clinical integration in evolving payment models. For each step, both clinical intelligence and AI are seen as jointly needed for success, with a relentless focus on patient-centered care. Ultimately, cardiology will benefit from AI-guided care, augmenting CV clinicians in terms of both efficiency and improving health outcomes, but we need to look past the hype and embrace the work needed to realize this potential.

References

[1] *Health IT Analytics*.Post in Tools and strategies. Artificial Intelligence in Healthcare spending to hit 36B. December 28, 2018.<https://healthitanalytics.com/news/artificial-intelligence-in-healthcare-spending-to-hit-36b> [accessed 16.06.19].

[2] Statistical thinking. Is medicine mesmerized by machine learning? *Statistical thinking*.December 12, 2019.< https://www.fharrell.com/post/medml/> [accessed 16.06.19].

[3] Verghese A, Shah NH, Harrington RA. What this computer needs is a physician: humanism and artificial intelligence. JAMA 2018;319(1):19—20.

[4] Zhang J, Gajjala S, Agrawal P, Tison GH, Hallock LA, Beussink-Nelson L, et al. Fully automated echocardiogram interpretation in clinical practice. Circulation 2018;138(16):1623—35.

[5] Rajkomar A, Oren E, Chen K, et al. Scalable and accurate deep learning with electronic health records. NPJ Digital Med 2018;1 Article number: 18. Available at: https://www.nature.com/articles/s41746-018-0029-1.

[6] Bhavnani SP, Parakh K, Atreja A, Druz R, Graham GN, Hayek SS, et al. Roadmap for innovation-ACC health policy statement on healthcare transformation in the era of digital health, big data, and precision health. J Am Coll Cardiol 2017;70(21):2696—718.

[7] Walsh MN, Rumsfeld JS. Leading the digital transformation of healthcare: the ACC innovation strategy. J Am Coll Cardiol 2017;70(21):2719—22.

Several reviews have focused on ML in the context of modalities of cardiac imaging that included not only MRI and echocardiograms but also coronary artery calcium scoring and coronary computed tomography angiography [17—20]. In addition, a review on decision support systems in cardiology with 41 relevant publications revealed that knowledge base as well as artificial neural network (ANN) and fuzzy logic are the most commonly used approaches to diagnosis and prediction [21] while a review of data mining papers over a 15-year-period yielded neural networks and support vector machines (SVMs) to be the most accurate and more efficient compared to other techniques in the development of data-mining models in cardiology [22].

There is relatively increased use of AI and data science in the domain of cardiac imaging ranging from ECG to echocardiography for both diagnosis and prognosis. For signal modalities an arrhythmia detection end-to-end DL strategy (accomplished with a 34-layer CNN) with a single-lead ECG (with more than 90,000 single-lead ECGs from over 50,000 patients with 12 rhythm disturbance categories such as atrial fibrillation and ventricular tachycardia) accomplished interpretation at a cardiologist level [23]. Recently, Attia and his group reported a study that paired 12-lead ECGs with echocardiograms to predict eventual LV dysfunction in patients with asymptomatic LV dysfunction with the use of CNN [24].

Deep learning (DL) for medical ultrasound

Rima Arnaout[1]

Rima Arnaout, a cardiologist with a passion for using computational science in medical imaging, authored this commentary about how the domains of DL and medical ultrasound can converge and be brought into practice and research.

[1]*Bakar Institute for Computational Health Sciences, University of California, San Francisco, San Francisco, CA, United States*

Seeing an unmet need in medicine, one physician hypothesized that recent technological innovations outside of medicine could be adapted for biomedical use. His ideas were met with trepidation and disbelief from many physician colleagues, but he found a scientist to help develop and test his ideas, and together they succeeded.

It was 1953; the technology was sonar; the physician was Inge Edler and the scientist was Carl Hertz. That was the story of the birth of medical ultrasound.

But that story could just as easily describe the present day. Now, ultrasound is part of the medical establishment, one of the most widely used medical imaging modalities worldwide. And the disruptive technology sweeping in from the computer science and engineering fields is DL.

DL and related ML techniques hold great promise for improving patient care through better medical ultrasound, one of the highest volume medical imaging tests in the world. To do this right, innovators must make use of both clinical and technical expertise.

Ultrasound has maintained its utility for medical diagnosis and management even as more sophisticated imaging modalities such as CT and MRI have earned their place in medical imaging, because of several key advantages. Ultrasound can show not just organ structure but organ *function*, such as the beating of the heart or the flow of blood in the liver, and it does so in real time. Ultrasound exams are radiation-free, relatively inexpensive, and portable, which means that there is relatively little selection bias among the types of patients, and the types of diseases, represented in ultrasound datasets. For rich and poor, young and old, healthy or sick, ultrasound imaging provides an excellent balance of information richness and ease of use (Fig. 1).

However, ultrasound has its challenges. Compared to other imaging modalities, ultrasound images have poor spatial resolution, can be difficult to interpret accurately and reproducibly [1], and are difficult to acquire, because a human hand must guide the ultrasound probe at just the

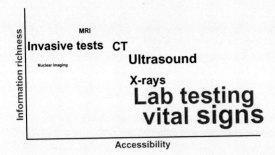

FIGURE 1 **Qualitative summary of testing modalities in medicine by information richness and accessibility.** Size of each word estimates volume in the United States.

right angle and with just the right pressure to get a good image. The ease of ultrasound is in a way deceptive; both acquiring and interpreting it take minutes to learn, but a lifetime to master. DL promises to accelerate and democratize that mastery. DL is revolutionizing pattern recognition and analysis for images across many fields, including in medicine [2,3]. In ultrasound, DL has been used to recognize views of the heart from different angles to aid in image acquisition [4]. It has also begun to power screening and diagnosis for rare diseases [5]. Through this type of work, DL can help improve accuracy and reproducibility for ultrasound image acquisition and interpretation.

As in the days of Edler and Hertz, this goal will be reached only with multidisciplinary collaboration. A DL model is a function of the selection and tuning of a particular neural network architecture, and the annotation and curation of data used to train it. Data science experts have led research into neural networks [6–8]. Similarly, we want the best clinical experts participating in data curation and validation, so that trained models represent the best clinical expertise available. These models can then be deployed worldwide to improve diagnosis and management wherever ultrasound imaging is used.

At the birth of medical ultrasound in the 1950s, there were more naysayers than proponents [9]. In contrast, today it seems the potential for DL to revolutionize medicine is being overhyped [10]. This is especially dangerous as ultrasound presents several challenges from a machine-learning perspective. As mentioned, in addition to being noisy, ultrasound includes multiple different views of each structure of interest; it can include multiple videos and still images in a single exam, presenting the challenge of a large amount of multimodal data input. Fundamental research in how to apply DL to ultrasound is, therefore, required, as is the participation of physicians as both stakeholders and innovators for medical ultrasound's continued success [11].

References

[1] Thavendiranathan P, et al. Improved interobserver variability and accuracy of echocardiographic visual left ventricular ejection fraction assessment through a self-directed learning program using cardiac magnetic resonance images. J Am Soc Echocardiogr 2013;26(11):1267–73.

[2] Esteva A, et al. Dermatologist-level classification of skin cancer with deep neural networks. Nature 2017;542 (7639):115–18.

[3] Gulshan V, et al. Development and validation of a deep learning algorithm for detection of diabetic retinopathy in retinal fundus photographs. JAMA 2016;316(22):2402–10.

[4] Madani A, Arnaout R, Mofrad M, Arnaout R. Fast and accurate view classification of echocardiograms using deep learning. NPJ Digital Med 2018;1(1):6.

[5] Arnaout R, et al. Deep-learning models improve on community-level diagnosis for common congenital heart disease lesions. ArXi e-prints [Internet]. 2018. Available from: <https://arxiv.org/abs/1809.06993> [cited 19.09.18].

[6] Hinton G. Deep learning—a technology with the potential to transform health care. JAMA 2018;320(11):1101−2.

[7] Karpathy A. The unreasonable effectiveness of recurrent neural networks. Available from: <http://karpathy.github.io/2015/05/21/rnn-effectiveness/>; 2015

[8] Montavon G, Samek W, Müller K-R. Methods for interpreting and understanding deep neural networks. ArXiv e-prints [Internet]. Available from: <https://ui.adsabs.harvard.edu/#abs/2017arXiv170607979M>; 2017.

[9] Singh S, Goyal A. The origin of echocardiography: a tribute to Inge Edler. Tex Heart Inst J 2007;34(4):431−8.

[10] de Saint Laurent C. In defence of machine learning: debunking the myths of artificial intelligence. Eur J Psychol 2018;14(4):734−47.

[11] Lindner JR. The importance of understanding the technology that serves us. J Am Soc Echocardiogr 2018;31(7): A27−8.

While a prior study described work on automated quantification in echocardiography (chamber quantification for left ventricular function and assessment of valve disease) [25], more recent works delineated how CNN was used for automated real-time standard view classification and image segmentation to improve workflow [26,27] (see Fig. 8.2).

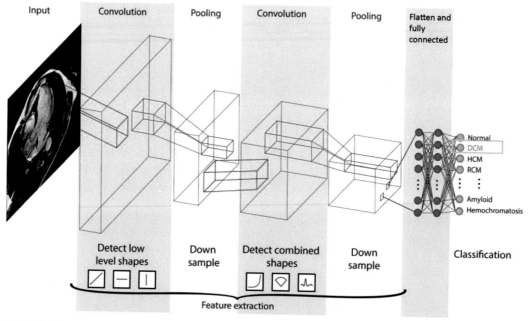

FIGURE 8.2   Schematic diagram of CNN in cardiac imaging. The convolutional layers act as filters to detect shapes or objects in the previous layer (in *red*). The complexity of these shapes progresses from the first layers to the last year. The data are downsampled in a pooling layer (in *blue*) that results in a higher level of complexity. After this series of convolutional and pooling layers the layers are flattened to a single layer that passes some fully connected layers to rest on a predicted output. In this case the image is consistent with DCM. *CNN*, Convolutional neural network; *DCM*, dilated cardiomyopathy; *HCM*, hypertrophic cardiomyopathy; *RCM*, restrictive cardiomyopathy.

In addition, ML has been applied to the ever so challenging heart failure patients with preserved ejection fraction and helped to set up a new phenotypic risk assessment system for heart failure [28] and also patients with either hypertrophic cardiomyopathy or athletes LV hypertrophy based on expert-annotated, speckle-tracking of echocardiograms [29]. Interestingly, a cognitive ML strategy via associative memory classifier was deployed to differentiate patients with constrictive pericarditis from restrictive cardiomyopathy [30]. One study even described using NLP for large-scale, automated, and accurate extracting of structured, semistructured, and unstructured data from echocardiography reports [31]. DL algorithms have also been applied to cardiac MRI as a prognosis prediction tool in patients with pulmonary hypertension and shown to be superior to clinicians' assessment [32]. There is even a report on CNN analyzing retinal fundus photographs (specifically optic disc or blood vessels) as a tool for prediction of cardiovascular risk factors;[33] this is a good example of crossing the boundaries of specialists to attain a higher level of care. In short, while ECGs and static images such as cardiac MRI are relatively straight forward for ML and DL, more complicated cardiac imaging such as echocardiograms as well as 3D and 4D images are also being studied with ML and DL.

There are also published reports of using AI for CDS in various settings in both adult and pediatric cardiology. Decision-making in cardiology is often complex and is particularly vulnerable to many heuristics and biases but also traps such as group think, sunk cost trap, and status quo trap [34]. A contribution from AI and data science is the detection of a clinical event in EHRs: one study used a supervised CNN model enhanced with bidirectional **long short-term memory (**LSTM)-based autoencoders to detect a bleeding event [35]. Similarly, a DL algorithm (called DL—based early warning system, or DEWS) that used only four vital signs had a high sensitivity and a low false positive rate for detection of patients with cardiac arrest in a multicenter study [36]. Choi reported the use of RNN with gated recurrent units to DL models to leverage temporal relations appear to improve the performance of these models to detect incident heart failure even with a short observation window of 12–18 months [37]. Finally, ML in the form of SVM devised an effective risk calculator that was shown to be superior (less recommended drug therapy with less adverse events) to the existing accepted ACC/American Heart Association cardiovascular disease risk calculator [38].

There are relatively few reports of AI-focused publications in children and adults with congenital heart disease although a very early report by Warner delineated a mathematical approach to the diagnosis of congenital heart disease [39]. An early report of utilizing an AI strategy for screening tests such as ECGs for sudden cardiac death with a ML/DL approach to maximize accurate diagnosis [40]. An AI-assisted auscultation algorithm performed well in a virtual clinical trial but may be difficult to become a routine approach given readily available echocardiographic assessments; in short, an AI strategy for an older technology (the stethoscope) may or may not be adopted by clinicians [41]. ML algorithms were deployed to train a large dataset of adults with congenital heart disease to prognosticate and facilitate clinical management [42,43]. A similar study to the aforementioned DEWS study revealed that predictive models created by AI can lead to earlier detection of patients at risk for clinical deterioration and thereby improves care for pediatric patients in the pediatric cardiac intensive care setting [44]. In addition, four AI-based algorithms were employed to facilitate a CDS system (CDSS) for estimating risk in congenital heart surgery [45]. One innovative report was described using ML and system modeling to facilitate a multicenter collaborative learning project for rapid structured fact-finding and dissemination of expertise; this forward thinking approach can provide a complement (and perhaps render less necessary) the traditional multicenter, randomized clinical trials that are sometimes challenging to execute [46].

Present assessment and future strategy

Overall, as cardiology is both a perceptual or image intensive field as well as a cognitive or decision-making subspecialty with a myriad of procedural tasks, AI is a particularly valuable technology for cardiology with potentially very rich dividends that are vastly underexplored at present but has great promise. Cardiologists presently are curious about AI but have not adopted AI methodologies as have the radiologists even in the image sector. There has been increasing activity in both conferences as well as the publishing domain on AI in cardiology, and there are already algorithms deployed to ECG monitoring on an outpatient basis. In the medical image domain, DL with CNN is being deployed for not only static images such as EKG, CT, and MRI but CNN with modifications (with LSTM) or in combination with RNN can also now interpret dynamic images such as angiograms and echocardiograms. These methodologies are direly needed in order to deploy much more sophisticated serial echocardiographic determinations of heart performance than the current conventional metrics of systolic and diastolic function.

For the future, cardiology and cardiologists can benefit greatly from AI as a technological resource. First, the full continuum of cardiac imaging presents serious but solvable challenges for AI deployment if a cardiologist wishes to integrate all of these studies for hybrid images or image fusion (echocardiogram and cardiac MRI which focuses on physiology and anatomy, respectively). In essence, there will be the emergence of a "super scan" in which all imaging modalities lead to a single set of 4D images that can be easily moved and manipulated. AI can also be integrated into the entire continuum, an "AI image continuum," from acquisition via operators (AI-enabled or "intelligent" image acquisition) to image interpretation as discussed previously and finally on to integration of images into individualized precision health assessment. The cardiologist can also be liberated from the long list of relatively mundane tasks to higher level of medical decision-making with full deployment of the various AI tools available. Second, complex decision-making with AI and the use of deep reinforcement learning can be particularly useful for the ever increasingly complex nature of diagnostic and therapeutic precision cardiovascular medicine in both the intensive care (precision intensive care) or hospital setting as well as the outpatient arena (precision cardiovascular care). This type of individualized medicine will need the many layers of data and information all integrated into an AI-enabled strategy for delivery of key information for knowledge and treatment. Routine individualized precision cardiovascular medicine will be practiced in which pharmacogenomic profiles and all aspects of biomedical data (including wearable devices) are combined to yield a precise individual diagnostic and therapeutic strategy as well as risk stratification for cardiovascular disease supported by population data. Third, the application of altered reality in cardiology and cardiac surgery for education, training, and trial procedures and interventions prior to the live procedure in heart disease will be made available with AI as a resource. Given the complexity and nuances of operative and interventional procedures in cardiology, perhaps it will be quite some time before a robot will be performing a cardiac procedure in its entirety even with human oversight. Lastly, the

TABLE 8.3 Cardiology and cardiac surgery.

Category of AI application	Clinical relevance	Current AI availability
Medical imaging	+ + +	+ + +
Altered reality	+ + +	+
Decision support	+ + +	+ +
Biomedical diagnostics	+ + +	+
Precision medicine	+ + +	+
Drug discovery	+ +	+
Digital health	+ + +	+
Wearable technology	+ + +	+
Robotic technology	+ +	+ +
Virtual assistants	+ +	+

AI, Artificial intelligence.

administration aspects of a busy and complicated heart program can be better managed with some of the RPA tools that are already available. Prediction models can also be in place to prevent unnecessary readmissions and complications. In short, AI is a much needed and very timely resource in cardiology especially since the cardiovascular disease burden remains singularly the largest and continues to climb in an aging population worldwide in both the developed as well as underdeveloped worlds (Table 8.3).

Critical Care Medicine

The intensive care physician has significant overlap with the anesthesiologist in skill set and tasks, and often have had anesthesiology training or board certification. These physicians usually care for patients that are housed in the intensive care setting and usually not in the hospital ward or outpatient setting. The ICU physician and team can work in many types of ICUs, include adult-focused ICUs such as the medical ICU, the coronary care unit, and the surgical ICU as well as the ICUs that are devoted to children such as the pediatric ICU (PICU) and the neonatal ICU (NICU). These physicians are multifunctional within the ICU in which they look at medical images (such as chest X-rays or head CT); perform procedures such as central line or chest tube placements; interpret vital signs and integrate all of this data; and make a medical decision, sometimes on an urgent and continual basis.

Published reviews and selected works

Like anesthesiology, critical care medicine is data-rich, but there has been only a relatively moderate level of publishing about utilizing AI in the ICU setting (for the same reasons as described earlier for anesthesiology). A review of AI applications in the ICU that was published close to two decades ago, including several older methodologies, is still worth reviewing [47]. Another good review focused on the integration of AI into complex decision-making in the acute care medical arena [48]. The author very astutely pointed out that the future physician and nurse will need to adapt to the AI at the bedside with a new paradigm of time-pattern recognition (vs the traditional threshold decision-making). This new AI-human synergistic thinking requires teaching archives of time pattern phenotypes to be taught and medical education to reflect this paradigm shift. An excellent and thoughtful review discussed not only the potential but also the limitations and pitfalls of big data in critical care medicine; the authors strongly advocate an open and collaborative environment for future ICU data efforts [49]. Lastly, Johnson reviewed the limitations of the current ICU databases for precision medicine [50]. These authors contend that the three challenges are compartmentalization, corruption (erroneous, missing, and imprecise data), and complexity (multimodal data).

An example of application of AI in the hospital setting is the use of ML algorithms for training expert-labeled vital sign data streams to automatically classify vital sign alerts as real or artifact in order to clean such data for future modeling [51]. Another robust illustration of the value of databases coupled with analytics is the Medical Information Mart for Intensive Care (MIMIC) accessible critical care database [52] (see Fig. 8.3). MIMIC is a

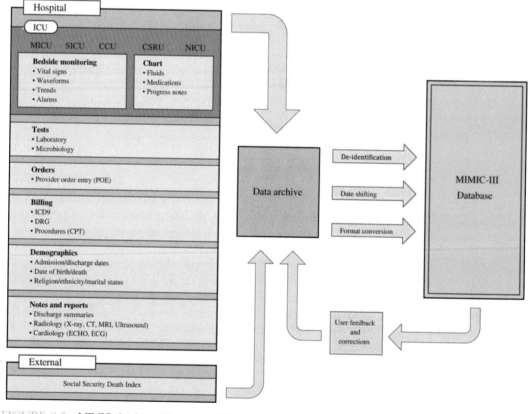

FIGURE 8.3   MIMIC database. The updated MIMIC-III database integrates deidentified clinical data from a data archive so that it is widely accessible to researchers and students under a data user agreement. The data archive is a data repository that includes a large amount of hospital data (from several different ICUs) and also includes an external data source (social security death index). *ICU*, Intensive care unit; *MIMIC*, Medical Information Mart for Intensive Care.

large, single-center (Mass General Hospital) database comprising of information in the critical care unit from tens of thousands of admitted patients (over 50,000 adults and close to 8000 neonates) for data mining and modeling of conditions that resulted in many publications. The data include general data (such as patient demographic, hospital admissions and discharge dates, death dates, and ICD-9 codes), physiological data (such as hourly vital signs and ventilator settings), medications (including IV medications), lab tests (including imaging data), fluid balance, and notes and reports (imaging reports, progress notes, discharge summary, etc.). The data can be downloaded as flat files an imported into a database system. The code that underpins the MIMIC-III database is openly available and a Jupyter notebook containing the code is also available. MIMIC-III has been used for "datathons" around the world by the MIT group to enable multidisciplinary groups that consist of clinicians and data scientists to answer and solve clinically relevant research queries and clinical problems [53].

Multidisciplinary datathons and lessons learned

Christina Chen

Christina Chen, a nephrologist with a strong background in data science and a faculty member for numerous datathons, authored this commentary on the value of the datathon hands-on experience for both clinicians and data scientists to answer clinical queries together.

Increasing the volume of digital health data and growing complexity of "big data" research has led to higher demand for team-based collaborative efforts. One such collaborative effort is the datathon. A "datathon" builds from the computer industry's "hackathon" model where a diverse group of professionals are brought together to identify a problem area and develop an innovative software solution over a focused period of time (typically a continuous 24–48 hours) [1]. Each group's solution is then presented at the close of the event. In particular, a datathon applies the hackathon model to data analytics and serves as a platform for real-time effective exchange of ideas among specialists. These events provide an excellent opportunity to reach a diverse community of people, introduce them to collaborative data science projects, generate new ideas, and pave the path for future collaborations. Although these events are short in duration, they can have long lasting benefits.

Developing new networks and creating collaborations

Researchers are often siloed into disparate groups with limited exposures to other departments, institutions, and industries. To challenge this limitation, datathons provide a place and opportunity for specialists to have real-time respectful and effective exchange of ideas. Furthermore, there is often field-specific jargon that is difficult for other people to understand. Working in small groups provides a more intimate setting to communicate effectively which can be difficult to accomplish in larger more anonymous and less collaborative events. This close knit collaboration capitalizes on the power of bringing together diverse expertise to effectively address problems immediately that command a deeper understanding of subject matter and design resulting in solutions that are both clinically interesting, technically feasible, and statistically sound. The long-term goal is for groups to continue working on their projects following the event and potentially collaborate in future projects.

Promoting public-access data and code sharing

MIMIC-III is a public-access high-resolution clinical ICU database that has yielded many publications and has been used in numerous datathons [2–4]. MIMIC-III is hosted on a platform that allows communities to view and publish shared problems. Using this database during datathons demonstrates what is possible when data can be explored. Furthermore, the platform allows for direct querying of questions to the dataset and returns the query specific results. These functionalities make it easy for communities to collaborate and contribute to shared concepts [5]. The high-yield in publications from using this public access dataset highlights the impact of open data and aims to encourage others to publicly release their datasets as well.

Lack of reproducibility in medical studies has been a barrier to producing consistently reliable findings. Transparent research processes can help to improve the quality of research and allow for complete understanding of methods and research design. Datathons provide an opportunity to advocate for behaviors that promote reproducibility in research such as open source code and sharing concepts [5,6].

Words of wisdom for a successful datathon

Typical datathons are short events, often 24–48 hours in duration. To ensure the event runs smoothly and productively, here are some suggestions:

1. Accessing data: There is no datathon without data. Providing a platform to access data during the datathon is essential, and in the past it has been done in different ways such as providing physical servers or using cloud services. Since datathons are short, it is more productive if the time is used to work on their projects instead of figuring out connectivity. Forms for data release are often filled out prior to the event.
2. Well defined questions: Meeting with clinicians prior to the event can help teams identify the areas of interests and develop a question that would be interesting and suitable given the time restraints and dataset used in datathons.
3. Balance of data scientists to clinicians: Taking some effort to ensure a balanced ratio of data scientists and clinicians creates more balanced groups. In some datathons, groups were randomly assigned ahead of time. In other datathons, active recruitment of clinicians was necessary.
4. Availability of mentors: Having a group of mentors, both clinicians and data scientists, who are both familiar with the data and clinical questions was a helpful resource for datathon groups.
5. Predatathon workshops: Bootcamps or workshops prior to datathons prepare participants for conducting research and informs them about data available. Some workshops that have been helpful include the ones that address asking a good research questions (such as PICO: P—patient, problem, or population; I—intervention; C—comparison, control or comparator; O—outcome), statistics and epidemiology, ML methods, code sharing concurrency control platforms (such as GitHub) and use of programs that promote documentation (such as Juptyer notebooks). Ideally these workshops can make use of the dataset being used for the event.

Datathons provide an excellent opportunity to investigate interesting questions, become familiar with the data, promote good habits, and build collaborations. The upfront collaborative interdisciplinary review may help identify ideas that are worth additional investigation and ones that are unlikely to be successful. The short time investment also allows for exploration of unusual innovative ideas. In addition to the technical and research advantages, these events have the potential to lead to long lasting friendships and provide inspiration for a common goal of responsibly improving healthcare.

References

[1] Aboab J, Celi LA, Charlton P, Feng M, Ghassemi M, Marshall DC, et al. A "datathon" model to support cross-disciplinary collaboration. Sci Transl Med 2016;8(333):333ps8.
[2] Johnson AEW, Pollard TJ, Shen L, Lehman L-WH, Feng M, Ghassemi M, et al. MIMIC-III, a freely accessible critical care database. Sci Data 2016;3:160035.
[3] Li P, Xie C, Pollard T, Johnson AEW, Cao D, Kang H, et al. Promoting secondary analysis of electronic medical records in China: summary of the PLAGH-MIT critical data conference and health datathon. JMIR Med Inform 2017;5(4):e43.
[4] Serpa Neto A, Kugener G, Bulgarelli L, Rabello Filho R, de la Hoz MÁA, Johnson AE, et al. First Brazilian datathon in critical care. Rev Bras Ter Intensiva 2018;30(1).
[5] Johnson AE, Stone DJ, Celi LA, Pollard TJ. The MIMIC Code Repository: enabling reproducibility in critical care research. J Am Med Inform Assoc 2018;25(1):32–9.
[6] Moseley ET, Hsu DJ, Stone DJ, Celi LA. Beyond open big data: addressing unreliable research. J Med Internet Res 2014;16(11):e259.

AI is deployed in the form of an intelligent reinforcement learning agent (the AI Clinician) that extracted implicit knowledge form patient data that exceeds the experience of a human clinician by many fold [54]. This AI model was able to provide individualized treatment decisions in sepsis that could improve patient outcomes. In addition, NLP-driven prediction models with laboratory data and vital signs were successful in yielding better predictive performance for mortality [55] while another ML model emphasized the importance of time in outcome predictions [56]. AI has also been used to enhance the simulation-based experience and training in resuscitation medicine [57]. Lastly, there is ongoing interest in forming a dyad between telemedicine ICU (tele-ICU) and ML-supported CDSSs (for sepsis prediction and ventilator management) especially with the enormous quantities of data from the tele-ICUs for a generalizable ML-CDSS algorithms [58].

Pediatric and neonatal intensive care

In the pediatric intensive care setting, there is a discussion about using AI more robustly toward "precision intensive care" where there is real-time DL with cognitive architecture to bring about precision care for each individual ICU patient [59]. One study used AI (in the form of logistic regression, random forests, and CNN) to identify physio-markers in predicting sepsis in the PICU 8 hours earlier than a screening algorithm [60]. In addition, there is ongoing work on the use of AI and predictive modeling for earlier detection of patients at risk for clinical decompensation in the pediatric cardiac intensive care setting. This AI strategy used neural network and decision tree classification as well as logistic regression to provide a real-time individualized risk assessment [44]. Another report by Williams described the use of *k*-means clustering unsupervised learning to discover 10 clusters with varying prognostic information [61]. Overall, The PICU community appears to be interested in pursuing AI tools for this complex environment [62]. There is less academic activity in the data-rich milieu of the NICU, but Singh et al. described a futuristic i neonatal intensive care unit with real-time analytics with a big data hub and a rule-based engine and DL [63].

The virtual pediatric intensive care unit (ICU)

Randall C. Wetzel

Randall C. Wetzel, a pediatric intensivist and the progenitor of a virtual pediatric ICU (PICU) concept, authored this commentary on the virtual PICU and its many dividends, from quality improvement, education, to data analytics for the PICU community.

In the era before Facebook and Google, when Yahoo was in its infancy and before our modern world was connected and created, a handful of pediatric intensivists met in St. Louis in 1997 to discuss how best to harness information technologies, the newfangled thing the Internet and the exciting potential of a connected world for the care of critically ill children. After 2 days a consensus plan emerged to improve the quality of the care provided for critically ill children by sharing data, supporting a wider Internet presence for pediatric critical care, and exploring telemedicine and distance learning. At the core of the virtual PICU (VPICU) was the hope of gathering digital data about our patients to discover how critical illness happens to children and to make more rapid progress sharing that knowledge. The primary impetus was the realization that medical data would become increasingly digital and that we would have improving digital copies of our patients, leading to the concept that *the patient is data* and that these data can be analyzed and inform the care of the next patient. These ideas led to the vision of creating a common information space for the practice of pediatric critical care that would enable us to care for children more than one at a time. After some discussion the CyberPICU was rejected and the VPICU was chosen to indicate that we would all, someday in the hopefully not too distant future, be practicing in one connected, sharing, collaborative, "virtual" PICU bringing our collective expertise to help children ([1], www.vpicu.org).

Initially, the VPICU focused on gaining a better understanding of the practice of critical care, from outcomes to types of ICUs, to the frequency of patient types, demographics and really defining our practice. Working with the Critical Care Focus Group established by the National Association of Children's Hospital and Related Institutions (NACHRI) the VPICU developed a database to serve the need to better understand pediatric critical care and to support research. This grew into a data registry to collect information about every ICU admission, how critical care was practiced, what sorts of patients were cared for, and what the severity-adjusted outcomes were. This subsequently became the independent partnership between the VPICU at Children's Hospital Los Angeles, NACHRI and subsequently The Children's Hospital Association, and The Children's Hospital of Wisconsin, now operating as Virtual Pediatric Systems, LLC (VPS, www.myvps.org).

VPS provides detailed severity of illness adjusted admission data on over 1.7 million ICU episodes from over 160 hospitals predominately in North America but also globally serving PICUs, cardiothoracic intensive care units, and NICUs. The high-quality data is collected by clinicians who are specifically trained and maintain an inter-rater reliability in excess of 0.90 with numerous electronic validations. In the past 15 years, VPS has supported hundreds of local quality improvement projects with data and customized benchmarking. In addition, VPS has provided high quality, validated data for over 150 research publications. VPS, LLC continues to provide improving, increasingly detailed customized and comparative reports to support intensivists and hospitals providing care for critically ill children [2,3].

Concurrently, the VPICU has supported online education through several projects, including supporting the largest online presence for pediatric critical care (www.pedsccm.org). Exploring telemedicine the VPICU supported a Southern California telemedicine network and collaborated in developing a telemedicine pediatric trauma network to provide critical care in the golden hour for mass casualties [4]. Educationally, the VPICU has sponsored fellowships to support education around providing telemedicine and research within the VPS database. Currently, the VPICU continues this educational mission training intensivists and data scientist in mutual understanding and collaboration around data science and AI.

At the core of the VPICU is data analytics. The VPICU is currently focused on applying ML and AI to provide CDS in the PICU. Over the years the VPICU has developed an approach to gathering clinical data, data architecture, data principals, and data curation techniques to facilitate this goal (https://ieeexplore.ieee.org/abstract/document/5999031/citations?tabFilter = papers#citations).

The VPICU founded the largest national meeting bringing together ML expertise with clinicians to foster the application of DL and AI to health-care data held annually around the country (www.mucmd.org). The VPICU's data scientists continue to explore the application of DL to critical care data [5,6] on the way to AI CDS. An important VPICU project is the training data scientists and intensivists to work together to provide the workforce for the application of AI methodologies to critical care. In addition, one of the largest barriers to data research is the availability of high-quality medical data [7]. The VPICU is leading a data collaborative among PICUs to create a large, data repository to support data science research in pediatrics, just as MIMIC has led to the widespread application of ML for adult critical care (www.vpicu.org/projects/datacollaborative).

Future challenges will be the continued need for high quality, well-curated data, and well-trained data scientists and clinicians familiar with AI techniques. We will need to decrease the time from discovery to the bedside and support near real-time CDS at the bedside. The ethical application of AI to critical care, data democratization, and the increasingly detailed capture of digital copies of our patients must also be addressed. The VPICU continues to have the ethical imperative to capture and learn from every breath, every heartbeat, and response of our patients to improve their care.

References

[1] Wetzel Randall C. The virtual pediatric intensive care unit: practice in the new millennium. Pediatr Clin North Am 2001;48:795−814.

[2] Wetzel RC, Sachedeva R, Rice TB. Are all ICUs the same? Paediatr Anaesth 2011;21(7):787−93. Available from: https://doi.org/10.1111/j.1460-9592.2011.03595.x Epub 2011.

[3] Wetzel RC. Pediatric intensive care databases for quality improvement. J Pediatr Intensive Care 2016;5:81−8.

[4] Burke RV, Berg BM, Vee P, Morton I, Nager A, Neches R, et al. Using robotic telecommunications to triage pediatric disaster victims. J Pediatr Surg 2012;47(1):221−4. Available from: https://doi.org/10.1016/j.jpedsurg.2011.10.046.

[5] Marlin BM, Kale DC, Khemani RG, Wetzel RC. Unsupervised pattern discovery in electronic health care data using probabilistic clustering models. Proceedings of the 2nd ACM SIGHIT international health informatics symposium (IHI '12). New York: ACM; 2012. p. 389−98. Available from: http://doi.acm.org/10.1145/2110363.2110408.

[6] Carlin CS, Ho LV, Ledbetter DR, Aczon MD, Wetzel RC. Predicting individual physiologically acceptable states at discharge from a pediatric intensive care unit. J Am Med Inf Assoc 2018;25(12):1600−7. Available from: https://doi.org/10.1093/jamia/ocy122.

[7] Wetzel RC. First get the data, then do the science. Pediatr Crit Care Med 2018;19(4):382−3. Available from: https://doi.org/10.1097/PCC.0000000000001482.

Time-series continuous physiologic data: artificial intelligence to manage risk and uncertainty

Peter C. Laussen[1,2]

Peter C. Laussen, a pediatric cardiac intensivist with much research experience in intensive care unit physiologic data, authored this commentary on the concept of time-series, or data in motion, and its challenge in attaining precision critical care.

[1]Department of Critical Care Medicine, The Hospital for Sick Children, Toronto, ON, Canada

[2]University of Toronto, Toronto, ON, Canada

Critical care units are dynamic, complex, and resource intense environments, where humans directly interface with technology. Decisions are time-sensitive and often occur in circumstances where clinicians need to quickly aggregate and integrate data from multiple sources. In turn, there can be variability in practice and uncertainty in management, whether it be related to the patient disease and diagnosis, the patient physiologic state and unpredictable responses to management, and to the ability of clinical teams to interpret data correctly. In addition, there are numerous competing pressures inherent to any ICU environment around work flow, communication, distraction, and resource utilization.

Typically, patients are surrounded by an array of monitors detecting physiologic signals such as heart rate, BP, and oxygen level among many others, as well as devices to support organ function such as the mechanical ventilator and infusion pumps administering medicines. There is an increasing amount of research in both adult and pediatric critical care to develop CDSSs. There are numerous examples of successful data capture of critical care data [1], but there are compromises and challenges when it comes to using the higher frequency waveform data.

Challenge of time-series data

The continuous physiologic data streaming from devices and monitors are time-series data, *data in motion*. They are characterized by a number of features: huge in volume and velocity, signals are variable in frequency and subject to considerable artifacts. No question that these data

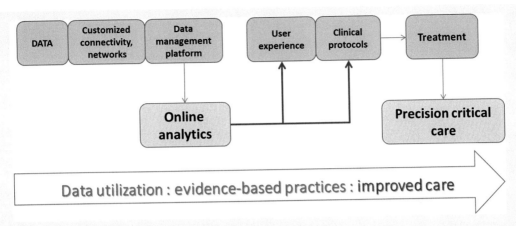

FIGURE 1   Meaningful use of complex physiologic data.

are essential for supporting decision at the bedside, but when it comes to modeling for the purpose of real-time (bedside) predictive analyses, the ability to capture, label, and integrate these data is limited. The data are messy and hard to manage, store and retrieve.

The problem of volume has placed constraints around data storage, and there are bottlenecks as the data are input/output (I/O) bound, and there needs to be an efficient compression/decompression and file indexing architecture to enable retrieval and analysis. Because of these challenges, there have been a number of approaches to handling higher frequency waveform data, including (1) relying on low frequency (1 Hz) data as recorded in the patient electronic record although such lower frequency sampling is unable to adequately capture the variability of the underlying physiological system, or (2) collect and then purge waveform data after a period of time; purging data assumes that all the features of a particular dataset are known at the time it is collected or for the duration stored, and that after purging it, there is no additional information that can be gleaned from this data.

Another approach is to capture and store low and high frequency data permanently, acknowledging that the data is "owned" by the patient, and as such, is discoverable. There are no off--the-shelf products currently available to manage the volume and variety of high frequency physiologic signals. In our department, we have built and deployed since 2016 a bespoke data management platform to facilitate collection, file indexing, compression, and decompression, Fig. 1 (see www.laussenlabs.ca). As an indication of the scale of continuous physiologic data that is measured and can be collected, and depending on the complexity of disease and treatment, there are routinely 500 + signals per hour and between 70 and 150 million physiologic data points per day generated in the 42 beds of our pediatric ICU. Over 700,000 hours and 2 trillion data points from over 4000 patients is currently stored in the database.

Data analysis

Once the data have been obtained and continuity and quality ensured, the next step is to make sure that the data can be readily analyzed. There are broad categories for using these data in critical care. Some of these include the following:

1. Describing the physiologic phenotype and individualized physiology according to age, disease, treatment, and time [2].
2. Understand the physiologic state, such as risk for low oxygen delivery, hemodynamic instability [3], effectiveness of mechanical ventilation, risk for neurologic injury, and metabolic state.
3. Develop early warning systems and the risk for an event within a physiologic state through recognizing patterns in the data, such as risk for sepsis or cardiac arrest [4,5].
4. Track the trajectory of a patient in response to treatment protocols, and direct specific interventions according to a change in the expected trajectory.
5. Develop decision support and business analytic tools to ensure the efficiencies of care against validated outcome metrics, such as length of stay and risk for readmission.
6. Enhanced signal processing and waveform analysis such as to diagnose changes in heart rhythm [6] and uncover previously hidden signals that may be embedded within composite waveforms.
7. Develop new insights into the underlying physiology, tease out subpopulations of patients that may respond to a particular treatment [7] and develop prognostic and predictive enrichment strategies that can lead to individualized and precise critical care management.

The promise and problem with physiologic data

There is a disconnect between how we make decisions and the trust we might otherwise have in a model or algorithm [8]. Models can be opaque, containing weighted features, components, and simplifications with blind spots related to the inputs and the priorities and judgment of their creators. They are mathematical outputs that may not take into consideration of conditions and behaviors. The risk therefore may be incorrect assumptions and spurious correlations, reinforced and contaminated by bias. Models need feedback of mistakes and of the results and outcomes; they need to be explainable, scalable and context sensitive.

It is possible to utilize the data generated by continuous physiologic signals at the bedside to help us understand physiologic states and phenotypes in critical care. At the same time, it is important to also understand that using big physiologic data to determine these states will not replace the clinician at the bedside, rather it will augment our decision-making, improve communication and information transfer, and to a large extent, level the playing field and reduce learning curves with respect to our experience and capability as clinicians.

References

[1] Johnson AE, Pollard TJ, Shen L, Lehman LW, Feng M, Ghassemi M, et al. MIMIC-III, a freely accessible critical care database. Nat Sci Data 2016;3:160035.
[2] Eytan D, Goodwin AJ, Greer R, Guerguerian AM, Mazwi M, Laussen PC. Distributions and behavior of vital signs in critically ill children by admission diagnosis. Pediatr Crit Care Med 2018;19(2):115—24.
[3] Potes C, Conroy B, Xu-Wilson M, Newth C, Inwald D, Frassica J. A clinical prediction model to identify patients at high risk of hemodynamic instability in the pediatric intensive care unit. Crit Care 2017;21(1):282.
[4] Meyer A, Zverinski D, Pfahringer B, Kempfert J, Kuehne T, Sundermann SH, et al. Machine learning for real-time prediction of complications in critical care: a retrospective study. Lancet Respir Med 2018;6(12):905—14.
[5] Tonekaboni S, Mazwi M, Laussen PC, Eytan D, Greer R, Goodfellow S, et al. Prediction of cardiac arrest from physiologic signals in the pediatric ICU. In: The proceedings of machine learning for healthcare conference, 85, ISSN 1938-7228, Palo Alto, CA.

[6] Goodfellow S, Goodwin A, Greer R, Laussen PC, Mazwi M, Eytan D. Atrial fibrillation classification using step-by-step machine learning. Biomed Phys Eng Express 2018;4(4):045005.
[7] Wong HR, Atkinson SJ, Cvijanovich NZ, Anas N, Allen GL, Thomas NJ, et al. Combining prognostic and predictive enrichment strategies to identify children with septic shock responsive to corticosteroids. Crit Care Med 2016;44(10) e1000-3.
[8] Topol EJ. High-performance medicine: the convergence of human and artificial intelligence. Nat Med 2019;25 (1):44−56.

Barriers to implementation of artificial intelligence (AI) in the critical care setting

Mjaye L. Mazwi[1], Danny Eytan[1,2], Sebastian D. Goodfellow1, Robert W. Greer[1] and Andrew J. Goodwin[1,3]

Mjaye L. Mazwi, Danny Eytan, Sebastian D. Goodfellow, Robert W. Greer, and Andrew J. Goodwin authored this commentary on the potential barriers to clinicians adopting AI in the intensive care unit setting, and these range from lack of domain expertise to data curation and data integrity issues.

[1]The Hospital for Sick Children, University of Toronto, Toronto, ON, Canada

[2]Faculty of Medicine, The Technion, Israel Institute of Technology, Haifa, Israel

[3]The University of Sydney, Sydney, NSW, Australia

The modern practice of critical care medicine relies on medical technologies to support patients. This technologically rich environment is very rich in data, sampled and displayed at frequencies of up to 2000 Hz. The clinical behavior of critically ill patients is characterized by unpredictable nonlinear responses to disease processes and therapeutic interventions [1]. These highly variable responses in conjunction with a lack of evidence backed by research trials leads to a great deal of uncertainty about optimal therapy at the level of the individual patient. This dynamic, data-rich environment characterized by uncertainty represents a uniquely compelling use case for AI both to cognitively offload clinicians and to augment their decision-making [2].

In spite of interest in the application of techniques such as ML to problems in critical care medicine, there has been very limited translation of ML models to the bedside [3]. This deficiency is problematic because translation should be considered the true test of model design and efficacy. Major drivers of this failure to translate models to the bedside include the following:

1. Lack of domain knowledge amongst researchers—This hampers model development by attenuating insight into both the nature of the clinical problem and the limitations associated with data types being used in model development. Integration of clinicians in multidisciplinary teams can provide this context, assist in feature engineering, and identify mechanisms for integration of model output into patient care.
2. Heterogenous and sparsely sampled data—Clinical data are heterogenous in type and missing data are a prevalent problem even in different patients with the same condition. This creates challenges in model development often requiring definition of minimum datasets for accurate predictions. In addition, a lack of accepted data collection standards between institutions limits the feasibility of transfer of models been institutions.

3. Data curation—Models developed using highly curated retrospective datasets are unsuited to the noisy, artifact-laden data generated in the process of patient care. This limits the ability to deploy such models to make prospective predictions on actual patient data at the point of care.

4. The compartmentalized nature of medical data—Heterogenous data generated in medical environments typically resides in a variety of different repositories in different formats (e.g., imaging studies in an imaging application, demographic data and outcomes in an EHR application, and biomarker data in a laboratory information system) [4]. This creates barriers to access for both development and deployment of models that integrate data from a variety of sources. This problem is compounded by a limited ability to integrate data from medical devices as a result of a lack of interoperability or proprietary data formats. We believe that a central aspect of the appraisal of any technology being purchased to support patients' needs to include assessment of how to access data displayed by the device and integrate and store it with other potentially useful data types.

5. Explainability—Classically, medical research has valued hypothesis-driven research and mechanistic understanding of disease processes. This mechanistic understanding is required to identify therapeutic targets. Data-driven model development is a new paradigm in medical research. Uncertainty about the inclusion and weighting of variables in model development or analytical techniques used can create misgivings amongst clinicians that limit incorporation of model predictions into care. Incorporating explainability into model development may mitigate this.

6. Regulatory barriers—The typical institutional review board process for approving research is unsuited to data-driven modeling techniques that require a flexible approach to identifying model inputs and iteration to improve performance over time. Patient privacy concerns can also hamper data access and the creation of the sort of large, detailed training datasets best suited to robust model development. This is an important barrier to assessing the utility of AI in medicine and a fundamental question that the medical and research communities will have to address is how to navigate these competing imperatives.

7. Variable handling of time in medical datasets—Different time protocols in medical datasets pose challenges to effective data integration. This is important because there is information embedded in the temporal sequence of disease progression that is only discoverable if data reflecting on the evolution of the disease maintains accurate alignment in time. For model implementation with the goal of real-time or near real-time prediction, this alignment needs to be maintained both in the retrospective data used to train the model as well as prospective data that serves as model inputs at the point of care.

8. Integration of prediction into clinical care—An important challenge associated with model implementation is how to most effectively incorporate model output into clinical care. This may take the form of data visualization for more complex models or simple alerts [5]. Integration needs to be undertaken with the goal of minimizing unnecessary disruption of care as well as defining thresholds for notification that minimize false positives and alarm fatigue and unnecessary cognitive loading of clinicians.

9. Ethical and legal considerations—Model implementation should be undertaken with an understanding of potential ethical and legal implications of modifying care. An example

would be the implementation of models that predict an impending event. Decisions about the responsibility to disclose model prediction to patients and families as well as expected clinician responses to predictions need to be characterized before model deployment.

Ideally, model implementation should occur in an improvement science framework with prospective identification of outcome, process, and balancing measures. This measurement framework needs to be dynamic as model performance is not static and ideally improves over time.

Although the challenges enumerated here are substantial, we believe that none of these barriers are insurmountable. The community of clinicians and researchers working in this area will need to collaborate as closely on solutions to barriers to implementation as they will on methodology and model development.

References

[1] Buchman TG. Nonlinear dynamics, complex systems, and the pathobiology of critical illness. Curr Opin Crit Care 2004;10(5):378−82.

[2] Workman M, Lesser MF, Kim J. An exploratory study of cognitive load in diagnosing patient conditions. Int J Qual Health Care 2007;19(3):127−33.

[3] Patel VL, Shortliffe EH, Stefanelli M, Szolovits P, Berthold MR, Bellazzi R, et al. The coming of age of artificial intelligence in medicine. Artif Intell Med 2009;46(1):5−17.

[4] Dimitrov DV. Medical Internet of Things and big data in healthcare. Healthc Inform Res 2016;22(3):156−63.

[5] Harrison AM, Herasevich V, Gajic O. Automated sepsis detection, alert, and clinical decision support: act on it or silence the alarm? Crit Care Med 2015;43(8):1776−7.

Present assessment and future strategy

Overall, there is now increasing academic and clinical activity compared to a similar field of anesthesiology but critical care medicine is still under leveraging the entire AI portfolio of tools. Just as in anesthesiology, one major reason is that the AI tools for real-time, complex decision support remains suboptimal but can be more sophisticated and mature in the near future. The work by the MIT group and its MIMIC effort has been instrumental in maintaining an AI and ML/DL presence in the ICU domain.

For the future the clinical relevance and current AI availability profile is similar to that of anesthesiology. Once deep reinforcement learning or other DL methodologies are sophisticated enough for the real-time, complex decision-making challenges of ICU medicine, there will be predictably an even higher level of AI adoption by the physicians in this group. The true value will be a universal ICU data repository for an enriched data source for ML/DL and deep reinforcement learning. In addition, AI and altered reality variations can also be an important aspect of education and training the clinicians in critical care unit setting with virtual and simulated resuscitations. From the administrative aspects of a busy ICU, robotic and automated assistance can be extremely useful as well to mitigate the escalating burden of documentation and workflow (Table 8.4).

TABLE 8.4 Critical care medicine.

Category of AI application	Clinical relevance	Current AI availability
Medical imaging	+ +	+ + +
Altered reality	+ + +	+
Decision support	+ + +	+ +
Biomedical diagnostics	+ + +	+
Precision medicine	+ +	+
Drug discovery	+ +	+
Digital health	+	+
Wearable technology	+	+
Robotic technology	+ +	+ +
Virtual assistants	+ +	+

AI, Artificial intelligence.

Dermatology

A dermatologist deals with the disorders of the skin. The pathology ranges from a simple sunburn or rashes to skin lesions that can be manifestation of systemic diseases (such as lupus) or cancer (such as malignant melanoma). A dermatologist, therefore, spends much of his/her time interpreting these skin lesions with their observational skills, usually at higher volumes in the clinic compared to other clinicians. A dermoscopy is an examination of skin lesion with a dermascope (also called incident light microscopy or epiluminescence microscopy). The dermatologist also performs minor surgeries such as laser therapy, excision, or biopsy of a lesion.

Published reviews and selected works

There is a relatively low number of publications on AI in dermatology and skin diseases despite the proliferation of DL/CNN in medical image interpretation that is observed in radiology, ophthalmology, pathology, and now in cardiology. There is an excellent review on ML and application of AI to imaging and diagnosis with discussion of dermatology and skin lesions [64]. The relatively comprehensive paper reviews the more relevant aspects of ML and image interpretation, including the bias and variance tradeoff, under and overfitting, and loss functions.

A landmark paper was the work published by Esteva et al. on the application of DL in skin cancer in 2017 [65]. This paper showed that the authors trained a CNN using a dataset of close to 130,000 clinical images that consisted of over 2000 different diseases. This CNN was tested against 21 dermatologists in two critical binary classification use cases and performed on par with the experts across both tasks (see Fig. 8.4). This achievement renders it possible to have cell phones be the low-cost universal access to this important aspect of medical care. Following this publication, a Cochrane Database Systematic Review concluded that prospective comparative studies are needed to evaluate the use of

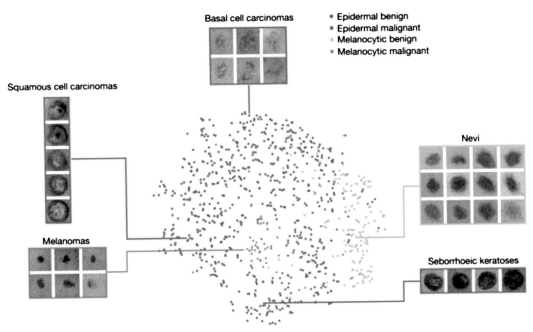

FIGURE 8.4   Classification of skin lesions. The colored clusters represent different skin disease categories after a CNN (trained end-to-end from the images directly) was applied. The CNN training consisted of using a dataset of over 120,000 clinical images and over 2000 diseases. The images correspond to various points in the clusters. A method for visualizing high-dimensional data called t-distributed stochastic neighbor embedding, or t-SNE, is applied to the last hidden layer.

computer-assisted diagnosis (CAD) systems as diagnostic aids versus dermoscopy even though in highly selected patient populations, CADs demonstrated high sensitivity and would serve well as a back-up for specialists to minimize the risk of missing melanomas [66]. Another "human versus machine" comparison was reported by Haenssle, who described a CNN versus an international group of 58 dermatologists with most dermatologists outperformed by the CNN [67]. Lastly, Zhang et al. reminded us that the best situation actually entails a combined effort from CNN and human expertise in improving the diagnosis of skin diseases [68].

Present assessment and future strategy

Overall, there is little AI utilization in dermatology but increasing clinical curiosity and interest for AI to be a second pair of eyes for diagnosis of skin lesions. Based on the dermatologists' meeting agendas and the absence of discussions on AI and image interpretation, this clinical interest is presently perhaps more curiosity than genuine inclination to incorporate AI into the practice (as observed with radiologists and other subspecialties). There are a few efforts to have available an AI-enabled dermatology visual diagnostic aid, and these efforts are likely to increase referrals to the dermatologists as these screening efforts will lead to diagnoses that will need human oversight and possible intervention.

TABLE 8.5 Dermatology.

Category of potential AI application	Clinical relevance	Current AI availability
Medical imaging	+++	+++
Altered reality	++	+
Decision support	++	++
Biomedical diagnostics	+	+
Precision medicine	+	+
Drug discovery	+	+
Digital health	+++	+
Wearable technology	+	+
Robotic technology	++	++
Virtual assistants	++	+

AI, Artificial intelligence.

For the future an AI-enabled dermatologist will have routine CNN-enabled dermoscopy in the office on a routine basis, and this dyad will correctly diagnose skin conditions almost perfectly. This human–machine synergy can also be combined with digital health capabilities to transmit photographs for not only routine screening but also follow-up; this capability can, therefore, obviate the need for an excessive number of clinic visits as well as make available a valuable service in global health. A universal visual repository of all skin lesions coupled with accurate labeling can enhance diagnostic capability immensely. The abovementioned strategy can also be coupled to a sophisticated decision support tool to achieve precision dermatology in terms of diagnosis and therapy. Most if not all procedures performed by the dermatologist can also be enhanced with AI and altered reality as to minimize skin trauma and injury. Lastly, AI can also decrease the workflow and administrative burden of a typical busy dermatologist office and clinic (Table 8.5).

Emergency Medicine

The emergency room clinician sees a large and heterogeneous group of patients in a short timeframe and often makes many fast medical decisions within this time. Similar to anesthesiologists and ICU physicians, these clinicians often have to make these decisions with inadequate data and information. The emergency room physician often performs minor procedures such as stitching of wounds or insertion of catheters. Like the ICU clinician, emergency medicine clinicians often feel like a real-time strategy game player having to make fast decisions with many complex situations in many patients without all the data. The typical emergency medicine clinician does not see patients in the hospital nor the outpatient clinic, so sometimes the eventual outcome (including death or serious morbidity) of his/her interventions is not known.

Published reviews and selected works

There is little academic or publication activity in emergency medicine in the realm of AI. A recent overview of how AI and ML can impact of emergency medicine discussed how essential it is for AI to be integrated into emergency medicine but concomitantly delineated the concerns such as algorithm opacity and data security [69]. Areas of AI incorporation include public health surveillance, clinical image analysis (including real-time ultrasound), clinical monitoring (with reduction of false alarms), clinical outcome predictions, population and social media analysis, vital signs and warnings, and finally home monitoring; the greatest challenge remains adoption of AI into the complex milieu of the health-care system. Another recent review introduced basic AI concepts such as ML and NLP in the context of emergency medicine and health informatics as well as clinical and operational scenarios [70].

A report delineated just exactly how AI can benefit emergency room operations: on arrival to accurately triage and risk-stratify in order to reduce waste; during visit to accelerate time to diagnosis in order to assign appropriate level of care; and at discharge to predict adverse events in order to individualize follow-up planning [71]. In addition, one study compared ML to traditional ED triage and demonstrated that ML (using random forest) was superior in predicting a higher severity level that has need for critical care, emergency procedure, or inpatient hospitalization [72]. There was also a report on using ML to predict, detect, and intervene older adults vulnerable for an adverse drug event in the emergency department [73]. Finally, Goto reported a ML-based prediction tools (lasso regression, random forest, gradient-boosted decision tree, and deep neural network) for clinical outcomes of children during emergency department triage [74].

Augmented intelligence in delivery of trauma care

Mustafa Kabeer and David Gibbs
Mustafa Kabeer and David Gibbs, pediatric surgeons focused on trauma care and innovations, authored this commentary to imagine just how artificial intelligence can add value to a common complicated, real-time medical condition.

The exponential increase in the amount of information available to health-care practitioners has created opportunities for improved care but has markedly increased the complexity of decision-making. Simultaneously, the fact that patients have access to an ever-increasing amount of information has increased expectations of health-care providers to make rapid, accurate decisions in a consumer-friendly fashion. The consequences of failure to navigate this landscape can be severe for the patient, the practitioner, and the organization providing care. At the same time, augmented intelligence offers the opportunity to assimilate individual clinical information into proven effective care algorithms in real time with decreased variability, improved quality, and increased patient and provider satisfaction.

Perhaps nowhere in healthcare has such a major reduction in variability and improved access to optimum treatment protocols been shown to save lives as in trauma. Prior to the advent of improved protocols in the 1970s, trauma care was fragmented and consisted of individual practitioners doing their best to provide care with resources that were intended for general medical and surgical care. With the implementation of Advanced Trauma Life Support training, verification of trauma centers, a systems approach to care, and a specialty with a significant emphasis on research, trauma care has seen marked improvements in survival and reduced morbidity.

However, as expertise has improved and knowledge has grown, expectations of perfection and transparency of outcomes has served to increase the challenge facing trauma care providers. Protocols for care are increasingly numerous and complex and often change based on the latest research outcomes and consensus papers. A system intended to promote standardization is again at risk on becoming reliant on individual provider knowledge rather than a well-functioning seamless system of care. Furthermore, many trauma patients enter the trauma system far from the tertiary trauma centers with the greatest capability. First responders and local hospital providers may have the ability to query higher level centers for advice but may not know which questions to ask and may not recognize a problem before valuable time is lost.

The capabilities of AI in its current and near future state offer great promise in providing the highest level of knowledge and standardization of care from the moment the patient enters the trauma system until their rehabilitation and recovery is complete. The best outcomes in trauma are noted when the initial care from the onset of the patient encounter is at its highest level of performance. Initial assessment and stabilization of the injured patient is based on the ABCDE (Airway, Breathing, Circulation, Disability, Exposure) algorithm, but within this there are a number of decision points. Mechanisms of airway support, control of bleeding, spinal and extremity stabilization, and fluid administration all represent opportunities to either advance or delay care. Upon arrival to the hospital, multiple assessments and interventions must be undertaken, including full assessment and examination, radiologic imaging, and laboratory analysis. Simultaneous with this assessment, notification of needed consultants and preliminary intervention decisions must be made. The sooner the proper assessment and intervention can be undertaken, the better the outcome for the patient. At the same time the trauma system must cope with finite resources and ensure that the proper care is directed to the proper patient in a timely fashion without unnecessary or even harmful interventions.

Augmented intelligence may have an important and beneficial role in complex injuries by alerting any level provider to the potential benefits of certain interventions given particular findings. Currently, less common findings such as scapular and first rib fractures should alert a provider to the possibility of a major vessel injury and the utility of obtaining a CT angiogram. Complex head injuries with suspicion of increased intracranial pressure in the initial care may require the physician to remember the benefits of mannitol or hyperventilation or even the basic requirements of optimizing oxygen delivery including oxygen supplementation and adequate perfusion pressure. Complex lung contusions resulting from pulmonary injury may benefit from specially tailored ventilatory strategies.

It is important for this technology to be easy to use and to integrate seamlessly within a complex, high-stress medical environment. Trauma care occurs in very specific locations with many team members within an ambulance or trauma bay in an emergency room setting, often followed by the ICU or operating room. Given the dynamic and fast paced setting of trauma care delivery, the requirements for AI intervention would require a voice based approach with a trained algorithm based on key word recognition. Alexa, or similar systems, can be utilized and integrated into a smart intervention algorithm that can be trained over time with various scenarios ranging from head injury to orthopedic injury to solid organ injury with very specific requirements for imaging, intervention, lab analysis and length of time for bedrest. These currently require a hard copy guideline or memory to remember the current recommendations. The potential for harmful variability in the current trauma care system remains high.

AI with voice recognition could first be deployed in an ambulance and could begin its algorithm based on utilizing the radio call from the ambulance to the base hospital to make recommendations or could respond to queries real time from the crew. Also based on initial call-in voice report, AI based at the receiving hospital could ensure that the appropriate resources are notified in timely fashion such as radiology, pharmacy, blood bank, and specialists. An AI-augmented system could also allow individualized notification of specialists based on their preference (phone vs pager) rather than the mass paging systems currently utilized at most centers. Once the patient arrives, an integrated AI system with access to the EMR could assist in recommending additional follow-up tests based on a patient's initial clinical, imaging, or laboratory findings. Particularly in the case of unusual injuries, this would ensure that an important step of the workup in an uncommonly seen injury is not neglected. Finally, a robust AI-augmented system would facilitate communication between providers or when care is transitioned from one unit to another, ensuring that checklists are not neglected.

Beyond improved care for individual patients, AI-augmented systems offer substantially improved opportunities for quality improvement and research initiatives. Most current research efforts require the investigator to have a clearly formed question prior to the initiation of the project. AI allows the unasked question to be identified and potentially answered. Trauma care is the archetypal clinical endeavor that will both benefit from and advance the success of efforts to deploy AI to the benefit of patients and care systems.

Present assessment and future strategy

Overall, there is very little academic or clinical activity for AI-related topics in this domain (based on recent meeting agendas and personal communications), even though emergency room has a large number of complex decisions and workflow challenges that would benefit from AI tools and methodologies (similar to aforementioned anesthesiologists and critical care medicine clinicians). The medical imaging area in the emergency room setting is now occasionally augmented by CNN and computer vision so this support is exceedingly helpful for the emergency room clinician (who are not as experienced with more sophisticated medical image interpretations such as head CT or echocardiography).

For the future, like the other subspecialties that often require fast thinking and decision-making (ICU, surgery, and anesthesiology), emergency medicine can greatly benefit from an AI-enabled strategy of deep reinforcement learning designed for real-time decision-making. This AI methodology can augment the clinicians' capabilities to make the best decisions in a busy environment. In addition, the emergency room is similar to the OR for the anesthesiologist and so can benefit from an AI-enabled workflow augmentation that can include history taking and triaging as well as discharge planning and follow-up (Table 8.6).

TABLE 8.6   Emergency medicine.

Category of AI application	Clinical relevance	Current AI availability
Medical imaging	+ +	+ + +
Altered reality	+ +	+
Decision support	+ + +	+ +
Biomedical diagnostics	+ + +	+
Precision medicine	+ +	+
Drug discovery	+ +	+
Digital health	+	+
Wearable technology	+	+
Robotic technology	+ +	+ +
Virtual assistants	+ +	+

AI, Artificial intelligence.

Endocrinology

The endocrinologist deals with the body's endocrine systems and glands that secret hormones (such as thyroid hormone from the thyroid gland and insulin from the pancreas); in essence the metabolism or biochemical processes of the body. The most common diseases seen by the endocrinologist include diabetes, thyroid disorders, and reproductive organs in adults and diabetes and growth disorders in children. The endocrinologist usually has a large outpatient volume but does not typically perform procedures nor look at large volumes of medical images (but can consult radiologists for various medical image studies such as bone age and certain scans). Outpatient management is a large part of the endocrinologist practice.

Published reviews and selected works

There are relatively few publications on AI utilization in endocrinology, and most of the existing published references are focused on diabetes, especially in the context of diabetic retinopathy. A recent opinion article by Gubbi et al. [75] discussed the AI areas relevant to an endocrinologist, including medical image interpretation (especially with diabetic retinopathy), preemptive medicine in which the onset of chronic diseases (such as diabetes) is delayed, and proactive diagnosis of endocrine diseases based on phenotypic expressions of these disorders. In addition, AI has been involved in the functional capacity of the artificial pancreas to manage diabetes. In a similar report, AI methodologies and their application to diabetes as a disease entity was reviewed; tools ranged from expert systems to ML as well as fuzzy logic in closed-loop systems were discussed [76]. Lastly, a thorough review of the AI in diabetes care literature are grouped into four areas: automated retinal screening, CDS, predictive population risk stratification, and patient self-management tools [77].

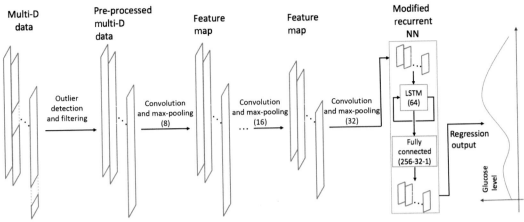

FIGURE 8.5   RCNN architecture for blood glucose predictions. The data on the left is the concatenated time series data of glucose levels as well as other data. The multidimensional data is first sent to the CNN followed by a modified RNN (seen in *red box*). The RNN contains LSTM cells and a fully connected layer and provides the output of glucose level. *CNN,* Convolutional neural network; *LSTM,* long short-term memory; *RNN,* recurrent neural network.

A report described the patient experience with the AI-based diabetic retinopathy screening tool in an endocrinology outpatient setting [78]. The fundus photographs were automatically read by a DL algorithm and results were provided real-time (vs 2 weeks by traditional retinal grading center); most patients (78%) preferred the automated screening process. In addition, a study examined expert algorithm-based identification of type 2 diabetes patients versus a novel data informed framework via feature engineering and ML (including *k*-nearest neighbor, Naive Bayes, decision tree, random forest, and SVM along with logistic regression). The latter performed much higher with an AUC of 0.98 (vs 0.71 for the expert algorithm) [79]. Lastly, Li et al. reported a clever approach to forecasting glucose levels by using an RCNN model with very short turnaround time for a future artificial pancreas [80] (see Fig. 8.5). An example of AI in endocrinology outside of the diabetic patient is a study of ML (rank regression and random forest) and transcriptomics in patients with growth hormone deficiency in disease classification [81].

Present assessment and future strategy

Overall, aside from some AI-related work on diabetes and disease management with fundus imaging and follow-up, there is little AI work in endocrinology as a subspecialty. This lack of clinical activity outside of diabetes is understandable as most of the attention for AI adoption has been centered on medical imaging interpretation with DL and CNN, including some significant work with diabetic retinopathy. There is little imaging focus for endocrinologists so present AI utilization will need to evolve into other areas, such as diabetes and glucose control with ML/DL and fuzzy logic.

For the future, exciting innovative advances with closed-loop systems and fuzzy logic as well as CRNN AI methodologies will all be extremely useful for essentially an artificial pancreas to treat diabetes. This will be a highly significant advance in the management of one of

TABLE 8.7   Endocrinology.

Category of AI application	Clinical relevance	Current AI availability
Medical imaging	++	+++
Altered reality	+	+
Decision support	+++	++
Biomedical diagnostics	+++	+
Precision medicine	++	+
Drug discovery	+	+
Digital health	+++	+
Wearable technology	+++	+
Robotic technology	++	++
Virtual assistants	++	+

AI, Artificial intelligence.

the most significant disease burdens in the future. In addition, AI-enabled population health with precision medicine will converge for diabetes to minimize population morbidity and mortality by not only improved treatment of the disease but also proactive prevention of the disease. Lastly, as diabetes is a complex disease with many complicated follow-up strategies necessary to adequately treat this multifaceted disease, automated tools as well as chatbots can be part of the AI portfolio to manage these patients (Table 8.7).

Gastroenterology

The gastroenterologist treats patients with not only gastrointestinal disorders (from esophagus to stomach and intestines), but also conditions of the liver gallbladder, and pancreas. The gastroenterologist sees medical imaging studies (such as CT/MRI but also contrast studies of the intestines and bowels) and performs procedures such as endoscopy (upper gastrointestinal tract) and colonoscopy (lower gastrointestinal tract). Common diseases include peptic ulcer, gastritis, gastroesophageal reflux, gastrointestinal bleeding, hepatitis, various types of malabsorption syndromes as well as inflammatory bowel diseases (ulcerative colitis and Crohn's disease) and cancers. In children, diseases include feeding and/or nutritional disorders (including reflux and food intolerances), liver disorders, and diarrhea/constipation.

Published reviews and selected works

An earlier review of ANN in gastroenterology discussed the added value of this AI technology to classification accuracy and survival prediction of diseases [82]. There is recent increased interest in AI in gastroenterology mainly because of the medical imaging work with DL. There is review of the use of AI, specifically use of DL, for gastroenterology and gastrointestinal endoscopy [83]. The use of computer-aided diagnosis as a second observer is better than prior experiences for conditions such as polyps as well as bleeding,

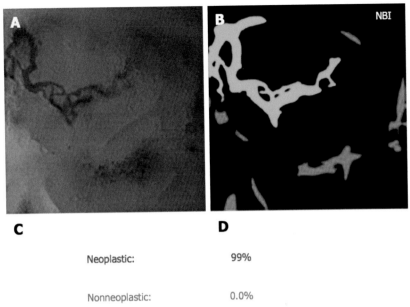

C	D
Neoplastic:	99%
Nonneoplastic:	0.0%

FIGURE 8.6   AI-assisted endocytoscopy system. (A) Endocytoscopy with narrow band imaging and (B) vessel image is extracted (where green light is the vessel image). (C) The AI system outputs the diagnosis (neoplastic vs nonneoplastic) and the probability (in D) is calculated by SVM. *AI*, Artificial intelligence; *SVM*, support vector machine.

inflammation, and infections. The concept of "optical biopsy" is presented as a methodology to achieve the diagnosis without biopsy (or at least more precise biopsy with real-time AI-enabled imaging to determine relative health of the observed tissue area) (see Fig. 8.6). The possible applications of AI in gastroenterology can be summarized as (1) technical integration of AI with EMR and endoscopy platforms will be essential to optimize clinical workflow; (2) AI systems will need to expand their library of clinical applications to include earlier diagnoses of other disease such as inflammatory bowel disease; and (3) AI systems will need regulatory and ethical considerations.

One report discussed AI as a useful methodology for gastroenterology in clinical decision-making with predictive models and in endoscopy with DL [84]. Another report on imaging and use of AI also discussed a long list of studies of AI (mainly ANN) methodologies for both the upper as well as the lower GI tract, including one also for the capsule endoscopy [85]. Lastly, a pediatric paper focused on the use of ML for classifying Crohn's disease and ulcerative colitis in pediatric inflammatory bowel disease [86]. This ML approach used endoscopic and histological data from 287 children and applied both supervised and unsupervised learning for delineating disease categories.

Present assessment and future strategy

Overall, based on meeting agendas and discussion with gastroenterologists, there is relatively little clinical AI in gastroenterology activity outside of the use of DL and CNN in

TABLE 8.8   Gastroenterology.

Category of AI application	Clinical relevance	Current AI availability
Medical imaging	+ + +	+ + +
Altered reality	+ +	+
Decision support	+ + +	+ +
Biomedical diagnostics	+ +	+
Precision medicine	+ +	+
Drug discovery	+ +	+
Digital health	+	+
Wearable technology	+	+
Robotic technology	+ + +	+ +
Virtual assistants	+ +	+

AI, Artificial intelligence.

endoscopic imaging, including real-time support during the procedure. As gastroenterology (along with other subspecialties such as cardiology and pulmonology) can have a very heterogeneous group of patients, AI-enabled precision medicine is underutilized in gastroenterology.

For the future, AI-enabled endoscopic examinations should be routinely embedded with CNN as an augmentation for the gastroenterologist. This procedure should involve an entire continuum from acquisition of the image to CNN interpretation of suspicious lesions and decision to biopsy or not. In addition, the examination can also include all data from at home monitoring to EHR from the hospital. In addition, capsule endoscopy with miniaturized cameras could play an important role in the future in a strategy to obviate the need for endoscopy or colonoscopy. In the near future the convergence of AI in the form of CNN (coupled with RNN) and miniaturized capsule technology will finally make these examinations almost entirely noninvasive (perhaps even at home). Lastly, precision gastroenterology medicine and care should be provided for all patients with these disorders with an AI-centric strategy that will incorporate all imaging and EMR as well as health data from outside the clinic and hospital settings (Table 8.8).

Global and Public Health/Epidemiology

Clinicians who deal with public health around the world is someone who specializes in global health. These clinicians are usually very knowledgeable of infections and chronic diseases (such as diabetes and mental health) as well as basic concepts of epidemiology (including vaccinations and epidemics, public sanitation, and environmental health). Common diseases seen by this type of clinician include respiratory diseases, diarrheal diseases, and malnutrition as well as common infections indigent to the area: malaria, tuberculosis, and HIV. The clinician in the field may view simple medical images (like chest X-rays) as well as perform relatively easy procedures such as minor surgeries and placement of IV lines. The global public health agenda follows the United Nations audacious Sustainable Development Goals.

Published reviews and selected works

There is a paucity of both reviews and published reports utilizing AI in global or public health or epidemiology. Wahl et al. [87] published a review with an appropriate question: How can AI contribute to health in resource-poor settings? This paper informs us that the United Nations convened a global meeting in 2017 to discuss the development and deployment of AI technologies to reduce poverty and improve public health. Although the review has minimal AI substance in the discussion, the paper delineates just how AI can be used in a resource-poor setting: expert systems can be used to support health programs, ML can model certain infectious diseases and their patterns, and NLP can be leveraged for surveillance and outbreak predictions using data from EHR and social media sources. Another review focused on AI and big data in public health [88]. In this work the authors delineated the fundamentals of big data and big data and their relevance to public health as well as highlighted the issues that would impact on medical professionals. The authors also conclude that AI and automation are two key elements in a specialized AI in the form of diagnostics but will need to be configured in the future to be generalized AI. Thiebaut et al. reported an increase in the use of AI in public health with a review of more than 800 papers in public health and epidemiology in a recent year that included a paper on the use of ANN to deidentify patient notes in EHR that outperformed existing methods [89]. The big data hurdles in precision public health mentioned possible solutions as informatics-oriented formalization of study design and interoperability throughout all levels of knowledge inference process [90]. Lastly, a positive discussion on the use of ML in epidemiology provided several applications, including predicting risk of nosocomial infections or predicting patients at greatest risk of developing septic shock [91].

Big data and a global public health intelligence network is described by a Canadian group; this effort and its adoption of big data and capability to detect international infection disease outbreaks can provide an early warning system that is life-saving [92]. In addition, Guo reported the ample use of ML methodologies in a dengue forecast model in China and discovered that support vector regression had the smallest prediction error rates compared to other methods [93]. Lastly, an interesting new methodology of using unsupervised ML to measure health system proved to be superior to traditional method of health facility surveys and can potentially save resources [94].

Global health and artificial intelligence (AI)

Sanjay Joshi[1], Sowmya Viswanathan[2,3], Weike Mo[4,5] and Sunita Nadhamuni[6]
Sanjay Joshi, Sowmya Viswanathan, Weike Mo and Sunita Nadhamuni, all committed to global health, authored this commentary to elucidate two large health care systems in Asia and the challenges for AI to be deployed and scaled in global health in general.

[1]Industry CTO Healthcare, Global CTO Office, Dell EMC, Hopkinton, MA, United States

[2]Tenet Healthcare, Dallas, TX, United States

[3]Harvard TH Chan School of Public Health, Boston, MA, United States

[4]Digital China Health Technologies Corp Ltd, Beijing, PR China

[5]Shanghai East Hospital, Tongji University, Shanghai, PR China

[6]Bangalore CoE, Dell EMC, Bengaluru, India

According to OECD and census records, ~42% of the population in China and ~65% in India live in rural areas. Combined, this is a total of about 1.2 billion people. Patient access, care, follow-up, and outcomes at scale are not possible with current physician and allied health worker throughput. This article provides an introduction to the scale of public health in China and India, identifies key clinical challenges, and provides a summary of the components and use-cases for AI.

China's public health at scale

As the nation with the largest population, China scale is massive both for the number of patients and hospitals. In the first 11 months of 2018, more than 7.5 billion people visited hospitals or clinics [1]. Of the 32,476 hospitals in China, most are entry level hospitals while only 2498 are tier 3 (the highest level according to Chinese government) [2]. The average annual number of patients admitted to each hospital is more than 100,000 (Fig. 1). As one of the biggest tier 3 hospitals in China, the First Affiliated Hospital of Zhengzhou University system has reported more than 6.5 million outpatient visits in 2017 [3]. A physician has <5 minutes for each patient. In addition, the overall expense for public hospitals is increasing quickly over the past few years (Fig. 1). It is a challenge and opportunity for the government and companies to meet the demanding needs with limited health-care resources.

The Chinese government is tackling this public health scale using a big data and AI. China's largest effort in health-care data is the National Cancer Big Data Platform. Led by the National Health Commission and National Reform and Development Commission, the Chinese government, in collaboration with partners, is building the world's biggest data warehouse from over 200 cancer specialty hospitals in China [4]. Most are top-tier hospitals with electronic data since early 2010s. Two years ago, the project reached a total of 36 million patients in the system. Terminology systems such as SNOMED-CT (systemized nomenclature of medicine—current

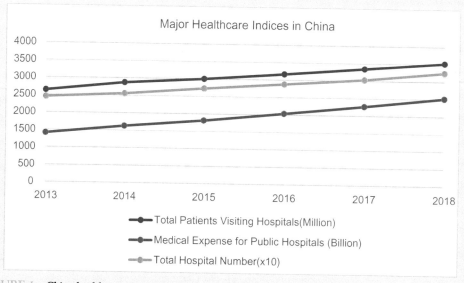

FIGURE 1   China health-care statistics.

terminology) has been used to standardize data from different resources; NLP technology has been applied to structure medical records; a common data model has been modified from the OHDSI/OMOP (observational health data sciences and informatics) model to unify data between hospitals. Although only a small portion of the millions have sufficient longitudinal medical record, the goal is to apply ML to gain in-depth understanding these data.

India's public health at scale

Most developing/emerging economies go through an epidemiological transition from dealing with communicable diseases such as typhoid and cholera to noncommunicable diseases (NCDs) such as diabetes, cancer, cardiovascular disease, and hypertension. India is no exception. But the scale of this challenge is what makes India stand out. Lifetime changes and rapid urbanization are resulting in NCDs contributing to 60% or 5.8 million of India's deaths annually [5]—often premature deaths in a 30–70-year bracket.

Counterintuitively rural India is facing a disproportionately large burden of NCDs due to malnutrition in early years which paradoxically increases the risk of NCDs, low awareness on the importance of preventive care and the need for continuous management of these chronic conditions, lack of access to quality health-care facilities and providers, and inability to bear the above costs which results in 55 million being pushed into medical poverty annually [6].

The Indian Government has identified the need to strengthen health systems to address the gaps in infrastructure, equipment and supplies, human resource availability and capability, governance and performance of health programs. To address these gaps and achieve the vision of Universal Health Care, *Ayushman Bharat* or "Healthy India" was realized as the part of National Health Policy in 2017. Ayushman Bharat adopts a continuum of care approach, through the establishment of 150,000 health and wellness centers which will bring healthcare closer to the homes of mostly rural India. These centers will provide Comprehensive Primary Health Care, covering both maternal and child health services and NCDs, including free essential drugs and diagnostic services, quality standards, new health cadres, financing models, and standard treatment protocols. This is preventive telemedicine at massive scale which will drive the necessary scale and adoption of AI. The increase in the percentage of NCD in India from 2007 to 2017 is shown in Fig. 2:

Clinical challenges in global health

With globalization occurring at a more rapid pace than society can fathom, clinical challenges in global health have come to the forefront in several areas. Advances in transportation systems globally have led to more dissemination of rare and communicable diseases across continents; while simultaneous advancements in health-care technology tools have endowed practicing clinicians with immediate availability of patient care data while providing better access to healthcare for all, amid the emergence of noncommunicable and chronic diseases.

About 60 million people in the United States are uninsured or underinsured; about half of them avoided getting access to healthcare due to costs [8]. The rising costs of healthcare and lack of access to providers has left them not seeking care. Driven by this patient demand and advents in technological advancements, telemedicine services offering video visits, remote patient monitoring, and data analytics solutions have cropped up. Precision medicine using genetic testing is

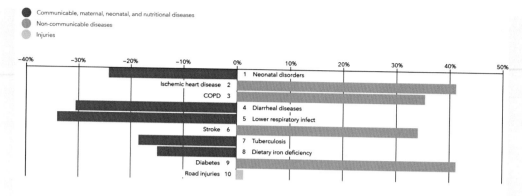

Top 10 causes of disability-adjusted life years (DALYs) in 2017 and percent change, 2007-2017, all ages, number

FIGURE 2   What causes the most death and disability combined in India? [7].

on the rise. The exponential growth of wearables and in-home diagnostic tools has led to the need for patient-centric research to be tailored to evidence-based methods of digital health education. Bringing health professionals who have traditionally been taught that the physician—patient rapport is built with the initial hand-shake and "touch," have had to learn the nuances of EHRs documentation and are now asked to absorb mobile-health, virtual health, online and offline digital courses, and simulations [9].

All of these rapid advancements have led to a great degree of physician burnout especially amongst the baby-boomers and Generation X: $\sim 32\%$ of physicians state this "burnout" is relating to computerization in their practices [10]. Health-care leaders and CEO's are not at par with investments in global health. According to Ashish Jha, Dean of Global strategy at Harvard T.H. Chan school of Public health, only 9% of Fortune 500 companies have a global health strategy, while 74% of them have an environmental strategy with clearly defined metrics [11].

For Global Health and AI to converge, digital education for clinicians and the public on technology and AI inferences, evidence-based policy making and a remarkable shift in corporate attitude needs to happen.

Artificial intelligence and machine learning in global health

There have been three previous "winters" (or "hype-failure cycles") in AI: the ALPAC report [12] in 1966 for "machine translation" from Russian to English; the 1973 Lighthill Report [13] which mentioned the "combinatorial explosion of data" for "speech understanding"; and "Expert Systems" including LISP and MUMPS for medical records in the late 1980s. In spite of the over-promise and current hype in 2019, ML and DL stand a much better chance in Global Health and Rural Telemedicine. Why? Physician and allied health worker shortages ($\sim 2/1000$ each in China and India with $>50\%$ unqualified in rural areas) [14] and underserved rural populations.

Clinical knowledge has traditionally been imparted via empirical learning and internship. While probabilistic inference is the current backbone of AI and ML, techniques such as General

Causal Inference (GCI) [15] need to be developed further, since random assignment of clinical treatments is difficult to achieve in practice. Predictive models connect inputs and outcomes via patterns and models. To be prescriptive the input needs to be changed for which no data may exist or be understood. The context of GCI is within the scope of the following clinical decision processes:

1. Will the clinical method work?
2. Why did it work?
3. What were the processes and actions?
4. What were the effects and side-effects?

Following are 10 real-world global health use-cases for AI:

1. healthy mother and child (OBGYN)
2. antimicrobial resistance and infectious disease origin and spread (influenza, malaria, encephalitis, meningitis, zika, chikungunya, cholera, typhoid, TB, MRSA, etc.)
3. NCD preventive care
4. the roaming hospital and the connected ambulance
5. teledermatology, teleradiology, telepathology
6. mental health
7. quality and regulatory data capture and reporting via CDS systems
8. outcomes and RWE (real world evidence) at scale
9. patient follow-up and calendaring
10. false information, misreporting, and fraud

AI architectures have four major components:

1. data aggregation visualization and transformation
2. modeling and model tuning
3. model interpretation and validation
4. implementation

Scaling this architecture, for both hardware and protocols, in resource constrained areas (power, networking, and environment) to provide analyses, decisions, process automation, reporting and outcomes at scale will remain a challenge in the near future. Telemedicine is slated to grow at >20% CAGR in China and in India through 2025 with a total global market of $130 billion by then [16]. We need a thoughtful, practical, real-world, and a policy- and outcome-driven approach to implement AI in global health.

References

[1] Chinese government statistical data. <http://www.nhc.gov.cn/mohwsbwstjxxzx/s7967/201901/57dec69d2c8c4e669864b067d2a1fb2e.shtml> [accessed March 2019].
[2] Chinese government statistical data. <http://www.nhc.gov.cn/mohwsbwstjxxzx/s7967/201901/94fcf9be64b84ccca2f94e3efead7965.shtml> [accessed March 2019].
[3] Reported by PharmNet. <http://news.pharmnet.com.cn/news/2018/02/12/491165.html> [Accessed March 2019].

[4] Shi WZ. Progress of China National Cancer big data platform. Beijing, China: International Cancer Big Data Application Form; 2018.

[5] WHO report 2015.

[6] Selvaraj S, Farooqui HH, Karan A. Quantifying the financial burden of households' out-of-pocket payments on medicines in India: a repeated cross-sectional analysis of National Sample Survey data, 1994—2014. BMJ Open 2018;8(5):e018020.

[7] "What causes the most death and disability combined?", India top 10 and 2010 to 2017 change, IHME. <http://www.healthdata.org/india> [Last accessed March 2019].

[8] <http://www.commonwealthfund.org/publications/press-releases/2017/oct/underinsured-press-release>.

[9] <https://www.jmir.org/2019/2/e12913/#Discussion>.

[10] <https://www.healthexec.com/topics/quality/more-half-physicians-burned-out-or-depressed>.

[11] <http://fortune.com/2019/03/25/private-sector-global-health-strategy/>.

[12] <http://www.sts.rpi.edu/public_html/nirens/SergeiPapers/Readings%20in%20Machine%20Translation%20Book%20Chapters/13.pdf> [last accessed March 2019].

[13] <http://www.chilton-computing.org.uk/inf/literature/reports/lighthill_report/contents.htm> [last accessed March 2019].

[14] WHO report 2018.

[15] Kosuke IK, van Dyk DA. Causal inference with general treatment regimes. J Am Stat Assoc 2004;99 (467):854—66. Available from: https://doi.org/10.1198/016214504000001187.

[16] Global telemedicine market: global market insights. 2019.

Present assessment and future strategy

Overall, there is only an early appreciation for the possible benefit of AI and related methodologies especially as it relates to big data, analytics, and perhaps image interpretation. There is little evidence of interest, understandably, beyond these topics based on a review of academic programs in public and global health.

For the future the fields of global and public health as well as epidemiology are ideally structured for use of ML/DL and other methodologies in studying and managing epidemics and natural disasters in the context of large populations. For instance, epidemics could be entirely followed and proactively contained with deep reinforcement learning. In addition, as more population health and precision medicine experience is gathered with the assistance of AI, these methods can be directly utilized in public and global health as well. Lastly, use of robotic technology including drones and orchestrate public health interventions in difficult geographical areas or natural disaster relief (Table 8.9).

Hematology

A hematologist is a specialist in disorders of the blood such as anemia (including sickle cell anemia), bleeding or clotting disorders, and malignancies such as leukemia, lymphoma, and myeloma. Hematologists often work with oncologists especially in caring for patients with cancers of the blood and bone marrow. Hematologists usually see patients in the hospital as well as outpatient settings.

Published reviews and selected works

There are very few publications on AI in hematology. A brief review by Sivapalaratnam discussed the impact and future of AI in hematology [95]. Convolutional deep neural

TABLE 8.9   Global and public health/epidemiology.

Category of AI application	Clinical relevance	Current AI availability
Medical imaging	+	+ + +
Altered reality	+	+
Decision support	+ + +	+ +
Biomedical diagnostics	+ + +	+
Precision medicine	+ + +	+
Drug discovery	+ +	+
Digital health	+ +	+
Wearable technology	+ +	+
Robotic technology	+ +	+ +
Virtual assistants	+ +	+

AI, Artificial intelligence.

networks was deployed for white blood cells identification system that also included the use of transfer learning [96]. A novel end-to-end convolutional deep architecture called "WBCsNet" was built and this system achieved an accuracy of 96%. Another use of AI in hematology was in the domain of hematopoietic cell transplantation and patient registries for efficient matching [97].

Present assessment and future strategy

Overall, there is very little present activity in the use of AI in hematology outside of imaging; but as with other subspecialties without a heavy emphasis on image interpretation, the use of AI in this field will be steady and in other domains such as decision support and biomedical diagnostics.

For the future, creative use of AI in hematology can be in precision hematology to better manage individual hematology patients based on many layers of data and pharmacogenomic profiles. This is particularly valuable as many hematology patients are particularly challenging in terms of precise diagnosis and eventual therapy. A future direction in the area of nanotechnology can offer an additional dimension to anticoagulation therapy in the future (Table 8.10).

Infectious Disease

The infectious disease clinicians focus on all aspects of infection, from symptoms and signs of infection to selected medical images as well as laboratory data and culture results from body fluids and/or tissue samples. These clinicians then integrate all the data to derive at a treatment plan, either antibiotic therapy (for bacterial infections) or antiviral or antifungal therapy. The deadliest infections include HIV/AIDS, tuberculosis, pneumonias,

TABLE 8.10 Hematology.

Category of AI application	Clinical relevance	Current AI availability
Medical imaging	+ +	+ + +
Altered reality	+	+
Decision support	+ + +	+ +
Biomedical diagnostics	+ +	+
Precision medicine	+ +	+
Drug discovery	+ +	+
Digital health	+	+
Wearable technology	+	+
Robotic technology	+ +	+ +
Virtual assistants	+ +	+

AI, Artificial intelligence.

diarrheal diseases, and malaria. The infectious disease expert can see patients in any clinical venue, from the ICU to the hospital bed and then also outpatient facility.

Published reviews and selected works

With so much data and possible dividends from ML and DL, there is at present relatively few published works in this domain of AI applications in infectious diseases. The work by Wong et al. reviewed the advent of big data analytics in infectious disease and public health surveillance due to the advancement of information and communications technologies and data collections systems in place [98]. The authors also feel that AI methodologies will be instrumental in supporting government agencies as well as health-care services analytics of massive infectious diseases to provide a proactive strategy. The authors conclude that an integrated big data conceptual model for infectious disease will be vital with shared data repository, epidemic models with projections of potential effects of disease spread, and AI for emergency response. Valleron commented on the need for a data science focus and priority as well as a strategy to include data science manpower for a university hospital-based institute of infectious diseases, especially with the advent of big data and computational power [99].

One Korean report from the Korean CDC described the use of a DL and long short-term memory models with optimization of parameters as well as inclusion of big data (including social media data) for prediction of infectious diseases 1 week into the future [100]. These methodologies were superior to autoregressive integrated moving average model by about 20%. A report reminiscent of the historical MYCIN project is an ontology-driven CDSS for infectious disease diagnosis and antibiotic prescription [101]. The system has over 500 infectious diseases and much more complete than any existing infectious diseases-relevant ontologies in the field of knowledge comprehension and had a receiver operating curve (ROC) result of nearly 90%. This methodology is a good example of human−machine synergy for the best medical decision-making tool. Recently, a study of sepsis patients with the use of unsupervised learning technique of cluster analysis yielded four novel subtypes that were distinctly different [102]. This study heralded perhaps the future of a "bottom-up" approach to patient data and their characterization (vs the traditional top-down clinical criteria paradigm that may be displaced in the future). In spite of this significant discovery, however, whether such use of AI in complicated patients and their EHR can affect better outcomes remains to be seen. Lastly, ANN and AI (with fuzzy logic) can even be utilized to not only increase efficacy and specificity for diagnosis of tuberculosis but also improve therapy [103].

Present assessment and future strategy

Overall, there are some clinical and organizational activities in deploying AI and ML/DL in infectious diseases but not in an organized global manner. The field of infectious disease with its many sources of data is ideally suited for big data and ML/DL with some expert knowledge oversight for many dividends.

For the future, new developments in infectious diseases and epidemiology can include continual real-time modeling of epidemics for much higher level of strategic resource allocation and disease containment. If this effort can be at the international level with every country contributing its data on infectious diseases, the full benefit of AI and its capabilities would be even higher. In addition, more precise treatment of infectious diseases with the appropriate antibiotic or antiviral/antifungal therapy and pharmacogenomic matching can minimize resistances from organisms and increase survival of infected patients (Table 8.11).

Internal and Family Medicine/Primary Care

The primary care physicians (internal medicine and family medicine for adults and pediatrics for children) see the patients for their common and usual medical needs.

TABLE 8.11   Infectious disease.

Category of AI application	Clinical relevance	Current AI availability
Medical imaging	+ +	+ + +
Altered reality	+	+
Decision support	+ + +	+ +
Biomedical diagnostics	+ + +	+
Precision medicine	+ +	+
Drug discovery	+ +	+
Digital health	+ +	+
Wearable technology	+ +	+
Robotic technology	+	+ +
Virtual assistants	+	+

AI, Artificial intelligence.

Common adult chronic diseases include heart failure, chronic obstructive pulmonary disease (or COPD), renal failure, diabetes, cancers, Alzheimer's, and infectious diseases. Pediatrics will be discussed separately under pediatrics.

Published reviews and selected works

Sidey-Gibbons authored an excellent comprehensive review on ML and medicine with a review of ML and its process, including examples of R computer code that accompanied this paper [104]. In a letter to the editor, Krittanawong outlines the ways AI can help physicians: medical imaging, triage, administrative burden, decision support, clinical prediction scores, etc. [105]. A review on the use of AI in medical practice delineates the applications of AI in internal medicine as not only image interpretation but also cognitive programs and NLP to extract knowledge from the exponentially increasing amount of published works in medicine [106]. In addition, AI can optimize the trajectory of care of chronic disease patients, suggest precision therapies for complex conditions, and reduce medical errors. This review ends with an astute reminder that AI will need humans for intelligent use of AI in medical practice. A more recent review on AI and health-care reverberates the theme of high potential dividend of AI in medicine and healthcare, including in silico clinical trials [107]. Application and exploration of big data mining in clinical medicine was reviewed to include methodologies such as fuzzy theory, rough set theory, genetic algorithm, and ANN[108]; these techniques can be used for assessing disease risks and supporting clinical decisions as well as practical drug use guidance. Lastly, one insightful and comprehensive review specifically addressed the 10 ways AI can transform primary care [109] (see Fig. 8.7).

According to the authors, the ways that AI will transform primary care will need to augment the patient–physician relationship while freeing up the physicians' cognitive and emotional space for their patients. The 10 ways AI will transform primary care include the following:

1. Risk prediction and intervention ($100 billion)—AI technology can be used to make better predictions and save money by avoiding preventable conditions such as unplanned readmissions or prolonged length of stay.
2. Population health management ($89 billion)—AI tools can be utilized as a smart platform to manage and execute population health projects.
3. Medical advice and triage ($27 billion)—AI methodologies can be used to serve as a medical advice resource for patients and augment the human clinician workforce.
4. Risk-adjusted paneling and resourcing ($17 billion)—AI tools can be utilized to determine proper staffing and resource allocation for clinicians and nursing staff.
5. Device integration ($52 billion)—AI in its embedded form will need to be part of wearable medical and health devices to enable practitioners to gain insight from this new resource.
6. Digital health coaching ($6 billion)—AI-driven fully automated text-based health coaching can increase health and decrease office visits and hospitalizations.
7. Chart review and documentation ($90 billion)—AI tools are being deployed to automate chart documentation and reduce the EHR burden of clinicians.

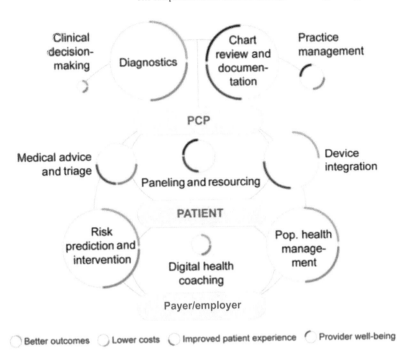

8. Diagnostics ($100 billion)—AI-enabled diagnostic tools for a myriad of diseases such as diabetic retinopathy, cardiac disease, and even Parkinson's disease can improve healthcare and quality of life.

9. Clinical decision-making ($1–2 billion)—AI-supported decision support tools will be embedded into EHR in the near future and take EHR beyond simply alerts.

10. Practice management ($10 billion)—AI-powered automated algorithms will reduce the human burden of repetitive tasks such as prior authorizations and eligibility checks.

In addition, ML has been compared to traditional prediction score for predicting hospital readmissions and proved to be superior in even three different hospital settings [110]. This automated strategy in more than 16,000 discharges can efficiently target patients at highest risk for readmission. Lastly, even palliative care has been demonstrated to improve with DL and EHR via automated screening and notification; this support is justifiably needed since physicians tend to over-estimate prognoses, and there is a shortage of palliative care staff [111]. Of course the human and humane aspects of palliative care will need to remain an essential part of the decision-making process.

Present assessment and future strategy

Overall, the hype of AI in medicine is mainly focused on medical imaging with DL and somewhat focused on decision support as well, so the primary care physician understandably may not perceive AI as very relevant in primary care. As the aforementioned article

TABLE 8.12 Internal and family medicine/primary care.

Category of AI application	Clinical relevance	Current AI availability
Medical imaging	++	+++
Altered reality	+	+
Decision support	+++	++
Biomedical diagnostics	+++	+
Precision medicine	++	+
Drug discovery	+	+
Digital health	++	+
Wearable technology	++	+
Robotic technology	++	++
Virtual assistants	++	+

AI, Artificial intelligence.

delineated, however, there are many ways that AI can help the primary care physicians outside of medical image interpretation and decision support. The areas of chart review, practice management, risk prediction, and population health management are large market potentials with growing number of AI-enabled solutions.

For the future the AI-enabled primary care physician will be in a better position to deliver precision primary care with individualized health-care planning and disease monitoring. In addition, routine screening using AI-enabled medical imaging (such as fundoscopic or dermoscopic examinations) as well as biomedical diagnostics will be automated with high level interpretation but without the requisite sub specialist referral that usually delay the screening process. These screening examinations will be accompanied by the appropriate recommendations for follow-up. The advent of wearable technology with eAI will also be a routine part of the health-care portfolio for the primary care physician. Finally, a health avatar will be available to coach each individual for the health issues that need to be addressed (weight management, mental health issues, smoking cessation, etc.) (Table 8.12).

Nephrology

A nephrologist looks after disease of the kidneys and urinary tract. This subspecialist can see patients in both the hospital or outpatient setting. Medical imaging of the kidneys as well as the urinary tract is important as is laboratory data that pertain to the kidneys. Renal dialysis, of which there are several types, is used to treat renal failure, temporary or permanent. Some of the adult patients with chronic diseases such as diabetes or heart failure can present complex scenarios as they can also have renal failure.

TABLE 8.13 Nephrology.

Category of AI application	Clinical relevance	Current AI availability
Medical imaging	+ +	+ + +
Altered reality	+ +	+
Decision support	+ + +	+ +
Biomedical diagnostics	+ + +	+
Precision medicine	+ + +	+
Drug discovery	+ +	+
Digital health	+ + +	+
Wearable technology	+ + +	+
Robotic technology	+ +	+ +
Virtual assistants	+ +	+

AI, Artificial intelligence.

Published reviews and selected works

There are not many published works on the use of AI in nephrology. There is no general review of AI use in nephrology but there is a review focused on the AI for the artificial kidney concept with emphasis on a personalized hemodialysis therapy [112]. Similar to the artificial pancreas discussed under endocrinology, the artificial kidney will need real-time decision support and predictive modeling with biofeedback and eAI in the device. Elements need to be incorporated in this model will include anemia, total body water, and intradialysis hypotension.

The use of AI for optimal anemia management in end stage renal disease showed that the models used were accurate and decreased the use of erythropoiesis-stimulating agents as well as the number of transfusions [113]. One report from pediatric nephrologists studied the use of AI for predicting dry weight in children on chronic hemodialysis [114]. The authors utilized neural network predictions and these outperformed experienced nephrologists in most cases in predicting dry weight.

Present assessment and future strategy

Overall, there is little clinical or academic activity in the use of AI in nephrology as it is not an image-intensive field. There is some interest, however, in developing AI for decision-making in nephrology.

For the future, there can be contribution from AI for precision nephrology with individualized decisions for patients with renal disorders, especially chronic renal failure. There is also the possibility of developing an artificial kidney with eAI in the device that will serve as a dialysis unit (Table 8.13).

Neurosciences (Neurology/Neurosurgery and Psychiatry/Psychology)

The neurologist sees disorders of the central and peripheral nervous systems. The list of diseases include seizures, peripheral neuropathies, muscular dystrophy, and brain tumors. The neurologist often views electroencephalogram (EEG) as well as various types of brain imaging (MRI including functional MRI, CT, and positron emission tomography, are the major modalities) with the radiologist. As with cardiologists and cardiac surgeons, neurologists have a special working relationship with neurosurgeons as often medical conditions warrant surgical intervention(s). The psychiatrist sees patients with mental disorders, ranging from mania depression to schizophrenia and can prescribe medications; a psychologist can also follow these patients but usually do not prescribe medications. The Diagnostic and Statistical Manual of Mental Disorders, or DSM-V, classifies mental disorders with well over 200 mental conditions in total.

Published reviews and selected works

There are a few important reviews on the impact of AI on this range of subspecialties, which is ironic given that AI is closest to these fields from a philosophical and scientific perspective. An overview by Ganapathy et al. described that application of AI in the neurosciences mandates a better understanding of the intelligent functioning of the human brain [115]. The authors describe AI applications in neurosciences to include upskilling in neurosurgical procedures, predicting outcome of neurosurgery for seizure disorders, and AI-assisted functional registration for outcome prediction, and other areas (neuro-oncology, neuro-traumatology, imaging services, strokes, and neurorehabilitation). Another review focused on ML and NLP as AI tools that are relevant for the early detection and diagnosis as well as outcome prediction and prognosis evaluation for stroke [116]. An excellent review of AI (as well as natural intelligence) in neurosurgery summarizes neurosurgical implications of ML as compared with clinical expertise in several areas (diagnosis, preoperative planning, and outcome prediction) [117]. The authors conclude that ML models have potential to augment the decision-making capacity of clinicians, but there are several obstacles in this review of ML models (publication bias, ground truth definitions, and interpretability). Another review on ANN in neurosurgery in the realm of clinical decision-making showed several applications, including diagnosis, prognosis, and outcome prediction [118]. Finally, in the realm of psychiatry, there are two reviews of significance: one review of ML and its role in psychiatry is explored as this field continues to become more and more complex (high-dimensional) and traditional statistical analyses will no longer suffice [119] and the other review focused on DL with CNN for brain imaging as well as RNN for mobile devices and projected future concepts (such as embedding semantically interpretable models of brain dynamics into a statistical ML context) [120].

In children with autism spectrum disorder (ASD) a robot-based approach showed that children with this disorder are more engaged in the several learning task and seem to enjoy more the task when interacting with the robot compared with the interaction with the adult [121]. In addition, electromechanical and robot-assisted arm training for improving activities after a stroke showed improvement of these activities in randomized controlled trials [122]. Much is expected of robots and virtual assistants in the future for both physical rehabilitation and psychiatric therapy as well as health-care education and chronic disease management. If accompanied by robust AI tools, these supportive services will be particularly useful in delivering value for the patients.

A call for mobile artificial intelligence (AI) solutions for unanimous and global mental health

Dennis Wall

Dennis Wall, a brilliant data scientist with a penchant for neurosciences, authored this commentary on the deployment of portable AI tools to favorably change the expected standard of care in children and adults with mental health.

Brain-mediated neurological and psychiatric disorders together account for 28% of disability due to NCD [1], and for 40% of premature deaths in the United States [2]. Current standards of mental healthcare suffer from issues with subjectivity and scalability, yet many can and should give way to mobile solutions that can function outside of the clinic in a more consistent and continuous fashion. Given the importance of early cognitive development on long-term mental well-being, a priority focus of precision mobile health should be on developmental pediatrics. Top among them, ASD is a developmental disability that affects at least 214 million children worldwide, including 1 million children in the United States at or under the age of 10 [3,4], with prevalence up from 1 in 125 to 1 in 40 US children within the last 5 years alone [3,5,6]. It has become one of the most pressing pediatric health concerns globally [7,8]. Children with ASD experience repetitive and restricted behaviors and interests, as well as social communication deficits that include difficulty with language, social attention, recognizing facial expressions, and engaging in social interactions [9–16], all of which require support throughout life if left untreated in early childhood [9].

Early intervention can greatly reduce the health-care burden across the lifespan [9,17]. The standard of care (SOC) for autism is early behavioral intervention, specifically, applied behavioral analysis (ABA) therapy, which can be effective at increasing IQ, improving eye contact, face-to-face gaze, and emotion recognition in children with ASD. While this SOC promotes improvements in several core deficits of children with autism, these therapies have quickly become unattainable by the population of families in need [3]. Wait times for therapy can exceed 12 months, and delays in access to care are far worse for children from ethnic minorities [18] and for those in rural and underserved areas [19–23]. These delays leave many untreated until after the time when early intervention can have its greatest impact [9]. To make matters worse the average annual health-care cost per family is more than $30,000 per year, and the annual health-care costs associated with ASD in the United States will reach approximately $500 billion by 2025 [24].

Delays in access to the SOC, coupled with the powerful potential of mobile AI, make development of new tools essential. A move toward mobile ubiquitous computing solutions that function at home and in the hands of the primary stakeholders (e.g., the children and the families with autism) will be among the best ways to meet demand affordably. More, such approaches will be among the best way to create the feedback loop necessary for adaptive and personalized intervention, all while amassing a rich and growing collection of mobile device data. As a step in this direction, we have deployed a novel wearable therapy using Google Glass that provides dynamic social training through a form of AI that finds and translates faces into core human emotions (e.g., happy, sad, angry, surprised, afraid) for the child to understand immediately, and in their natural moments of social interaction [25–27]. Our studies suggest that the system provides significant increases in eye contact, emotional understanding, and socialization, key deficits with which all children on the autism spectrum struggle [28,29]. But more, this tool shows the potential to not only treat but to simultaneously generate new computer vision libraries that can enable advanced models for detection and treatment.

As a second example, we have constructed an expandable prototype mobile app (GuessWhat.stanford.edu) that engages children through a fluid social interaction with his/her social partner, turning the focus of the camera on the child and reinforcing prosocial learning while simultaneously measuring the child's progress. At its simplest level the *GuessWhat* is in its most basic form, mobile charades. After selecting a card deck, namely, an array of images associated with a theme, such as "farm animals," the autistic child's social interacting partner, for example, Mom or Dad, holds the smartphone on his or her forehead to display a sequence of images that the child must imitate. The parent flips the phone upward/downward if able/unable to correctly guess what the child is acting, switching to the next image in the deck with each flip. Meanwhile, the roughly 90-second game sessions are recorded by the phone's outward facing camera, and dynamically passed to computer vision algorithms and emotion classifiers integrated into the gameplay to detect emotion in the child's face, automatically capturing features such as gaze, eye contact, and joint attention, while building entirely novel and personalized computer vision libraries. This simple game can serve as a complete data science solution to treat and track autism while automatically building labeled computer vision libraries to train new models with greater precision for detection of prosocial behaviors, social reciprocity, eye contact, and emotion recognition. Indeed, the more you play it the better the libraries, and the more precise the AI models, get.

These two examples, which focus on early developmental brain health to promote longer term mental wellness are just two small steps to showcase the potential of mobile AI. Advances in mental healthcare require mobile solutions that scale *and* domain specific AI that can personalize to the problem and the person. The previous two solutions show that handheld or wearable games to treat and track while growing domain specific and context dependent computer vision libraries—a holy grail for the field. Advances in ubiquitous computing including mobile and wearable technologies, together with advances in AI and computer vision, make viable the deployment of tools to large patient populations that go beyond simple wearable health consumer tools such as Fitbit to ones that can not only deliver diagnostic, scientific, and therapeutic value but can establish a communication portal among patients and clinicians. Indeed, a new paradigm in medicine is possible, one that uses ubiquitous computing tools to offer value to consumers in return for data that will validate and deploy new models for healthcare, create a fluid patient-to-clinician dialogue, and concomitantly enable a feedback loop of improved science and preventative intervention for unanimous and global mental health.

References

[1] Prince M, Patel V, Saxena S, et al. Global mental health 1—no health without mental health. Lancet 2007;370 (9590):859—77.

[2] Schroeder SA. We can do better—improving the health of the American people. N Engl J Med 2007;357 (12):1221—8.

[3] Baio J, Wiggins L, Christensen DL, et al. Prevalence of autism spectrum disorder among children aged 8 years—Autism and Developmental Disabilities Monitoring Network, 11 Sites, United States, 2014. MMWR Surveill Summ 2018;67(6):1.

[4] Boyle CA, Boulet S, Schieve LA, et al. Trends in the prevalence of developmental disabilities in US children, 1997-2008. Pediatrics 2011;127(6):1034—42.

[5] Nicholas JS, Charles JM, Carpenter LA, King LB, Jenner W, Spratt EG. Prevalence and characteristics of children with autism-spectrum disorders. Ann Epidemiol 2008;18(2):130—6.

[6] Xu G, Strathearn L, Liu B, et al. Prevalence and treatment patterns of autism spectrum disorder in the United States, 2016. JAMA Pediatr 2019;173(2):153—9.

[7] Murray CJ, Lopez AD. Alternative projections of mortality and disability by cause 1990-2020: Global Burden of Disease Study. Lancet 1997;349(9064):1498—504.

[8] Whiteford HA, Degenhardt L, Rehm J, et al. Global burden of disease attributable to mental and substance use disorders: findings from the Global Burden of Disease Study 2010. Lancet 2013;382(9904):1575—86.

[9] Dawson G. Early behavioral intervention, brain plasticity, and the prevention of autism spectrum disorder. Dev Psychopathol 2008;20(03):775—803.

[10] Dawson G, Rogers S, Munson J, et al. Randomized, controlled trial of an intervention for toddlers with autism: the Early Start Denver Model. Pediatrics 2010;125(1):e17—23.

[11] Dawson G, Webb SJ, McPartland J. Understanding the nature of face processing impairment in autism: insights from behavioral and electrophysiological studies. Dev Neuropsychol 2005;27(3):403—24.

[12] Howlin P, Goode S, Hutton J, Rutter M. Adult outcome for children with autism. J Child Psychol Psychiatry 2004;45(2):212—29.

[13] Landa RJ, Holman KC, Garrett-Mayer E. Social and communication development in toddlers with early and later diagnosis of autism spectrum disorders. Arch Gen Psychiatry 2007;64(7):853—64.

[14] Palumbo L, Burnett HG, Jellema T. Atypical emotional anticipation in high-functioning autism. Mol Autism 2015;6(1):47.

[15] Sasson NJ, Pinkham AE, Weittenhiller LP, Faso DJ, Simpson C. Context effects on facial affect recognition in schizophrenia and autism: behavioral and eye-tracking evidence. Schizophr Bull 2016;42(3):675—83.

[16] Xavier J, Vignaud V, Ruggiero R, Bodeau N, Cohen D, Chaby L. A multidimensional approach to the study of emotion recognition in autism spectrum disorders. Front Psychol 2015;6:1954.

[17] Dawson G, Jones EJH, Merkle K, et al. Early behavioral intervention is associated with normalized brain activity in young children with autism. J Am Acad Child Adolesc Psychiatry 2012;51(11):1150—9.

[18] Angell AM, Empey A, Zuckerman KE. Chapter four—A review of diagnosis and service disparities among children with autism from racial and ethnic minority groups in the United States. In: Hodapp RM, Fidler DJ, editors. International review of research in developmental disabilities. Academic Press; 2018. p. 145—80.

[19] Chiri G, Warfield ME. Unmet need and problems accessing core health care services for children with autism spectrum disorder. Matern Child Health J 2012;16(5):1081—91.

[20] Dawson G, Bernier R. A quarter century of progress on the early detection and treatment of autism spectrum disorder. Dev Psychopathol 2013;25(4 Pt 2):1455—72.

[21] Gordon-Lipkin E, Foster J, Peacock G. Whittling down the wait time: exploring models to minimize the delay from initial concern to diagnosis and treatment of autism spectrum disorder. Pediatr Clin North Am 2016;63 (5):851—9.

[22] Siklos S, Kerns KA. Assessing the diagnostic experiences of a small sample of parents of children with autism spectrum disorders. Res Dev Disabil 2007;28(1):9—22.

[23] Ning M, Daniels J, Schwartz J, et al. Identification and quantification of gaps in access to autism resources in the U.S. J Med Internet Res 2019;21.

[24] Leigh JP, Du J. Brief report: forecasting the economic burden of autism in 2015 and 2025 in the United States. J Autism Dev Disord 2015;45(12):4135—9.

[25] Voss C, Washington P, Haber N, et al. Superpower glass: delivering unobtrusive real-time social cues in wearable systems. Proceedings of the 2016 ACM international joint conference on pervasive and ubiquitous computing: adjunct; 2016. ACM; 2016. p. 1218–26.

[26] Washington P, Voss C, Haber N, et al. A wearable social interaction aid for children with autism. Proceedings of the 2016 CHI conference extended abstracts on human factors in computing systems; 2016. ACM; 2016. p. 2348–54.

[27] Washington P, Voss C, Kline A, et al. SuperpowerGlass: a wearable aid for the at-home therapy of children with autism. Proc ACM Interact Mob Wearable Ubiquitous Technol 2017;1(3):112.

[28] Daniels J, Haber N, Voss C, et al. Feasibility testing of a wearable behavioral aid for social learning in children with autism. Appl Clin Inf 2018;9(1):129–40.

[29] Daniels J, Schwartz JN, Voss C, et al. Exploratory study examining the at-home feasibility of a wearable tool for social-affective learning in children with autism. NPJ Digital Med 2018;1(1):32.

Present assessment and future strategy

Overall, the area of neurosciences is starting to be aware of and is focused on AI and its full range of capabilities. The academic meetings in neurosciences have an occasional speaker in the AI domain, and the presence of AI in these academic gatherings as well as clinical settings has continued to increase. The neurosciences can advance AI by incorporating cognitive elements and architecture that is so well understood by its members into the current DL paradigm.

For the future, with emerging altered reality as well as robotic and virtual assistance technologies, the entire spectrum of neurosciences from neurology and neurosurgery to psychiatry will benefit greatly from these technologies in the therapeutic areas of precision rehabilitation. Medical imaging of the brain will advance with multiple modalities converging into a "superscan" similar to what was discussed under cardiology. In addition, "precision psychiatry" with all the dimensions of precision medicine will mandate strategic use of AI and all its capabilities similar for rehabilitation.

Lastly, there will be a necessary convergence and synergy between AI and the neurosciences so that this dyadic relationship will increase the knowledge and capabilities of both of these intimately related sciences (Table 8.14).

Obstetrics/Gynecology

These clinicians see female patients at all ages during or after adolescence for regular physical examinations as well as for births or medical conditions that involve the female-reproductive system. These physicians are the only clinicians who go from operating in the OR to seeing patients in the clinic as a dually trained surgeon and medical doctor. These clinicians often will follow the fetuses prior to their birth, so fetal monitoring (fetal ECG used to track heart rate and specifically heart rate abnormalities), and fetal echocardiograms are an essential part of fetal follow-up. Maternal-fetal medicine, or perinatology, involves the medical and surgical management of high-risk pregnancies. Other subareas of focus include reproductive endocrinology and fertility, gynecological oncology, and family planning.

TABLE 8.14 Neurosciences.

Category of AI application	Clinical relevance	Current AI availability
Medical imaging	+ + +	+ + +
Altered reality	+ + +	+
Decision support	+ + +	+ +
Biomedical diagnostics	+ + +	+
Precision medicine	+ + +	+
Drug discovery	+ +	+
Digital health	+ +	+
Wearable technology	+ +	+
Robotic technology	+ + +	+ +
Virtual assistants	+ + +	+

AI, Artificial intelligence.

Published reviews and selected works

There are a few published reviews on AI in obstetrics and gynecology. A short commentary mentioned AI and big data to make impact in obstetrics and gynecology and was accompanied by a review of the literature [123]. Wang et al. discussed AI in the context of assisted reproductive technology as well as its limitations and challenges [124]. AI applications in reproductive medicine include evaluation and selection of oocytes; sperm selection and semen analysis; embryo selection; and finally prediction of in vitro fertilization (IVF) (see Fig. 8.8).

The various aspects of AI applications are place in this figure. (A) Decision tree model is shown to make predictions for embryos with the model first separating cumulus cells samples with high or low AMHR2 expression followed by high or low LIF expression. (B) The human semen samples and their quantitative phase maps and features are trained with a two-class SVM classifier. (C) An ANN model is used to produce a decision support system that can help predict the semen parameters.

One manuscript focused on AI and its use in the interpretation of intrapartum fetal heart rate tracings [125]. The analysis concluded with the observation that the use of AI for fetal heart rate monitoring during labor showed that the agreement between the AI tool and human observers was moderate but AI assistance did not improve neonatal outcomes (neonatal acidosis, APGAR scores, or death), which is worth remembering as an important aspect of AI (outcome vs performance of algorithm). Another review published prior to proliferation of DL networks focused on the use of ANNs in gynecological diseases [126]. Even in 2010, the authors appreciated the robust ANN in evaluating multifactorial data from multiple sources to derive at subtle and complex patterns and nonlinear statistical modeling that was superior to conventional logistic regression. Categories of ANN application in gynecological diseases include gynecological oncology (especially early detection and prognosis), assisted reproduction, and reproductive endocrinology,

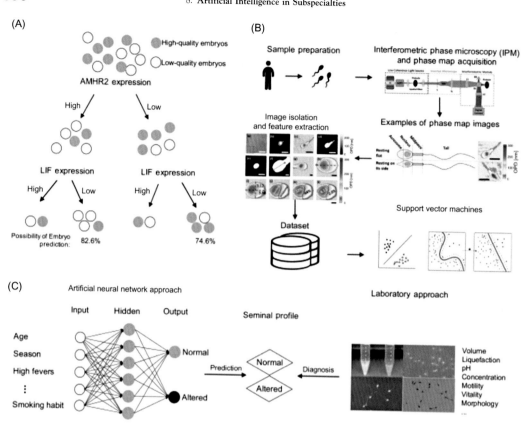

FIGURE 8.8   AI applications in reproductive medicine. *AI*, Artificial intelligence.

and gynecological urogynecology (prediction of surgical outcome). Another area of interest in exploring computational science in reproductive medicine is IVF.

Simopoulou et al. reviewed the necessary leap of evolution from basic mathematics to bioinformatics for analyzing complex models for embryo selection in IVF [127]. Lastly, an interesting reference to the use of fuzzy cognitive maps, which combines fuzzy logic with neural networks, in obstetrics is suggested as a medical decision support system that includes cause and effect relationships among concepts [128].

Present assessment and future strategy

Overall, much of the interest in the use of AI in obstetrics and gynecology has been in the areas of fetal monitoring, which as a very small signal-to-noise ratio (similar to ECGs and EEGs), and IVF. Similar to many of the aforementioned clinical areas, obstetrics and gynecology has a very heterogeneous group of patients and even include fetuses as an essential part of the clinical scope. This level of complexity is well suited for more sophisticated ML/DL to decrease the burden and stress of decision-making for the clinician.

For the future an AI-enabled fetal monitoring strategy will reduce the human burden and concomitantly increase diagnostic accuracy for fetal events in a proactive manner.

TABLE 8.15   Obstetrics and gynecology.

Category of AI application	Clinical relevance	Current AI availability
Medical imaging	+ +	+ + +
Altered reality	+ +	+
Decision support	+ + +	+ +
Biomedical diagnostics	+ + +	+
Precision medicine	+ + +	+
Drug discovery	+ +	+
Digital health	+ +	+
Wearable technology	+ +	+
Robotic technology	+	+ +
Virtual assistants	+	+

AI, Artificial intelligence.

Advances in monitoring technology will lead to continuous at-home monitoring that is both accurate and noninvasive. For all aspects of gynecological as well as obstetrical conditions and situations, support from AI tools will enable the clinicians to render better decisions in precision medicine format (Table 8.15).

Oncology

These physicians follow patients who are diagnosed with cancer and set up their diagnostic as well as therapeutic strategies; primary care physicians (internists, pediatricians, and family practitioners) usually are the clinicians that screen patients for cancer. The oncologists are essentially specialized internists who focus on the specific cancer and all that is required of that cancer in terms of continued surveillance and therapy (including complications and sequelae). Radiation oncologists treat cancer with the use of high energy radiation therapy. There are also pediatric oncologists who focus on children and young adolescents with cancer. There is sometimes clinical overlap between oncology and hematology (as mentioned previously).

Published reviews and selected works

There are a large number of publications on the use of AI in oncology; this may be due to the interest in precision medicine and therapy for cancer patients as well as in cognitive computing (IBM Watson) and its application in cancer. A discussion of AI in oncology discussed this topic in terms of virtual (ML and algorithms) and physical (robots and medical equipment) AI [129]. Areas for AI and its impact in oncology include tumor segmentation, histopathological diagnosis, tracking tumor development, and prognosis prediction. The authors concluded by discussing the power of cancer-related platforms to track tumor progression and treatment effects. Another good review of ML and imaging discussed also AI in oncology [130]. This review covered not only ML but also DL with discussion on hand-crafted versus machine-engineered feature extraction in radiomics; three cases were presented at the end of the review. The domain of cancer genomics and precision medicine with AI applications was reviewed by Xu; there is a myriad of ways AI can be applied to this area [131]. There are also many review papers on radiation oncology, and one review paper that focused on radiation oncology discussed the various aspects of how AI can be applied to this domain: image segmentation, treatment plan generation and optimization, normal tissue complication probability modeling, quality assurance, and adaptive replanning [132]. As NLP is essential in AI projects in oncology, a review on NLP appeared to encourage oncologists to automate unstructured data from their practice [133]. Lastly, a commentary by Kantarjian and Yu succinctly summarized potential applications of AI in cancer care and research, ranging from developing national/international cancer registries and pathways to uncovering genomic or molecular events that render certain cancer patients more or less sensitive to treatments [134].

Important papers were published on the experience (good and less than good) with Watson for Oncology (WFO). The system was demonstrated to be effective in having high concordance with a tumor board after incorporating data from more than 500 breast cancer cases [135]. In addition, the Watson for Genomics tool was used to demonstrate that human curation is not sufficient to interpret somatic next-generation sequencing and that a molecular tumor board empowered by cognitive computing can improve patient care by being more expedient [136]. Lastly, WFO had significant issues with the system's inability to solve data quality problems in unstructured data (such as doctors' notes and written case reports) and departed from the well-known MD Anderson Hospital [137]. In breast cancer the combination of computer-aided diagnosis of image-omics (pathological images) and functional genomic features improved the classification accuracy by 3% [138]. In this study, SVM for differentiating stage I breast cancer from other stages are learned with the use of computer-aided diagnosis that enables joint analysis of functional genomic information and image from pathological images. In another study the entire biomedical imaging informatics framework consisted of image extraction, feature combination, and classification. In addition, a DL set of algorithms were used to examine hematoxylin and eosin-stained tissue sections for the detection of lymph node metastases and achieved better diagnostic performance than a panel of 11 pathologists who interpreted without time constraints [139]. The fascinating aspect of this study is that it was part of an international contest (Cancer Metastases in Lymph Nodes Challenge, or CAMELYON16) with 23 teams competing, with most teams using a deep CNN for the image interpretation. Lastly, an exciting development in the area of precision oncology is dynamic risk profiling using serial tumor biomarkers for personal outcome prediction in the strategy called continuous individualized risk index [140].

Present assessment and future strategy

Overall, there remains to be relatively little academic or clinical activity in AI in oncology other than the activity generated by WFO. The relatively negative publicity that promulgated from the MD Anderson/IBM WFO project did have significant repercussions and offered an important lesson in AI adoption and accountability. Launched in 2013, this partnership garnered much publicity in the media, and the hyped promise to cure cancer was never realized. Issues involved in this failure include lack of competitive bidding,

TABLE 8.16 Oncology.

Category of AI application	Clinical relevance	Current AI availability
Medical imaging	+ +	+ + +
Altered reality	+ +	+
Decision support	+ + +	+ +
Biomedical diagnostics	+ + +	+
Precision medicine	+ + +	+
Drug discovery	+ +	+
Digital health	+	+
Wearable technology	+	+
Robotic technology	+	+ +
Virtual assistants	+	+

AI, Artificial intelligence.

insufficient due diligence, and decision made without IT department buy-in. The seemingly obvious questions regarding whether it would improve patient outcomes and lower costs were not adequately answered by the institution.

For the future, AI promises to be an essential resource for oncology in many ways since oncology is multisystem and multidimensional as a subspecialty. Using NLP, key discoveries can be made via EHR data for early cancer detection as part of population health management as well as for ascertaining patient-chemotherapy coupling to be maximally efficient and efficacious. Medical imaging and radiology, an important component of cancer care, will become extremely sophisticated and perhaps be even closer with pathology. This will form an AI-enabled image continuum (radiology to pathology) that will even include multiple layers of additional information such as response to therapy and outcomes. As with most other subspecialties, a precision medicine approach to cancer (precision and translational oncology) to include pharmacogenomic profiles including immunotherapy is even more critical in oncology as it can provide a differential advantage toward survival. Finally, there is also altered reality-enabled strategies for treating therapy-related side effects in patients receiving chemotherapy, which may change in the future as well (Table 8.16).

Ophthalmology

An ophthalmologist examines the eyes (including a fundoscopic examination which is a detailed examination of the retina) and cares for the diseases that pertain to the eye. Optical coherence tomography is another imaging in ophthalmology. There is a myriad of diseases that affect the eye, with most common ones being glaucoma, cataracts, diabetic retinopathy seen in patients with diabetes, and macular degeneration seen in older patients. In addition to corrections for near sightedness (myopia), far sightedness

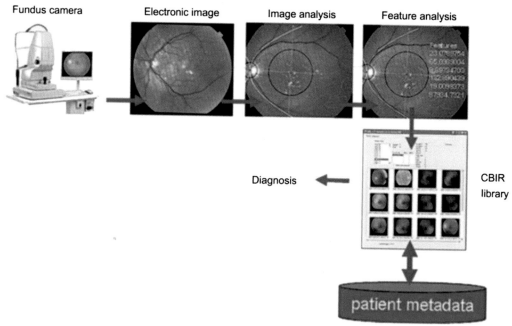

FIGURE 8.9   Fundus photograph and artificial intelligence. The fundus photograph is taken with the camera and image analysis with feature extraction and analysis is performed. The diagnosis is made with comparisons with other fundus photographs in the CBIR library with coupling to the patient metadata.

(presbyopia), and astigmatism, an ophthalmologist can perform operations that include laser surgery and other procedures that deal with eye pathology.

Published reviews and selected works

With the advent of DL and CNN in medical imaging, there is a relative increase in the number of publications in this area; most of the works focus on the DL aspect of fundoscopic examination. A review by Kapoor et al. reviewed the current state-of-the-art in AI in ophthalmology [141]. The review not only focused on ophthalmology but briefly reviewed AI and its impact in a few other fields in medicine as well. The major areas of impact for AI in ophthalmology include teleophthalmology for various eye diseases such as retinopathy of prematurity and glaucoma screening as well as the use of AI and DL in eye diseases. Another review discussed the use of AI and patient metadata for rapid diagnosis and follow-up with an ophthalmologist (see Fig. 8.9) [142]. There are additional reviews on AI in ophthalmology [143−145].

One of the most significant studies published in the application of AI in medicine was the study by Gulshan et al. that validated the accuracy of a DL algorithm for the detection of diabetic retinopathy in retinal fundus photographs [146]. The study showed that the algorithm is as good as board-certified ophthalmologists in making the diagnosis with high specificity and sensitivity and an area under the ROC of 0.99 (see Fig. 8.10). The

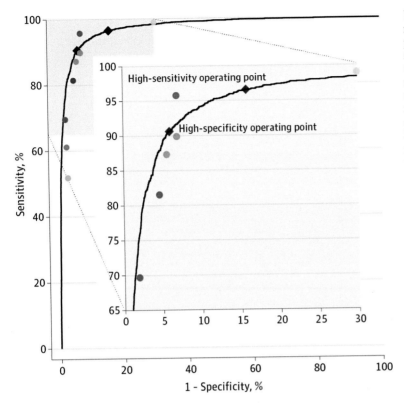

FIGURE 8.10 Diabetic retinopathy ROC curve. The performance of the algorithm for automated diagnoses (*black curve* with *black diamonds*) is compared with eight ophthalmologists with manual grading (*colored circles*) on an ROC curve (upper left corner is highest level of performance). Notice that there is one ophthalmologist (in purple above the ROC curve) who performed better than the algorithm. *ROC*, Receiver operating characteristic.

deep CNN algorithm used over 100,000 retinal images for its data set and it had high sensitivity and specificity for detecting diabetic retinopathy in test sets. In 2018 Abramoff et al. published a study of his autonomous diagnostic system (IDx-DR), which was later approved by the FDA in 2019 as the first fully autonomous AI-based diagnostic tool in biomedicine [147].

Present assessment and future strategy

Overall, the public and the subspecialty have been enthusiastic about the first autonomous AI-enabled diagnostic tool (as mentioned previously) and, therefore, screening of certain common eye conditions such as diabetic retinopathy can be achieved with good accuracy. The global health implications are sizable: the DL algorithm improves in the scope of eye conditions it can help diagnose, and the referrals may increase for the ophthalmologist to treat eye disorders.

For the future the AI-enabled ophthalmologist may need to adopt the same strategy as the radiologist in that the nonperceptive as well as the procedural parts of the job will become even more important. The automated screening eye examination may be done at local centers with automated interpretation rather than in the clinical office of the ophthalmologist with delayed results. Precision ophthalmology care will involve appropriate follow-up of significant findings on the fundus examination that will

TABLE 8.17   Ophthalmology.

Category of AI application	Clinical relevance	Current AI availability
Medical imaging	+ + +	+ + +
Altered reality	+ +	+
Decision support	+ + +	+ +
Biomedical diagnostics	+	+
Precision medicine	+ + +	+
Drug discovery	+	+
Digital health	+ +	+
Wearable technology	+ +	+
Robotic technology	+	+ +
Virtual assistants	+	+

AI, Artificial intelligence.

routinely include patient metadata. It is possible that with widespread screening for common eye disorders that we will have a significant increase in the incidence of eye disorders; the increase in prevalence will need to be configured into national and international health budgets as well as treatment strategies to avoid over diagnosis and over treatment (Table 8.17).

Pathology

A pathologist typically reviews many types of medical images of biopsies or sections of tissues in the form of microscopic slides (microscopic morphology remains the gold standard in diagnostic pathology). In addition to image interpretation, a pathologist also performs autopsies (called "gross" pathology). Lastly, many aspects of laboratory testing fall under the auspices of the pathologist and his/her division. The recent advent of digital pathology has expedited the workflow of the pathologist: whole-slide imaging, faster networks, and cheaper storage solutions have enabled pathologists to manage digital slide images.

Published reviews and selected works

There is a robust body of literature on the use of AI in the form of DL and computer vision in pathology as it is an image-focused domain. A recent review of AI and relevance to pathology recognized the significance of DL in incorporating clinical, radiological, and genomic data to pathology data [148]. Much of the review focuses on CNN and computer vision for pathological specimens. In addition, a review of automation and AI in laboratory medicine emphasize the robust coupling of these two technologies can increase efficiency and lead to personalized medicine as well as precision public health [149]. Different levels of automation in the laboratory ranges from inoculation to partial and complete laboratory automation. In addition, reviews of digital pathology and AI delineated the integration of digital slides into the pathology workflow coupled with advanced algorithms and computer-aided diagnostic techniques [150,151].

There is also a report on the use of combined automation and AI in the clinical laboratory [149]. The authors pointed out that both of these technologies will disrupt the clinical laboratories with development of new diagnostic and prognostic models that will be essential aspects of personalized medicine. Another review on use of AI in laboratory medicine focused on not only workflow but also regulation [152]. A third review focused on basic concepts of ML as well as how these ML models can relate to laboratory medicine [153]. Similar to a "human versus machine" landmark papers in other areas, a report compared DL algorithms to a group of 11 pathologists in a simulated workflow interpreting whole-slide images of lymph nodes in women with breast cancer; some of the ML algorithms proved better [139] (see Fig. 8.11). In addition, automated interpretation of blood culture gram stains by use of CNN can be extended to all Gram stain interpretive activities in the clinical laboratory [154]. Furthermore, a report on prostate cancer risk stratification and treatment section pointed out that multiparametric MRI and digital pathology should together can enable advanced characterization of disease through a combined AI-enabled pathology—radiology assessment [155]. Nir et al. reported recently that patch-based training and evaluation of CNN models may be flawed and that multiexpert data should be used to obtain a more realistic performance evaluation of the model [156]. Finally, O'Sullivan et al. suggested the interesting integration of AI and autopsy via autonomous robots that can use a trained algorithm for a robotic autopsy [157]. This development can have valuable insight into surgical procedures that are partly robotic as well.

Present assessment and future strategy

Overall, the present situation warrants the pathologist to accommodate DL as a partner in image interpretation perhaps as much as the radiologist, if not more so. This accommodation will need all the pathology images to be digitized and stored in the cloud. It is entirely possible that the pathologist job description is even more vulnerable than that of the radiologist if the pathologist is not open to adoption of AI (as much of the pathologist focus is medical images). The laboratory data part of the pathologist portfolio can be improved with education for both the patient as well as the caretaker; in addition, the laboratory data should be one of the essential layers of data used for precision medicine.

For the future, while there is perception that some or even most of the pathologist work can be displaced by computer vision, there are several strategies. Perhaps the most intriguing is to have a new medical image subspecialist in the future who has a strong background in computer vision and DL that will entail the work of the present day pathologist and radiologist combined. This will be a sub specialist who will study the medical image continuum from the molecular and microscopic to the human-sized anatomic images. In addition, it is conceivable that the future laboratory will be entirely devoid of humans as full automation and AI will be an end-to-end AI-enabled laboratory (Table 8.18).

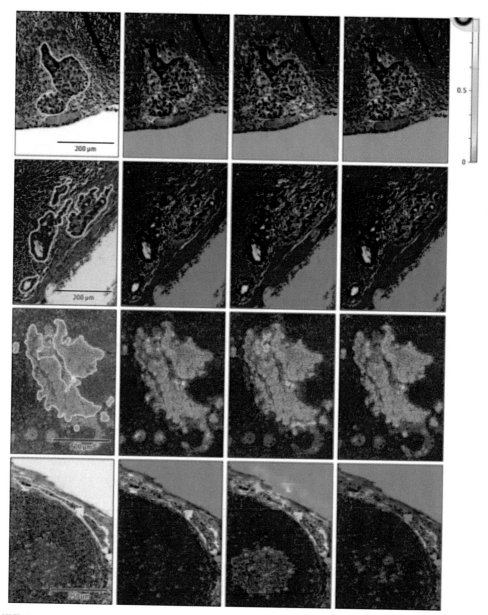

FIGURE 8.11   Probability map of CAMELYON16 competition. The color slide bar at top right signifies probability of each pixel to be part of a metastatic region. In (A), four annotated micrometastatic regions of hematoxylin and eosin-stained lymph node tissue sections are taken from the dataset of the test set of the CAMELYON16 dataset. In (B–D) the probability map from each competing team is overlaid on the original images.

TABLE 8.18   Pathology.

Category of AI application	Clinical relevance	Current AI availability
Medical imaging	+ + +	+ + +
Altered reality	+ + +	+
Decision support	+ +	+ +
Biomedical diagnostics	+ + +	+
Precision medicine	+ + +	+
Drug discovery	+	+
Digital health	+	+
Wearable technology	+	+
Robotic technology	+ +	+ +
Virtual assistants	+	+

AI, Artificial intelligence.

Pediatrics

Pediatric physicians are either primary care pediatricians or a sub specialist in a specific field such as pediatric cardiology or pediatric infectious disease. Pediatric patients are challenging due to the heterogenous patient population (size and age) as well as the prevalence of pediatric rare diseases. Some pediatric subspecialties also accommodate adult patients with pediatric diseases.

Published reviews and selected works

There has been only a few published works in pediatrics and AI. There is, however, a growing number of AI-related publications predominantly within certain pediatric subspecialties such as radiology, cardiology, oncology, neurology, ophthalmology, and pathology. In addition, there are publications in the realm of global health and infectious disease that are good realms for data mining as well as precision medicine which is ideal for certain pediatric diseases. A recent concise "synthetic" mini review by Kokol et al. showed the evolution of AI-related pediatric medicine papers through several time periods and concluded that AI still has not been widely adopted in pediatrics due to a myriad of factors [158]. The major AI themes as it related to children were brain mapping, pattern recognition, developmental disorders, emergency care, ML, and oncology and gene profiling. In addition, a more comprehensive review by Shu et al. covered a myriad of topics in the use of AI in pediatrics: decision support and hospital monitoring, medical imaging and biomedical diagnostics, precision medicine and drug discovery, cloud computing and big data, digital medicine and wearable technology, and robot technology and virtual assistants [159]. Lastly, a review of the importance of data science in child health discusses the three unique features of child health for impact by data science: imperative for data sharing given less volume of relevant data, rareness of congenital diseases, and importance of

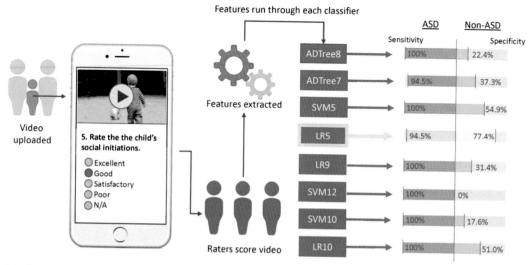

FIGURE 8.12   Rapid and mobile classification of ASD versus non-ASD. Performance of various models are seen in this figure, starting with ADTree8 (decision tree). Study participants can directly upload video clips or a YouTube link. The video raters via crowdsourcing tagged all features and generated the feature vectors to run each of the eight classifiers automatically. The best performing model turned out to be LR5, a five-feature logistic regression classifier. *ASD*, Autism spectrum disorder.

potentially sensitive temporal information. This review also raises the interesting prospect of utilizing transfer learning with models developed on nonpediatric data [160].

A few important works are worth noting in AI in pediatric medicine. A recent publication from China focused on the use of AI and analysis of diverse and massive EHRs (symptoms and signs, history, laboratory data, and PACS report) in over 1 million children visits. This paper revealed that ML classifiers can query medical records in a similar fashion to hypothetico-deductive reasoning of clinicians to make a diagnosis by using a DL-based NLP information extraction model that also includes a disease hierarchical logistic regression classifier in order to predict a clinical diagnosis for the encounter [161]. This model was capable of out-performing junior pediatricians in diagnosing common pediatric diseases and may have a role for triaging based on severity of illness. An earlier study of ML (with 9 ML algorithms) used for medical decision support for the early detection of neonatal sepsis found that 8/9 algorithms exceeded the physicians in terms of treatment sensitivity and specificity [162]. Another good illustrative example of innovative application of AI tools is the mobile detection of autism through ML on home video [163]. In this study, the use of ML on 162 two-minute home videos speed the diagnosis without compromising the accuracy of the autism diagnosis. The home video is uploaded and three raters (via a web portal) tagged all features (such as eye contact, stereotyped speech, or echolalia) to generate feature vectors to run each of the eight classifiers automatically for diagnosis of autism (see Fig. 8.12). The five-feature logistic regression classifier showed to be superior to other models that used decision trees or SVMs.

Artificial intelligence (AI) for pediatric medicine

Sudhen Desai

Sudhen Desai, a pediatric radiologist at a large children's hospital, authored this commentary to envision the multiinstitutional collaboration that will be needed for AI projects in pediatrics to take on rare diseases.

The world of pediatric medicine historically tends to be a few steps behind adult medicine, whether discussing medical device development, pharmaceutical development, or technological advancements. The introduction of AI to the armamentarium of pediatric providers offers a mechanism to level the playing field with potential for profound impact on care provisions for the pediatric population.

There are unique challenges in the utilization of AI for the pediatric sphere that do not have to be considered with adults. In example, normal heart rates vary with age, showing variability between a neonate and a teenager. In my field of imaging, normal growth centers also vary depending on the patient's age (e.g., an elbow radiograph of a 3-year old has differing growth centers than that of a 14-year old). The very definition of normal is a "moving target."

Therefore pediatric AI algorithms must be more robust than those used for the evaluation of adult pathology, to handle the inherent differences between the clinical and imaging data of pediatric "normal" and "pathology". These algorithms are being developed. Already, models exist to map central lines in pediatric radiographs [1], identify pediatric elbow fractures [2], and determine bone age [3]. Others are currently in development, as well as the use of AI to streamline imaging workflow and efficiencies.

These technologies have great potential to help disrupt care models of the pediatric population, to allow AI algorithms to serve as trusted adjuncts to the current workflow of the medical providers at pediatric institutions. However, they require deep support of nonclinical hospital decision makers, as this support requires time and resources.

Participant institutions in the pediatric AI sphere are diverse. Free-standing pediatric hospitals (FSPH) and academic pediatric hospitals are involved in these efforts. In spite of lacking some of the infrastructure that more traditional academic institutions provide their affiliated children's hospitals, FSPH are leaders in the development of pediatric AI technologies. They are willing to invest in infrastructure that mandates hardware, software, and intellectual upgrades from the current standards, as most readers of this text are familiar with.

These FSPH face some unique challenges. Compared to their academic counterparts, FSPH generally lack patient volumes to allow adequate training and refinement of the algorithms/models. They tend not to have staff computer and data scientists, or ready access to such personnel. As with most medical care environments, hardware and software set-ups are not designed to accommodate the necessary large data transfers that permit robust training of the models and seamless integration of the outputs of these models.

However, as more pediatric hospitals embark on AI development programs, we foresee a future of collaboration and data-sharing. Our expectation is that with time, we will be able to create site-neutral databases and infrastructure of pediatric-specific algorithms. The hope is that this will allow North American and international hospitals unrestricted access to cutting edge algorithms that will improve their ability to treat their sickest children in the most efficient and expedient manner. Diagnostic, and by extrapolation, therapeutic, improvements will occur. Pediatric healthcare delivery can improve with these technological augments. In addition, we envision embedding algorithms within currently available technologies to help improve imaging throughput (e.g., CT, MRI), thus reducing scan times (and anesthetic times) as well as contrast and radiation exposure.

While these might seem like flights-of-fancy right now, it is clear that AI represents a space in which the pediatric population need not fall behind advances in the adult space. In fact, the unique challenges of our pediatric populations might even force AI in the pediatric space to be far more robust than our adult algorithms need be.

References

[1] Yi X, Adams S, Babyn P, et al. Automatic catheter and tube detection in pediatric X-ray images using a scale-recurrent network and synthetic data. J Digit Imaging 2019;. Available from: https://doi.org/10.1007/s10278-019-00201-7.

[2] Rayan J, Reddy N, Kan H, et al. Binomial classification of pediatric elbow fractures using a deep learning multiview approach emulating radiologist decision making. Radiology 2019;. Available from: https://doi.org/10.1148/ryai.2019180015.

[3] M. Cicero and A. Bilbily. Machine learning and the future of radiology: How we won the 2017 RSNA ML challenge. November 23, 2017. <https://www.16bit.ai/blog/ml-and-future-of-radiology>.

Recently, a report on girls with suspected central precious puberty and their response to the gonadotropin-releasing hormone stimulation test with the use of extreme boosting and random forest classifiers proved to be useful [164] to render the algorithm more interpretable, a LIME modification was deployed. Lastly, the use of NLP to recognize Kawasaki disease showed this methodology to be superior to manual review of charts by physicians and the authors astutely pointed out that this tool should be embedded in EHR as a support mechanism [165].

Present assessment and future strategy

Overall, the field of pediatrics, both primary care and subspecialty care, underutilizes AI but certain areas are starting to adopt AI (cardiology, PICU, etc.). Major obstacles to adoption include lack of understanding of AI and appreciation of AI tools amongst particularly pediatricians. The combination of heterogeneity of the population with the enigma of rare diseases renders pediatrics and the care of children ideal for deployment of AI for faster diagnoses and more accurate decision-making process.

For the future, there can be several AI-related projects that will be impactful: (1) with AI and facial recognition capabilities as well as genomic sequencing, undiagnosed genetic syndromes in pediatrics can be a phenomenon of the past; (2) with lifelong precision medicine as a goal, multilayered information such as genomic profile, family history, imaging data, and laboratory data will not only be the strategic for diagnosis and therapy in childhood, but throughout the entire lifespan of the individual; and (3) sophisticated AI tools can be deployed to accelerate speed of AI-enabled diagnosis of common childhood diseases (such as pneumonia, heart disease, and diarrheal illnesses) in areas of clinician shortage. In addition, medical images of children are often low in number, especially for certain rare diseases; a convergence of medical image types as well as pooling of such medical images can

TABLE 8.19   Pediatrics.

Category of AI application	Clinical relevance	Current AI availability
Medical imaging	+ +	+ + +
Altered reality	+ +	+
Decision support	+ + +	+ +
Biomedical diagnostics	+ + +	+
Precision medicine	+ + +	+
Drug discovery	+ +	+
Digital health	+	+
Wearable technology	+	+
Robotic technology	+ +	+ +
Virtual assistants	+	+

AI, Artificial intelligence.

create a universal DL library for rapid diagnosis and interpretation. Lastly, altered reality as well as robotic technology can be useful AI tools in the therapy for children (Table 8.19).

Pulmonology

These sub specialists are educated and trained to diagnose and treat diseases of the lungs and chest. Common disorders that they treat include reactive airway disease, lung cancer, and chronic obstructive lung disease (COPD). These clinicians often perform and/or overlook pulmonary function tests that measure lung function and capacity. In addition, they also perform bronchoscopy, a procedure that involves endoscopic examination of the airways and lungs. These clinicians work closely with the ICU physicians, and sometimes are even boarded in ICU medicine.

Published reviews and selected works

There is a paucity of systematic reviews as well as reports of AI in pulmonary medicine. There is no comprehensive review that focuses on the use of AI in pulmonology.

A commentary on the use of AI and chest imaging (in the form of chest CT) discusses the enormous potential of AI by incorporating algorithms into standardized workflows and estimating COPD burden, exacerbation risk, and mortality in any given geographical area [166]. Gonzalez et al. [167] in this same issue reported on their experience with DL-based analysis of CT scans of the chest can directly predict outcomes including respiratory events and mortality. Interestingly, this group also utilized transfer learning to have another cohort use the same methodology. Lastly, a recent report used AI for predicting prolonged mechanical ventilation and tracheostomy placement in adults [168]. The AI methodology used was gradient-boosted decision trees algorithm for the classifier. Lastly, a pediatric study using physiological data and ML showed that this pediatric automated asthma respiratory score

TABLE 8.20 Pulmonology.

Category of AI application	Clinical relevance	Current AI availability
Medical imaging	+ +	+ + +
Altered reality	+ +	+
Decision support	+ + +	+ +
Biomedical diagnostics	+ + +	+
Precision medicine	+ + +	+
Drug discovery	+ +	+
Digital health	+ +	+
Wearable technology	+ +	+
Robotic technology	+ +	+ +
Virtual assistants	+	+

AI, Artificial intelligence.

with ANN had a good predictive accuracy [169] and another pediatric study discovered pediatric asthma phenotypes based on response to controller medications [170].

Present assessment and future strategy

Overall, there is little academic or clinical activity on the use of AI in pulmonology. In addition to utilizing AI and DL/CNN for CT and MRI of the chest, pulmonologists can also leverage AI more for its image-related procedures, such as bronchoscopy.

For the future, there are many areas that can be explored for AI applications in pulmonary medicine. Even though there are not as many types of imaging for the pulmonologist (as compared to radiologist or cardiologist), there are many possible advances from precision pulmonology and testing. For instance, an easy at home pulmonary function testing capability can be interpreted with an AI-embedded pulmonary function device (Table 8.20).

Radiology

Radiology is a field dedicated to interpretation of medical images ranging from an X-ray to more advanced images (such as CT and MRI) as well as some types of ultrasound (the exception being an echocardiogram, or an ultrasound of the heart, which is usually interpreted by a cardiologist). Some radiologists also perform procedures (called "invasive" as these procedures involve entering the vasculature or body organ) such as biopsies or interventions (such as vascular occlusions of certain blood vessels or aneurysms or insertion of catheters into veins). The field of nuclear medicine involves use of radioactive materials to diagnose or treat diseases.

Published reviews and selected works

Radiologists are very familiar with computer-aided detection and diagnosis (CADe and CADx, respectively) software that are explicitly programmed to detect specific

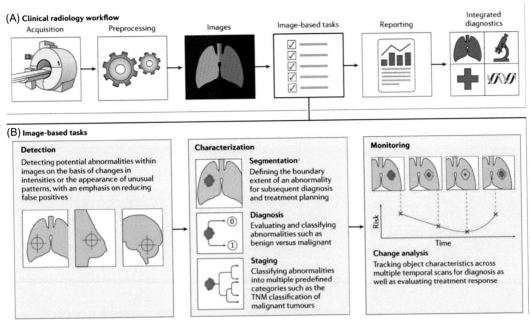

FIGURE 8.13 AI and imaging. This schematic diagram outlines the AI and clinical work flow (A) and AI and the image-based tasks (B). The workflow starts with acquisition of the image, to preprocessing, and then image-based tasks (quantification of features), followed by the reporting and integration of diagnostics from other sources. AI can have an impact on all of these steps to facilitate and enable a faster and more accurate process. The image-based tasks (seen in B) include detection of abnormal findings; characterization (segmentation, diagnosis, and staging as in cancer); and monitoring with change analysis. *AI,* Artificial intelligence.

presentations of a disease on the medical image. As this field has been the most enthusiastic about AI especially as it relates to CNN with medical image interpretation, this is reflected in the number of publications as it has been exponential in its increase. In short, radiology is now the leading subspecialty in the number of new publications in the AI in biomedicine realm per annum (followed by oncology, surgery, and pathology in order). There are a number of review articles as well as substantive commentaries on AI in radiology and related subareas, but only a few will be mentioned here.

A recent review of AI and ML in radiology reviewed not only the imaging aspects of AI but also workflow and surveillance [171]. The paper defined success of AI in radiology as increased diagnostic certainty, faster turnaround, better outcomes for patients, and better quality of work life for radiologists. Another review discussed the evolution of AI versus human intelligence and put this in the context of image-based tasks with excellent diagrams for these processes (see Fig. 8.13) [172]. The basic differences between traditional ML (with feature engineering and classification) and DL for radiology image interpretation is also shown (see Fig. 8.14). Yet another review in the same year focused on ML and its impact in workflow, specifically scheduling and triage, CDSSs, detection and interpretation of findings, postprocessing and dose estimation, examination quality control, and radiology reporting [173]. The review piece by Liew added the legal and ethical issues within health policy applied to AI systems [174].

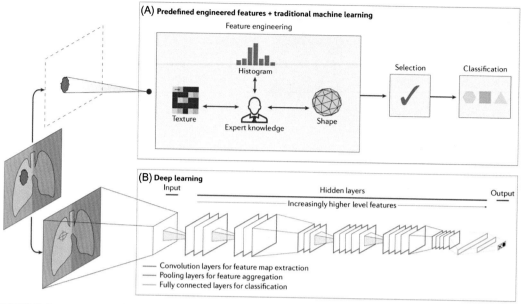

FIGURE 8.14   Machine versus deep learning in radiology. For the diagnosis of a suspicious finding on an image, the traditional machine learning is compared with deep learning. In (A), feature engineering consists of expert knowledge and feature extraction (with features such as tumor shape or location), followed by selection and eventual classification. In (B) deep learning does not require annotation but convolution, pooling, and fully connected layers with increasingly higher level features interpret the image.

Turning data into actionable insights—the artificial intelligence (AI) continuum

Jörg Aumuller[1]

Jörg Aumuller with a strong background in AI and advanced technologies, authored this commentary on the concept of an AI continuum in medical imaging that covers the range from image data acquisition to population health management.

[1]*Digitalizing Healthcare Marketing, Siemens Healthineers, Erlangen, Germany*

What is AI in medicine? Most people will answer with specific examples: Back in the 1970s, it would have been associated with something like MYCIN, a rule-based expert system for infectious diseases from Stanford University. Today, most people will mention software to detect pulmonary nodules on chest X-rays and CT studies, chatbots that are used in triage situations, or algorithms that predict which patient will suffer from heart failure, experience a seizure, or develop postoperative bleeding.

All these are excellent examples, but what about the broader picture? Let us do some localizing. The illustration given later can be thought of as providing an expedition map into a world that is only partly explored and that could be called the Artificial Intelligence Continuum. A useful first step is to give present-day medical AI its proper place within the "grand narrative" of AI, that is, the evolution from task-specific artificial narrow intelligence (ANI) toward artificial super intelligence (ASI).

Artificial narrow intelligence—not narrow at all

Medicine would clearly benefit from ASI—who would not want some intelligence that is technically, psychosocially, and emotionally superior to every doctor who has ever lived? Attractive as it is, it will not become a reality in any foreseeable future. As for now, in medicine, we are in the realm of ANI, characterized by AI solutions that perform single tasks based on specific datasets within a predefined range. However, within this area, there are many advantageous applications to be explored. Many ANI solutions can perform tasks much faster and far more accurately than a human being ever could—and a machine will never tire, no matter how narrow and repetitive a task is.

In addition, the technology suffers from a mislabeling. ANI might sound narrow, but it is pretty diverse, and it offers plenty of opportunities for innovators. On closer inspection, it becomes clear that we have an overarching trajectory from ANI to ASI. There is another trajectory within ANI, marked by increasing complexity, thanks in large parts to more data being integrated and aggregated from more and more data sources.

With this in mind, we can map different application fields of AI in medicine along the ANI trajectory, starting with fields in which algorithms are chiefly being used for automation and quantification, often embedded into medical devices and software solutions. Further along the curve, we approach more complex fields in which AI algorithms perform advanced analytics and predictions, be it on the level of the individual, of a certain population, or of some other cohort.

Turning data into actionable insights – the artificial intelligence continuum

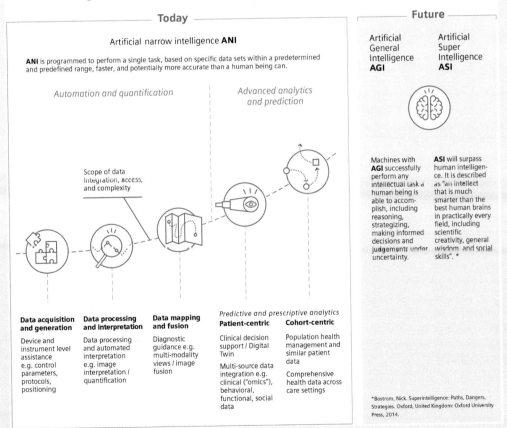

III. The current era of artificial intelligence in medicine

Automation and quantification: the foundation of the artificial intelligence continuum

Let me briefly outline five application fields on the ANI trajectory. Even at the very beginning of the digitalization process, we mainly talk about improving data acquisition and data generation. Take for instance CT imaging: a DL-powered 3D camera can improve patient positioning and can thus help to reduce radiation exposure while achieving higher quality CT scans even for stressed or novice operators. Similarly, a camera-based AI system can optimize the placement or routing of lab tubes in laboratory analyzers, which will improve workflows.

Algorithms like these can enhance the average quality of the digital data collected by any medical device. With the AI Continuum in mind, there is an interesting aspect here: improved and standardized data quality will likely make it easier to use more complex algorithms further up the curve. In other words, data acquisition algorithms have immediate benefits to patients and user. But, they can also be seen as a prerequisite for making progress in higher ANI applications.

With increasing data complexity, we reach the fields of data processing/interpretation and data mapping/fusion. Many groups around the globe conduct research on data processing and interpretation. In imaging, we talk about digital segmentation and characterization tools, and about algorithms that automatically visualize, measure and classify. Beyond imaging, think of medical data mining in plain text documents, such as reports, or of interpreting other medical data of almost any kind.

Data mapping and fusion is where we leave single datasets and start to aggregate and apply AI algorithms on data from different sources, albeit from sources that still belong to a similar category. The most obvious example is imaging: fusing live ultrasound with 3D MRI datasets in order to visualize a catheter during an intervention in real-time. It can be an enormous gain for both doctors and patients in terms of workflow optimization and reducing unwarranted variations.

The higher altitudes: advanced analytics and prediction

This becomes different when we ascend the ANI trajectory further, toward where AI is being used for advanced analytics and prediction. This is happening now—it is the frontier of AI in medicine of our time, the big data realm of patient-centric predictions and cohort analytics.

Patient-centric predictive simulations are colloquially called "digital twins." High-level digital twins are lifelong physiological data models. They make use of all kind of available patient data—imaging, clinical records, and lab data including omics, and they might also draw on behavioral data and on social determinants of health. These data sets can be integrated and analyzed, not only to provide a multidimensional risk model but also to run simulations on the course of a disease and on the outcomes of treatments. Digital twins will also feed into sophisticated decision support systems which apply guideline recommendations and the most recent clinical trial knowledge to the individual patient data set, providing clinical guidance that is as tailored and as precise as possible.

Complexity further increases when predictive algorithms are used to compare the digital twin data set of an individual with those of similar patients. However, in order for this scenario to become a reality, there are still some technical as well as legal hurdles to be cleared. What we are talking about today are low-level digital twin models of individual organs like the heart. These are based on a manageable number of data sources—in the example of the heart, for instance, there is MRI data for dynamic mechanical and fluid-mechanical modeling, electrophysiological data, and vital data such as BP. Models like these are currently being evaluated in clinical trials.

In summary the concept of an AI continuum outlined previously helps us to classify medical applications of AI along a trajectory marked by rising complexity, increasing number of data sources, and closer patient involvement. On the lower end of the continuum, AI tools embedded into medical technology are available today and ready to increase quality of care and patient safety. With R&D ongoing, they will be supplemented by increasingly sophisticated predictive tools that help us to come closer to a future of healthcare in which precision medicine will be the new SOC.

There are several excellent overviews of ML/DL in imaging. A comprehensive primer on DL for radiologists by Chartrand et al. is both comprehensive as well as easy to understand, with exquisite and concise explanations of key concepts as well as excellent diagrams to illustrate the many nuances in this domain [175]. Other good reviews of varying lengths and details in DL in radiology have been published since 2015 [176–178]. Finally, an excellent AI guide of medical image analysis for the user (as well as authors and reviewers) offers excellent best practices, covering technical, statistical, and other aspects of evaluation and application of AI tools in radiology [179].

Another key area for radiologists is NLP as much of the workflow is focused on reporting and communicating the image findings. Cai et al. [180] provided an overall primer on NLP as it pertains to radiology research and clinical applications (see previous citation) and Pons provided a similar review of NLP with focus on workflow elements such as diagnostic surveillance, case recall, and quality of practice [180].

A landmark paper in radiology was the CheXNet study that developed an algorithm for detecting pneumonia on chest X-rays with performance that proved superior to radiologists [181]. The algorithm was a 121-layer CNN trained on the largest publicly available chest X-ray dataset called ChestX-ray14 (with over 100,000 frontal-view chest X-ray images with 14 diseases) (see Fig. 8.15). Although there have been many appropriate criticisms of this study (such as accuracy of ChestXray 14 dataset, and medical meaning of labels), it raised awareness amongst radiologists that a new era of AI-enabled medical image interpretation has arrived [182].

Titano et al. reported use of CNN and weakly supervised classification for screening head CT images for acute neurologic events and was able to reduce the time to diagnosis from minutes to seconds [183]. GE Healthcare partnered with Boston Children's Hospital to develop an AI-backed decision support platform with cloud computing for the diagnosis and treatment of pediatric brain diseases [184]. This AI strategy of having a reference library of normal pediatric MRI brain scans coupled with DL is particularly helpful for pediatric brain imaging as disorders can be misinterpreted as normal brain maturation while the opposite is also true (normal brain maturation can be misdiagnosed as abnormal changes and lead to misdiagnoses and inappropriate treatment). This decision support platform will be available worldwide for all

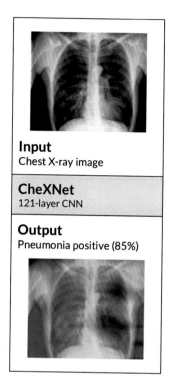

Input
Chest X-ray image

CheXNet
121-layer CNN

Output
Pneumonia positive (85%)

FIGURE 8.15   CheXNet CNN. This is a CNN with 121 layers with chest X-ray image as input and probability of a pathology is the output. In this case, there was an 85% probability that the image has pneumonia (area highlighted). *CNN*, Convolutional neural network.

pediatric brain imaging as pediatric neuroradiologists are scarce. In addition, a recent report described a DL language modeling approach to attain a radiology–pathology correlation by taking reports of each domain and use.

Universal Language Model Fine-Tuning for Text Classification methodology [185]. A recent study in JAMty -20

A showed that in diagnostic study of intracranial aneurysms with augmentation with a DL segmentation model, clinicians showed significant increases in sensitivity, accuracy, and interrater agreement when augmented with neural network model-generated segmentations [186] while a study of cerebral arteriovenous malformations showed ML was able to provide best possible predictions of AVM radio surgery outcomes and to identify a novel radiobiological feature [187]. A brief paper described the innovative Data Science Institute and AI Advisory Group under the American College of Radiology [188]. Finally, Jha and Topol suggested that radiologists (as well as pathologists) learn to adapt to AI to become "information specialists" [189].

Artificial intelligence (AI) in radiology: the road to augmented radiology

Imad Nijim

Imad Nijim, a CIO in radiology and advocate for AI implementation, authored this commentary to elucidate his IT perspective on how AI can be effectively deployed in radiology to augment not only interpretation of medical images but also workflow in radiology.

Machine vision and allure of artificial intelligence in radiology

The story of AI in radiology begins with machine vision. The availability of extremely large data sets, access to relatively cheap and massively scalable compute power, and no-cost access to sophisticated open-source development packages has led to many breakthroughs in computer science. Most relevant to radiology is the use of CNNs for computer or machine vision. Simply put, machine vision is the ability for computer programs to "see" images. With computer algorithms able to identify a picture of a simple object, such as a chair or a cat, why not build an algorithm to identify pneumonia? Can a computer software generate a diagnostic report from a medical image? This is the fundamental promise and allure of AI in radiology.

Complications unique to radiology

The potential for AI in radiology generated excitement, which grew into substantial hype that has finally peaked. Innovators, software developers, academics, radiologists, and the industry collectively recognized the complexities of using AI in radiology.

The first complication is that medical images contain health information protected by an act of Congress we all know as HIPAA. Medical images are not as readily available for software developers and data scientists as are nonmedical images. For example, a simple Internet search using the keywords "white cat" will instantly produce a large dataset of white cat images. In contrast, an Internet search for pneumonia will produce a variety of results, none of which are specifically an image of pneumonia.

The second major complication for AI in Radiology is the problem of "lay-person classification." A software developer can glance at an image of a cat and confirm that the object in the image is indeed a cat. In contrast, most software developers cannot look at an X-ray and identify the presence of pneumonia. Expert radiologists using specific software and equipment are needed to make this determination. A properly annotated and carefully curated dataset specific to pneumonia is necessary to build an AI algorithm that can identify the presence of pneumonia.

Current commercial development efforts in radiology AI are not focused on creating a diagnostic report from images. Most development efforts are focused on improving patient care through physician efficiency, severity prioritization, and quality initiatives. The most common AI models in radiology are classifiers. Classifiers identify the existence of an object with a certain level of statistical confidence. These models can classify pathology, segment a body part, or identify a disease in images such as x-ray or computed tomography. Models are implemented in a variety of workflows or packaged as products based on their performance. Performance is typically measured by sensitivity and specificity and each intended use has a unique tolerance for false positives and false negatives.

Expanding on the white cat versus pneumonia analogy, any person or even a child can quickly inspect a picture of a white cat and further classify the image. With very high levels of accuracy, a person can classify the cat as having blue eyes, green eyes, being old, skinny, fluffy, or having a flat, round face. However, a lay-person cannot identify pneumonia and further classify the disease. Can a model that identifies the presence of pneumonia also determine opacity, severity, or fluid characterization?

Phenotype characterization is the next wave of innovation in radiology AI. With the hype of AI in radiology behind us, and the groundwork in place, we are moving to an exciting phase of practical implementations, expanding use cases, commercialization, and a complete overhaul of the informatics systems of radiology.

Practical implementations of artificial intelligence radiology

Once a model is developed with satisfactory performance and acceptable validation, integration to existing radiology workflows becomes the last mile in delivering AI to radiologists. Nonacademic AI developers can be organized into three group: equipment and modality manufacturers, medical informatics vendors, and independent software developers.

Equipment manufacturers can deliver AI models directly into the modality and utilize model insights at the point of patient care. For example, mammography imaging equipment can apply advanced detection models to breast cancer screening exams and display the results on the modality console. X-ray manufacturers can display statistical likelihood of the presence of pneumothorax in near real-time right on the console.

Medical informatics vendors are one step removed from the point of patient care but have the advantage of broad access to historic patient data, prior exams, and radiology reports. In some cases, additional patient history is available in the patient's medical record. For example, a pulmonary nodule detection model can also evaluate prior exams and quickly calculate change in volumes.

Independent software vendors tend to be new entrepreneurs focusing on AI development by model or by body part. They tend to deliver model insights through cloud services. Independent AI software vendors are producing broad innovations and model developments that cover detection algorithms for intracranial hemorrhage, spine fracture, lung nodule, and vessel laceration.

This is only a glimpse of what is possible with AI in radiology.

The future is augmented radiology

Radiology and radiologists have always embraced technology, innovation, and early adoption. AI, specifically CNNs, that allow us to build image analysis models, are the most disruptive technology to the informatics of radiology since the digitization of the printed film.

We have entered a new and exciting time that is engaging radiologists in rethinking technology's role in diagnostic imaging. The next chapter in this rapidly evolving landscape will be how AI will augment the radiologist in the continuum of patient care.

Present assessment and future strategy

Overall, AI in radiology is extremely robust with some if not most of the radiologists very cognizant of the significance of AI in their future work. A milestone was reached when FDA granted approval of the AI radiology software called ContaCT (Viz.ai) in 2018; this software is able to decrease the time to make diagnosis of a major stroke. Even though there had been some discussion regarding whether radiology will be an attractive field for future students and trainees, the prevailing thinking is that radiology remains a secure field but with AI being an ever important partner in the practice. Despite what was said about not training any more radiologists, it is entirely conceivable that AI will render radiology even more enticing as a medical field. The American College of Radiology's Data Science Institute deserves much praise for its conceptualization and agenda as an AI-centric center of resources and education for radiologists. There is an ongoing discussion or even debate about the performance of single-site AI tools in medical image interpretation and whether these studies are applicable to other sites. In addition to the present work on CNN and medical image interpretation, there is also progress on NLP applications in radiology for less prose in reports and data mining. Some focus is now directed to workflow inadequacies in radiology that can potentially be neutralized by AI techniques.

For the future a key concept for radiologists and medical image interpretation is the creation of a "AI-enabled, end-to-end imaging" that creates a continuum from image acquisition to reconstruction and segmentation and then on to image interpretation so that AI is embedded in all aspects of image acquisition and interpretation. Of note is the AI-enabled acquisition of images so that a poor image or study will potentially be an observation of the past; in addition, the future acquisition of images will be accompanied by labels that will be already in place. A concept that was discussed earlier under cardiology is the emergence of a "super scan" in which static images such as CT or MRI as well as dynamic images as seen in ultrasound can converge into one modality with an AI-enabled strategy. A future concept also entails incorporation of this image interpretation into precision medicine and population health (precision radiology or imaging) so that follow-up studies will be based on data rather than intuition alone. It is possible that there will be a closer dyadic relationship between radiology and pathology so a future field of medical image specialist could emerge. In areas outside of imaging, AI can decrease the dosage of nuclear materials needed in nuclear medicine. Lastly, AI can also mitigate the administrative burden of a busy program by automating certain aspects of the radiology administration (such as deployment of RPA) (Table 8.21).

Rheumatology

The rheumatologist deals with musculoskeletal diseases and systemic autoimmune diseases such as rheumatoid arthritis, osteoarthritis, gout, and lupus. The rheumatologist does not typically perform procedures but are often involved in difficult diagnoses and physical rehabilitation of joint and musculoskeletal disorders.

Published reviews and selected works

There are no searchable lengthy reviews on the use of AI in rheumatology. One editorial discusses that the advent of bioinformatics and AI and the explosion of patient data along with image analysis will enable the clinicians in rheumatology to make timely diagnosis and accurate prognosis for diseases such as rheumatoid arthritis [190]. Another editorial similarly advocated for AI and its potential to help rheumatologists in their practice [191].

TABLE 8.21   Radiology.

Category of AI application	Clinical relevance	Current AI availability
Medical imaging	+++	+++
Altered reality	+++	+
Decision support	++	++
Biomedical diagnostics	+	+
Precision medicine	+++	+
Drug discovery	+	+
Digital health	+	+
Wearable technology	+	+
Robotic technology	++	++
Virtual assistants	+	+

AI, Artificial intelligence.

One report described the published literature using either automatic image segmentation and/or classification or regression methodologies for prediction of fracture risk in osteoporosis [192]. Another recent report discussed the impact of big data and AI on a disease such as Sjogren's syndrome via and NLP-based program called BIBOT [193]. Lastly, clinical outcome forecasting using longitudinal DL model in patients with EHR data showed that it is possible to accurately predict the outcome in these complicated patients with rheumatoid arthritis [194].

Present assessment and future strategy

Overall, there is little academic or clinical activity in the use of AI in rheumatology to date. There is, however, indication that some practitioners can foresee the value of AI in deciphering some of the enigmas as well as augmenting some of the chronic disease challenges in rheumatology.

For the future, AI and its portfolio of tools can be potentially very useful for this subspecialty. For instance, wearable technology and eAI can provide valuable real time and daily information for AI-enabled chronic disease management that can be designed to be patient-centric. In addition, diagnostic dilemmas that frequently encountered by the rheumatologist can be aided by an AI resource in the form of updated information as well as an able cognitive partner. Lastly, as many of these patients have chronic disability, the AI tools in robotic technology and virtual assistants can be valuable as assisting dimensions for patient care (Table 8.22).

Surgery

While there are general surgeons, surgery is a large domain that encompasses many subareas, such as plastic surgery, orthopedic surgery, thoracic surgery, and neurosurgery.

TABLE 8.22   Rheumatology.

Category of AI application	Clinical relevance	Current AI availability
Medical imaging	+	+ + +
Altered reality	+ +	+
Decision support	+ + +	+ +
Biomedical diagnostics	+ +	+
Precision medicine	+ + +	+
Drug discovery	+ +	+
Digital health	+ +	+
Wearable technology	+ +	+
Robotic technology	+ +	+ +
Virtual assistants	+	+

AI, Artificial intelligence.

These subareas of surgery focus on a particular body area or organ (such as the heart for cardiac surgery) or a specific system (such as urologic surgery for the genitourinary system). A surgeon often interprets medical images although often with consultation with a radiologist but at times without such a support (such as during emergencies).

Published reviews and selected works

The majority of current published works on AI and related topics concentrate on aspects of robotic surgery, but there are few reviews and published reports on AI in surgery (neurosurgery and cardiac surgery will be discussed separately under neurosciences and cardiology, respectively). One recent review on AI in surgery emphasized that four AI areas outside of robotics are particularly relevant to surgeons: ML, NLP, ANN, and computer vision [195]. The implications for surgeons include more precise selection of patients for procedures, especially biopsies; higher level of precision care as part of preoperative care; and increased surgical community collaboration via video and EMR data sharing and analytics. The integration of multimodal data with AI can significantly augment the surgeons' capability to improve care: preoperative comprehensive risk score calculation, intraoperative event detection and predictive analytics, and postoperative morbidity and mortality detection and prediction. Another general review in surgery but primarily focused on the transition from laparoscopic surgery to robotic surgery that is integrated with AI [196]. These robots will need to be able to perceive surroundings, recognize problems, implement appropriate action plans and finally, and produce solutions to new problems.

A similar review in otolaryngology encourages surgeons to put forth more efforts to collaborate with data scientists [197]. The author correctly pointed out the pitfalls of AI in healthcare, such as the inflated expectation of Watson in oncology as well as the somewhat comical nuances in medical image interpretation (such as chest tubes in pneumothorax and rulers in malignant lesions). In addition, a focused review of AI in plastic surgery emphasized the importance of precision medicine and AI in plastic surgery in order to use patient data to formulate an individualized plan of intervention [198] while another review of AI in the same surgical subspecialty focused on not only surgical applications but also resident training [199]. Patients discussed in this review include those undergoing breast surgery (and risk for associated anaplastic large cell lymphoma as well as over diagnosis of breast cancer), wound care (via use of imaging and CNN as well as AI-assisted evaluation for surgical flaps), and craniofacial surgery (perioperative and intraoperative surgical planning for craniosynostosis). A comprehensive review of AI and ML in orthopedic surgery is a thorough systematic literature review from the past two decades of 70 journal articles with the caveat that AI, like any other technology, needs to adhere to the tenets of health technology assessment [200] while a word of caution was raised as well to balance the positive outlook [201]. An excellent review of AI in spine research also covered a wide range of topics, from ML to image segmentation and prediction of outcomes [202].

Artificial intelligence (AI) and its potential in surgery

Daniel A. Hashimoto[1]

Daniel A. Hashimoto, a young surgeon with a technology background, envisions a promising future with a myriad of AI tools that can be deployed for surgeons that range from the more obvious intraoperative application but also perioperative risk assessment and intervention for patients.

[1]Surgical AI & Innovation Laboratory, Department of Surgery, Massachusetts General Hospital, Boston, MA, United States

In the United States, over 48 million operations are performed annually [1]. We know this because hospitals, insurers, medicare, and other groups keep track of such data. It seems pretty obvious that if we are doing something as invasive as surgery, we should probably keep track of data about the operation and see what is working and what is not. Perhaps surprisingly, recording and tracking data about patient outcomes was not always considered by everyone to be the right thing to do.

In 1914 a surgeon by the name of Ernest Amory Codman worked at the Massachusetts General Hospital (MGH) and suggested that the hospital and its surgeons systematically track outcomes on their patients. However, his colleagues and the hospital's administration were reluctant to incorporate his ideas and even discouraged them, leading to his resignation from the MGH staff [2]. Perhaps some may have held a fear of what he might find and what we know to be true today: complications happen after surgery, and some complications are avoidable.

In modern surgery, we recognize that Codman was on the right track: studying outcomes makes us better surgeons because we can learn from the data. National and international efforts, such as the American College of Surgeons National Safety and Quality Improvement Program, have been underway to systematically collect preoperative, operative, and postoperative data on surgical patients to help us better understand how to improve surgical care. MGH, the hospital that had resisted Codman's ideas over a century ago, now even has a research center named after him—the Codman Center for Clinical Effectiveness in Surgery.

Techniques in AI serve as additional tools with which we can study and explore the data that we currently have in such large databases. However, most of that data is based on insurance claims data or patient registries. These datasets can describe the operation, usually in the form of Current Procedural Terminology or ICD codes but not the specifics of what happened in any individual case. Thus the more exciting promise of AI is its potential to allow us to unlock previously untapped sources of data through which to improve surgical care.

Multiple groups, such as our Surgical AI & Innovation Laboratory at MGH the CAMMA group in Strasbourg and the International Centre for Surgical Safety in Toronto, are developing and researching AI systems that can analyze surgical video to accomplish tasks such as identifying steps of a procedure [3] or tracking instruments [4]. Such systems hold the promise of providing more specific, granular data about the events that occur during the course of an operation that may correlate to poor outcomes. As this type of data increases in its availability, it could then be utilized to help quantitatively compare different operative techniques and approaches as well as provide greater insight into what pre- and postoperative pathways are best paired with specific operative approaches.

Imagine a scenario where an AI system is able to augment the surgeon's ability to make decisions. Drawing on a database of hundreds of thousands of previously recorded cases, the AI could warn the surgeon that they are deviating off course. The surgeon can alter their operative plan and prevent a potential injury or complication from ever occurring. The potential here is an AI for real-time CDS—a GPS for surgery that, instead of avoiding traffic, helps surgeons avoid complications and saves lives.

A well-designed system that utilizes AI (and state-of-the-art technology for the protection of patient privacy) could allow for data to be collected, analyzed, and shared in real-time across multiple surgeons. For this to be possible, efforts are needed to build a surgical database that includes video of the intraoperative phase of care. Importantly, such a database must be shared across institutions in a collaborative manner. Isolating the data within specific groups defeats the purpose of democratizing surgical data and knowledge. With a worldwide database of video and outcomes, one can imagine an AI with the collective experience of hundreds of surgeons—a "collective surgical consciousness" that pools the experience of a multitude of surgeons. Such a database could facilitate the development of a GPS for surgery [5].

The collective surgical consciousness is built on the lofty premise that we can use AI to create the highest volume "surgeon"—one who has learned techniques from the best surgeons across the world. This means that every patient could have access to the same expertise, translating not only to lives saved but also to reduced cost by reducing complications. It could make surgical care more affordable and thus more accessible to patients who need it.

This concept is early in development, and researchers, engineers, and surgeons have a lot of difficult challenges to solve in order to achieve this potential that AI holds. Given the promise of improving the care of and impact on surgical patients worldwide, I believe AI's potential is worth chasing.

References

[1] Hall MJ, et al. *Ambulatory surgery data from hospitals and ambulatory surgery centers: United States, 2010*. National Center for Health Statistics; 2017. p. 1–15.
[2] Brand RA. Ernest Amory Codman, MD, 1869-1940. Clin Orthop Relat Res 2009;467(11):2763–5.
[3] D.A.Hashimoto, G. Rosman, E.R. Witkowski et al. Computer vision analysis of intraoperative video: automated recognition of operative steps in laparoscopic sleeve gastrectomy, *Ann Surg* **270** (3), 2019, 414–421.

[4] Twinanda AP, et al. EndoNet: a deep architecture for recognition tasks on laparoscopic videos. IEEE Trans Med Imaging 2017;36(1):86–97.
[5] Hashimoto DA, et al. Artificial intelligence in surgery: promises and perils. Ann Surg 2018;268(1):70–6.

Artificial intelligence in healthcare from an academic reconstructive surgeon's perspective

Brian Pridgen and James Chang

Brian Pridgen and James Chang, academic plastic surgeons with a penchant for innovation, delineates how a surgical subspecialty like plastic surgeon can adopt for a myriad of areas beyond diagnosis and screening, such as preoperative planning and surgical training.

Surgeons have been slow to embrace the use of AI. This is not surprising, and is in part due to the dearth of structured databases in surgery. Most importantly, surgeons have not imagined how AI would aid intraoperative decision-making and technique. Particularly in reconstructive surgery, the unique nature of each case and each tissue defect relies on individualized, bespoke reconstruction. This has not been amenable to analysis and guidance by machines.

However, surgeons are now looking toward AI to aid diagnosis and triage of complicated patients. Similar to the work flow of other physicians, a reconstructive surgeon's care for a patient begins with screening and referral from nonspecialists, and then continues to surgical decision-making and technical execution of an operation. Using AI to aid screening and diagnosis for reconstructive surgeons mirrors the approach of other specialties such as pathology and radiology. The relatively few number of patients undergoing reconstructive surgery and the scarcity of structured databases present challenges for training data-hungry AI algorithms. Despite this, there exists a tremendous opportunity for AI to bridge the knowledge gap between reconstructive surgeons and their colleagues outside of the specialty, particularly as healthcare becomes increasingly specialized.

AI in reconstructive surgery is in its nascent phase. Using a needs-based approach to designing computer vision AI projects, we have focused on clinical scenarios that meet three criteria:

1. The specialist can reliably make a diagnosis based on an image.
2. There is a disparity between the specialist and nonspecialist's diagnostic accuracy.
3. Timely diagnosis is critical for appropriate management.

We used these criteria to identify two areas of need in reconstructive surgery. Both projects are ongoing areas of research and are presented as illustrative examples of opportunities to apply AI to academic reconstructive surgery.

The first project addresses timely and accurate diagnosis of acute burn wounds. Acute burns often present to emergency departments without on-site burn specialists. Due to the relative infrequency of burns, the initial assessment by nonspecialists is prone to error [1]. This can lead to under- or overtriage, or inappropriate initial fluid resuscitation. Our team built an annotated image database of prior patients with acute burns. Using this database, we are developing a computer vision algorithm that has achieved similar accuracy to burn specialists (Fig. 1). With

 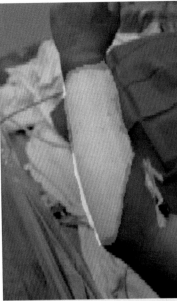

FIGURE 1 Photograph of a patient with an acute forearm burn (left) and AI-predicted mask overlaid on the burn (right) (unpublished results). *AI*, Artificial intelligence.

smartphone accessibility to this algorithm, this AI tool would allow nonspecialists (emergency room physicians and family practitioners) to provide expert-level initial evaluation and triage of burns.

The second project attempts to reduce the rate of missed wrist perilunate dislocations. Delayed presentation of a missed perilunate dislocation, an injury that should be promptly identified and treated, is a clinical problem encountered by hand surgeons [2]. Because these injuries are apparent to a specialist on radiographs, we sought to build a computer vision algorithm that could provide this expertise to nonspecialists in the emergency room or clinic. We have assembled a database of images and are developing an algorithm that accurately identifies perilunate dislocations. This could be applied as a screening tool for wrist trauma to avoid a missed injury and to allow prompt treatment by a hand surgeon.

While our needs-based criteria guided the identification of these initial projects, these should not constrain the development of AI. Other examples within reconstructive surgery that close the experience gap between specialists and nonspecialists include identification of children with craniofacial features suggestive of craniosynostosis and outpatient monitoring of chronic wounds. We predict there are many more.

Beyond diagnosis and screening, there are other possibilities for reconstructive surgeons to utilize AI. When planning an operation, the clinical judgment of a senior surgeon guides decision-making for fracture stabilization or flap perforator artery selection. AI programs trained on the senior surgeon's analysis of scans and angiograms may be used by less experienced surgeons to obtain an "AI expert consult" to guide surgical planning.

Using AI to support the technical execution of an operation will require further research, including advances in surgical robotics. However, a closely related area of opportunity is in surgical education, for both initial training and ongoing skill-based maintenance of certification for practicing surgeons. Prior research has shown that observing videos of laparoscopic cases can accurately rate a surgeon's skill, which correlates with outcomes [3]. Early work has shown that AI similarly may be capable of rating the technical ability of a surgeon [4]. Such an AI program would obviate the expense and labor of experienced surgeons having to perform these evaluations.

We are on the verge of seeing the impact of AI on screening and diagnosis throughout healthcare. More effort will be placed toward building the labor-intensive databases needed to train algorithms as their value becomes more evident. In surgery the key will be to translate the experience of expert senior surgeons to AI algorithms. Once these databases mature, AI will guide us through every phase of a patient's care, even within specialized fields such as reconstructive surgery. We look forward to this revolution in reconstructive surgery!

References

[1] Armstrong JR, Willand L, Gonzalez B, Sandhu J, Mosier MJ. Quantitative analysis of estimated burn size accuracy for transfer patients. J Burn Care Res 2017;38(1):e30–5.
[2] Herzberg G, Comtet JJ, Linscheid RL, Amadio PC, Cooney WP, Stalder J. Perilunate dislocations and fracture-dislocations: a multicenter study. J Hand Surg Am 1993;18(5):768–79.
[3] Birkmeyer JD, Finks JF, O'Reilly A, Oerline M, Carlin AM, Nunn AR, et al. Surgical skill and complication rates after bariatric surgery. N Engl J Med 2013;369(15):1434–42.
[4] Jin A, Yeung S, Jopling J, Krause J, Azagury D, Milstein A, et al. Tool detection and operative skill assessment in surgical videos using region-based convolutional neural networks. In: 2018 IEEE winter conference on applications of computer vision (WACV) [Internet]. 2018. p. 691–9. Available from: <doi.ieeecomputersociety.org/10.1109/WACV.2018.00081> [cited 27.01.19].

Present assessment and future strategy

Overall, with the exception of neurosurgery, most of the surgical disciplines remain relatively dormant in the area of AI. The field of robotic surgery does involve AI and is the sector that has garnered the most attention by the surgeons in general.

For the future, in addition to the evolution of higher sophistication of robotic surgery, there are many areas that can be developed in the surgical areas. The surgeons would benefit from computer vision and medical image interpretation especially in the absence of a qualified radiologist. In addition, preoperative planning with the use of AI and virtual or augmented reality could reduce suboptimal surgical results and/or unnecessary complications. Data mining can be used for risk assessment and stratification to provide both the patients and payors a more precise prediction of outcome and resource allocation. Lastly, altered reality and AI can provide a resource for preoperative planning as well as education and training in surgery (Table 8.23).

In addition to the aforementioned subspecialties the presence and influence of AI in other health-care—related areas are discussed later:

Dentistry

A dentist performs diagnosis and treatment that pertain to the teeth and also oral structures (such as the mucosa) and craniofacial complex. Diagnostic tests consist mainly of X-

TABLE 8.23   Surgery.

Category of AI application	Clinical relevance	Current AI availability
Medical imaging	+++	+++
Altered reality	+++	+
Decision support	+++	++
Biomedical diagnostics	++	+
Precision medicine	+++	+
Drug discovery	++	+
Digital health	++	+
Wearable technology	++	+
Robotic technology	+++	++
Virtual assistants	++	+

AI, Artificial intelligence.

rays and treatments range from restorative (filling of caries) to prosthodontics (replacement of teeth). Some dentists specialize in endodontics (root canal therapy) or oral and maxillofacial surgery.

Published reviews and selected works

An earlier review on applications of AI for dentistry discussed aim to reduce cost, time, human expertise, and error with special focus on global needs [203].

Present assessment and future strategy

Overall, a dentist has a balanced perception, cognition, and operation portfolio of tasks, so there can be many AI applications in these areas that would be very similar to the aforementioned areas for medical subspecialties. For the future an AI-enabled dentist can triage from dental images (even conveniently from cell phones) and history and schedule appointments appropriately from this strategy. Some aspect of robotic surgery may be feasible, and this resource can be utilized, along with even AI-enabled 3D printing of prosthetics (which currently take weeks).

Digital Health

Digital health heralds the era of technological advances such as apps, wearable technology and remote monitoring, telemedicine and communication tools, and other diagnostic devices to affect a more optimal quality of care as well as a more timely response to any situation. While it is not a subspecialty per se, there are concentrated efforts and meetings in this domain to promote the use of existing technologies for population health and personalized medicine.

Published reviews and selected works

There is a number of reports in digital health coupled with AI that clearly demonstrates not only the proof of concept in applying AI to an app or device but also clinical benefit. A recent editorial in *Lancet* cautions the use of AI in digital medicine and strongly recommends a continual evaluation of digital health interventions for both clinical effectiveness and economic impact [204]. A more positive review by Fogel discussed how AI in digital medicine can improve not only basic health screening and prevention as well as medication adherence but also the human-to-human experience of healthcare. Another review in this domain focused on the concept of a medical Internet of Things (IoT) in digital healthcare that is imbued with AI-related tools [205]. In order to reduce overall costs for both prevention and management of chronic diseases, devices are needed to execute this strategy: to monitor health biometrics, to auto-administer therapies, and to track real-time health data during therapy. Along with these devices, mobile applications for access to medical records as well as tools for telemedicine and telehealth for this new paradigm of medical IoT. All of these devices and equipment will need an AI-centric strategy for data integration and interpretation for delivering optimal health-care advice and direction.

While chronic diseases such as diabetes care can benefit greatly from a coordinated and efficient strategy, use of technology including AI remains fragmented at present due to a myriad of issues: lack of supportive policy and regulation, unsustainable reimbursement, inefficient business models, and concerns regarding data security and privacy [206]. The advent of wearable devices and sensors to continuously track physiologic parameters can provide an overall patient care strategy that will improve outcome and lower health-care costs in cardiac patients with heart failure [207]. This new paradigm of cardiovascular disease management can also improve the physician—patient relationship. ML algorithms have also been applied to large-scale wearable sensor data in neurological disorders such as Parkinson's disease to significantly improve both clinical diagnosis and management [208]. This sensor-based, quantitative, objective, and easy-to-use system for assessing Parkinson's disease has potential to replace traditional qualitative and subjective ratings by human interpretation.

Present assessment and future strategy

Overall, the present digital medicine sector is gradually being developed with interaction with basic analytics and AI tools. Most of the wearable devices are not embedded with analytics. As this domain also requires data infrastructure to be well organized, some of the efforts in this area have been also challenging. An overall strategy for preliminary and continual evaluation of AI applications in digital medicine has been ongoing as the barrier to entry may continue to be low for some apps and devices. This evaluation process will need insight from not only organizations such as the American Medical Association or the FDA but perhaps also by an international consortium of multidisciplinary experts. Finally, attention is being directed toward the cybersecurity of these intelligent devices to mitigate the risk of data breaches and, therefore, intentional harm to patients and caretakers.

For the future, eAI and ML algorithms evolve toward the Internet of Everything and will bring together people, process, data, and things; this strategy will allow the accrued data be streamlined and organized in the cloud proactively in an overall paradigm of personalized precision medicine. As these devices become more intelligent, increasingly higher levels of sophistication in decision support can also be part of both (1) preventive medicine (such as retinal images for retinopathy screening or skin lesions for melanoma detection) and (2) chronic disease care management (such as diabetes, hypertension, or heart failure).

Genomic Medicine and Precision/Personalized Medicine

Genomic medicine involves using genomic information of an individual as part of the clinical diagnosis and therapy and has had impact on oncology, pharmacology, rare and undiagnosed diseases, and infectious disease. Genomic medicine is also an important part of precision or personalized medicine. There is some understandable confusion between personalized medicine and precision medicine: while the former is an older term, the latter is a more current term to reflect an approach to medicine that incorporates genetic, environmental, and lifestyle factors. Pharmacogenomics, the study of how genes can affect a person's response to drugs, is an important part of precision medicine.

Published reviews and selected works

There is significant academic activity focusing on AI in this burgeoning domain. A recent review on AI utilization in precision medicine discussed the importance of data quality and relevance [209]. The authors contend that much of the effort to advance AI in precision medicine has been focused on algorithms and generation of genomic sequence data and EHR but should also be on physiological genomic readouts in disease-relevant tissues as well. Another review discussed advances in ML and AI are vital for the understanding of epigenetic processes, specifically DL for the generation and simultaneous computation of novel genomic features [210]. Grapov et al. reviewed DL in the context of omics and EHR and astutely pointed out that the challenges of DL is akin to those observed in biological message relay systems such as gene, protein, and metabolite networks [211]. In biomedical diagnostics, medical geneticists are often frustrated by the tedious nature of genotype-phenotype interrelationships among syndromes, especially for extremely rare syndromes. Now, medical geneticists are able to use a visual diagnostic system that employs ML algorithms and digital imaging processing techniques in a hybrid approach for automated diagnosis in medical genetics, especially in rare diseases [212]. One such proposal is the BioIntelligence Framework proposed by Farley et al. [213]. In this model a scalable computational framework

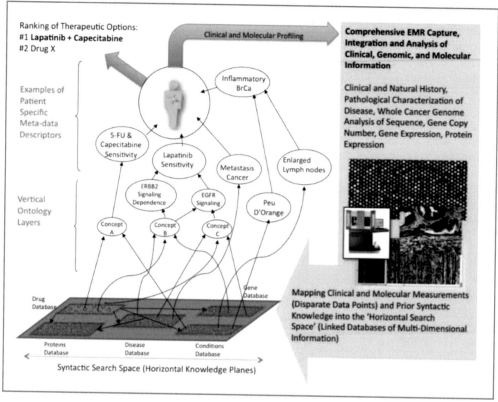

FIGURE 1   BioIntelligence framework. The figure shows the multidimensional genomic and clinical data can be configured (mapped and projected though an ontology graph data structure) to search for individualized therapy. Clinical and molecular profiles from individuals are used along with their EHR data for a three-dimensional approach (horizontal knowledge planes or search space and vertical mapping with ontology layers) to recover concepts to infer therapeutic options. The basis for this framework is a hierarchically organized and ontologically based knowledge representation schema.

leverages a hypergraph-based data model and query language that may be suited for representing complex multilateral, multiscalar, and multidimensional relationships. This hypergraph-like store of public knowledge is coupled with an individual's genomic and other patient information (such as imaging data) to drive a personalized genome-based knowledge store for clinical translation and discovery. Patients of very similar genomic and clinical elements can be discovered and matched for diagnostic and therapeutic strategies (see Fig. 8.16) [214].

Artificial intelligence (AI) and nutrition—a personalized diet strategy

Minhua He[1] and Cindy Crowninshield[2]
Minhua He and Cindy Crowninshield both with a background in nutrition authored this
commentary on the innovative concept of applying AI concepts to attain a personalized diet
strategy that includes personalized meal planning.

[1]*Shenzhen Yihua Investment Management Co., Ltd., Shenzhen, P.R. China*

[2]*Eat4yourGenes and Eat2beWell, Ashland, MA, United States*

Since the proverb "An apple a day keeps the doctor away" originated in the 19th century [1], almost every child has come across it since parents use it as a rule of thumb to encourage fruit and vegetable consumption. To promote consumption the United States Department of Agriculture issues food guides. Eight guides have released since 1916 [2], including the well-known "Food Guide Pyramid" published in 1992. The most recent food guide "MyPlate" issued in 2011 illustrates fruits, grains, vegetables, protein, and dairy as five food groups that are building blocks for a healthy diet using a familiar mealtime symbol. "My" in "MyPlate" emphasizes the personalization approach to finding a lifelong healthy, balanced eating style shaped by many factors and choices. Other popular diet and nutrition planning approaches exist including a reduced calorie diet, ketogenic diet, intermittent fasting, Whole30, and Paleo [3]. Some of these approaches require elimination of entire food groups that may cause serious nutrient deficiencies over time. The emergence of meal plan varieties and personal diet planning tools point to an increasing awareness that no universal diet plan fits all. AI has started to play a significant role in this field. Recent findings suggest the way we build models and collect data can push the edge of diet planning to be more personal than ever.

AI strengthens the ability of scientific studies in gathering, analyzing, interpreting, and eventually predicting the best diet plan a person needs to achieve a certain health goal. While we know a model is not a perfect description or a prediction of reality, we are getting closer to reality with data science, ML, and more thoughtful interpretation. For example, Habit is a company that looks at 70 + health markers and uses ML algorithms to inform users how their body handles macronutrients. Users learn what their ideal plate looks like and receive a personalized food guide and list of recipes. Passio, an ML company that uses image recognition to provide real-time on-device food recognition, enables users to have seamless food tracking and nutrition insights.

Currently, most diet planning service providers use AI technologies that quantify each participant who contributes to their research as a number in a dataset; the sample of one participant is likely to be evened out by thousands of other participants. Predictions from participant data will be made based on general results of the pros and cons of a food type to a group of people with a certain biomarker type. Data predictions cannot include every factor or measure some factors that may influence the user's actual reaction to different food types. Or, how a user's health condition changes over time, including lifestyle, medical conditions, immune system, anatomy, physiology, medications, and environment. It is possible that a typical diet plan would miss out on these factors. One example of a medical condition in a study shows a postmeal evaluation can be as important as premeal planning.

Research conducted by Weizmann Institute "Personalized Nutrition by Prediction of Glycemic Responses" [4] found people eating identical meals present high variability in postmeal blood glucose response. Personalized diets created with the help of an accurate predictor of blood glucose

From: Zeevi D, Korem T, Zmora N, et al. Personalized nutrition by prediction of glycemic responses, Cell 2015 163, 1079-1094DOI: (10.1016/j.cell.2015.11.001)

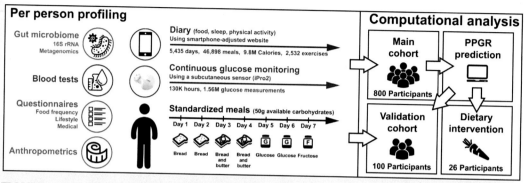

FIGURE 8.16 Profiling of postprandial glycemic responses, clinical data, and gut microbiome. (A) Illustration of our experimental design. Source: *From Zeevi D, Korem T, Zmora N, et al. Personalized nutrition by prediction of glycemic responses, Cell 2015;163:1079–94. doi:10.1016/j.cell.2015.11.001.*

response that integrate parameters such as dietary, habits, physical activity, and gut microbiota may successfully lower postmeal blood glucose and its long-term metabolic consequences. In this study, 800 healthy and prediabetic individuals were continuously monitored and responses measured to 46,898 meals. Participants were also measured with other blood parameters, anthropometrics, physical activity, self-reported lifestyle behaviors, and gut microbiota composition and function. The research group devised a machine-learning algorithm that accurately predicted personalized postprandial glycemic response to real-life meals. Weizmann Institute validated these predictions in an independent 100-person cohort. Finally, a blinded randomized controlled dietary intervention based on this algorithm resulted in significantly lower postprandial responses and consistent alterations to gut microbiota configuration (Fig. 1). A recent study "Assessment of a Personalized Approach to Predicting Postprandial Glycemic Responses to Food Among Individuals Without Diabetes" [5] conducted by Mayo Clinic echoed this finding.

In these studies, one person is no longer a single data in a dataset; they become the center of new data generated over time. By using a key indicator that reflects potential factors that may affect one person's unique relationship to a type of food, feeding this second set of user's individual data into the model after predictions based on a dataset from the population sample, diet planning for each user is differentiated. In other words the same usage of ML can generate a more personalized selection of diet components if we add a tier of algorithm modification with the user's personal response data over time.

Many AI-based diet planning services are trying to improve their algorithms by collecting more data from more users. However, for each individual, a person's unique connection to foods may be more effective in building diet planning. For example, Whisk's Culinary Coach uses AI to provide personalized food recommendations based on flavor preferences and food avoidance. Another example is Plant Jammer, a recipe-generating app, that uses AI to help users build personalized recipes and improve low kitchen confidence by pairing ingredients together based on factors (e.g., food likes/dislikes) and setting different filters (e.g., season and region). Although

AI is smart enough to perform calculations that humans cannot compete with, its utility highly depends on where it was built in a hypothesis. AI may give a better explanation about a view of how a mechanism runs, but it cannot create the view for us. In diet planning, volume and quality of data are critical, but another dimension to consider is data validity to the individual case. The time horizon of each data generator may be too valuable to ignore.

In 100 years, we have evolved from static food guides to automated nutrition planning. AI's role and value as a tool in the future of personalized nutrition and meal planning to help people live healthier and balanced lifestyles will increase.

References

[1] Margaret E. History behind 'An apple a day'. Washington Post. 2013. [Internet]. Available from: <https://www.washingtonpost.com/lifestyle/wellness/history-behind-an-apple-a-day/2013/09/24/aac3e79c-1f0e-11e3-94a2-6c66b668ea55_story.html?utm_term = .b4dfc41f6fae> [cited 11.03.19].
[2] USDA ChooseMyPlate. A brief history of USDA food guides. United States Department of Agriculture. [Internet]. Available from: <https://choosemyplate-prod.azureedge.net/sites/default/files/relatedresources/ABriefHistoryOfUSDAFood
Guides.pdf> [cited 11.03.19].
[3] Kotecki P. The most popular diets millennials want to try in 2019. Business Insider. 2019. [Internet]. Available from: <https://www.businessinsider.com/most-popular-diets-millennials-want-to-try-2019-2018-12> [cited 11.03.19].
[4] Zeevi D, Korem T, Zmora N, et al. Personalized nutrition by prediction of glycemic responses. Cell 2015;163 (5):1079—94. Available from: https://doi.org/10.1016/j.cell.2015.11.001.
[5] Mendes-Soares H, Raveh-Sadka T, Azulay S, et al. Assessment of a personalized approach to predicting postprandial glycemic responses to food among individuals without diabetes. JAMA Netw Open 2019;2(2): e188102. Available from: https://doi.org/10.1001/jamanetworkopen.2018.8102.

Present assessment and future strategy

Overall, the field of genomic medicine with its extension into precision medicine has increased its focus on AI as it embodies the principles of big data and integration of this data for knowledge discovery in a multiomics data integration strategy. The necessary complex work for all the disparate data to converge and for the prediction models to work creates a daunting challenge for the data scientist in this domain but concomitantly an exciting opportunity. For the future, there is optimism that precision medicine that is enabled by AI can lead to precision medicine from birth. This lifelong precision medicine continuum means that we can perhaps one day prevent diabetes starting from birth. In short, precision medicine as a paradigm for the future of medicine is attainable with full deployment of AI methodologies.

Physical Medicine and Rehabilitation

This is a subspecialty that focuses on physical rehabilitation of usually medical and surgical patients after procedures and/or with disabilities (temporary or permanent). This field is also known as physiatry and the physician who practices physiatry a physiatrist. Common diagnoses of patients seen by physiatrists include those with brain or spinal cord injuries, neuromuscular disorders, strokes, multiple sclerosis, burn injury, and others. The overarching goal of this sub specialist is to restore functional capacity of and quality of life of any individual in need of interventions to achieve that goal. A subspecialist in this domain needs to have working knowledge of not only musculoskeletal system but also other systems, such as neurological, circulatory, and rheumatological.

Published reviews and selected works

There are very few papers on the use of AI in this subspecialty. Barry reviewed the use of AI in this subspecialty and emphasized that adaptation, cooperation, and trust are at the center of rehabilitation and that AI and use of robots and other equipment can, therefore, enhance adaptation with guidance for movement as well as other elements such as cues for sensation and control of environment [215]. Applications for AI in physical medicine and rehabilitation include exoskeletons and neuroprosthetics; exercise and movement control with robots; telepresence, social robots, and smart environments (including smart homes). In addition, use of AI along with wearable technology and patient biometric measurements can improve efficiency and efficacy of musculoskeletal physiotherapy [216]. Other papers focused on the use of robotics in rehabilitation but will not be covered in detail here.

Present assessment and future strategy

Overall, there are some activities in this domain but limited to robotics but little in combining robotics and advanced AI methodologies or other emerging technologies. For the future, since AI-enabled robotics loom large in the coming decades, this subspecialty will see a large panoply of robots and equipment that will be made available for rehabilitation of patients with handicap and/or functional impairment. Additional technologies such as altered reality will also provide an additional innovative dimension to rehabilitation strategies.

Regenerative Medicine

This field of medicine, such as genomic medicine, can be an element in some if not most subspecialties. Regenerative medicine is the discipline of creating living and functional tissue to repair or replace the patient's own tissue (due to congenital defect, injury, or disease). Progress has been made in bone, soft tissue, and corneas, as well as organ transplantation but the future will bring many more possible therapies (such as tissue-engineered vascular grafts, stem and precursor cells for myocardial infarction, and even artificial pancreas, kidney, and even spinal cord).

Published reviews and selected works

There is a moderate amount of published works on AI in regenerative medicine. One review in pediatric cellular therapy and regenerative medicine discussed the use of predictive modeling for personalized treatment in children [217].

Present assessment and future strategy

Overall, there is some activity in the area of regenerative medicine and AI in that this is a necessary cornerstone of precision medicine for the future. Regenerative medicine can be enabled by AI as computational modeling in both cellular immunotherapy and genetic engineering in addition to regenerative medicine can yield major dividends in the decade to come for precision medicine. For the future, promises loom large for this branch of medicine and many improvements in diagnosis will necessarily lead to therapy. An AI-enabled strategy for regenerative medicine will complement precision medicine with individualized medical and surgical therapy from the converging sciences of *tissue engineering, 3D-printing, and AI as "organ printing." The AI strategy can be extended into the artificial organ such as pancreas and kidneys as AI will be a necessary coupling for not only its morphological genesis but also functional maintenance in the form of fuzzy logic and deep learning.*

Veterinary Medicine

This branch of veterinary medicine involves diagnosis and treatment of diseases and conditions that pertain to animals. A veterinarian often not only a medical doctor but also a surgeon for these animals. Some of these caretakers specialize in certain groups of animals or subareas such as surgery or dermatology just like human clinicians. Other than the obvious difference in patients, these doctors have the same qualifications and needs as their counterparts in pediatric or adult medicine.

Published reviews and selected works

There are a few published works on AI in veterinary medicine. One excellent and comprehensive review on medical decision support and ML methods delineated the importance of data and examined three of the common methods in detail [218]. Other than this comprehensive review, there are relatively few reports in veterinary medicine.

Present assessment and future strategy

Overall, much of the decision-making process in veterinary medicine and parallel the process in human medicine and much information and insight can be extracted from the medical literature. With more champions for this new paradigm, AI use in veterinary medicine can develop further and faster. For the future, most if not all aspects of AI applications in human medicine can be applied in veterinary medicine. In some ways, since animals are not able to communicate symptoms (as in pediatric patients), perhaps AI methodologies can be even more helpful in decision support and other clinical areas.

Medical Education and Training

Medical school education consists of a 4-year-curriculum that consists of 2 years of basic sciences prior to 2 years of subspecialty exposure. Medical education consists of preclinical and clinical parts, with the former involving courses such as anatomy, physiology, biochemistry, pharmacology, and pathology and the later rotations such as surgery, internal medicine, pediatrics, and radiology. After one graduates from medical school, one enters a residency program for subspecialty training and further on to a fellowship program for some areas such as cardiology, critical care medicine, or specialized surgical subspecialties. The length of residency and fellowship training after graduation from medical school depends on the subspecialty; it can be as short as 3 years for primary care (pediatrics or family practice) to as long as seven or more years for (1) some surgical subspecialties such as neurosurgery or cardiothoracic surgery; (2) some medical subspecialties that require additional fellowship training (such as pediatric cardiology); or (3) someone who desires multiple subspecialty board certification (such as cardiology and intensive care). After their specialized training the typical physician is required to take a board examination to be certified in that particular subspecialty and then is required to take a recertification examination every 5–10 years. Physicians attain board certification and recertification as well as continue their education with continuing medical education. In the United States the Association of American Medical Colleges leads the academic medicine effort for the education of medical school students. With the exponentially increasing medical knowledge, the doubling time for medical knowledge has substantially shortened to the point where the average clinician does not have enough time to keep up with knowledge even in his/her own field.

A future child neurologist's perspective

Adam Z. Kalawi[1]

Adam Z. Kalawi, a soon-to-be pediatric resident and neurology fellow as well as a data science enthusiast, authored this commentary to elucidate the importance of data science and artificial intelligence early in a physician's medical education and clinical training.

[1]University of California San Diego, Rady Children's Hospital, San Diego, CA, United States

Introduction

We are born into this world as nearly blank slates, programmed with primitive reflexes centered around the ability to feed and the most basic abilities of self-preservation. Through the decades of complex environmental and interpersonal interactions, we have the potential to develop into organisms with such complexity of thought that we are not only able to study our own consciousness, but also have the audacity to program machines to learn and further guide our collective pursuit of truth. The statements discussed previously regarding human development and the pursuit of AI encapsulate the underlying motivations behind my life's biggest passions—the former serving as motivation for my clinical study of the developing human brain and the latter serving as motivation for my role as a technology and data science enthusiast.

Through serendipity and great fortune I have come to know Dr. Anthony C. Chang as a mentor and friend for the better part of a decade, and I was filled with joy for the AI community when he announced that he would be publishing a comprehensive book on *Artificial Intelligence in Medicine: Principles and Applications*. With so much momentum in the AI community right now, this book aims to crystalize and capture the energy of this era to serve as a unifying guide for the coming decades of change. Never before have data been so readily available in such large volumes, never before has computing been so powerful, and never before have so many clinicians been so actively present at the forefront of AI. The future for AI in medicine is bright.

Applications of artificial intelligence in neurology

In bridging my two passions, it is always amazing to see how much the fields of AI and neurology overlap. Many ML algorithms use processes completely foreign to human learning, however some—such as CNNs—have processes that mirror the layered cortical networks that occurs within the human brain. In a full circle story, these and other ML algorithms have the potential to improve the diagnosis and treatment of an array of neurologic disorders, some of which I will review later.

Neonatal seizure detection

In many parts of the world, amplitude-integrated electroencephalography has been substituted in the place of formal routine encephalography (rEEG) and video encephalography in community-based NICUs for the monitoring of brainwave activity in premature infants. Though this option for brainwave monitoring and seizure-detection is more simple than its alternatives due to its easier application and bedside interpretability, studies have found that it is prone to artifacts and suggest that it is less reliable than rEEG with a neurologist's interpretation [1,2]. Advances in rEEG signal analysis and seizure detection with AI will prove invaluable in bringing a higher SOC to some of the most vulnerable patients in our communities and decrease time to seizure detection for neonatologists.

Autism diagnosis and screening

It was announced in February 2019 that Cognoa devices for autism screening obtained US FDA breakthrough device designation for its AI driven autism screening technology [3]. I was excited to hear this excellent news from my colleagues and friends and feel that this is a large step forward for the AI in medicine community in demonstrating proof of concept for operationalization within the US health-care system. This technology applies ML algorithms to analyze parental input and videos of children to help predict children that warrant further autism diagnostic testing. This approach enables earlier detection of autism, and, in turn, earlier access to ABA therapy which is associated with improved long-term outcomes for patients with autism.

Neuroimaging and neuropathology

In this most recent wave of enthusiasm for AI in medicine analysis of imaging and tissue pathology have been viewed as low hanging fruit for ML algorithms. In fact, some authors posit that radiology and pathology are becoming data science driven fields [4]. Indeed, studies have demonstrated that adult chest radiography can be triaged with AI systems based on deep CNNs and that this method is both operational feasible and technically able to reduce delays in critical imaging findings [5]. Similar success and promise has been demonstrated in the imaging analysis of breast tissue pathology for cancer diagnosis [6]. As these technologies mature, I am optimistic that they will aid in the more complex imaging (MRI) and tissue diagnosis of many neurologic conditions such as pediatric brain tumors. As personalized therapeutics become more advanced, AI techniques may prove beneficial in genomic analysis of tumor cells for targeted therapies.

Closing words

I look forward to joining you on the forefront of this clinical journey of AI in medicine and congratulate you for owning this book, because in doing so you are identifying yourself as part of the AI-aware generation of clinicians and medical data scientists that will bring about massive breakthroughs in patient safety, workflow efficiency, diagnostic accuracy, and medical advancement. If I may leave you with some closing wisdom before you return to your studies, it is this: mentorship will be critical in advancing AI in medicine—involve your young trainees and you may find that one day, the mentorship goes both ways.

References

[1] Glass HC, et al. Amplitude-integrated electro-encephalography: the child neurologist's perspective. J Child Neurol 2013;28(10):1342–50. Available from: https://doi.org/10.1177/0883073813488663. Available from: https://www.ncbi.nlm.nih.gov/pmc/articles/PMC4091988/.

[2] Suk D, et al. Amplitude-integrated electroencephalography in the NICU: frequent artifacts in premature infants may limit its utility as a monitoring device. Pediatrics 2009;123(2). Available from: https://doi.org/10.1542/peds.2008-2850. Available from: https://pediatrics.aappublications.org/content/123/2/e328?download = true.

[3] Ndivya. Cognoa autism devices obtain FDA breakthrough status. Verdict Med Devices 2019;8. Available from: www.medicaldevice-network.com/news/cognoa-autism-devices/. Available from: https://www.medicaldevice-network.com/news/cognoa-autism-devices/.

[4] Jha S, Topol EJ. Adapting to artificial intelligence. JAMA 2016;316(22):2353. Available from: https://doi.org/10.1001/jama.2016.17438. Available from: https://jamanetwork.com/journals/jama/article-abstract/2588764.

[5] Annarumma M, et al. Automated triaging of adult chest radiographs with deep artificial neural networks. Radiology 2019;291(1):272. Available from: https://doi.org/10.1148/radiol.2019194005. Available from: https://pubs.rsna.org/doi/10.1148/radiol.2018180921.

[6] Robertson S, et al. Digital image analysis in breast pathology—from image processing techniques to artificial intelligence. Transl Res 2018;194:19–35. Available from: https://doi.org/10.1016/j.trsl.2017.10.010. Available from: https://www.sciencedirect.com/science/article/pii/S1931524417302955.

The artificial intelligence (AI) and digital disruption of surgical knowledge and skill

Todd A. Ponsky[1]

Todd A. Ponsky, a pediatric surgeon who is always envisioning innovations in medicine, authored this commentary on the many possible innovations in medical and surgical education and training that can be enabled with AI.

[1]Clinical Growth and Transformation, Cincinnati Children's Hospital, Cincinnati, OH, United States

Surgeons should start preparing for digital disruption. Medical knowledge is growing exponentially. There are approximately 2.5 million scientific publications per year and it is anticipated that the number of medical publications will double every 73 days by the year 2020 [1,2]. Furthermore, new surgical technology and techniques are also growing at an unprecedented pace. It is becoming impossible to stay current. Traditional methods of staying current in surgery included textbooks, journals, and medical society meetings. Newer digital trends are gradually disrupting these methods of knowledge sharing and skills transfer and the future will likely involve ML, crowd sourcing, digital learning platforms, and virtual mentoring.

One limitation to staying current in surgery is the exponential growth of new publications and the lack of a good filter for quality content. ML may help apply NLP to scientific publication databases to help identify high quality, important, publications that surgeons should pay attention to. Although journal editors seem "wise," the wisdom can likely be distilled down to the recognition of key elements that could be taught to NLP algorithms such as high power, study design, and topic relevance. Another tool that will likely help us tease out the best articles from the abyss of publications will be crowd sourcing. Utilizing the wisdom of the crowd, the articles of greatest interest that have been viewed or "liked" the most can rise to the top of search engines similar to YouTube or Google.

Traditional methods of learning new information in surgery are national meetings, textbooks, and journals. Meetings are becoming too expensive and time away from work is becoming less practical. Textbooks are 5 years old by the time they are published and the information is static until a new edition is published. This is no longer practical in the era of exponentially changing medical knowledge. With the advent of digital publishing, journals are no longer limited by paper and have drastically increased the number of articles published each year. This, along with the explosion of predatory, open access, journals has created an overwhelming litany of new, low quality publications making it near impossible to stay current on important information. The solution will be novel platforms which curate content, apply ML filters to the content, apply crowd sourcing to the content, and provide rich, micro-multimedia, digestible, shareable, content in a mobile device that can always be available at a surgeon's fingertips. These platforms must provide the ability for surgeons to easily "pull" just-in-time information and also receive "push" notifications of new, important, need-to-know information.

Finally, it is likely that virtual presence and ML will help practicing surgeons learn new skills and stay current in evolving techniques. Practicing surgeons are encountering a never-before seen exponential growth in surgical technology and techniques that often are developed after they have completes residency training. It is challenging for surgeons to be trained in these new skills during their busy clinical practice. The introduction of "surgical telementoring" can now allow experts at one institution to train other surgeons at remote locations through virtual presence technologies [3]. Through these platforms the expert surgeon can watch the operation remotely and guide the learning surgeon through verbal guidance and on-screen telestration (Fig. 1). However, the real

FIGURE 1 Surgical telementoring. On the right, a surgeon is set up to mentor another surgeon in the operating room. On the left, the mentoring surgeon is using an on-screen marking pen to illustrate anatomic details for the operation. In the future, deep learning and cognition can take place of the mentoring surgeon in certain cases.

III. The current era of artificial intelligence in medicine

needle-movement will come when real time, ML, and image recognition can be used to predict false moves and complications and help with hazard prevention. Similar to a driver who is driving a self-driving car that warns the driver about lane departures, potential blind spot crashes, or head on collisions, a surgeon could be warned that they are approaching a major blood vessel.

The new age of exponential growth brings new problems to the surgical world but these can likely be addressed by applying digital solutions such as ML and novel platforms.

Here one surgeon is telementoring another surgeon through a lung resection in an infant and using live telestration to point out the anatomy.

References

[1] The STM report: an overview of scientific and scholarly journal publishing. 2015 STM: International Association of Scientific, Technical and Medical Publishers Fourth Edition published March 2015; updated with minor revisions November 2015. Published by International Association of Scientific, Technical and Medical Publishers Prins Willem Alexanderhof 5, The Hague, 2595BE, The Netherlands. <https://www.stm-assoc.org/2015_02_20_STM_Report_2015.pdf>.

[2] Densen P. Challenges and opportunities facing medical education. Trans Am Clin Climatol Assoc 2011;122:48—58.

[3] Ponsky TA, Schwachter M, Parry J, Rothenberg S, Augestad KM. Telementoring: the surgical tool of the future. Eur J Pediatr Surg 2014;24(4):287—94.

Published reviews and selected works

It is more than 100 years since the Flexner report that shaped our present medical school education strategy, and it is now more important than ever to reassess our medical educational strategy. A report by Wartman and Combs emphasized the timeliness of AI in medical education and its role in the future of clinical work [219]. In addition, some subspecialties have also discussed the need for AI education also during clinical training as residents [220] as well as the use of virtual reality and AI in medical education [221]. On the other hand, a review showed that AI can also be effectively utilized for assessing physician competence at different levels; surgeons and radiologists seem to be the subspecialties that used this strategy the most [222]. Lastly, Johnston astutely pointed out that the training of the physicians of the future needs both information technology and analytics knowledge but also the humanistic aspects of medicine (the art of caring) is more essential than ever before [223].

Present assessment and future strategy

Overall, there is very little education of data science or AI in medical education or training; in addition, much more use of AI and altered reality would also be ideal for both education and training (such as use of altered reality for virtual dissection of anatomy). For the future, medical education and training is ripe for a major disruption to maintain pace with the exponentially increasing medical knowledge as well as the rapid rise in modern technologies. The altered reality technologies with gaming and deep reinforcement learning can radically change the medical education experience as well as the clinical learning and training effectiveness. The advent of AI is a precious gift from our technological colleagues, and while AI is not necessarily going to replace clinicians, it should be part of every medical student's educational curriculum as well as every physician's clinical portfolio from this point forward.

Nursing

Nurses carry the main burden of health-care delivery and range from bedside nursing to outpatient nursing, and even home nursing. A nurse practitioner has obtained additional training and education and is able to write orders and serve as a physician's partner. A physician's assistant is a similar partner to the physician but may have additional capability to perform clinical and procedural tasks.

Published reviews and selected works

The academic interest in AI for nursing is fairly robust. One paper details the potential of augmented intelligence in nursing [224]. This review of AI focused on cognitive computing and IBM Watson but also discussed other AI tools in use. The essence of the review is on AI as a resource to place the locus of meaning back at home and with the patient. Another report reviews applying AI technology to support decision-making in nursing [225].

Present assessment and future strategy

Overall, the interest in AI from the nursing colleagues is relatively high. One potential reason is the number of challenges that nurses face on a day-to-day basis can have AI-enabled solutions. For the future the domain of robotics and robotic assistance can be very robust in the future for nursing. In addition, the domain of chronic disease management and the possible role of virtual assistants will also be valuable for nursing care.

Healthcare Administration

Like medical education and training, there is very little use of AI in hospital administration with the exception of some analytics (see earlier section). The typical hospital administrator is usually not enlightened on aspects of AI with the exception of an occasional CIO or CMIO who has had experience working with a data science team.

Published reviews and selected works

There is a paucity of published works on the use of current AI tools in health-care administration. One review of AI in health-care delivery has a section on health-care administration [226]. This review and section acknowledges that the complex nature of healthcare with its administrative burdens and resource constraints are in need of AI tools. New approaches such as transfer learning, contextual analysis, knowledge injection, and distillation can be proposed to mitigate the health-care imbroglio that has persisted and escalated the past few decades. In addition, a report discussed the use of ML-enabled complex high-dimensional models to assess and predict hospital attendance and found that gradient boosting machine-based models were the most accurate [227]. In addition, there are many applications for RPA, which are intelligent software robots that can automate most repetitive tasks. These include processes such as physician credentialing, enrollment and patient eligibility, coding, claims administration, accounts receivable, and secondary claims management.

Business intelligence (BI) in medicine and healthcare

Jiban Khuntia and Xue Ning

Jiban Khuntia specializes in health information technology and his colleague Xue Ning, authored this commentary on using artificial intelligence more effectively for BI in the health care arena by providing more accurate and timely insight.

Business intelligence (BI) refers to a process of data analysis and information generation through strategies and technologies. BI has emerged as an essential strategic imperative in many sectors. BI is helpful for managers in enterprises to make informed business decisions in various levels and domains. BI technologies can handle large structured and unstructured (big) data in the health-care industry.

Medicine and healthcare have been in the forefronts of applying data and analytics in many areas, such as evidence-based science, clinical trials, and disease surveillance. However, top organizational level strategic decisions based on BI is a niche area in medicine and health-care industry. Some entities have moved toward this frontier, in various subsets of the health-care industry, such as hospitals and health insurance organizations.

The scope of BI in health-care medicine involves at least five aspects. First, recognize and understand how BI tools and applications can be used to make decisions. Second, apply different mining, modeling, and analytics techniques to medicine and health-care data. Third, articulate the value of significant volumes of data to medicine and healthcare and take effective decisions. Fourth, stay in the forefront of acquiring and evaluating different BI tools, generate application ideas, and explore emerging use of BI in different areas of healthcare, starting from quality management to patient empowerment to population health. Finally, develop a workforce who are not only eager to use BI in healthcare but should also be able to critically analyze different aspects of the BI in day-to-day use.

Supporting and improving decision-making: Medicine and healthcare, by nature, has been a data- and knowledge-centric discipline. With the advent of information technology and subsequent applications in recent years, a plethora of data collection opportunities are available. Data are growing day by day. Data sources are becoming abundant. Complex data generated in organizations should not just remain idle, but the value proposition of generated data will increase by using the data for better decision-making. Such a value proposition necessitates advanced analytics and intelligence to support decision-making capabilities is growing than before. BI tools such as predictive modeling, data visualization, and dashboards can inform hospitals and other health-care organizations regarding patient care, workforce distribution, clinical operations, daily practices of physician and nurses, and administration and management. Most important outcomes may be improvements in patient satisfaction through improved care while keeping the cost intact.

Issues related to operations, equipment and facilities, diagnostic and care are abundant in healthcare. Efficient and effective coordination across practices, units, and external agencies are playing significant roles in delivering better care. Health-care organizations, with the support of BI, can help to navigate the complexities of health-care governance. In addition, advanced analytical capabilities are necessary for solving key challenges and connecting patient, clinical, and operational data. Administrators can track performance indicators to analyze, manage, and help health-care organizations. Data and analytics based insights can help in operations to workflow improvement, and align strategic goals to the organization. Varied data sources can obtain patient-related insights, and connected data and ideas can assist in providing the best care.

Need of BI to improve patient care: Health-care context and health-care data, both, are involved. The decision taken in this context in one way has a simple objective, that is, to provide better care. However, the strategy to leverage from data for better care remains a significant shortcoming in healthcare. As much as healthcare is increasingly adopting digital technologies to support operations and workflow, the BI for healthcare would provide the critical avenues to manage the massive amounts of both structured and unstructured data, the health-care organizations deal in a daily basis. Embedding intelligence tools in the everyday practices of physicians, administrators, and other personnel in healthcare will enable them to make more informed decisions—that in turn will help in effective patient care.

Need for BI for cost-management in healthcare: A widely known issue in healthcare is the high-cost problem, that is, highly increasing cost against moderate quality and efficiency of care. BI platforms can increase organizational exchange and use of data to address a variety of cost-challenges involved in operations, finance management, care regime and decisions, and clinical practices.

When health-care professionals are trained to follow data-analysis-decision path, they will be able to apply deep and profound insights in all the fields of healthcare. They can address risks, forecast some future events, avoid errors, and take appropriate precautions. Then, the successful application of analytics-based insights is increased. Health-care organizations can provide better levels of patient and clinical care based on well-analyzed data. Subsequently, the outcome will be improving personnel distribution, decreasing readmissions, and managing expenses using the smart application of BI in healthcare. Also, using BI applications in medicine and healthcare is not only part of the evolving and emerging health-care industry, but a movement toward sharing and establishing evidence-based best practices across all dimensions of care and management.

Takeaways: Because of the complex nature of health-care data and the significant impacts of healthcare data analysis, it is essential for health-care professionals to understand the role of BI in medicine and healthcare. BI in health-care provides support to the process from data, information, analysis, to decision-making. It is critical for the path from operations to intelligence generation, and then to the strategy development. The possible benefit of BI spans from practical outcomes such as patient safety, readmissions, patient satisfaction, wait times, scheduling issues, hospital stays, cost management, to informed strategic planning and managing population health.

As a new area in the developing stage, BI in healthcare is facing both opportunities and challenges. More professionals are needed in this area. Nonetheless, the established and emerging usage of BI in other industries should motivate the health-care sector to be proactive toward assimilation of BI and look forward to future trends with an open mindset.

Superpowered operational decisions in the age of artificial intelligence (AI)

Benjamin Fine[1,2,3,4]

Benjamin Fine, an engineer-turned-radiologist who is an expert in improvement science and machine learning, authored this commentary on the concept of an AI-enabled health care manager to monitor real-time health system data to execute the best course of action.

[1]Quality and Informatics, Diagnostic Imaging, Trillium Health Partners, Mississauga, ON, Canada

[2]Department of Medical Imaging, University of Toronto, Toronto, ON, Canada

[3]Institute for Better Health, Santa Rosa, CA, United States

[4]Centre for Healthcare Engineering, University of Toronto, Toronto, ON, Canada

(Article inspired by the late Doris Gorthy and her passionate pursuit of a data-driven health-care system.)

Management guru Peter Drucker described a hospital as the most complex system humans have ever attempted to manage. One on hand, it has a single compelling mission to unite its unwaveringly caring people: provide excellent care for each patient it sees. Yet, coordinating the orchestra of specialized staff, equipment, and facilities to deliver the right care to the right patient at the right time and in the right place at the lowest possible cost is elusive. The net result is that healthcare is eating GDP—hospitals alone eat over a 1/3 of health-care spending [1]—while delivering, at best, variable quality of care, and patient experience.

To understand why this is the situation requires peering deep into every day hospital decision-making. Consider the following scenario:

> Doris is the manager of a hospital diagnostic imaging department. Her oncology referring providers are complaining patients are waiting too long for CT scans. Her emergency department is complaining about long turn around times on their CTs. Her CEO wants to have the best access for outpatient imaging in town. Some days CT seems overbooked; others it seems there is underutilized equipment. How does Doris best allocate her CT resources to maximize throughput, balance wait time while maintaining staff satisfaction at the lowest possible cost. Oh, and she has 5 minutes to solve this problem before the next fire lands on her desk.

Can we empower Doris to harness available data to deliver better care. In 2019 the answer is yes.

Superpowering operational decision-making

Superpowering managers such as Doris with machines to enable better operational decisions is one of the greatest opportunities for data science to improve health-care delivery. This is accomplished by letting modern algorithms loose on health system big data powered on cheap low cost computing infrastructure. The machines provide two important inputs to decision-making. First an accurate "prediction" of demand at a granular level (how many CTs of which type will be needed for all my patient types) and second "prescription" of which management decision will yield desired outcomes (how should I allocate my CT time for different test types to minimize overall wait time). These improved predictions and prescriptions are now possible because of exponential increases in data, computing power, and accompanying algorithmic improvement in the last decade.

Enabling modern hospital operations data science

Instead of understanding problems with observational studies, manual data collection or chart reviews, exponentially increasing machine-readable operational data—rich with patient, provider, equipment information and associated time-stamped activities—is available in near real time from now-prevalent hospital EMRs, IoT, and other sources. Algorithms have advanced rapidly that can handle complex health-care data; ones that assume, force, or require knowledge of an underlying data model to facilitate calculation—such as the Poisson distributions of queuing theory, best-fit/guess equations representing objectives in optimization—can be replaced or augmented by black box ML models that learn to approximate the underlying system (without a priori knowledge) and provide the prediction necessary to inform better operational decisions. Lastly, instead of simplifying algorithms, limiting data to make solutions computationally tractable, exponentially increasing computing power (2000s era supercomputer power in a $3000 2019 GPU) means large volumes of data can be analyzed in near real time. This means better, cheaper predictions [2], more insights, faster, and more frequently.

An example: "precision" hospital care delivery

Consider the flow of patients across the hospital from emergency department through various wards, ICU, imaging tests, through to discharge home or another facility. Each patient encounter with the health system now leaves a digital footprint—time stamps, patient measurement, a medical report—that can be mined to gather deep understanding of the "personalized" care of each patient, that can fuel future learning. Combine this operational data with socioeconomic, geospatial, weather, traffic, etc., and you can start to imagine possibilities along the course of care. Predicting emergency department arrivals—even by patient type, diagnosis—can allow for adequate staffing, which can even be simulated at the beginning of each shift [3]. Predicting admission at ED arrival would allow preparation of nursing and ward bed allocation [4]. Same could be said for ICU admission [5] or imaging use. While currently resource planning for, say, "stroke" patients is uniform, mining EHRs for features that predict length of stay would find clusters of patients who may benefit from a targeted intervention to facilitate early discharge [6]. Outcomes of decisions can be tracked in a "learning" ideal care system [7]. Applying data science techniques to enable predictions and better operational decisions along the patient journey can decrease costs (from idle or overused equipment and staff), patient wait times, and improve the patient experience.

Looking forward

Superpowered operational decision-making is in fact already happening. Some of the problems described here have been solved by commercial products available today. Others are described in research journals. But we are still early along the adoption curve. There is massive return on investment here by deploying expensive specialized health-care resources precisely. And massive opportunity by delivering improved value for patients. My prediction is that this is where more and more health-care leaders first apply predictive and prescriptive techniques to superpower operational decision-making and improve health-care delivery.

As Drucker also opined, hospitals must coordinate their information amongst various specialists to deliver care with precision akin to a symphony. The alignment of care resources should, ideally, appear just where the patient needs them (right care, right place, and right time). This is a very complicated symphony—and one that can be enabled by a data-driven "AI conductor" assisting people like Doris to deliver each patient precisely the care they need, when they need it (Fig. 1).

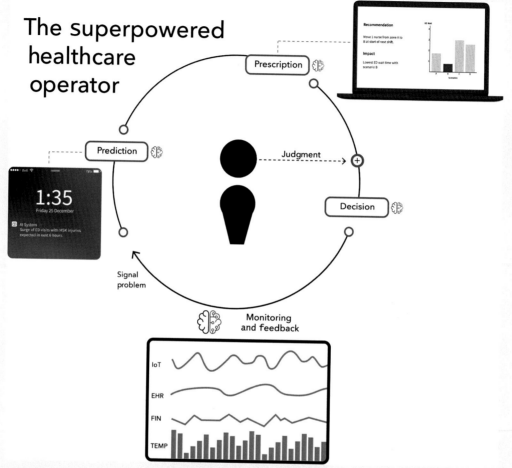

FIGURE 1   A superpowered health-care manager of the future. The "AI" system will monitor real-time health system data, predict deviations from normal requiring operator action, delivering suggested pre-scribed actions, and monitoring the impact of those changes to feed ongoing learning. In this example, icy conditions at night predict a surge of falls and suggests an appropriate shift of ED resources in response. *AI*, Artificial intelligence.

References

[1] Papanicolas I, Woskie LR, Jha AK. Health care spending in the United States and other high-income countries. JAMA 2018;319(10):1024–39.

[2] Agrawal A. Prediction machines: the simple economics of artificial intelligence. Boston, MA: Harvard Bus Review Press; 2018.

[3] Chan CW, Sarhangian V, editors. Dynamic server assignment in multiclass queues with shifts, with Application to Nurse Staffing in Emergency Departments. 2017.

[4] Hong WS, Haimovich AD, Taylor RA. Predicting hospital admission at emergency department triage using machine learning. PLoS One 2018;13(7):e0201016.

[5] Wellner B, Grand J, Canzone E, Coarr M, Brady PW, Simmons J, et al. Predicting unplanned transfers to the intensive care unit: a machine learning approach leveraging diverse clinical elements. JMIR Med Inf 2017;5(4):e45.

[6] El-Darzi E, Abbi R, Vasilakis C, Gorunescu F, Gorunescu M, Millard P. Length of stay-based clustering methods for patient grouping, vol. 189. 2009. p. 39−56.

[7] Stone D. Secondary analysis of electronic health records. Cham: Springer; 2016.

Present assessment and future strategy

Overall, there is much improvement that can be done in the administrative aspects of the hospital with deployment of AI. One main resource is the use of AI in the form of RPA to increase operational efficiency. For the future, AI in a hospital should be the formation of an embedded intelligence unit that encompasses all aspects of data analytics at a cognitive level and in real-time mode. There will no longer be a necessary difference between BI and clinical analytics as both will be part of the intelligence in the hospital.

References

[1] Ching T, Himmelstein DS, Beaulieu-Jones BK, et al. Opportunities and obstacles for deep learning in biology and medicine. J R Soc Interface 2018;15:20170387.

[2] Connor CW. Artificial intelligence and machine learning in anesthesiology. Anesthesiology 2019;131.

[3] Alexander JC, Joshi GP. Anesthesiology, automation, and artificial intelligence. Proc (Bayl Univ Med Cent) 2018;31(1):117−19.

[4] Gambus P, Shafer SL. Artificial intelligence for everyone. Anesthesiology 2018;128:431−3.

[5] Mathur P, Burns ML. Artificial intelligence in critical care. Int Anesth Clin 2019;57(2):89−102.

[6] Patriarca R, Falegnami A, Bilotta F. Embracing simplexity: the role of artificial intelligence in peri-procedural medical safety. Expert Rev Med Devices 2019;16(2):77−9.

[7] Corey KM, Kashyap S, Lorenzi E, et al. Development and validation of machine learning models to identify high-risk surgical patients using automatically curated electronic health record data (Pythia): a retrospective, single-site study. PLoS Med 2018;15(11):e1002701.

[8] Lee HC, Ryu HG, Chung EJ, et al. Prediction of bispectral index during target-controlled infusion of propofol and remifentanil: a deep learning approach. Anesthesiology 2018;128:492−501.

[9] Weintraub WS, Fahed AC, Rumsfeld JS. Translational medicine in the era of big data and machine learning. Circ Res 2018;123:1202−4.

[10] Bonderman D. Artificial intelligence in cardiology. Cent Eur J Med 2017;866−8.

[11] Johnson KW, et al. Artificial intelligence in cardiology. J Am Coll Cardiol 2018;71(23):2668−79.

[12] Benjamins JW, Hendriks T, Knuuti J, et al. A primer in artificial intelligence in cardiovascular medicine. Neth Heart J 2019;27:392−402.

[13] Shameer K, Johnson KW, Glicksberg BS, et al. Machine learning in cardiovascular medicine: are we there yet? Heart 2018;104:1156−64.

[14] Krittanawong C, et al. Artificial intelligence in precision cardiovascular medicine. J Am Coll Cardiol 2017;69 (21):2657−64.

[15] Bizopoulos P, Koutsouris D. Deep learning in cardiology. IEEE Rev Biomed Eng 2018;12:168−93.

[16] Dey D, Slomka PJ, Leeson P, et al. Artificial intelligence in cardiovascular imaging: JACC state-of-the-art review. J Am Coll Cardiol 2019;73(11):1317−35.

[17] Slomka PJ, Dey D, Sitek A, et al. Cardiac imaging: working towards fully-automated machine analysis and interpretation. Expert Rev Med Devices 2017;14(3):197−212.

[18] Alsharqi M, Woodward WJ, Mumith JA, et al. Artificial intelligence and echocardiography. Echo Res Pract 2018;5(4):R115−25.

[19] Al'Aref SJ, Anchouche K, Singh G, et al. Clinical applications of machine learning in cardiovascular disease and its relevance to cardiac imaging. Eur Soc Cardiol 2018;40:1975−86.

[20] Massalha S, Clarkin O, Thornhill R, et al. Decision support tools, systems, and artificial intelligence in cardiac imaging. Can J Cardiol 2018;34:827−38.

[21] Dudchenko A, Kopanitsa G. Decision support systems in cardiology: a systematic review. Stud Health Technol Inform 2017;237:209−14.

[22] Kadi I, Idri A, Fernandez-Aleman JL. Knowledge discovery in cardiology: a systematic literature review. Int J Med Inform 2017;97:12−32.

[23] Rajpurkar P, Hannun AY, Haghpanahi M, et al. Cardiologist-level arrhythmia detection with convolutional neural networks. arXiv:1707.01836v1[cs.CV]; 2017.

[24] Attia ZI, Kapa S, Lopez-Jimenez F, et al. Screening for cardiac contractile dysfunction using an artificial intelligence-enabled electrocardiogram. Nat Med 2019;25:70−4.

[25] Gandhi S, Mosleh W, Shen J, et al. Automation, machine learning, and artificial intelligence in echocardiography: a brave new world. Echocardiography 2018;35:1402−18.

[26] Zhang J, Gajjala S, Agrawal P, et al. Fully automated echocardiogram interpretation in clinical practice: feasibility and diagnostic accuracy. Circulation 2018;138:1623−35.

[27] Ostvik A, Smistad E, Aase SA, et al. Real-time standard view classification in transthoracic echocardiography using convolutional neural networks. Ultrasound Med Biol 2019;45(2):374−84.

[28] Shah SJ, Katz DH, Selvaraj S, et al. Phenomapping for novel classification of heart failure with preserved ejection fraction. Circulation 2015;131(3):269−79.

[29] Naurla S, et al. Machine-learning algorithms to automate morphological and functional assessments in 2D echocardiography. J Am Coll Cardiol 2016;68:2287−95.

[30] Sengupta P, et al. A cognitive machine learning algorithm for cardiac imaging: a pilot study for differentiating constrictive pericarditis from restrictive cardiomyopathy. Circ Cardiovasc Imaging 2016;9(6) https://doi.org/10.111/10.1161.

[31] Nath C, Albaghdadi MS, Jonnalagadda SR. A natural language processing tool for large-scale data extraction from echocardiography reports. PLoS One 2016;11(4):e0153749.

[32] Dawes TJW, de Marvao A, Shi W, et al. Machine learning of three-dimensional right ventricular motion enables outcome prediction in pulmonary hypertension: a cardiac MR imaging study. Radiology 2017;283(2):381−90.

[33] Poplin R, Varadarajan AV, Blumer K, et al. Prediction of cardiovascular risk factors from retinal fundus photographs via deep learning. Nat Biomed Eng 2018;2:158−64.

[34] Ryan A, Duignan S, Kenny D, et al. Decision making in pediatric cardiology: are we prone to heuristics, biases, and traps? Pediatr Cardiol 2018;39(1):160−7.

[35] Li R, Hu B, Liu F, et al. Detection of bleeding events in electronic health record notes using convolutional neural network models enhanced with recurrent neural network autoencoders: deep learning approach. JMIR Med Inform 2019;7(1):e10788.

[36] Kwon JM, Lee Y, Lee Y, et al. An algorithm based on deep learning for predicting in-hospital cardiac arrest. J Am Heart Assoc 2018;7:e008678.

[37] Choi E, Schuetz A, Stewart WF, et al. Using recurrent neural network models for early detection of heart failure onset. AMIA 2017;24(2):361−70.

[38] Kakadiaris IA, Vrigkas M, Yen AA, et al. Machine learning outperforms ACC/AHA CVD risk calculator in MESA. J Am Heart Assoc 2018;7:e009476.

[39] Warner HR, Toronto AF, Veasey LG, et al. A mathematical approach to medical diagnosis: application to congenital heart disease. JAMA 1961;177(3):177−83.

[40] Chang AC. Primary prevention of sudden cardiac death of the young athlete: the controversy about the screening electrocardiogram and its innovative artificial intelligence solution. Pediatr Cardiol 2012;33:428−33.

[41] Thompson WR, Reinisch AJ, Unterberger MJ, et al. Artificial intelligence-assisted auscultation of heart murmurs: validation by virtual clinical trial. Pediatr Cardiol 2018;40.

[42] Diller GP, Kempny A, Babu-Narayan SV, et al. Machine learning algorithms estimating prognosis and guiding therapy in adult congenital heart disease: data from a single tertiary centre including 10,019 patients. Eur Heart J 2019;40.

[43] Diller GP, Babu-Narayan SV, Li W, et al. Utility of machine learning algorithms in assessing patients with a systemic right ventricle. Eur Heart J 2018;20.

[44] Olive MK, Owens GE. Current monitoring and innovative predictive modeling to improve care in the pediatric cardiac intensive care unit. Transl Pediatr 2018;7(2):120—8.

[45] Ruiz-Fernandez D, Torra AM, Soriano-Paya A, et al. Aid decision algorithms to estimate the risk in congenital heart surgery. Comput Methods Prog Biomed 2016;126:118—27.

[46] Wolf MJ, Lee EK, Nicolson SC, et al. Rationale and methodology of a collaborative learning project in congenital cardiac care. Am Heart J 2016;174:129—37.

[47] Hanson WC, Marshall BE. Artificial intelligence applications in the intensive care unit. Crit Care Med 2001;29(2):427—35.

[48] Lynn LA. Artificial intelligence systems for complex decision-making in acute care medicine: a review. Patient Saf Surg 2019;13:6—14.

[49] Ghassemi M, Celi LA, Stone DJ. State of the art review: the data revolution in critical care. Crit Care 2015;19:118—27.

[50] Johnson AEW, Ghassemi MM, Nemati S, et al. Machine learning and decision support in critical care. Proc IEEE Inst Electr Electron Eng 2016;104(2):444—66.

[51] Hravnak M, Chen L, Dubrawski A, et al. Real alerts and artifact classification in archived multi-signal vital sign monitoring data: implications for mining big data. J Clin Monit Comput 2016;30(6):875—88.

[52] Johnson AE, Pollard TJ, Shen L, et al. MIMIC-III, a freely accessible critical care database. Sci Data 2016;3:160036.

[53] Personal communication with Dr. Leo Celi.

[54] Komorowski M, Celi LA, Badawi O, et al. The artificial intelligence clinician learns optimal treatment strategies for sepsis in intensive care. Nat Med 2018;24(11):1716—20.

[55] Marafino BJ, Park M, Davies JM, et al. Validation of prediction models for critical care outcomes using natural language processing of electronic health record data. JAMA Netw Open 2018;1(8):e185097.

[56] Meiring C, Dixit A, Harris S, et al. Optimal intensive care outcome prediction over time using machine learning. PLoS One 2018;13(11):e0206862.

[57] Brisk R, Bond R, Liu J, et al. AI to enhance interactive simulation-based training in resuscitation. Med Proc Br HCI **64**,2018;1—4.

[58] Kindle RD, Badawi O, Celi LA, et al. Intensive care unit telemedicine in the era of big data, artificial intelligence, and computer clinical decision support systems. Crit Care Clin 2019;3593:483—95.

[59] Chang AC. Precision intensive care: a real-time artificial intelligence strategy for the future. Pediatr Crit Care Med 2019;20(2):194—5.

[60] Kamaleswaran R, Akbilgic O, Hallman MA, et al. Applying artificial intelligence to identify physiomarkers predicting severe sepsis in the PICU. Pediatr Crit Care Med 2018;19:e495—503.

[61] Williams JB, Ghosh D, Wetzel RC. Applying machine learning to pediatric critical care data. Pediatr Crit Care Med 2018;19(7):599—608.

[62] Chang AC. Artificial intelligence in pediatric critical care medicine: are we (finally) ready? Pediatr Crit Care Med 2018;19.

[63] Singh H, Yadav G, Mallaiah R, et al. iNICU-integrated neonatal care unit: capturing neonatal journey in an intelligent data way. J Med Syst 2017;41:132.

[64] Nichols JA, Herbert Chan HW, Baker MAB. Machine learning: applications of artificial intelligence to imaging and diagnosis. Biophys Rev 2019;11(1):111—18.

[65] Esteva A, Kuprel B, Novoa RA, et al. Dermatologist-level classification of skin cancer with deep neural networks. Nature 2017;542(7639):115—18.

[66] Ruffano F, Takwoingi Y, Dinnes J, et al. Computer-assisted diagnosis techniques (dermoscopy and spectroscopy-based) for diagnosing skin cancer in adults. Cochrane Database Syst Rev 2018;(12) 12:CD013186.

III. The current era of artificial intelligence in medicine

[67] Haenssle HA, Fink C, Schneiderbauer R, et al. Man against machine: diagnostic performance of a deep learning convolutional neural network for dermoscopic melanoma recognition in comparison to 58 dermatologists. Ann Oncol 2018;29:1836—42.

[68] Zhang X, Wang S, Liu J, et al. Towards improving diagnosis of skin diseases by combining neural network and human knowledge. BMC Med Inform Decis Mak 2018;18(Suppl 2):69—76.

[69] Stewart J, Sprivulis P, Dwivedi G. Artificial intelligence and machine learning in emergency medicine. Emerg Med Australas 2018;. Available from: https://doi.org/10.1111/1742-6723.13145 [Epub ahead of print].

[70] Lee S, Mohr NM, Street WN, et al. Machine learning in relation to emergency medicine clinical and operational scenarios: an overview. West J Emerg Med 2019;20(2):219—27.

[71] Berlyand Y, Raja AS, Dorner SC, et al. How artificial intelligence could transform emergency department operations. Am J Emerg Med 2018;36(8):1515—17.

[72] Levin S, Toerper M, Hamrock E, et al. Machine learning-based electronic triage more accurately differentiates patients with respect to clinical outcomes compared with the emergency severity index. Ann Emerg Med 2018;71(5):565—74.

[73] Ouchi K, Lindvall C, Chai PR, et al. Machine learning to predict, detect, and intervene older adults vulnerable for adverse drug events in the emergency department. J Med Toxicol 2018;14:248—52.

[74] Goto T, Camargo CA, Faridi MK, et al. Machine learning-based prediction of clinical outcomes for children during emergency department triage. JAMA Netw Open 2019;2(1):e186937.

[75] Gubbi S, Hamet P, Tremblay J, et al. Artificial intelligence and machine learning in endocrinology and metabolism: the dawn of a new era. Front Endocrinol 2019;. Available from: https://doi.org/10.3389/fendo.2019.00185.

[76] Rigla M, Garcia-Saez G, Pons B, et al. Artificial intelligence methodologies and their application to diabetes. J Diabetes Sci Technol 2018;12(2):303—10.

[77] Dankwa-Mullan I, Rivo M, Sepulveda M, et al. Transforming diabetes care through artificial intelligence: the future is here. Popul Health Manag 2019;22(3):229—42.

[78] Keel S, Lee PY, Scheetz J, et al. Feasibility and patient acceptability of a novel artificial intelligence-based screening model for diabetic retinopathy at endocrinology outpatient services: a pilot study. Sci Rep 2018;8 (1):4330.

[79] Zheng T, Xie W, Xu L, et al. A machine learning-based framework to identify type 2 diabetes through electronic health records. Int J Med Inf 2017;97:120—7.

[80] Li K, Daniels J, Liu C, et al. Convolutional recurrent neural networks for glucose prediction. IEEE J Biomed Health Inform 2019;. Available from: https://doi.org/10.1109/JBHI.2019.2908488 [Epub ahead of print].

[81] Murray PG, Stevens A, De Leonibus C, et al. Transcriptomics and machine learning predict diagnosis and severity of growth hormone deficiency. JCI Insight 2018;3(7):e93247.

[82] Grossi E, Mancini A, Buscema M. International experience on the use of artificial neural networks in gastroenterology. Dig Liver Dis 2007;39:278—85.

[83] Alagappan M, Glissen Brown JR, Mori Y, et al. Artificial intelligence in gastrointestinal endoscopy: the future is almost here. World J Gastrointest Endosc 2018;10(10):239—49.

[84] Ruffle JK, Farmer AD, Aziz Q. Artificial intelligence-assisted gastroenterology—promises and pitfalls. Am J Gastroenterol 2019;114(3):422—8.

[85] Yang YJ, Bang CS. Application of artificial intelligence in gastroenterology. World J Gastroenterol 2019;25 (14):1666—83.

[86] Mossotto E, Ashton JJ, Coelho T, et al. Classification of pediatric inflammatory bowel disease using machine learning.

[87] Wahl B, Cossy-Gantner A, Germann S, et al. Artificial intelligence and global health: how can AI contribute to health in resource-poor settings? BMJ Global Health 2018;3:e000798.

[88] Benke K, Benke G. Artificial intelligence and big data in public health. Int J Environ Res Public Health 2018;15:2796—805.

[89] Thiebaut R, Thiessard F. Artificial intelligence in public health and epidemiology. Yearb Med Inf 2018;27:207—10.

[90] Prosperi M, Min JS, Bian J, et al. Big data hurdles in precision medicine and precision public health. BMC Med Inform Decis Mak 2018;18:139—52.

III. The current era of artificial intelligence in medicine

[91] Wiens J, Shenoy ES. Machine learning for healthcare: on the verge of a major shift in healthcare epidemiology. Clin Inform Dis 2018;66(1):149−53.

[92] Dion M, AbdelMalik P, Mawudeku A. Big data and the global public health intelligence network (GPHIN). Can Commun Dis Rep 2015;41(9):209−14.

[93] Guo P, Liu T, Zhang Q, et al. Developing a dengue forecast model using machine learning: a case study in China. PLoS Negl Trop Dis 2017;11(10):e0005973.

[94] Leslie HH, Zhou Z, Spiegelman D, et al. Health system measurement: harnessing machine learning to advance global health. PLoS One 2018;13(10):s0204958.

[95] Sivapalaratnam S. Artificial intelligence and machine learning in hematology. Br J Haematol 2019;185(2):207−8.

[96] Shahin AI, Guo Y, Amin KM, et al. White blood cells identification system based on convolutional deep neural learning. Comput Methods Programs Biomed 2019;168:69−80.

[97] Muhsen IN, Jagasia M, Toor AA, et al. Registries and artificial intelligence: investing in the future of hematopoietic cell transplantation. Bone Marrow Transplant 2019;54:477−80.

[98] Wong ZAY, Zhou J, Zhang Q. Artificial intelligence for infectious disease big data analytics. Infect Dis Health 2019;24:44−8.

[99] Valleron AJ. Data science priorities for a university hospital-based institute of infectious diseases: a viewpoint. Clin Infect Dis 2017;65(1):S84−8.

[100] Chae S, Kwon S, Lee D. Predicting infectious disease using deep learning and big data. Int J Environ Res Public Health 2018;15(8).

[101] Shen Y, Yuan K, Chen D, et al. Ontology-driven clinical decision support system for infectious disease diagnosis and antibiotic prescription. Artif Intell Med 2018;86:20−32.

[102] Seymour CW, Kennedy JN, Wang S, et al. Derivation, validation, and potential treatment implications of novel clinical phenotypes for sepsis. JAMA 2019;321.

[103] Dande P, Samant P. Acquaintance to artificial neural networks and use of artificial intelligence as a diagnostic tool for tuberculosis: a review. Tuberculosis 2018;108:1−9.

[104] Sidey-Gibbons JAM, Sidey-Gibbons CJ. Machine learning in medicine: a practical introduction. BMC Med Res Methodol 2019;19:64.

[105] Krittanawong C. The rise of artificial intelligence and the uncertain future for physicians. Eur J Int Med 2018;48:e13−14.

[106] Miller DD, Brown EW. Artificial intelligence in medical practice: the question to the answer? Am J Med 2018;131:129−33.

[107] Noorbakhsh-Sabet N, Zand R, Zhang Y, et al. Artificial intelligence transforms the future of health care. Am J Med 2019;132.

[108] Zhang Y, Guo SL, Han LN, et al. Application and exploration of big data mining in clinical medicine. Chin Med J 2016;129:731−8.

[109] Lin SY, Mahoney MR, Sinsky CA. Ten ways artificial intelligence will transform primary care. J Gen Intern Med 2019;34.

[110] Morgan DJ, Bame B, Zimand P, et al. Assessment of machine learning vs standard prediction rules for predicting hospital readmissions. JAMA Netw Open 2019;2(3):e190348.

[111] Avati A, Jung K, Harman S, et al. Improving palliative care with deep learning. BMC Med Inform Decis Mak 2018;18(Suppl. 4):122.

[112] Hueso M, Vellido A, Montero N, et al. Artificial intelligence for the artificial kidney: pointers to the future of a personalized hemodialysis therapy. Kidney Dis 2018;4(1):1−9.

[113] Brier ME, Gaweda AE. Artificial intelligence for optimal anemia management in end-stage renal disease. Kidney Int 2016;90(2):259−61.

[114] Niel O, Bastard P, Boussard C, et al. Artificial intelligence outperforms experienced nephrologists to assess dry weight in pediatric patients on chronic hemodialysis. Pediatr Nephrol 2018;33(10):1799−803.

[115] Ganapathy K, Abdul SS, Nursetyo AA. Artificial intelligence in neurosciences: a clinician's perspective. Neurol India 2018;66:934−9.

[116] Jiang F, Jiang Y, Zhi H, et al. Artificial intelligence in health care: past, present, and future. Stroke Vasc Neurol 2017;2:e000101.

[117] Senders JT, Arnaout O, Karhade AV, et al. Natural and artificial intelligence in neurosurgery: a systematic review. Neurosurgery 2018;83:181−92.

III. The current era of artificial intelligence in medicine

[118] Azimi P, Mohammadi HR, Benzel ED, et al. Artificial neural network in neurosurgery. J Neurol Neurosurg Psychiatry 2015;86:251−6.

[119] Iniesta R, Stahl D, McGuffin P. Machine learning, statistical learning, and the future of biological research in psychiatry. Psychological Med 2016;46:2455−65.

[120] Durstewirz D, Koppe G, Meyer-Lindenberg A. Deep neural networks in psychiatry. Mol Psychiatry 2019;24:1583−98 [Epub ahead of print].

[121] Costescu CA, Vanderborought B, David DO. Reversal learning task in children with autism spectrum disorder: a robot-based approach. J Autism Dev Disord 2015;45(11):3715−25.

[122] Mehrholz J, Pohl M, Platz T, et al. Electromechanical and robot-assisted arm training for improving activities of daily living, arm function, and arm muscle strength after stroke. Cochrane Database Syst Rev 2015;11 CD006876.

[123] Desai GS. Artificial intelligence: the future of obstetrics and gynecology. J Obstet Gynecol India 2018;68 (4):326−7.

[124] Wang R, Pan W, Jin L, et al. Artificial intelligence in reproductive medicine. Soc Reprod Fertil 2019;158: R139−54 REP-18-0523.R1.

[125] Balayla J, Shrem G. Use of artificial intelligence in the interpretation of intrapartum fetal heart rate tracings: a systematic review and meta-analysis. Arch Gynecol Obstet 2019;300:7−14.

[126] Siristatidis CS, Chrellas C, Pouliakis A, et al. Artificial neural networks in gynaecological diseases: current and potential future applications. Med Sci Monit 2010;16(10):RA231−6.

[127] Simopoulou M, Sfakianoudis K, Maziotis E, et al. Are computational applications the "crystal ball" in the IVF laboratory? The evolution from mathematics to artificial intelligence. J Assist Reprod Genet 2018;35 (9):1545−57.

[128] Stylios CS, Georgopoulos VC. Fuzzy cognitive maps for medical decision support—a paradigm from obstetrics. In: Annual International Conference of the IEEE Engineering in Medicine and Biology. 2010.

[129] Londhe VY, Bhasin B. Artificial intelligence and its potential in oncology. Drug Discov Today 2019;24 (1):228−32.

[130] Tseng HH, Wei L, Cui S, et al. Machine learning and imaging informatics in oncology. Oncology 2018;. Available from: https://doi.org/10.1159/000493575.

[131] Xu J, Yang P, Xue S, et al. Translating cancer genomics into precision medicine with artificial intelligence: applications, challenges, and future perspectives. Hum Genet 2019;138:109−24.

[132] Kiser KJ, Fuller CD, Reed VK. Artificial intelligence in radiation oncology treatment planning: a brief overview. J Med Artif Intell 2019;2:9−19.

[133] Yim WW, Yetisgen M, Harris WP, et al. Natural language processing in oncology: a review. JAMA Oncol 2016;2(6):797−804.

[134] Kantarjian H, Yu PP. Artificial intelligence, big data, and cancer. JAMA Oncol 2015;1(5):573−4.

[135] Somashekhar SP, Sepulveda MJ, Puglielli S, et al. Watson for oncology and breast cancer treatment recommendations: agreement with and expert multidisciplinary tumor board. Ann Oncol 2018;29(2):418−23.

[136] N.M. Patel, V.V. Michelini, J.M. Snell, et al., Enhancing next-generation sequencing-guided cancer care through cognitive computing, Oncologist2392, 2018, 179−185.

[137] Schmidt C. MD Anderson Breaks with IBM Watson, raising questions about artificial intelligence in oncology. J Natl Cancer Inst 2017;109(5):4−5.

[138] Su H, Shen Y, Xing F, et al. Robust automatic breast cancer staging using a combination of functional genomics and image-omics. Conf Proc IEEE Eng Med Biol Soc 2015;2015:7226−9.

[139] Bejnordi BE, Veta M, van Diest PJ, et al. Diagnostic assessment of deep learning algorithms for detection of lymph node metastases in women with breast cancer. JAMA 2017;318(22):2199−210.

[140] Kurtz DM, Esfahani MS, Scherer F, et al. Dynamic risk profiling using serial tumor biomarkers for personalized outcome prediction. Cell 2019;178:699−713.e19 S0092-8674(19)30639-7.

[141] Kapoor R, Walters SP, Al-Aswad LA. The current state of artificial intelligence in ophthalmology. Surv Ophthalmol 2019;64:233−40.

[142] Du XL, Li WB, Hu BJ. Application of artificial intelligence in ophthalmology. Int J Ophthalmol 2018;11 (9):1555−61.

[143] Lu W, Tong Y, Yu Y, et al. Applications of artificial intelligence in ophthalmology: general overview. J Ophthalmol 2018;1−15.

[144] Ting DSW, Pasquale LR, Peng L, et al. Artificial intelligence and deep learning in ophthalmology. Br J Ophthalmol 2019;103(2):167−75.

[145] Hogarty DT, Mackey DA, Hewitt AW. Current state and future prospects of artificial intelligence in ophthalmology: a review. Clin Exp Ophthalmol 2019;47(1):128−39.

[146] Gulshan V, Peng L, Coram M, et al. Development and validation of a deep learning algorithm for detection of diabetic retinopathy in retinal fundus photographs. JAMA 2016;316:2402−10.

[147] Abramoff MD, Lavin PT, Brich M, et al. Pivotal trial of an autonomous AI-based diagnostic system for detection of diabetic retinopathy in primary care offices. NPJ Digital Med 2018;1:39.

[148] Chang HY, Jung CK, Woo JI, et al. Artificial intelligence in pathology. J Pathol Transl Med 2019;53:1−12.

[149] Naugler C, Church DL. Automation and artificial intelligence in the clinical laboratory. Crit Rev Clin Lab Sci 2019;56(2):98−110.

[150] Janowczyk A, Madabhushi A. Deep learning for digital pathology image analysis: a comprehensive tutorial with selected use cases. J Pathol Inf 2016;7:29.

[151] Niazi MKK, Parwani AV, Gurcan MN. Digital pathology and artificial intelligence. Lancet Oncol 2019;20(5): e253−61.

[152] Gruson D, Helleputte T, Rousseau P, et al. Data science, artificial intelligence, and machine learning: opportunities for laboratory medicine and the value of positive regulation. Clin Biochem 2019;69:1−7.

[153] Cabitza F, Banfi G. Machine learning in laboratory medicine: waiting for the flood? Clin Chem Lab Med 2018;56(4):516−24.

[154] Smith KP, Kang AD, Kirby JE. Automated interpretation of blood culture Gram stains by use of a deep convolutional neural network. J Clin Microbiol 2018;56(3):e01521−17.

[155] Harmon SA, Tuncer S, Sanford T, et al. Artificial intelligence at the intersection of pathology and radiology in prostate cancer. Diagn Interv Radiol 2019;25(3):183−8.

[156] Nir G, Karimi D, Goldenberg L, et al. Comparison of artificial intelligence techniques to evaluate performance of a classifier for automatic grading of prostate cancer from digitized histopathologic images. JAMA Netw Open 2019;2(3):e190442.

[157] O'Sullivan S, Leonard S, Holzinger A, et al. Anatomy 101 for AI-driven robotics: explanatory, ethical, and legal frameworks for development of cadaveric skills training standards in autonomous robotic surgery/ autopsy.. Int J Med Robot 2019;e2020 [ePub].

[158] Kokol P, Zavrsnik J, Vosner HB. Artificial intelligence and pediatrics: a synthetic mini review. Pediatr Dimens 2017;2(40):2−5.

[159] Shu LQ, Sun YK, Tan LH, et al. Application of artificial intelligence in pediatrics: past, present, and future. World J Pediatr 2019;15(2):105−8.

[160] Bennett TD, Callahan TJ, Feinstein JA, et al. Data science for child health. J Pediatr 2018;208:12−22.

[161] Liang H, Tsui B, Ni H, et al. Evaluation and accurate diagnoses of pediatric disease using artificial intelligence. Nat Med 2019;25:433−8.

[162] Mani S, Ozdas A, Aliferis C, et al. Medical decision support using machine learning for early detection of late-onset neonatal sepsis. J Am Med Inf Assoc 2014;21(2):326−36.

[163] Tariq Q, Daniels J, Schwartz JN, et al. Mobile detection of autism through machine learning on home video: a development and prospective validation study. PLoS Med 2018;15(11):e1002705.

[164] Pan L, Liu G, Mao X, et al. Development of prediction models using machine learning algorithms for girls with suspected central precocious puberty: retrospective study. JMIR Med Inf 2019;7(1):e11728.

[165] Doan S, Maehara CK, Chaparro JD, et al. Building a natural language processing tool to identify patients with high clinical suspicion for Kawasaki disease from emergency department notes. Acad Emerg Med 2016;23(5):628−36.

[166] Labaki WW, Han MK. Artificial intelligence and chest imaging: will deep learning make us smarter? Am J Respire Crit Care Med 2018;197(2):148−50.

[167] Gonzalez G, Ash SY, Vegas-Sanchez-Ferrero G, et al. Disease staging and prognosis in smokers using deep learning in chest computed tomography. Am J Respire Crit Care Med 2018;197(2):193−203.

[168] Parreco J, Hidalgo A, Parks JJ, et al. Using artificial intelligence to predict prolonged mechanical ventilation and tracheostomy placement. J Surg Res 2018;228:179−87.

[169] Messinger AI, Bui N, Wagner BD, et al. Novel pediatric automated respiratory score using physiologic data and machine learning in asthma. Pediatric Pulmonol 2019;54:1149−55.

[170] Ross MK, Yoon J, van der Schaar A, et al. Discovering pediatric asthma phenotypes on the basis of response to controller medication using machine learning. Ann Am Thorac Soc 2018;15(1):49–58.

[171] Thrall JH, Li X, Li Q, et al. Artificial intelligence and machine learning in radiology: opportunities, challenges, pitfalls, and criteria for success. J Am Coll Radiol 2018;15(3):504–8.

[172] Hosny A, Parmar C, Quackenbush J, et al. Artificial intelligence in radiology. Nat Rev Cancer 2018;18 (8):500–10.

[173] Choy G, Khalilzadeh O, Michalski M, et al. Current applications and future impact of machine learning in radiology. Radiology 2018;288(2):318–28.

[174] Liew C. The future of radiology augmented with artificial intelligence: a strategy for success. Euro J Radiol 2018;102:152–6.

[175] Chartrand G, Cheng PM, Vorontsov E, et al. Deep learning: a primer for radiologists. RadioGraphics 2017;37(7):2113–31.

[176] Lee JG, Jun S, Gho YW, et al. Deep learning in medical imaging: general overview. Korean J Radiol 2017;18 (4):570–84.

[177] McBee MP, Awan OA, Colucci AT, et al. Deep learning in radiology. Acad Radiol 2018;25:1472–80.

[178] Yasaka K, Akai H, Kunimatsu A, et al. Deep learning with convolutional neural network in radiology. Jpn J Radiol 2018;36:257–72.

[179] England JR, Cheng PM. Artificial intelligence for medical image analysis: a guide for authors and reviewers. AJR 2019;212:513–19.

[180] Cai T, Giannopoulos AA, Yu S, et al. Natural language processing technologies in radiology research and clinical applications. Radiographics 2016;36(1):176–91.

[181] Pons E, Braun LMM, Hunick MGM, et al. Natural language processing in radiology: a systematic review. Radiology 2016;279(2):329–43.

[182] Rajpurkar P, Irvin J, Zhu K et al. CheXNet: Radiologist-level pneumonia detection on chest X-rays with deep learning, 2017, arXiv:1711.05225.

[183] Personal communication with Luke Oaken-Rayner, May 2019.

[184] Titano JJ, Badgeley M, Schefflein J, et al. Automated deep neural network surveillance of cranial images for acute neurologic events. Nat Med 2018;24:1337–41.

[185] Al Idrus A. Boston children's to create deep learning tool for pediatric brain scans. FierceBiotech November 28, 2016.

[186] Filice R. Deep learning language modeling approach for automated, personalized, and iterative radiology-pathology correlation. J Am Coll Radiol 2019;16:1286–91 [Epub ahead of print].

[187] Park A, Chute C, Rajpurkar P, et al. Deep learning-assisted diagnosis of cerebral aneurysms using the HeadXNet Model. JAMA Netw Open 2019;2(6):e195600.

[188] Oermann EK, Rubinsteyn A, Ding D, et al. Using a machine learning approach to predict outcomes after radiosurgery for cerebral arteriovenous malformations. Sci Rep 2016;6:21161.

[189] McGinty GB, Allen B. The ACR data science institute and AI advisory group: harnessing the power of artificial intelligence to improve patient care. J Am Coll Radiol 2018;15(3):577–9.

[190] Jha S, Topol EJ. Adapting to artificial intelligence: radiologists and pathologists as information specialists. JAMA 2016;316(22):2353–4.

[191] Kothari S, Gionfrida L, Bharath AA, et al. Artificial intelligence and rheumatology: a potential partnership. Rheumatology 2019;58:1894–5.

[192] Foulquier N, Redou P, Saraux A. How health information technologies and artificial intelligence may help rheumatologists in routine practice. Rheumatol Ther 2019;6(2):135–8.

[193] Ferizi U, Honig S, Chang G. Artificial intelligence, osteoporosis, and fragility fractures. Curr Opin Rheumatol 2019;31(4):368–75.

[194] Foulquier N, Redou P, Le Gal C, et al. Pathogenesis-based treatments in primary Sjogren's syndrome using artificial intelligence and advanced machine learning techniques: a systematic literature review. Hum Vaccin Immunother 2018;14(11):2553–8.

[195] Norgeot B, Glicksberg BS, Trupin L, et al. Assessment of a deep learning model based on electronic health record data to forecast clinical outcomes in patients with rheumatoid arthritis. JAMA Netw Open 2019;2(3):e190606.

[196] Hashimoto DA, Rosman G, Rus D, et al. Artificial intelligence in surgery: promises and perils. Ann Surg 2018;268(1):70–6.

[197] Kose E, Ozturk NN, Karahan SR. Artificial intelligence in surgery. Eur Arch Med Res 2018;34(Suppl. 1):54−6.

[198] Bur AM, Shew M, New J. Artificial intelligence for the otolaryngologist: a state of the art review. Otolaryngol Head Neck Surg 2019;160(4):603−11.

[199] Kim YJ, Kelley BP, Nasser JS, et al. Implementing precision medicine and artificial intelligence in plastic surgery: concepts and future prospects. Plast Reconstr Surg Glob Open 2019;7:e2113.

[200] Kanevsky J, Corban J, Gaster R, et al. Big data and machine learning in plastic surgery: a new frontier in surgical innovation. Plast Reconstr Surg 2016;137(5):890−7.

[201] Cabitza F, Locoro A, Banfi G. Machine learning in orthopedics: a literature review. Front Bioeng Biotechnol 2018;6:75.

[202] Jones LD, Golan D, Hanna SA, et al. Artificial intelligence, machine learning, and the evolution of health care: a bright future or cause for concern? Bone Joint Res 2018;7:223−5.

[203] Galbusera F, Casaroli G, Bassani T. Artificial intelligence and machine learning in spine research. JOR Spine 2018;2:e1044.

[204] Khanna S. Artificial intelligence: contemporary applications and future compass. Int Dental J 2010;60:269−72.

[205] The Lancet. Is digital medicine different?. Lancet 2018;392:95.

[206] Dimitrov D. Medical Internet of Things and big data in health care. Healthc Inf Res 2016;22(3):156−63.

[207] Fatehi F, Menon A, Bird D. Diabetes care in the digital era: a synoptic overview. Curr Diab Rep 2018;18 (7):38−47.

[208] Steinhubl SR, Topol EJ. Moving from digitalization to digitization in cardiovascular care: why is it important, and why could it mean for patients and providers? J Am Coll Cardiol 2015;66(13):1489−96.

[209] Kubota KJ, Chen JA, Little MA. Machine learning for large-scale wearable sensor data in Parkinson's disease: concepts, promises, pitfalls, and features. Mov Disord 2016;31(9):1314−26.

[210] Williams AM, Liu Y, Regner KR. Artificial intelligence, physiological genomics, and precision medicine. Physiol Genomics 2018;50(4):237−43.

[211] Holder LB, Haque MM, Skinner MK. Machine learning for epigenetics and future medical applications. Epigenetics 2017;12(7):505−14.

[212] Grapov D, Fahrmann J, Wanichthanarak K, et al. Rise of deep learning for genomic, proteomic, and metabolomic data integration in precision medicine. OMICS 2018;22(10):630−6.

[213] Kuru K, Niranjan M, Tunca Y, et al. Biomedical visual data analysis to build an intelligent diagnostic decision support system in medical genetics. Artif Intell Med 2014;62(2):105−18.

[214] Farley T, Kiefer J, Lee P, et al. The BioIntelligence framework: a new computational platform for biomedical knowledge computing. J Am Med Inf Assoc 2013;20(1):128−33.

[215] Mousses S, Kiefer J, Von Hoff D, et al. Using biointelligence to search the cancer genome: an epistemological perspective on knowledge recovery strategies to enable precision medical genomics. Oncogene 2008;27: S58−66.

[216] Barry DT. Adaptation, artificial intelligence, and physical medicine and rehabilitation. Phys Med Rehabil 2018;S131−4.

[217] Tack C. Artificial intelligence and machine learning: applications in musculoskeletal physiotherapy. Musculoskelet Sci Pract 2019;39:164−9

[218] Sniecinski I, Seghatchian J. Artificial intelligence: a joint narrative on potential use in pediatric stem and immune cell therapies and regenerative medicine. Transfus Apheresis Sci 2018;57:422−4.

[219] Awaysheh A, Wilcke J, Elvinger F, et al. Review of medical decision support and machine learning methods. Ve Pathol 2019;56:512−25.

[220] Wartman SA, Combs CD. Reimagining medical education in the age of AI. AMA J Ethics 2019;21:146−52.

[221] Boggs SD, Luedi MM. Nonoperating room anesthesia education: preparing our residents for the future. Curr Opin Anesthesiol 2019;32:490−7 [Epub ahead of print].

[222] Yu W, Wen L, Zhao LA, et al. The applications of virtual reality technology in medical education: a review and mini-research. J Phys Conf Ser 2019;1176:022055.

[223] Dias RD, Gupta A, Yule SJ. Using machine learning to assess physician competence: a systematic review. Acad Med 2019;94(3):427−39.

[224] Johnston SC. Anticipating and training the physician of the future: the importance of caring in an age of artificial intelligence. Acad Med 2018;93(8):1105−6.

III. The current era of artificial intelligence in medicine

[225] Skiba DJ. Augmented intelligence and nursing. Natl Leag Nurs 2017;108–9.
[226] Liao PH, Hsu PT, Chu W, et al. Applying artificial intelligence technology to support decision making in nursing: a case study in Taiwan. Health Inform J 2015;21(2):137–48.
[227] Reddy S, Fox J, Purohit MP. Artificial intelligence-enabled health care delivery. J R Soc Med 2019;112 (1):22–8.
[228] Nelson A, Herron D, Rees G, et al. Predicting scheduled hospital attendance with artificial intelligence. NPJ Digital Med 2019;2:26.

Implementation of Artificial Intelligence in Medicine

Extreme uncertainty and dissatisfaction in data and information exists in the imbroglio of the practice of medicine and the world of health care especially with the increased incorporation of electronic health records (EHR) into the hospital and clinics and limited input into the few EHR vendors. Some of the data issues facing the present day practitioner include escalating and missing data, exponentially increasing information, rigid regulatory policies, and decreasing access to information. The author described the main implementation obstacles of artificial intelligence (AI) in medicine as data sharing and standardization, transparency, patient safety, financial issues, and education [1]. Medical providers lack sufficient insight and education in the realm of data science and this ignorance results in an inadequate knowledge from the rich data that now exist in health care and medicine. To date, there has not been wide acceptance of data science or AI in the medical school or clinical training program curriculum. Finally, there is also a significant cultural and intellectual schism between the clinical world and data science domain: most medical meetings lack data science or AI discussions on application and gatherings of devotees of machine learning (ML) or data science in health care or medicine also lack strong clinician presence. Practitioners require accurate data as well as up to date information and sharing of ideas to ensure best outcomes for the patient population and not solely be distracted by high performance of these tools [2,3]. In the future the data domain in health care will also be more inclusive of nontraditional sources such as social media and home monitoring especially with the proliferation of applications. In addition to the aforementioned issues, Maddox posed several very relevant questions in this nascent domain [4]: (1) What are the right tasks for AI in health care? (2) What are the right data for AI? (3) What is the right evidence standard for AI? And finally (4) What are the right approaches for integrating AI into clinical care?

The challenge is in the acceptance of artificial intelligence (AI)

Matthieu Komorowski

Matthieu Komorowski, an intensivist and data scientist as well as a former AIMed abstract winner, authored this commentary on what he perceives as the main obstacles to adoption of AI for clinicians: trust and AI literacy.

Question: What are the next steps for getting AI into the hands of clinicians across the United Kingdom?

The challenge to deploying AI technologies in the National Health Service (NHS) is not so much technical but rather societal and cultural. Indeed, most of what we call AI actually relies on relatively simple computational models that are readily available, and many companies, large and small, are eager to roll out their technologies into the NHS with the vision of improving the care of potentially millions of patients.

If we want these tools to be used, the biggest hurdle will be to improve their acceptance by all the shareholders: patients, clinicians, and policy makers. Acceptability of a technology is strongly correlated with two factors: the trust and the literacy that users have in this technology [1,2]. Trust is able to improve acceptance of a new product by reducing perceived risk, which has been defined in the innovation literature as the likelihood that a product will fail [1]. Improving the trust implies demonstrating that the technology is both effective and safe. People use AI-based products such as navigation systems and home personal assistants because they perform well and they deliver value. In health care, demonstrating value and effectiveness of a new treatment classically happens in the form of randomized controlled trials [3]. For example, we can imagine comparing patient outcomes when receiving either standard "human" treatment or AI-augmented treatment.

Demonstrating safety will be more complex. First, regulators and policy makers need to define a framework for AI governance. In the United Kingdom the key principles of such a framework were outlined in the House of Commons *Science and Technology Committee* report from May 2018: "Algorithms in decision making" [4]. Undeniably, such technologies will have to be certified and approved prior to use, within the regulatory framework in place [1]. Next, AI must be developed in an ethical manner, it should augment the capabilities of humans and benefit society as a whole. This aspect is in general straightforward in health care, unlike other sectors such as defense. Safety also means protecting patient data privacy against breaches. The new General Data Protection Regulation (GDPR—https://www.eugdpr.org) recently brought some important changes, in an effort to unify regulation on data protection within the European Union. GDPR introduced the concept of a "right to explanation," meaning that the users affected by a model are due an accounting of how a particular decision was reached. In practice, this requirement will prove very challenging to implement for some branches of AI such as deep learning (DL), because the neural network's reasoning is embedded into thousands of complexly interconnected nodes [5]. Finally, safeguards should be put in place to improve further AI safety. In medicine the most obvious safeguard is to leave human physicians in charge, and to augment rather than attempt to replace their intelligence with computerized systems.

The second way to improve acceptability of AI is to improve literacy in AI and data science among shareholders. This can be achieved in a number of ways, ranging from formal education, public events, debates, to joint conferences on AI in medicine, datathons, and hackathons [6]. Education from top-tier institutions is now freely available online and accessible to clinicians without previous background in computer science, for example, in the form of massive open online courses.

Andrew Ng, one of the most influential minds in AI and DL, stated that "whoever wins AI will own the future." But further than simply creating genius machines, future developments in AI must include maturing the rules and the culture that will enable AI to function according to our common values. This, in return, will generate trust and boost acceptability, for the ultimate benefit of our NHS patients.

References

[1] Hengstler M, Enkel E, Duelli S. Applied artificial intelligence and trust—the case of autonomous vehicles and medical assistance devices. Technol Forecast Soc Change 2016;105:105—20. Available from: https://doi.org/10.1016/j.techfore.2015.12.014.
[2] Lee JD, See KA. Trust in automation: designing for appropriate reliance. Hum Factors 2004;46(1):50—80. Available from: https://doi.org/10.1518/hfes.46.1.50_30392.
[3] Murad MH, Asi N, Alsawas M, Alahdab F. New evidence pyramid. BMJ Evid Based Med 2016; ebmed-2016-110401. https://doi.org/10.1136/ebmed-2016-110401.
[4] The House of Commons. Algorithms in decision-making—Science and Technology Committee—House of Commons. Retrieved August 11, 2018, from: <https://publications.parliament.uk/pa/cm201719/cmselect/cmsctech/351/35102.htm>; 2018.
[5] Knight W. The dark secret at the heart of AI. MIT Technol Rev. April 11, 2017. Retrieved from: <https://www.technologyreview.com/s/604087/the-dark-secret-at-the-heart-of-ai/>.
[6] Aboab J, Celi LA, Charlton P, Feng M, Ghassemi M, Marshall DC, et al. A "datathon" model to support cross-disciplinary collaboration. Sci Transl Med 2016;8(333):333ps8. Available from: https://doi.org/10.1126/scitranslmed.aad9072.

Artificial intelligence (AI) strategy for health-care systems

Aziz Nazha

Aziz Nazha, a physician with an AI background and the leader of an AI center, authored this commentary on lessons learned in deploying AI in a large institution and strategies to assure longevity of such an enterprise.

AI already has a significant impact on our lives and its impact is expected to continue to rise exponentially. As AI is radically transforming each industry, the health-care industry is not far from that. In fact, it is expected that the health-care industry will be most the transformed industry by AI in the next decade. This increases fear and anxiety among physicians and hospital administrators especially on how to incorporate AI in hospitals workflow. This fear and anxiety are mainly driven by multiple factors that include (1) physicians are unfamiliar with AI and ML technologies especially that these technologies have always been viewed as a "black box," (2) the rapid change in algorithms and digital transformation that makes it difficult to keep up with the latest developments, and (3) the overselling and lack of understanding of the complexity of health care of AI products from big companies and startups. Thus hospitals and health-care systems must develop a strategy to adopt AI technologies and incorporate them into the workflow of daily operating activities. This strategy should focus on the application of AI in operation, financial decision-making, and more importantly research and development. Each of these aspects must have a different strategy to maximize the return on investment (ROI) of each of them. In a recent essay in JAMA journal authors, Edward H. Shortliffe, MD, PhD, Editor-in-Chief of the Journal of Biomedical Informatics, and Martin J. Sepulveda, MD, formerly with IBM's Watson Research Laboratory highlighted six elements that are needed for physician acceptance of an AI-supported decision support system that includes no black boxes "Transparency is required." The physicians and other health-care providers must fully understand the basis for any advice or recommendation given. Systems must save—not waste—time. Clinical support "must blend into the workflow," not complicate it. Systems should be easy to use so that no major training is needed to get results. Relevance is very important as clinical support should "reflect an understanding" of what physicians will be asking and how the result can change the patient outcome and finally the delivery of information must be respectful and advice should be given in a manner that respects the user's expertise, have a strong scientific foundation, and the advice should be reproducible, reliable, and usable and based on rigorously peer-reviewed scientific evidence. Understanding these elements can help the hospital's administrators understand how to implement AI in health care.

So how hospitals, in health-care systems, can build an AI strategy? Hospitals have to adopt a few steps. First, build a team of clinician–data scientists who understand and foster ML and AI capabilities in the system. Second, establish small projects that can yield higher impact value and can prove that the team is capable of delivering bigger projects with a higher ROI. Third, build internal talent that can enhance and increase the productivity of the team and finally expand the team capabilities to reach multiple AI technologies to be applied in multiple areas from diagnosis to prognosis and personalized treatment. These approaches can enhance the patient overall outcome and decrease cost. This path is not easy but can be achieved by getting all health-care providers together from physicians to researchers and statisticians to pharmacists with the main goal of embracing AI capabilities to aid health-care personals in providing state-of-the-art care and improve outcomes to all patients.

In conclusion, the convergence of Big Data, improved algorithms, computational power, and cloud storage in health care has started to yield robust ML projects and reasonable results in biomedicine and health care. There is a wide range of subspecialty focus and interest level in AI applications to date, with radiology surging ahead with particularly DL in medical image interpretation. There are also many nuances between clinicians and data scientists in culture and understanding in being partners in this domain. Best practice in the near future will involve the use of AI to answer the clinical questions (intelligence-based medicine) rather than current practice of solely relying on published reports and other resources (evidence-based medicine or expert groups).

Key Concepts

- There is currently an escalating interest and enthusiasm for artificial intelligence (AI) in medicine and health care, particularly in a few areas such as medical imaging and decision support; the origin of this surge in accommodation of AI in medicine has been mostly the rapid proliferation as well as adoption of machine and deep learning (DL) in many areas.

- Overall, the Food and Drug Administration and its Good Machine Learning Practices is a much more congruent regulatory strategy with the exponential increase in AI technologies in clinical medicine and health care.
- This dichotomy conveniently delineates some of the key differences between clinicians (more prone to System 1 thinking) and data scientists (with their affinity for System 2 thinking). Physicians, especially those in the acute care clinical setting [such as emergency room, intensive care units (ICUs), and operating and procedure rooms] often rely on a fast intuition-based System 1 thinking that is based on past experiences and judgments. Data scientists, on the other hand, more frequently approach problems with slower and more logical progressive thinking that is rationality-based System 2 thinking.
- Overall, perhaps the myriad of human biases and heuristics can potentially be neutralized with an objective AI-supported strategy in the decision-making process to reduce the proportion of human-related biases and heuristics that often lead to errors.
- Major criticisms of evidence-based medicine include a publication bias that refers to the clinician and investigator tendencies to publish only studies with a positive diagnostic or therapeutic result. In addition, the terms of the criteria for the levels of evidence are not always agreed upon and these imprecise definitions are often confusing and misleading. These guidelines or recommendations are very often out of date and do not accommodate more recent ideas and/or study results; in short, the information is not timely and therefore lack real-time relevance.
- Most, if not all physicians in their subspecialties, can be configured by three basic areas of thinking that involves different areas of the brain in which they perform their tasks: perception, cognition, and operation.
- There is much promise in the utilization of AI methodologies such as DL (in particular convolutional neural network, or CNN) for automated medical image interpretation and/or augmented medical imaging. The image interpretation tasks include classification, regression, localization, and segmentation.
- The area of augmented (AR), virtual (VR), and mixed reality will be able to leverage AI technology and use this resource to the fullest for a variety of purposes; these include education and training as well as simulation and immersive scenarios for all stakeholders (including patients and families) and preoperative and intraoperative imaging and planning for certain medical and surgical subspecialists.
- While it is laudable that AI was able to defeat the human champion in the game *Go*, the practice of medicine, especially in the chaotic domains of the emergency room, ICU, and operating rooms, are more akin to the real-time strategy games like *StarCraft*.
- Bedside biomedical monitoring has been unidirectional: displaying data such as vital signs in a continuous fashion but not analyzing and understanding data internally so therefore not at all "intelligent." AI has the potential to change this paradigm by deploying machine and DL to this rich data milieu (with recurrent neural network described previously) and deriving knowledge and intelligence in a real-time fashion.
- Human−robot interaction and relationship are being evaluated in a variety of clinical scenarios such as physical or psychiatric rehabilitation and education and training.

- AI is very involved in the evolution of virtual assistants as these are dividends of natural language processing (NLP) (including natural language understanding and generation).
- The paradigm of precision medicine with its complexity of decisions that can be made and enormity of data to be analyzed is particularly well suited for the portfolio of AI methodologies such as DL (especially its AI congener deep reinforcement learning) as similar patients can be identified and assessed.
- Diverse disciplines such as language, neurophysiology, chemistry, toxicology, biostatistics, and medicine can converge to leverage AI and machine learning (ML)/DL to design novel drug candidates. The cognitive solutions are designed to fully integrate and analyze relatively large data sets such as in life sciences for drug discovery.
- An essential part of digital health is the use of information and communication technologies as well as AI in the form of data mining of the incoming data as well as machine and DL for anomaly detection, prediction, and diagnosis/decision-making in a continuous manner.
- Wearable technology has also taken a different perspective in this time of AI, especially with the possibility of wide adoption of simple AI tools embedded in medical devices.
- AI technology, while maturing in the area of medical image interpretation, still needs to be more sophisticated in the realm of decision support; real-time, complex decision support with feedback mechanism is the crux of the need for AI-enabled anesthesia practice.
- Overall, as cardiology is both a perceptual or image intensive field as well as a cognitive or decision-making subspecialty with a myriad of procedural tasks, AI is a particularly valuable technology for cardiology with potentially very rich dividends that are vastly underexplored at present but has great promise.
- Once deep reinforcement learning or other DL methodologies are sophisticated enough for the real-time, complex decision-making challenges of ICU medicine, there will be predictably an even higher level of AI adoption by the physicians in this group. The true value will be a universal ICU data repository for an enriched data source for ML/DL and deep reinforcement learning.
- For the future an AI-enabled dermatologist will have routine CNN-enabled dermoscopy in the office on a routine basis and this dyad will correctly diagnose skin conditions almost perfectly. This human—machine synergy can also be combined with digital health capabilities to transmit photographs for not only routine screening but also for follow-up; this capability can therefore obviate the need for an excessive number of clinic visits as well as make available a valuable service in global health.
- For the future, like the other subspecialties that often require fast thinking and decision-making (ICU, surgery, and anesthesiology), emergency medicine can greatly benefit from an AI-enabled strategy of deep reinforcement learning designed for real-time decision-making.
- For the future, exciting innovative advances with closed-loop systems and fuzzy logic as well as CRNN AI methodologies will all be extremely useful for essentially an artificial pancreas to treat diabetes. This will be a highly significant advance in the management of one of the most significant disease burdens in the future.
- AI-enabled endoscopic examinations should be routinely embedded with CNN as an augmentation for the gastroenterologist. This procedure should involve an entire continuum from acquisition of the image to CNN interpretation of suspicious lesions and decision to biopsy or not.

- The fields of global and public health as well as epidemiology are ideally structured for the use of ML/DL and other methodologies in studying and managing epidemics and natural disasters in the context of large populations.
- There are some clinical and organizational activities in deploying AI and ML/DL in infectious diseases but not in an organized global manner. The field of infectious disease with its many sources of data is ideally suited for Big Data and ML/DL with some expert knowledge oversight for many dividends.
- There are many ways that AI can help the primary care physicians outside of medical image interpretation and decision support. The areas of chart review, practice management, risk prediction, and population health management are large market potentials with growing number of AI-enabled solutions.
- Similar to the artificial pancreas discussed under endocrinology, the artificial kidney will need real-time decision support and predictive modeling with biofeedback and embedded AI (eAI) in the device. Elements need to be incorporated in this model will include anemia, total body water, and intradialysis hypotension.
- There will be a necessary convergence and synergy between AI and the neurosciences so that this dyadic relationship will increase the knowledge and capabilities of both of these intimately related sciences.
- Much of the interest in the use of AI in obstetrics and gynecology has been in the areas of fetal monitoring, which as a very small signal-to-noise ratio (similar to electrocardiograms and electroencephalograms), and in vitro fertilization. Similar to many of the aforementioned clinical areas, obstetrics and gynecology has a very heterogeneous group of patients and even includes fetuses as an essential part of the clinical scope. This level of complexity is well suited for more sophisticated ML/DL to decrease the burden and stress of decision-making for the clinician.
- Overall, the public and the subspecialty have been enthusiastic about the first autonomous AI-enabled diagnostic tool and therefore screening of certain common eye conditions such as diabetic retinopathy can be achieved with good accuracy. The global health implications are sizable: the DL algorithm improves in the scope of eye conditions can help one to diagnose, and the referrals may increase for the ophthalmologist to treat eye disorders.
- The relatively negative publicity that promulgated from the MD Anderson/IBM Watson for Oncology project did have significant repercussions and offered an important lesson in AI adoption and accountability. Issues involved in this failure include lack of competitive bidding, insufficient due diligence, and decision made without information technology department buy-in. The seemingly obvious questions regarding whether it would improve patient outcomes and lower costs were not adequately answered by the institution.
- While there is perception that some or even most of the pathologist work can be displaced by computer vision, there are several strategies. Perhaps the most intriguing is to have a new medical image subspecialist in the future who has a strong background in computer vision and DL that will entail the work of the present day pathologist and radiologist combined. This will be a subspecialist who will study the medical image continuum from the molecular and microscopic to the human-sized anatomic images.

III. The current era of artificial intelligence in medicine

- The field of pediatrics, both primary care and subspecialty care, underutilizes AI but certain areas are starting to adopt AI (cardiology, pediatric intensive care unit, etc.). Major obstacles to adoption include lack of understanding of AI and appreciation of AI tools among particularly pediatricians. The combination of heterogeneity of the population with the enigma of rare diseases renders pediatrics and the care of children ideal for deployment of AI for faster diagnoses and more accurate decision-making process.
- Even though there are not as many types of imaging for the pulmonologist (as compared to radiologist or cardiologist), there are many possible advances from precision pulmonology and testing. For instance, an easy at home pulmonary function testing capability can be interpreted with an AI-embedded pulmonary function device.
- In addition to the present work on CNN and medical image interpretation, there is also progress on NLP applications in radiology for less prose in reports and data mining. Some focus is now directed to workflow inadequacies in radiology that can potentially be neutralized by AI techniques.
- Wearable technology and eAI can provide valuable real time and daily information for AI-enabled chronic disease management that can be designed to be patient-centric. In addition, diagnostic dilemmas that frequently encountered by the rheumatologist can be aided by an AI resource in the form of updated information as well as an able cognitive partner. Lastly, as many of these patients have chronic disability, the AI tools in robotic technology and virtual assistants can be valuable as assisting dimensions for patient care.
- For the future, in addition to evolution of higher sophistication of robotic surgery, there are many areas that can be developed in the surgical areas. The surgeons would benefit from computer vision and medical image interpretation especially in the absence of a qualified radiologist. In addition, preoperative planning with the use of AI and VR or AR could reduce suboptimal surgical results and/or unnecessary complications.
- eAI and ML algorithms evolve toward the Internet of everything and will bring together people, process, data, and things; this strategy will allow the accrued data be streamlined and organized in the cloud proactively in an overall paradigm of personalized precision medicine.
- The multidimensional genomic and clinical data can be configured (mapped and projected through an ontology graph data structure) to search for individualized therapy. Clinical and molecular profiles from individuals are used along with their electronic health records data for a three-dimensional approach (horizontal knowledge planes or search space and vertical mapping with ontology layers) to recover concepts to infer therapeutic options. The basis for this framework is a hierarchically organized and ontologically based knowledge representation schema.
- For the future, since AI-enabled robotics looms large in the coming decades, this subspecialty will see a large panoply of robots and equipment that will be made available for rehabilitation of patients with handicap and/or functional impairment. Additional technologies such as altered reality will also provide an additional innovative dimension to rehabilitation strategies.
- The altered reality technologies with gaming and deep reinforcement learning can radically change the medical education experience as well as the clinical learning and training effectiveness.

- The domain of robotics and robotic assistance can be very robust in the future for nursing. In addition, the domain of chronic disease management and the possible role of virtual assistants will also be valuable for nursing care.
- AI in a hospital should be the formation of an embedded intelligence unit that encompasses all aspects of data analytics at a cognitive level and in real-time mode. There will no longer be a necessary difference between business intelligence and clinical analytics as both will be part of the intelligence in the hospital.

Assessment of Artificial Intelligence Readiness in Health-care Organizations

The strategy for AI implementation in a health-care organization can be in a series of steps of as a programmatic change. The AI strategy should be customized to the institution on the basis of overall institutional strategy and financial flexibility. To both assess and increase AI readiness of a health-care organization, the following 10 pillars can be used to assess the situation and to execute an action plan:

	Hospital	Score	Notes
		1–5 (10 categories) × 2 = 100 points maximum	
Data	Quality	1. PACS and/or CPOE only	*Higher score indicates better data*
		2. Above + physician documentation	
		3. Complete EHR	
		4. Complete EHR with HIE and patient portal	
		5. Complete EHR (as previous) with enterprise data warehouse	
Science	Analytics	1. Basic analytics (reporting)	*Higher score indicates better analytics*
		2. Basic analytics (forecasting)	
		3. Advanced analytics (machine learning)	
		4. Predictive analytics	
		5. Real time and/or prescriptive analytics	
Technology	Infrastructure	1. Poor infrastructure	*Higher score indicates better infrastructure*
		2. Below average infrastructure	
		3. Average infrastructure	
		4. Above average infrastructure	
		5. Excellent infrastructure	
		(AI-enabled hybrid cloud, robotic process automation penetration, 25% IoT/IoE, mobile access)	
Security	Cybersecurity	1. Poor cybersecurity	*Higher score indicates better security*
		2. Below average cybersecurity	

III. The current era of artificial intelligence in medicine

		3. Average cybersecurity	
		4. Above average cybersecurity	
		5. Excellent cybersecurity	
		(Firewall, web security, training, dark web id monitoring, layered defense, and endpoint protection)	
		1–5	
Team	Competencies	1. One full-time data scientist or data champion	*Higher score indicates more expertise*
		2. More than one data scientist (at least 1 at PhD level)	
		3. More than above but no true team leader	
		4. Team + designated senior leader in data and/or AI	
		5. AI program or center	
Problem	Relevance	1. Organization feels AI is not relevant	*Higher score indicates more relevance*
		2. Organization feels AI is of little relevance	
		3. Organization feels AI is somewhat relevant	
		4. Organization feels AI is very relevant	
		5. Organization feels AI is extremely relevant and urgent	
		1–5	
Strategy	Significance	1. AI not mentioned in strategic plan	*Higher score indicates higher significance*
		2. AI mentioned in strategic plan	
		3. AI pilot project(s) planned	
		4. AI project(s) ongoing with defined metrics	
		5. AI a major transforming force within organization	
Financial	Resources	1. No financial support	*Higher score indicates more finances*
		2. Support for small project ($25k or less)	
		3. Support for larger project ($25k or more)	
		4. Part of IT budget designated for AI projects	
		5. Separate budget for AI projects/team	
Timing	Culture	1. No/little interest/knowledge within organization	*Higher score indicates better timing*
		2. Few (<5%) have engaged in AI education	
		3. Some (>5%) have engaged in AI education	
		4. Above + ongoing education within organization	
		5. Above + regional/national meeting	
Barrier	Adoption	1. No/little interest from leadership	*Higher score indicates lower barrier*
		2. Some (<50%) interest from leadership	

III. The current era of artificial intelligence in medicine

3. Major ($>$50%) interest from leadership

4. Above + commitment from leadership

5. AI playing a role in executive leadership decisions

| Intangibles | AI and data-related papers/books, startups, awards, presentations, meetings, partnerships, consultancies, news, environmental, community support, etc. | *State reason(s)*

1 Point per reason (up to 10) |

AI, Artificial intelligence; *CPOE*, computerized physician order entry; *EHR*, electronic health records; *HIE*, health information exchange; *IoE*, Internet of everything; *IoT*, Internet of things; *IT*, information technology; *PACS*, Picture Archive and Communication System.

Ten Elements for Successful Implementation of Artificial Intelligence in Medicine

These are 10 necessary elements and solutions to solve issues and problems for AI in medicine to be an ultimately successful paradigm shift:

Improving data access, storage, and sharing strategies in health care. Much of the data in health care is missing, inaccurate, and disorganized as well as fragmented. In addition, patients need to be empowered to own their data as often health-care data are sequestered in a hospital or clinic with virtually no access. Finally, all the stakeholders must come together and be willing to provide access and share health-care data. If AI is the rocket to launch us into orbit and onto moonshot projects, data are the fuel we need; yet this fuel is not centrally collected. Timing is key as major data sources are still yet being formed, such as genomic data, wearable technology data, and socioeconomic data. While AI methodologies are ahead of schedule, data in health care have been in disarray for decades and will require effort to be improved. One key strategy will be the deployment of a revolutionary data infrastructure (graph or hypergraph databases). All of this discussion of health care data being collected and stored will require close attention to cybersecurity and the emergence of blockchain technology may be timely. In short, good AI in medicine mandates good not merely basic data and database foundation but also some level of connectivity via Internet of things (IoT) and Internet of everything (IoE).

Fostering AI in medicine awareness and education. Even with escalating increases in venture capital in the area of AI in health care and medicine, particularly in the areas of decision support and medical imaging, hospital administrators and clinicians as well as investors still lack sufficient education about AI methodologies applied to health care and medicine. There is a paucity of data science classes in medical schools and residency programs.

In addition, computer and data scientists are also less than fully enlightened in the realm of clinical medicine and what clinicians actually need most for AI to be utilized in order to lessen their burden. An overall higher level of knowledge will also prevent procuring AI solutions that are "Mechanical Turks."

Understanding human-to-human (H2H) collaboration for AI-driven agenda. In the work of AI in medicine, it is vital that H2H interactions and relationships drive these agendas and projects. From sharing data to doing projects using AI in health care and medicine, human champions and leaders from a myriad of domains need to be the drivers of these joint agendas.

Increasing clinician-to-data scientist synergy. There is a general lack of awareness and education for clinicians about DL (try to find a clinically active physician who even understands what DL is) and concomitantly for data scientists about clinical medicine (try to find a data scientist who interfaces with a physician more than once in a while and who actually spends time in the clinical domain). This clinician-to-data scientist distance is further increased with at times human hubris on both sides. One example is the preliminary ML work in atrial fibrillation: this has the potential to lead to overdiagnosis if there is not enough discussion between the clinical and data science domains. A great fictional comparison is that of Sherlock Holmes (the data scientist) and Dr. Watson (the clinician), the dyad of inspectors who work well together as both neutralize each other's weaknesses. Lastly, the present state of imbroglio in medicine and its future solutions based on AI mandates a special duality and synergy of clinicians and data scientists (like the double helices of DNA) with their respective different modes of processing contributing without hubris. Take the integer of "5" and have the clinician and data scientist each be this integer: if the two parties are antagonistic, then it is 5/5 or 1; if the two parties are complementary but not synergistic, then it is $5 + 5 = 10$; and if the two parties are synergistic, then it is $5 \times 5 = 25$. The future clinician may benefit from a dual education in data science and AI. These dual-trained scientists can then serve as valuable liaisons between the clinical and data science domains. It is possible that DL and AI can render the clinicians less clinically astute so these clinician–data scientists can mitigate this loss of clinical acumen. Lack of a dual perspective can easily lead to false presumptions; overdiagnosis is a potential problem such as studies that have excluded a clinician input. This cohort will also be useful to demystify the "black box" issue of AI. The cohort is helpful to police false discovery with finding spurious associations in the training set that have little relevance to new data.

Appreciating Small (and not always Big) Data in biomedicine. Finally, for the future, if we have sustained H2H collaboration and if we have design DL in health care and medicine, we will need to utilize not only DL but also variants of DL for solutions as there cannot be always Big Data to satisfy the ideal situation for DL. We need to work with "Small" Data (individual patient's serial data, e.g.) that is extremely important for many clinicians in serial follow-up. We also need to think of innovative DL methods such as one-shot learning and deep reinforcement learning for less than Big Data.

Making the visible invisible and the invisible visible as well as being able to explain AI. The AI projects should render present day paraphernalia such as computers and biometric devices obsolete (making the visible invisible). One such AI-inspired tool is the intelligent agent (or chat bots) that will replace many humans. Conversely, the signal in the noise will be picked up by DL and other techniques to make the invisible knowledge and intelligence in health care and medicine more discernible. It is good to have someone knowledgeable to triage the situations to see which will be best suited for AI methodologies. Lastly, we need explainable AI, or explainable artificial intelligence, to minimize the "black box" perception of AI and, in particular, DL among stakeholders who do not have a data science education or background.

Utilizing all aspects of AI tools in the AI portfolio (not only DL). Despite the attention, it is not always about machine or DL in biomedicine. The lack of a meaningful continual human (clinician)-to-human (data scientist) interface then results in a paucity of best solutions to real problems in health care and medicine with a full understanding of all the nuances. In addition, we have the possibility of creating a myriad of overdiagnoses that can result in inappropriate therapy and its inherent complications. In short, we do not, in health care, have "design" thinking. It is prudent to implement basic techniques in statistical analysis rather than always thinking that DL is the only methodology. Some of the knowledge can be culled from good basic data analytics without heavy duty DL and it is important to appreciate when situations do not demand AI. Conversely, many future problems in biomedicine and health care will need to be solved with a cognitive solution and not simply DL.

Understanding the complex nature of biomedicine. In spite of the success of AlphaGo in defeating the human Go champion, biomedicine can arguably be even more complex than the Go game. The "deep" phenotype of biomedicine will include hundreds of layers of data, including genomic and pharmacogenomic profile, socioeconomic milieu, and psychological profile. This deep phenotyping necessary to demystify the imbroglio of clinical biomedicine and precision medicine will become far more complex as the years of AI in health care and medicine become more omnipresent (see figure). There is a potential mismatch between precision medicine and population health; AI in biomedicine may be useful to reconcile these two forces.

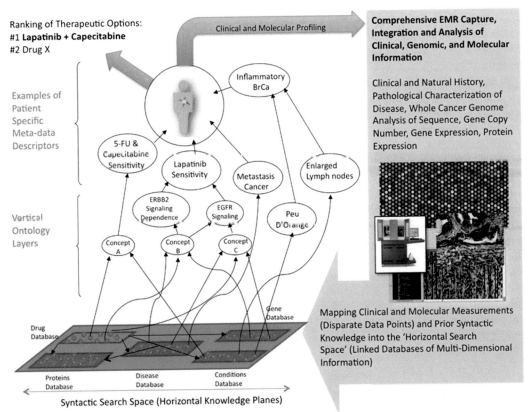

III. The current era of artificial intelligence in medicine

Introducing elements of cognition in AI in medicine. All stakeholders involved in AI in medicine and health care should have a rudimentary understanding of the types of AI and all their nuances and limitations. Appreciation of the third wave (neuroscience) of AI will be vital for all practitioners of AI in health care and medicine. The advent of IoT will evolve into the AI-inspired IoE in which devices will have some embedded AI capabilities; this is analogous to afferent peripheral nerves connecting to a central nervous system. With this capability, each person's health care could be provided a "clinical GPS" with illnesses being "traffic congestions" to be navigated around.

Executing more realistic projects as well as grand vision projects. It is essential to have a portfolio of AI in medicine projects especially ones that are field-proven automation projects (administrative tasks such as obtaining authorizations and credentialing health-care workers) using robotic process automation. The AI team should have a balanced portfolio of projects that include both easier to accomplish ones as well as more ambitious DL projects that focus on medical image interpretation or decision support in various clinical venues. This approach will increase adoption from both the administrative as well as the clinical leadership as there will be an ROI as well as value from the portfolio of projects.

Ten Obstacles to Overcome for Implementation of Artificial Intelligence in Medicine

Here are 10 potential obstacles (there are obviously more but these are some of the common ones observed) for successful implementation of AI projects or agendas in any organization or group:

Cultural differences between clinicians/hospital administrators and data scientists. Physicians, especially ones in acute care settings, prefer to make decisions expediently as they may have many such decisions to make within a short time (rounds or conferences). Data scientists usually have a less aggressive timeline and tend to work with more flexibility (although start-ups have a similar attention to timelines). There is also a difference in the hierarchical structure of these two groups.

Value proposition of AI projects and services. It is sometimes very difficult and/or tedious to convince all stakeholders the value proposition of individual AI projects from various services. This obstacle is usually easier to overcome when a previous AI service or project has been productive or has returned some value on investment. It is also helpful to have at least a few clinician champions in the organization to emphasize value in terms of quality of patient care delivery.

Knowledge deficit on both sides. Clinicians are mostly familiar with statistics from their medical education but are not usually well educated in data science. Data scientists, on the other hand, can search for knowledge in medicine but at times do not comprehend the many nuances and intrinsic complexities and uncertainties of clinical medicine nor how physicians think. This creates a sizable domain knowledge schism that can exaggerate the cultural differences.

Trust in new technology. The clinicians sometimes feel somewhat antagonistic toward certain technologies that had great promise but resulted in much lower delivery. One constant comparison is that of EHR and AI as the latter technology is sometimes bundled with the former. Clinicians naturally still harbor some discontent with EHR due to its increased burden and lack of perceived value. AI can focus on mitigating this burden or at least bring value as much as feasible.

Explainability of AI tools (and the "black box" perception). There is a perception that there is a lack transparency of AI tools. It may be unfair to expect AI tools to be fully elucidated as clinicians do not always have that expectation for all technologies (pacemakers, e.g.). If clinicians attain a basic understanding of AI and if the data scientists work on rendering these AI tools more understandable, perhaps an acceptable middle ground of interpretability can be reached.

Workflow affected by AI. There is low tolerance among clinicians for more burden of any kind due to a relatively high rate of professional burnout from many reasons (one major reason is the EHR burden). The ideal AI project would not only decrease the burden but also improve patient care and bring the cost of care down. The only sure way to get adoption for an AI project that would increase burden is if the project will clearly demonstrate improved quality of care.

Access to large volume of high integrity data. Biomedical data, in their cleanest and most complete form, are very difficult to secure in high volumes. Even very large biomedical datasets can have very inaccurate labeling as well as many other issues. Clinicians can help one to overcome this limitation by collaborating with other institutions, which is rarely done but hopefully will be more routine in the near future. Data scientists can also be more flexible with working with small data.

Interoperability and EHR infrastructure issues. Aside from the aforementioned data size and integrity issue, there are logistical hurdles of data access and gathering that can have high levels of time and resource demands. Lack of full interoperability between hospitals and health-care venues makes it difficult to have full access to the data as well as to gather all the data. There is also an understandable lack of full collaboration among EHR vendors that will facilitate data sharing.

Lack of clinician champion(s). Just as with any project, it is much more difficult to get AI adoption if there is no clear clinician champion who would be very committed to the AI project. Even with a clinician champion, that person will still need to gather several cochampions who will not only support but also sustain the project. The champion(s) will need to maintain constant communication with the AI stakeholders to have circular feedback throughout the project.

Fear of AI taking over and displacing clinicians (and others). There is a concern among health-care workers that part or most of their job descriptions can be displaced by automation and AI. The premature public declaration by a vocal few that doctors will be replaced was not widely accepted nor appreciated by clinicians. The more realistic concept is that AI will complement the skill sets of health-care workers and reduce the excessive burdens that they face on a daily basis.

In addition, here are a few more obstacles from Kevin Lyman of Enlitic:

1. Radiology data are highly nuanced and unintended bias is hard to avoid without proper education.
2. More time is spent building tools that enable us to build clinical AI than actually building clinical AI.

3. Most hospitals do not design their software infrastructure with ease of AI integration in mind.
4. Models do not get regulatory approvals, claims around very specific uses of a model do.
5. Privacy and security are subjective measures with moving targets, especially on a global scale.

References

[1] He J, Baxter SL, Xu J, et al. The practical implementation of artificial intelligence technologies in medicine. Nat Med 2019;25:30−6.
[2] Darcy AM, Louie AK, Roberts LW. Machine learning and the profession of medicine. JAMA 2016; 315(6):551−2.
[3] Nsoesie EO. Evaluating artificial intelligence applications in clinical settings. JAMA Netw Open 2018;1(5): e182658.
[4] Maddox TM, Rumsfeld JS, Payne PR. Questions for artificial intelligence in health care. JAMA 2019; 321(1):31−2.

PART IV

The Future of Artificial Intelligence and Application in Medicine

We can only see a short distance ahead, but we can see plenty that needs to be done. *Alan Turing, British mathematician*

The future of artificial intelligence (AI) and its applications in biomedicine is not only very promising but also full of challenges [1]. This premise of advances in AI is based on rapid improvement and accelerated deployment of advanced computer technology (such as quantum computing and neuromorphic chips) and rapid development and evolution of AI techniques (such as deep learning and its many variants) as well as cognition-influenced architecture that have led to a current Cambrian explosion of AI.

Peter Voss so aptly described the future of AI as the "third wave": the first wave of AI was good old fashioned AI that focused on traditional programming followed by the second wave of AI of the current deep learning so that this third wave will be reliant on many cognitive architectures (see Fig. 1). While this third wave of cognitive architecture is much more complex as it incorporates the relevant elements of human cognition, the former two waves do lack the biological completeness and integration of the third wave. Machine learning alone in medical decision-making can be dangerous if it is performed without human cognition and "common sense" [2]. In this third wave, cognitive architectures will need to possess: general learning ability; real-time and interactive learning; dynamic goals and context; transfer learning; and abstract reasoning in order to reach artificial general intelligence. In addition, human-like intuition AI developments designed to be between machine and deep learning and rule-based programming are helpful; these AI advances include interactive networks (a model that can reason about how objects in complex systems interact), neural physics engine (an object-based neural network architecture for learning predictive models), and recursive cortical networks (discussed next) [3].

(A)

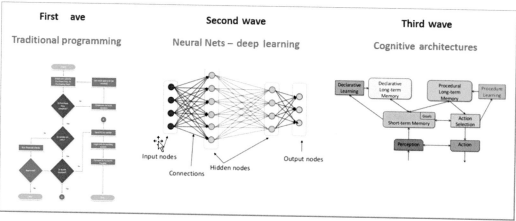

(B)

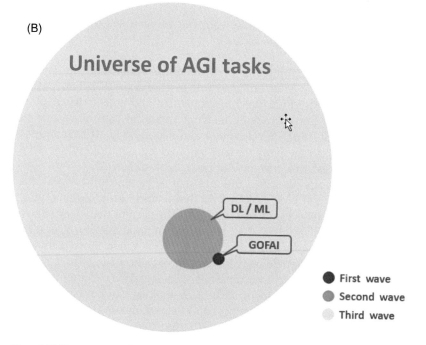

FIGURE 1   (A and B) Three waves of AI. The waves are seen in (A). The first wave of AI was programming, followed by the second of AI of which we are in now that of deep learning. The upcoming third wave of AI will be focused on cognitive architecture. In (B) the eventual AGI from the third wave is considerably larger than the two prior waves. *AGI*, Artificial general intelligence; *AI*, artificial intelligence.

References

[1] Stead WW. Clinical implications and challenges of artificial intelligence and deep learning. JAMA 2018;320 (11):1107−8.

[2] Cabitza F, Rasoini R, Gensini GF. Unintended consequences of machine learning in medicine. JAMA 2017;8318(6):517−18.

[3] How researchers are teaching AI to learn like a child (AAAS). <https://www.youtube.com/watch?v = 79zHbBuFHmw>.

10

Key Concepts of the Future of Artificial Intelligence

In addition to cognitive architecture, there are additional areas of development of artificial intelligence (AI) that are particularly relevant to clinical medicine and health care (in alphabetical order).

5G

This is the next-generation mobile Internet connectivity that will provide much faster speed (100 × faster than 4G) and more reliable connections on devices. This future capability will accommodate the exponential rise in Internet of Things (IoT) and medical devices with the infrastructure necessary for the large amount of data to come. Increased speed and data transfer as well as more bandwidth are the end result of having 5G capability; this will be essential as more devices become more sophisticated. The countries most active in bringing in 5G are the Unites States, China, and South Korea.

Augmented and Virtual Reality

The future of AI will enable both augmented and virtual reality (VR). Augmented reality is an enhanced reality that is a result of computer-generated enhancements atop of reality. VR, coded by a special language called VR modeling language, is a computer-generated artificial simulation or recreation of a situation mainly via vision and hearing (Facebook Oculus is an example). Both of these AI-enabled reality methodologies will become commonplace in the near future and will have many possible applications in medicine and health care, such as medical education (both clinicians and families), training (including simulation), and preoperative or preprocedural planning.

Blockchain and Cybersecurity

Blockchain is the use of cryptography to allow a collection of blocks or records to be maintained in such a way that is difficult to modify. This strategy was initially used for

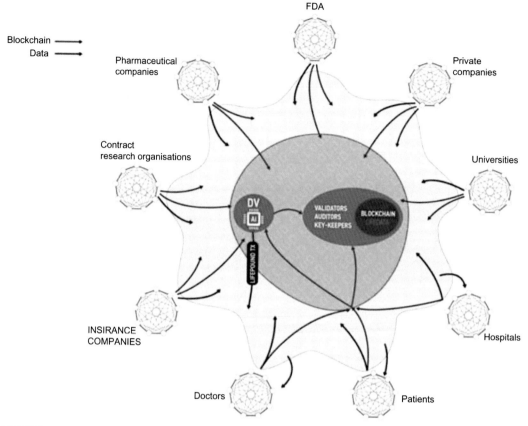

FIGURE 10.1   Personal data-driven ecosystem. Individuals have full autonomy of their own health-care data with various stakeholders (such as insurance companies, pharmaceutical companies, Food and Drug Administration, and doctors/hospitals) having access to this data.

bitcoin as a public ledger. Blockchain is therefore a disruptive innovation in information registration that utilizes three existing technologies: private key cryptography, peer-to-peer network, and the blockchain protocol. A successful deployment of technologies such as blockchain to improve cybersecurity in health care will facilitate data sharing among stakeholders in the near future. Additional future cloud and data security concepts to be adopted for the future will need to include novel concepts such as blockchain as well as (1) homomorphic encryption and (2) differential privacy. Homomorphic encryption is an encryption strategy that allows for certain computations to be performed on medical data while they are still encrypted [1]. One significant limitation for this security solution is that the processing speed is slowed during this process. On the other hand, differential privacy uses sophisticated algorithms to add sufficient "noise" to the data to render it less vulnerable to linkages to other databases for matching purposes. Mamoshina (see reference previously) proposed to create a large secure health-care data ecosystem that is enabled by a convergence of blockchain and AI (see Fig. 10.1).

Artificial intelligence (AI) and blockchain: stronger together

Sriram Vishwanath

Sriram Vishwanath, a professor of computer engineering, authored this commentary on the synergy between blockchain and AI, as these technologies are complimentary as blockchain can increase the transparency and interpretability of AI.

As highlighted in other sections of this book, AI is a powerful tool that will impact multiple health-care applications. Even though the use of AI in health care is still in its infancy, its central value in transforming health care remains undeniable. A few decades ago, AI was hyped as a magic solution for our needs across all aspects of life. We have since discerned the difference between fact and fiction in AI, and when used in a targeted manner, have realized its potential in improving our health-care system.

However, our understanding of blockchain and its benefits in health care is still in its very early stages, with blockchain remaining a somewhat enigmatic tool for many of us. Unlike AI and machine learning, which are much better understood and are considered to be significantly more mature, blockchain is undergoing a "hype-cycle" of its own as we speak, which is causing its potential impact on the health-care ecosystem to be obscured. Just as in the case of AI, targeted use of blockchain can prove to be immensely useful to multiple domains, including health care. To accomplish this, we must look past the hype and understand blockchain for what it actually is, and what societal need it addresses.

Blockchain enables decentralized exchange of value and incentivizes parties to participate in such an exchange in a trustless manner (providing a solution to a classical problem called the "Byzantine Generals Problem"). Blockchain can enable this to be accomplished without a central authority and can simultaneously provide privacy and compliance guarantees for the participants within its network. Much like AI, blockchain is an interdisciplinary, brilliantly devised solution to a major societal need: In the case of AI, it is the need to extract value from data; and in the case of blockchain, it is the need to transfer this value in a decentralized, trustless manner. A decentralized architecture is desirable across many multiparty settings, when power/decisions are not concentrated in the hands of one person, object, or entity. Blockchain is the first such system to enable such an interaction while providing provable guarantees for security and privacy.

There is a natural, synergistic relationship between AI and blockchain technologies. First, blockchain can increase interpretability and transparency within AI and, therefore, transform a "black-box" algorithm into a "glass box." This enables a deeper understanding of the features, factors, and reasons underlying AI's predictive algorithms and enables both the developer and user to comprehend the primary driving forces behind them. Second, blockchains enable private auditable ledgers, thus producing HIPAA compliant decentralized data structures on which AI algorithms can operate. This simultaneously enables compliance, auditability/traceability, and efficient decentralized learning.

AI and blockchain are two emerging technologies that are naturally complimentary, and together, they will change health care forever. We have a unique opportunity to embrace the benefits of these technologies and be part of this transformation, making lives better for each member of the health-care ecosystem through this process.

Brain—Computer Interface

The other terms for this concept include mind—machine interface, brain—machine interface (BMI), or direct neural interface. These are all communication pathways between the brain and an external device to augment natural intelligence. An example of this type of device is the one proposed by Elon Musk called "neural lace." This area of development is of particular interest to many areas of physical rehabilitation as these interfaces with AI capabilities can improve patients who are disabled in any way with augmentation of their capabilities.

Capsule Network

This is Hinton's recent description of neural network and a capsule [2] with an element of biomimicry that can be "smarter" with less input data. This aspect is potentially ideally suited for biomedical data. The conventional convolutional neural networks (CNNs) were instrumental in the popularity of deep learning but have significant limitations. One such drawback is the lack of spatial hierarchies between the objects. Capsules, which are groups of neurons that will encode spatial information, essentially can introduce the cognitive element of "intuition" to deep learning as these entities improves model's spatial information and hierarchical relationship.

An example in the very recent biomedical literature of this new development of capsule network with CNN is in Alzheimer's disease diagnosis with MRI and CNN with capsule network for superior performance (compared to 3D CNN with an autoencoder) [3].

Cloud Artificial Intelligence

Cloud computing has been thus far a public cloud model (exemplified by Amazon Web Services or Salesforce's CRM system), but the cloud of the future will enable virtualization and management of software-defined data services. Future AI applications, therefore, will be in the cloud so AI will be in the form of AI-as-a-service. The other aspect of cloud computing that will be essential for AI is its role in the formation of Internet of Everything (IoE) (see next) to create a ubiquitous decentralization of devices and sensors.

Edge Computing

As a counterbalance and a coupling to cloud computing, edge computing technology enables devices to collect and analyze data in real time locally so that it is processed outside the cloud datacenter. This is akin to having a peripheral nervous system with local signal processing (as opposed to having every signal going to the central nervous system for analysis). In addition to being more efficient, this process can also increase the level of security as the data can be processed without going to the cloud. Disadvantages of this edge computing include more sites that will need to be configured and monitored as well as issues with decentralization.

Embedded Artificial Intelligence (or Internet of Everything)

Data in health care will concomitantly escalate as well from the advent of wearable and home monitoring technology leading to IoE [4] (see Fig. 10.2). There are now an estimated 100,000 mHealth apps available [5]. While the IoT is the interconnection of billions of physiologic devices to the Internet, IoE will be essentially a "network of networks" to incorporate people, processes, and data to these devices to enable automation in data acquisition and analytics without human intervention [6]. All of these "smart" devices with wireless sensor networks (WSNs) will add to the collective intelligence of medical data and information. The IoE is basically intelligent connection of people, process, data, and things; it builds on IoT by adding network intelligence and turning information into actions. The future of AI will need to involve the IoE just as the brain needs a nervous system.

Embedded AI is the technological paradigm of integrating machine learning or other AI tools into the devices to attain a certain desired function. Embedded AI is advancing beyond the IoT, that is, devices communicating with each other and generating data. With biomedical devices, especially ones with continuous monitoring, the amount of physiological data received would be unmanageable by caretakers unless there is an "upstream" intelligent algorithm built in to the device to filter all the noise from the signal (for

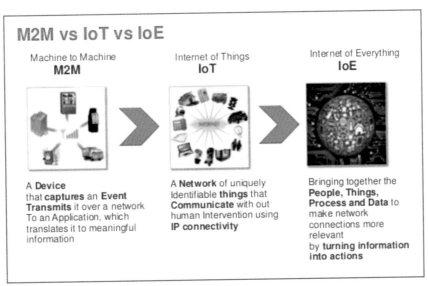

FIGURE 10.2 Internet of Everything. The current transition is from machine to machine to now Internet of Things, a state in which elements are connected. In the near future, more and more equipment and sensors will not only be interconnected but also embedded with AI tools to render these devices smart. *AI*, Artificial intelligence.

instance, filtering out normal sinus rhythm and sending an alert to the cardiologist only when there is an abnormal rhythm like atrial fibrillation or ventricular tachycardia).

Fuzzy Cognitive Maps

Fuzzy logic, combined with neural networks, can form fuzzy cognitive maps (FCMs) that are an efficient and robust AI technique for modeling complex systems in medicine [7]. FCM can be helpful in designing medical decision-support systems by focusing in four areas: decision-making, diagnosis, prediction, and classification [8]. The advantage of FCM is in its unique specification of integrating human knowledge and experience with computer-aided techniques so that one can achieve a human—machine synergy that is ideally suited for the complexities of medicine.

Generative Query Network

Generative query network is a very significant foray into how machines can learn like a child from Google's DeepMind. This is a framework for a machine to learn on its own (based on data) to perceive their surroundings without any human labeling. This is accomplished by having two different networks: a representation network (in which an agent makes observations) and a generation network (in which these observations are turned into predictions). This autonomous process will facilitate training for neural networks and have a sizable impact in medical imaging [9].

Hypergraph Database

As elucidated in an earlier section, traditional relational databases are weak in complex hierarchical data and processing graphical data structures. A graph database is designed to neutralize these disadvantages by processing data in a graphical (with nodes and edges) strategy and enables queries across the data network. In order to model even more complex and highly interconnected data, a new paradigm of data representation called hypergraphs will need to be implemented. A hypergraph is a graph model in which the relationships (called a hyperedge) can connect any number of nodes. Future applications of AI will become increasingly more sophisticated, and the data in biomedicine will need a paradigm shift to a graph and even hypergraph format with graph algorithms that will take graphs as inputs.

Low-shot Learning

Popularized by facial recognition work, one-shot learning is an advanced methodology of supervised learning algorithm that uses a siamese neural network to be able to learn from one or very few images. In addition, Fei-Fei of Stanford promulgated one-shot learning that can bring a special dimension to unique cases in medicine as it will not require the usual large dimensionality of data that the previous types of learning will need [10]. There is work on one-shot learning with both memory-augmented neural networks (see Neural Turing Machines, or NTM above under neural networks) and also siamese neural networks. Siamese neural network contains two or more identical subnetworks (same parameters and weights). One-shot learning would be particularly useful in biomedicine as CNN are based on large and labeled data sets that may not be scalable for some diseases.

Even more interesting, zero-shot learning is a supervised learning that is able to predict labels that are not in the training data by using embedding vectors called word embeddings. In other words, this type of learning is able to solve a task by not having received any training data. All of these low-shot learning methodologies that do not require big data can be particularly useful in clinical medicine and health care.

An example in the recent biomedical literature of one-shot learning is the deployment of this learning for cervical cell classification in histopathology tissue images in patients with cervical cancer with a 94.6% accuracy in detection [11].

Neuromorphic Computing

It is also known as neuromorphic engineering, this is a concept in which computer chips can mimic the brain by communicating in parallel using "spikes" that are bursts of electric current that the neuron can control. These neuromorphic chips have the advantage over traditional computer central processing units in that these chips require far less power to process AI-inspired algorithms. Advances in AI in medicine in the future will require technological improvements such as neuromorphic computing.

Quantum Computing

Another type of futuristic computing is a new approach to process information and is much more powerful than the conventional computer. A quantum computer utilizes quantum bits (or qubits) instead of the conventional bits (that is in 0 or 1 states) used in digital computing. Quantum computing takes advantage of the quantum phenomenon that subatomic particles can exist in more than one state at a time. A quantum computer such as the D-Wave computer is 100 million times faster than a conventional laptop. As the demand for computation exponentially increases in biomedicine as we head into the cognitive era of AI, quantum computing (and other types of computing such as DNA computing) will be a necessary technological tool.

Recursive Cortical Network

This is a generative model (structured probabilistic graphical model to be exact) that differs from deep learning in that it has a scaffolding (rather than learning from scratch, or *tabula rasa*, as in deep learning). Recursive cortical network is thus an object-based model that can be much more efficient in learning and was able to break the text-based

CAPTCHA (completely automated public turing text to tell computers and humans apart). The result of this model is that it has a contour hierarchy of features that is guided by neuroscience and the visual cortex.

Spiking Neural Network (SNN)

SNNs, called the third generation of neural networks, operate with events called spikes that are discrete events that occur at certain points in time; it therefore is more biological than machine learning in that the neural network often relies on the timing of individual action potential of neurons. This is work that is now more in the realm of brain-inspired AI.

An example in the recent biomedical literature of SNN is the Kasabov study: It is suggested that SNN, with its innovative spatiotemporal architecture, can serve as a framework to improve processing velocity as well as accuracy for spatiotemporal brain data (such as electroencephalogram) for earlier diagnosis of degenerative brain diseases such as Alzheimer's disease [12] (Fig. 10.3).

Swarm Intelligence

Swarm intelligence is intelligence derived from many individuals based on self-organizing group behavior. The collective behavior illustrates that unified systems

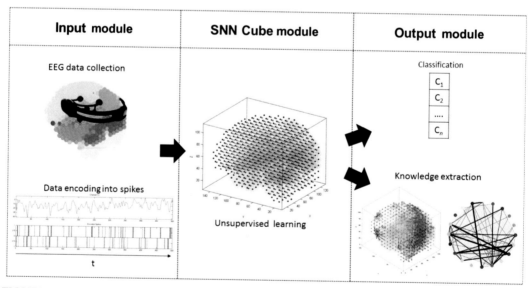

FIGURE 10.3   Brain-inspired architecture for EEG data classification and knowledge extraction. This architecture has three components: input module where EEG data are encoded into trains of spikes like an SNN into the SNN cube module; in the SNN cube module, the spatiotemporal characteristics of the EEG data are captured and learned; and the output module accommodates classification after unsupervised learning, and new knowledge discovery is obtained. *EEG*, Electroencephalogram; *SNN*, Spiking neural network.

outperform the majority of individual members, but since humans do not naturally have these connections as observed in ants or fish, swarm AI is executed by technology to provide feedback to human members. This group dynamic results in ad hoc information gathering and sharing, and consensus is based on dissemination of the pool of knowledge. In short, this type of intelligence leverages the "wisdom of the crowd," and this philosophy is direly needed to answer the numerous queries we have in medicine and health care.

Temporal Convolutional Nets

Up to now, recurrent neural networks (RNNs) have been involved with sequence problems such as language and speech. Temporal convolutional nets (TCN) are CNNs with some added new features and are challenging RNNs for the sequence problem category. In addition, video-based action segmentation has been previously dealt with in two steps: low-level features for each frame by using CNN followed by an RNN as a classifier for higher level temporal relationships. A unified approach with TCN and capture all levels of time scales [13]. This flexibility with TCN can be ideal for a myriad of situations in medicine.

An example in the recent biomedical literature of TCN is in its application in early recognition of sepsis with the TCN embedded with a multitask Gaussian process adapter framework so that it is applicable to irregularly spaced time series data [14].

Transfer Learning

Transfer learning involves adaptation of a trained model to predict examples from a different data set; this phenomenon is particularly favorable with deep learning networks since deep learning requires so much time and resources for its training. In other words, a model that is previously trained on a type of task is now repurposed on a different type of task. The type of learning is called inductive transfer. Transfer learning can be accomplished with both image and language data. Compared to traditional machine learning, transfer learning is very different as it accomplishes learning of a new task by relying on a previously learned task so that it has acquired knowledge (whereas machine learning does not retain that knowledge of a learned task). This knowledge of solving one problem is applied to a different (but related) problem (see Fig. 10.4). Finally, transfer learning can be accomplished in minutes compared to hours for deep learning and can be done with small data sets (vs large data sets for deep learning).

An example in the recent biomedical literature of transfer learning is the strategic use of transfer learning in classification of retinal images for macular degeneration and diabetic retinopathy based on pretrained weights in a CNN that was deployed for ImageNet [15]. Another notable example is the work of Wiens et al. on the use of transfer learning to leverage data from multiple hospitals for predictions on hospital-acquired infections [16].

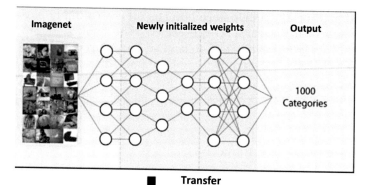

FIGURE 10.4 Transfer learning. Instead of training a deep neural network from the beginning, one can take a previously trained network on a different domain (such as ImageNet) for a source task and adapt it for a target task with smaller amount of data (retinal images). The transfer of learned knowledge renders the target model able to use small amount of data.

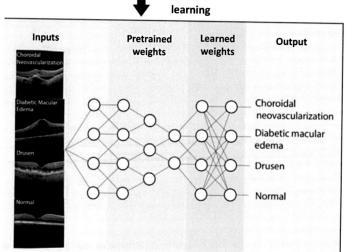

Artificial intelligence (AI) and natural language processing for rare diseases (RDs)

Mayur Saxena

Mayur Saxena, a biomedical engineer and the founder of a startup in AI in health care, authored this commentary on the strategic use of AI, specifically natural language processing, on diagnosis of RDs as many of these RDs.

Every physician is trained to be biased

Doctors are commonly taught a couple of truisms in medical school that help them approach real-world clinical situations: (1) "common conditions happen commonly" and (2) "when you hear hoofbeats, think horses not zebras." Years of medical practice only reinforce these truisms since doctors spend most of their time treating conditions that occur frequently.

Advancements in genetics have helped us identify more than 7000 RDs. Around 350 of these have some form of effective treatment. These conditions affect less than 1 person per 2000 individuals in the population [1]; on the extreme end of the spectrum, some diseases occur less frequently than one in a million people. Some of these conditions are well understood and have established treatment regimens. Without timely diagnosis, many of these patients will succumb to a disease that could have been managed with lifesaving interventions. Many of these RDs are themselves poorly understood in terms of disease presentation, progression, and management options. While finding successful treatments is difficult for any disease, this problem is compounded with RDs by the prohibitive challenge of finding enough patients to enroll into clinical trials.

Although physicians may have been exposed to some of these rare conditions in medical school, it is common for a doctor to go a lifetime of practice without experiencing many of these disease entities in person. Humans learn best from experience, and in the case of RD, experience is hard to come by. Moreover, many of these diseases have been only recently discovered and characterized; it is not uncommon for a physician to be completely unaware of the existence of many RDs. One of the unanswered questions in medicine is how can we help these well trained, experienced doctors to identify these zebras in addition to the horses they see and manage daily?

The potential of artificial intelligence for rare diseases

On average, patients with RD go undiagnosed for decades [2,3]. AI offers a means to help the physician find undiagnosed RD patients and get them the care that they need. In the era of digitized health data, it is possible to build a technology that sieves through massive repositories of patient data such as electronic medical record (EMR), claims, and labs and identifies patients that potentially have an undiagnosed RD. Such a system could assist the doctor that is trained to look for horses to confront the possibility that they are faced today with a zebra. In this commentary, we discuss how breakthroughs in AI and natural language processing (NLP) have the potential to revolutionize the diagnosis and management of RDs.

Why natural language processing is a key technology for rare diseases

Assuming you had a data repository the size of the US population, if one is searching for cases of a disease that has a prevalence of one in a million, one would expect to find approximately 350 such patients. Even though the widespread adoption of EMRs opens up the possibility of finding RD patients using data science, the actual EMR data is often messy and difficult to work with; the vast majority of this data exists in free-form clinical text. While problematic to analyze, this text is exceedingly valuable as it contains information that is found nowhere else in the EMR, including free-form medical history, observations, symptoms, nursing notes, and diagnoses that have not been billed (i.e., documented without ICD-10 codes).

One of the key AI technologies that can enable the use of EMR data is *NLP*. NLP encompasses AI methods that process, analyze, and make sense of natural language. Using NLP, it is possible to digest huge repositories of clinical text and extract *medical concepts* such as symptoms, conditions, and histories to come up with a representation of data that can be fed into algorithms that are able to detect patient cohorts suffering from a RD that has not yet been diagnosed. This can be accomplished by searching for patterns and connections between clinical concepts over huge data sets that contain enough information that even the hidden patterns of RD may be revealed.

The power of a combined big data and NLP approach can be illustrated through a simple toy example. Consider a fiction RD, which we call *RD1*. Of course by virtue of being a RD, all possible complaints and disease phenotypes for RD1 are not known. However, there are several underlying factors and associated symptoms that are known to be markers of RD1:

1. Chronic hemolysis (primary marker, occurs in ~100% of cases) symptoms of hemolysis include fatigue, rapid heartbeat, headache, chest pain, dyspnea on exertion, blood clots, and kidney damage
2. Venous thrombosis caused by blood clots (believed to occur in ~30% of cases)
   a. Liver: jaundice, abdominal pain, Budd–Chiari syndrome
   b. Lungs: dyspnea, heart palpitations
   c. ...
3. ....

Now consider a patient with a BMI of 32 that presents with venous thrombosis. However, the doctor failed to determine that chronic hemolysis was the cause of thrombosis and instead assumed it to be caused by the patient's obesity, which is a much more likely explanation. In this case an AI algorithm can compare this individual's EMR against the medical concepts extracted from similar patients (i.e., those suffering from obesity and venous thrombosis) and flag this patient if there are patterns that are either (1) atypical of venous thrombosis patients with obesity or (2) similar to reported cases of RD.

By no means can this pattern matching be conclusive on its own and be called a diagnosis of RD1. However, a trained provider might consider this information and potentially choose an appropriate diagnostic test for RD1.

Envisioning the future of rare disease diagnosis

With enough data access, such patterns of RD can be computed using massive amounts of data across systems, providers, EMRs, conditions, and geographies. To make this future a reality, hospitals, payers, and government bodies will have to come together and decide to create a system that can responsibly bring attention to these patients [4]. Each RD alone may only affect tens or hundreds of patients across the United States. But 7000 RDs together are estimated to affect 25–30 million US citizens. The authors strongly believe that AI and NLP with support from the right authorities can change the way RD care is dispensed globally [4].

References

[1] Rare diseases act of 2002.
[2] Black N, Martineau F, Manacorda T. Diagnostic odyssey for rare diseases: exploration of potential indicators. London: Policy Innovation Research Unit, LSHTM; 2015.
[3] Muir E. The rare reality – an insight into the patient and family experience of rare disease. In: Rare disease UK; 2016.
[4] The Pharmaceutical Research and Manufacturers of America (PhRMA). Progress in fighting rare diseases, <https://www.phrma.org/media/progress-in-fighting-rare-diseases>; 2019 [accessed 19.03.30].

Data and Databases

For all of these aforementioned advanced AI tools to effectively create a paradigm shift in medicine, we still need to vastly improve health-care data and databases.

First, health-care databases can utilize the above graph DMBS (see previously) mapped into networks of autonomous databases as a federated or virtual metadatabase management system for interoperability and collective intelligence [17,18]. Many subspecialties lack such a coordinated network of stakeholders but will benefit immensely from such a collaboration. The advantage of such as network is incorporation of a Web-enabled semantic search and global query capability with data discovery is that it is ideally suited for biomedicine especially with rare diseases (RDs) and with complex imaging data [19−21]. In addition, this federated approach using Internet-based networking technologies can provide excellent collaborative research in epidemiology and public health even at the international level [22]. Lastly, this federated system provides an excellent framework for the IoT networking paradigm of interconnected smart objects [23]. This data discovery capability can eventually mature into AI possibilities embedded into the database. One key future development of big data analytics is real-time analytic processing (RTAP). This is a process in which the data is captured and processed in a streaming fashion using online analytical processing and complex machine learning algorithms [24].

Second, each configurable cloud infrastructure in the biomedical system should retain the following essential characteristics of cloud computing: on-demand self-service, broad network access, resource pooling, rapid elasticity, and measured service [25]. The cloud infrastructure that is best suited for each situation will be configured to meet that need. Health care with its big data and data analytical needs that concomitantly demands privacy and security can use this flexible cloud infrastructure to meet the challenges.

Third, little in the biomedical data domain is truly virtualized at present [26]. There is, however, a report of software-defined network (SDN) in a hospital setting in Japan using *OpenFlow* as the hospital in-house LAN (one of major technical specifications of SDN published by Open Networking Foundation) [27]. In short, SDN decouples the control plane from the data plane and creates more dynamic resource utilization with better oversight of network bandwidth and with direct programmability. One limitation of SDN is its requirement for network architecture and design skills. In addition, the storage of heterogeneous health-care data sources can be abstracted into designated data pools, with this process being an app-centric policy-based automation (software-defined storage) (such as *IBM's Big Blue Elastic Storage*). This strategy is ideal for the biomedical system with its data-intensive applications that require immediate access and rapid analytics as well as matching the storage capability with the type of data [28]. As an example, this strategy has been implemented at Maimonides Medical Center with *DataCore's SANsymphony-V* software with savings in hardware and personnel costs and increased performance of applications [29]. The aforementioned SDN concept has evolved into a software-defined data center (SDDC) architecture (with server virtualization, software-defined networking, and storage hypervisor) that can be entirely virtualized so that all of the infrastructure and the management can be automated by software. In short, the SDDC frees the application layer from the hardware layer.

Fourth, an entirely virtualized infrastructure can include compute, network, security, and storage abstractions such that it is IT as a service (ITaaS) and its cloud infrastructure also managed by automation. One advantage of such a system is its remote programmability to render it agile and automated as well as global and continuous. All of the components are essentially decoupled from the hardware as an ubiquitous software for all users in various hospitals and programs at any given time. Another advantage is the accelerated service delivery (from weeks/days to even hours). The dynamic configuration of SDDC can optimize resource allocation and improve efficiency in health care.

The future of software-define data systems will be a federated system with more standardized network protocols and more automated interfaces for management. With emerging sophisticated database management systems and cloud and virtual computing technology, medical data can be efficiently organized into a virtual intelligent biomedical data "ecosystem" to better serve the needs of hospitals and health systems [30]. In addition, the WSNs as well as other patient-generated data will need to be virtualized to provide effective solutions in health care [31,32]. This virtual strategy will enable medical data to converge with AI methodologies to promulgate true medical intelligence in the cloud [33]. One major area of concern will always be data security in the cloud as present data shows 94% of health-care institutions having had breach of data [34]. The exponential convergence of existing biomedical data with both genomic and biophysiologic data will render medical data to be even more voluminous, complex, and heterogeneous. This explosion of medical data will need a more sophisticated database management strategy as well as cloud and virtual environments to enhance data discovery as well as ensure data security and privacy.

In summary, the future of biomedicine can include a proposal for an AI-inspired cloud continuum of data—information—knowledge—intelligence (a "medical intelligence" as a service, or "MIaaS"). First, future medical data can be managed in a graph-based metadatabase management system with RTAP for both its storage capability and its query flexibility to accommodate the large and complex medical data in the ensuing decades. Second, future medical data in the customized cloud infrastructure will be far more sophisticated than a simple public—private dichotomy and can be customized from a cloud infrastructure system based on customer versus supplier control, ownership, and responsibility as well as private versus shared infrastructure and operations; cloud security can be further enhanced by mechanisms such as homomorphic encryption and differential privacy. Third, a SDDC architecture can be entirely virtualized so that the infrastructure that includes compute, network, and storage abstractions will result in ITaaS. The future medical data system will be entirely in a virtual synergy with humans and contribute to medical intelligence.

By embedding intelligence into all aspects of medical data from graph database and metadatabase management system to customize cloud infrastructure and to SDDC and virtualization, the aforementioned strategies can accelerate this transformation in biomedicine from fragmented and unstructured data sets to cohesive and agile information imbued with medical intelligence.

References

[1] Kocaba O, et al. Medical data analytics in the cloud using homomorphic encryption. In: Chelliah PR, et al., (Eds.), Handbook of research on cloud infrastructures for big data analytics. ITI Global., Hershey, 2014, pp. 471−488.

[2] Sabour S, Frosst N, Hinton GE. Dynamic routing between capsules. arXiv:1710.09829v2.

[3] Rajeswari KRK, Maheshappa HD. CBIR system using capsule networks and 3D CNN for Alzheimer's disease diagnosis. Inform Med Unlocked 2019;14:59−68.

[4] Banaee H, et al. Data mining for wearable sensors in health monitoring systems: a review of recent trends and challenges. Sensors 2013;13:17472−500.

[5] Cortez NG, et al. FDA regulation of mobile health technologies. N Eng J Med 2014;171(4):372−9.

[6] Gubbi J, et al. Internet of things (IoT): a vision, architectural elements, and future directions. Future Gen Comput Syst 2013;29:1645−60.

[7] Amirkhani A, Papageorgiou EI, Mohseni A, et al. A review of fuzzy cognitive maps in medicine: taxonomy, methods, and applications. Comput Methods Prog Biomed 2017;142:129−45.

[8] Obiedat M, Samarasinghe SA. Novel semi-quantitative fuzzy cognitive map model for complex systems for addressing challenging participatory real life problems. Appl Soft Comput 2016;48:91−110.

[9] Eslami SMA, Jimenez Rezende D, Besse F, et al. Neural scene representation and rendering. Science 2018;360:1204−10.

[10] Ballinger B. Three challenges for artificial intelligence in medicine. Cardiogram in blog.cardiogr.am, September 19, 2016.

[11] Yarlagadda DVK, Rao P, Rao D, et al. A system for one-shot learning of cervical cancer cell classification in histopathology images. In: Proceedings; 2019. p. 1095611.

[12] Kasabov N, Capecci E. Spiking neural network methodology for modelling, classification, and understanding of EEG spatio-temporal data measuring cognitive processes. Inf Sci 2015;294:565−75.

[13] Lea C., Vidal R., Reiter A. et al. Temporal convolutional networks: A unified approach to action segmentation. arXiv:1608.08242 [cs.CV]. <www.cs.jhu.edu/'areiter/JHU/Publications_files/ColinLea_TCN_CameraReady.pdf>.

[14] Moor M, Horn M, Rieck B, et al. Early recognition of sepsis with Gaussian process temporal convolutional networks and dynamic time warping. Proc Mach Learn Res 2019;106:1-IX.

[15] Kermany DS, Goldbaum M, Cai W, et al. Identifying medical diagnoses and treatable diseases by image-based deep learning. Cell 2018;172:1122−31.

[16] Wiens J, Guttag J, Horvitz E. A study in transfer learning: leveraging data from multiple hospitals to enhance hospital-specific predictions. J Am Med Inf Assoc 2014;21:699−706.

[17] Sinaci AA, et al. A federated semantic metadata registry framework for enabling interoperability across clinical research and care domains. J Biomed Inf 2013;46(2013):784−94.

[18] Kim M, et al. An informatics framework for testing data integrity and correctness of federated biomedical databases. AMIA Jt Summits Transl Sci Proc 2011;2011:22−8.

[19] Krischer JP, et al. The Rare Diseases Clinical Research Network's organization and approach to observational research and health outcomes research. J Gen Intern Med 2014;29(Suppl 3):739−44.

[20] Forrest CB, et al. PEDSnet: how a prototype pediatric learning health system is being expanded into a national network. Health Aff 2014;7:1171−7.

[21] Ozyurt IB, et al. Federated web-accessible clinical data management within an extensible neuroimaging database. Neuroinformatics 2010;8(4):231−49.

[22] Doiron D, et al. Data harmonization and federated analysis of population-based studies: the BioSHaRE project. Emerg Themes Epidemiol 2013;10:12−20.

[23] Abu-Elkheir M, et al. Data management for the Internet of things: design primitives and solution. Sensors 2013;13(11):15582−612.

[24] Branescu I, et al. Solutions for medical databases optimal exploitation. J Med Life 2014;7(1):109−18.

[25] Barreto D. Lecture for MS&E 238 on July 11, 2014. Adopted from NIST, 10/09.

[26] Personal communication with Dr. Spyro Mousses, July 28, 2014.

[27] Nagase K. Software defined network application in hospital. J Innov Impact 2013;6(1):1−11.

[28] Personal communication with Dr. Marty Kohn (formerly of IBM), July 9, 2014.

[29] How software-defined storage brought Maimonides Medical Center to the forefront of health care IT. In: DataCore.com, July 30, 2014.

[30] Graschew G, et al. New trends in the virtualization of hospitals—tools for global e-Health. Stud Health Technol Inform 2006;121:168–75.

[31] Islam MM, et al. A survey on virtualization of wireless sensor networks. Sensors 2012;12(2):2175–207.

[32] Howie L, et al. Assessing the value of patient-generated data to comparative effective research. Health Aff 2014;7:1220–8.

[33] Scott DJ, et al. Accessing the public MIMIC-II intensive care relational database for clinical research. BMC Med Inform Decis Mak 2013;13:9.

[34] Perakslis ED. Cybersecurity in health care. N Engl J Med 2014;371(5):395–7.

The Future of Artificial Intelligence in Medicine

The myriad of issues in the application of artificial intelligence (AI) mentioned in the previous chapters will need to be addressed, perhaps on a grander scale, in the future.

One issue is the ethics of its use in the variety of sectors and the accompanying debates amongst scholars and scientists as well as the public. Elon Musk and Stephen Hawking both predict dire consequences while other Silicon Valley titans argue the other way. The truth may very well be in the middle: we need to be respectful of the power of AI and not be careless in its deployment. One approach is that suggested by Oren Etzioni to have three rules for AI that are inspired by Isaac Asimov's three rules for robotics: (1) an AI system must be subject to full gamut of laws that apply to its human operator; (2) an AI system must clearly disclose that it is not human; and (3) an AI system cannot retain or disclose confidential information without explicit approval from the source of that information [1].

Another issue is the economics of such a paradigm shift in work and compensation. AI is a prediction technology so the cost of goods and services that rely on prediction (such as inventory management and demand forecasting) will fall. But since all human activities rely on not just data and prediction but also judgment, action, and outcome, prices for the latter three can increase as the demand for these three capabilities go up. The issue of singularity (in which the computers will be more intelligent than all of mankind and replace its intellectual capacity) and how we accommodate this epoch will need to be discussed. The economic impact of such an event especially with countries that do not have large investments in AI needs to be examined.

The issue of bias in the formation and application of AI for all uses will need to be scrutinized. Biased data feeding into machine-learning algorithms can lead to biased systems in AI [2]. Bias can also be in the form of a patient population being too homogenous and therefore models become less valid for a more heterogenous population.

Another concern is that of data protection and privacy. The advent of the newly enacted European General Data Protection Regulation and its mandate to require specific informed consent for personal data use may herald an equivalent movement elsewhere in the world. In addition, Health Insurance Portability and Accountability Act does not regulate health care data generated outside the health system so new guidelines are needed for this upcoming data tsunami.

Intelligence-Based Medicine
DOI: https://doi.org/10.1016/B978-0-12-816462-4.00011-X

Data privacy and intellectual property in artificial intelligence (AI) in medicine

Gavin Bogle and James Silver

James Silver and Gavin Bogle, legal counsels within the AI realm including the challenging issue of privacy and intellectual property rights, authored this commentary on these issues with AI in medicine, specifically data privacy and intellectual property with regard to ownership and rights.

Magyar, Bogle & O'Hara LLP, Toronto, ON, Canada

"AI" is at the forefront of healthcare innovation and discovery, and signals a paradigm shift in the processes that inform new inventions, as well as the way that patients interact with their healthcare providers.

From a clinical standpoint the use and implementation of AI is revolutionary, as it is expected to yield greater efficiencies, deeper insights, and ultimately better patient outcomes. Indeed, despite considerable challenges, among regulatory bodies, the US Food and Drug Administration is at the forefront of the AI revolution and is working hard to keep up with advances [1]. However, other legal ramifications of AI, particularly in the areas of data privacy and intellectual property, do not currently benefit from robust statutory frameworks and jurisprudence as to define and guide AI development and use processes. A tension currently exists between pioneering AI technologies that do not fit cleanly within the molds of our current legal and regulatory systems, and the importance of our legal and regulatory systems to effectively govern innovation, research, development, and commerce in the public interest.

Data privacy

One prominent use of AI in healthcare involves collecting and mining big data. The use and sharing of such data are an integral part of informing developers and algorithms alike. The most glaring concern for AI's use of health information is data privacy, namely Health Insurance Portability and Accountability Act (HIPAA)'s Privacy Rule and the safeguards therein to protect patient information. The Privacy Rule does not go far enough to protect health information, as HIPAA does not apply to all entities, particularly to the many technology companies that are developing AI. The Privacy Rule is not at odds with AI per se but should serve as a benchmark for the protection of patient anonymity and should help the AI field self-regulate. If the healthcare AI industry is able to put in place its own conventions, limits, and accountability systems to govern the fair, compliant, and protective use of sensitive information, it may prevent less informed policymakers from stepping in and imposing unworkable restrictions on the use of data. The success of AI will largely depend on society's trust in it, and a misuse of patient information will greatly undermine AI in the public eye, and will likely lead to restrictive laws and regulations in response.

Intellectual property

Traditional device and pharmaceutical innovators have found great success, not just from their products saving lives but from patent protection which effectively gives inventors a market monopoly over their inventions. The ability to obtain patent protection is perhaps the most important commercial incentive driving healthcare innovation and development and justifies the huge research and development costs required to bring a product to market by derisking competition for sales for a limited time.

Patent eligible subject matter

Many aspects of AI do not fit neatly into conventional streams of patentability. In *Mayo Collaborative Services v. Prometheus Laboratories, Inc.* [2], the Supreme Court of the United States held that "laws of nature, natural phenomena, and abstract ideas are not patentable subject matter" and nor are "methods for making determinations that are well known in the art, that simply tell doctors to engage in well-understood, routine, conventional activity previously engaged in by scientists in the field." As such, many AI innovations, which use more efficient data aggregation, data structures, and data mining of biological information to guide medical diagnosis and treatment, face great hurdles to patentability, despite the breakthrough nature of such technology.

Inventorship and ownership

The US patent system only recognizes individuals as inventors, not machines [3]. This is at odds with AI systems, as the purpose of AI is for the machine to learn about, adapt to, and overcome challenges, which may yield independent inventions without the input of people. The current system is not set up to allow for invention by AI, and the often-expressed idea that the creator of the AI itself is the inventor of the AI's invention is not a satisfactory solution, as the software developer clearly did not conceive the AI's invention, which is a fundamental aspect of inventorship in patent law.

In addition, as AIs begin to invent, an issue of ownership also arises, as an AI does not presently have definable property rights. Legislators will have to craft a new statutory framework, or it will be left to the courts to decide who among the persons responsible for the AI should own the AI's invention. Common law jurisdictions such as the United States will emerge as the leaders of legal thought and policy regarding AI in healthcare, as the common law is agile and responsive to innovation and commerce, but this is a foreseeably hotly contested and undefined area of AI law.

References

[1] Bibb Allen, The role of the FDA in ensuring the safety and efficacy of artificial intelligence software and devices, J. Am. Coll. Radiol. 16 (2) (2019) 208—210.
[2] *Mayo Collaborative Services v. Prometheus Laboratories, Inc.* 566 U.S. 66. No. 10-1150 (argued December 7, 2011).
[3] Hattenback B, Glucoft J. Patents in an era of infinite monkeys and artificial intelligence. Stan Technol L Rev 2015;19(32).

There is also the issue of transparency for AI in medicine methodologies. The perception, fair or unfair, of AI techniques having a black box and lack of transparency breeds mistrust amongst the various stakeholders [3]. Some explainability will be necessary for wide adoption amongst clinicians as well as patients and families.

How will physicians know they can trust artificial intelligence (AI)?

Erik Lickerman

Erik Lickerman, a pathologist with a programming and data modeling background, authored this commentary on what physicians will be seeking for an inherent trust in AI and that we need to be patient in order to foster this trust.

CTO Sanguine Biosciences, Sherman Oaks, CA, United States

When will we, as physicians, trust AI to perform vital tasks in healthcare? There cannot be a purely rational answer. Trust is an emotionally driven state. We decide in our guts how much we can trust a person, and under which circumstances.

What would it even mean for a physician to "trust" AI?

Let us posit that "trust" would mean—have the same confidence in the output of AI that we would have in a consult from a human physician. If a primary care doctor sends a patient to a human cardiologist, he will not question if he should trust the cardiologist. To a great extent, this is because expert opinion provides a psychological and legal shield to the primary care doctor. To a lesser extent, it is because the consult report will contain an account of the cardiologist's clinical reasoning.

The lack of this latter element is one of the main barriers to "trusting" AI. Many AI algorithms are "black boxes." They can tell us that based on a large number of data, the current patient situation is most similar to those who were treated successfully by drug X, but it will not offer any step by step explanation. If the primary care physician cannot see the "reasoning," how can he question it? If he cannot question it, how can he trust it?

How often *would* he question it if the consultant were human? He might question, or ask a question, about a cardiology recommendation but he probably wouldn't question a pathologist's cancer diagnosis. "Oh, you think this is cancer. Really? Tell me why?". The pathologist could indeed tell him why. He would point to the dense blue color and irregularity of the nuclei and how they invade through a membrane etc.

An AI slide reader could do no such thing. The explanation would be "this slide pattern is similar to other patterns which were labeled as cancer by experts for my training set." If AI is just pattern matching, how do we know those patterns are valid? Without the scientific explanation, how do we know that past patterns will continue?

The explanation offered by the human pathologist is scientific. The dense blue indicates too much DNA, which means cancer. Science! Too much DNA causes the cells to reproduce too quickly and invade surrounding tissue and spread to other parts of the body because, well, you know, DNA and all that.

Except that isn't correct. Too much DNA *doesn't* cause cancer. The cells that *look like cancer* may actually be terminally differentiated cancer cells i.e. they are an *indication* that the really dangerous, probably more innocent-looking cells, lurk nearby. Have we demonstrated scientifically that the bizarre-looking cells are the driver cells? Have we scientifically demonstrated that a cell cannot look like that without being a cancer driver cell?

Of course not. Instead, we depend on the experience that shows when you remove a mass from a person and the cells look like that, in almost all cases the patient later dies of cancer.

Pathology diagnosis has *always* been a pattern-matching exercise. Most of the medicine is pattern matching, followed by scientific modeling, rarely the other way around.

Is cardiology different? Do we treat with digoxin because it inhibits adenosine triphosphatase which causes an increase in intracellular sodium, which in turn causes a decrease in the sodium–calcium exchanger which in turn causes an increase in intracellular calcium which increases the force of contraction while simultaneously lengthening phase 0 and 4 of the action potential, effectively slowing the heart rate?

No, we've been treating congestive heart failure with digoxin for centuries. We only figured out the scientific explanation for digoxin's effect in the 20th century. Back in the 18th century, William Withering discovered foxglove was effective for treating dropsy, the edema caused by congestive heart failure. How? Mother Hutton, a folk herbalist had it as one of 20 or so ingredients in her dropsy medicine. Withering experimented with the different components until he narrowed the effective one to foxglove. The herbalist knew because she was taught by someone who was taught by someone who. . .you get the picture.

After centuries of *pattern matching*, a scientist isolated it to a single plant. In the 1930s another scientist isolated it to a single chemical. Later in the 20th century, other scientists elucidated the mechanism.

The human body is a complex system and it is impossible to understand from the ground up. We tend to be taught the science first and the medicine later, which gives the false impression that medicine was developed from basic scientific knowledge. In reality, *most* of the medicine was developed the other way around. People noticed patterns, tried stuff and kept track of what worked. Later others looked for the basic building block level science that explained our previous habits.

In medicine experience trumps intellect. Experience is that which we use to predict behavior in systems which are too complex for deterministic reasoning; years of internalized observations working in the background of our minds and later given rationalization and verbalization.

Experience manifests as drawing conclusions based on combinations of similarities we see in the current situation to past situations. Is this very different from AI?

All of which is not to argue that we should trust AI because medicine is already mostly pattern matching. I'm only pointing out that our confidence in human medicine is somewhat emotional. A human in a white coat makes a pronouncement and we trust.

We cannot choose to overcome this emotional reaction. Even writing the above, I am not prepared to trust AI. Trust will come slowly after many successes and few terrifying failures.

What should we do, and expect of AI medicine, to make it most likely that physicians will trust it as they do human colleagues?

Lastly, AI and specific ethical issues in health care include not only bias, inequity, and other issues but also fiduciary relationship of patients and machine-learning systems and potential changes in the physician–patient dyadic relationship [4]. There are close to if not over 200 companies around the world in this domain of AI in clinical medicine and health care, so this aspect is looming larger by the month (see Fig. 11.1).

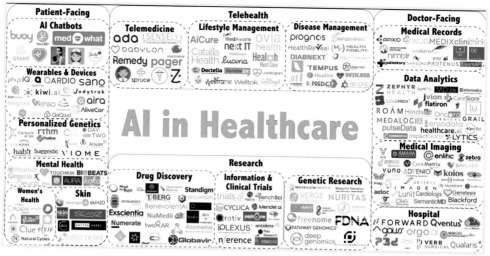

FIGURE 11.1 The artificial intelligenc (AI) in health-care industry landscape. The large number of AI in health-care companies (211 depicted here) is divided into patient-facing; telehealth; research; and doctor-facing categories. *Source:* Reproduced with permission from https://techburst.io/ai-in-healthcare-industry-landscape-c433829b320c (c) Emily Kuo.

The 100-Year Study on AI (AI100) is a long-term study of AI and its impact on people and society. The eight relevant areas that are considered most salient include health care (the other seven being transportation, service robots, education, low-resource communities, public safety and security, employment and workplace, and entertainment) [5].

Already used in areas such as radiology, pathology, genomic medicine, cardiology, outpatient services, and intensive care, AI will continue to have an escalating impact in medicine albeit with concomitant fear amongst stakeholders for an AI "takeover," especially their jobs.

In Nick McKeown's parlance, medical data and analytics need to be transformed from a "vertically integrated, closed, proprietary, slow innovating" data system to that of a "horizontally integrated, open interfaces, and rapid innovation" data ecosystem [6]. To put it into biological lexicon, the biomedical data system needs to transform from a rudimentary musculoskeletal system to an intelligent nervous system. In addition, cloud computing and storage will be vital to facilitate the panoply of AI techniques for multiinstitutional collaborations that will be essential for the future of AI in biomedicine and health care.

Perhaps clinicians and medical educators can embrace rather than distrust AI and allow its capacities to transform how we teach and deliver health care, especially in the changing health care milieu to fee-for-value systems. An effective AI in medicine strategy will liberate clinicians from the burgeoning burden of electronic health records and allow a return to an ideal physician−patient relationship. In addition, since computers are excellent at handling data and making predictions, the human judgment will become even more valuable. This philosophy of accommodating emerging technologies will need to be indoctrinated early in the clinician's career.

The education of dual clinician data cientist

King Li

King Li, the medical school dean of an exciting new medical school, authored this commentary on a new curriculum for medical students that will incorporate both innovation and engineering to create a cohort of dually educated clinician-data scientists.

Engineering and technology including data science if appropriately applied has the potential to revolutionize healthcare by increasing quality, decreasing cost and increasing accessibility and equity. Ideally, engineering, technology, and data science can be used to enhance the humanistic aspect of medicine by freeing healthcare providers from tasks that can more aptly and efficiently be done by machines. However, currently in the US technology has been blamed by some for increasing cost, dehumanizing healthcare, and increasing physician burnout. This mismatch between expectation and reality is due to a multitude of factors but the paucity of physicians who are familiar with and able to guide the appropriate deployment of engineering, technology and data science in healthcare is partly responsible. To remedy this situation a fundamental change in the way we educate healthcare professionals is required. In the past century, traditional medical education is heavily influenced by the Flexner report published in 1910 with focus mainly on basic and clinical sciences. In the current era, it may be more appropriate for healthcare education to be built on four pillars: basic science, clinical science, humanities related to healthcare, plus engineering and technology including data science.

At the Carle Illinois College of Medicine (CI Med) at the University of Illinois at Urbana-Champaign (UIUC) the entire medical school is dedicated to such a curriculum with the goal of developing physician-leaders and physician-innovators. All medical students in this engineering-focused college of medicine are required to have quantitative science background including high-level mathematics, statistics, and computer science. They are selected based on four essential characteristics, the 4Cs: compassion, competence, curiosity, and creativity. In phase I of the curriculum, which is about 18 months in duration, all courses are developed by three course directors, one from basic science, one from clinical science and one from engineering. The medical humanities and a 360-degree view of our complex healthcare system and other healthcare systems in the world are delivered as threads throughout all phases of the curriculum. The phase I courses are delivered using a problem-based learning approach with engineering, technology, and data science integrated into the organ system based courses rather than as standalone courses. There are two 3-week discovery periods when the students can have multiple options to choose from including a global experience program, research, community health in rural or urban underserved areas and industry exposures. In phase II, which is about 18 months in length, the students are required to do an IDEA (Innovation, Design, Engineering, and Analysis) project in each of the required clerkships. Each student is required to come up with a new idea to improve healthcare in each of these IDEA projects. To help the students refine their ideas engineering rounds are held once a week. During the engineering rounds an engineering faculty will make clinical rounds with a clinical faculty and the students. This is designed not just for educating the students but also for increasing the communication between the engineering and clinical faculty.

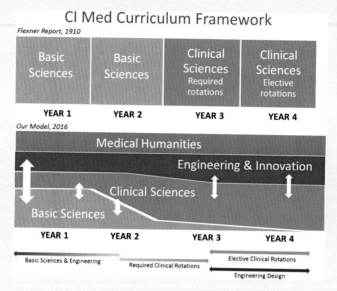

In phase III, essentially the final year, all students are required to select one idea from their seven IDEA projects to become their Capstone project. For the Capstone projects, each student is required to lead a team of other students from different disciplines on campus to help turn their ideas into prototypes. To assist in this effort each student is given financial support and the use of a Health Maker Lab network comprised of 16 core facilities on campus capable of fabrication at different scales from automated synthesis of molecules, micro- and nanofabrication, synthesis of new genes and cells, 3D printing, supercomputing all the way to design of new architecture. Some of these prototypes can be further developed in an incubator space in the Research Park with appropriate mentoring and other supports. All students are also required to finish a data science project using real-life data from the healthcare system and other sources which can potentially lead to improvement in healthcare processes. All graduates of CI Med will have experiential training in engineering, technology, data science, and innovation and can serve as catalysts for change in various parts of our complex healthcare ecosystem (Carrie, please put in a description of En Med here).

Artificial intelligence (AI) in medicine: all about the journey and not a destination

Tom Murickan

Tom Murickan, a high school student and an outstanding role model for his peers, authored this commentary about how he started a club at his high school to focus on AI in medicine and the future promise of medicine.

I am sure many people who will pick up this book will wonder why they are reading about AI in medicine from the perspective of a 17-year-old high school junior. The answer lies in a question Dr. Anthony Chang poses all the time: "The future of AI in medicine is already here, but how do we ensure future generations make the best and most ethical use of it?" When I was asked to contribute my thoughts on this subject, I was more than intimidated. However, something remarkable about everyone involved with this book, as well as with the entire AI-Med community, is that there are no egos involved, because there is a constant and unwavering belief that the next great AI breakthrough can come from anywhere or anyone. So here I am making my small mark in it, sharing my short path so far in the long journey I intend to take through AI in medicine.

From a young age, I was always fascinated by the medical field. As I grew older, many life experiences helped shape my interest in the medical field to be a combination of both a passion for learning constantly and a desire to help people in need. This desire led me to volunteer at both Pomona Valley Hospital Medical Center and Cedars-Sinai Medical Center, as well as with a local organization dedicated to helping special-needs children called Heart of Hope. While these were all amazing experiences and opportunities for me, I knew that becoming a doctor would take more. Then, in the summer of 2018, I applied and was selected for the MI3 Summer Internship Program at Children's Hospital of Orange County. The internship focused on the impact of AI on the medical field and allowed me to meet with and listen to a variety of professionals in medicine, data science, nursing, and many other fields. This completely changed my perspective on the medical field and on what kind of doctor I wanted to become. It also permeated my thoughts with a lot more questions about whether society, as a whole, and more specifically the medical education system, is ready to prepare a doctor, not only armed with AI but also using it ethically and effectively. As a partial solution to this issue, I had a vision of an AI-Med Club at my high school, whose purpose would be to spread awareness of the increasing prevalence of AI to my fellow classmates and their families. Through my work in the internship, I was officially introduced to the director of the program, Dr. Anthony Chang, who encouraged me to pursue this and, although I faced some significant headwind initially, I was finally able to accomplish my effort in establishing AI-Med Club at my high school with some great support from the AI-Med community. What was the initial skepticism? The main reason behind our school club approval committee's doubt was the lack of awareness of what AI in Medicine was and their uncertainty of my ability to deliver these concepts in a way high-schoolers could comprehend. My response? Their doubt was exactly the reason why I pursued establishing this club, because there needed to be an individual who could relay the vast quarry of information about the future of medicine in regards to AI to students, in order to prepare them for a future in any field affected by AI, and I believed that I was that individual. In fact, I knew I was the right individual, because in December of 2018, Dr. Chang invited my dad and me to attend AI-Med North America 2018 at Dana Point to provide some unique insights and perspectives to the AI-Med community. The 4-day conference featured a multitude of guest speakers from all different fields focused on increasing the advances of AI in Medicine. The discussions, feedback, and encouragement I received from the group in those 4 days emboldened my vision of what my future career as a medical professional would look like. I was given an opportunity to be a part of a panel on the future of AI-Med and speak to the audience about my outlook on AI-Med through my previous experiences. Among my many questions posed to the group was my growing concern on how well prepared the higher education system (undergraduate and medical schools) was to train me properly as a medical professional that will use AI in Medicine. This one issue sparked an idea within the AI-Med community that a student like me would be a good ambassador to launch AI-Med clubs countrywide, throughout high schools, universities, and medical schools, to create the awareness, the movement, the inertia needed for the education system to reevaluate how best to incorporate statistics, programming, data science and the like into the medical curriculum.

This has been an exciting 4 months for me as I have been able to launch the AI-Med Club at my high school and conduct the first few monthly meetings where we have discussed many of the concepts and also had incredible guest speakers who have delivered AI concepts in medicine at a high school level. I am glad to report that each meeting has had more and more attendees and I hope to keep the momentum going. In addition, I have been working with the CEO of AI-Med, Freddy White, to further define my role as the student ambassador and prepare to reach out to institutions across the United States to help start AI-Med clubs and organizations on their campuses. My goal this year is to be able to establish a dozen new high school AI-Med clubs across the United States and try to bring at least a few medical school program directors to AI-Med North America in December of 2019. By having these Medical Schools represented, I am hopeful that it will begin the process of bridging the gap between premedical school curriculums and the preparation of future medical students to be armed with deeper knowledge and understanding of AI in medicine. As I continue on this effort, one of my goals is to be able to reach out to the directors of any and all medical schools in the United States and obtain the answer to the question: "I am the future AI-assisted physician. Are you ready for me?" And with enough effort and the continued support from the AI-Med community, the answer will be a resounding "Yes." Thus when I look at AI in medicine, it is not a destination, rather a continuous journey that is unique for each and every individual on this path and I plan to meet a lot of incredible people and make a lot of exciting discoveries on my journey.

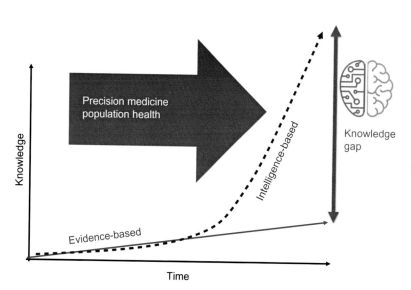

FIGURE 11.2　Intelligence-base medicine. With the advent of precision medicine and population health, an increasing knowledge gap needs to be made up by a paradigm shift from evidence-based medicine to intelligence-based medicine, the latter driven by data and AI tools.

In conclusion, the future of AI in medicine is extremely propitious with a myriad of advanced AI techniques such as deep reinforcement learning, generative adversarial network, one-shot learning, and cognitive methodologies that will need to be in synergy with clinicians with cognitive thinking to allow data to be an enabler of new knowledge and intelligence in biomedicine and health care. All health-care data will need to be liberated and shared without any obstacles so that AI can be ubiquitous and invisible in the future health care arena and discover new knowledge from all sources of data and information. In short, for us to fulfill our vision of precision medicine and population health for the next century, we need to change the paradigm of evidence-based medicine to that of a data science-driven intelligence-based medicine (see Fig. 11.2).

Key Concepts

- This premise of advances in AI is based on rapid improvement and accelerated deployment of advanced computer technology (such as quantum computing and neuromorphic chips) and rapid development and evolution of AI techniques (such as deep learning and its many variants) as well as cognition-influenced architecture that have led to a current Cambrian explosion of AI.
- Peter Voss so aptly described the future of AI as the "third wave": the first wave of AI was good old-fashioned AI that focused on traditional programming followed by the second wave of AI of the current deep learning so that this third wave will be reliant on many cognitive architectures.
- A successful deployment of technologies such as blockchain to improve cybersecurity in health care will facilitate data sharing amongst stakeholders in the near future.
- As a counterbalance and a coupling to cloud computing, edge computing technology enables devices to collect and analyze data in real time locally so that it is processed outside the cloud datacenter. With biomedical devices, especially ones with continuous monitoring, the amount of physiological data received would be unmanageable by caretakers unless there is an "upstream" intelligent algorithm built into the device to filter all the noise from the signal.
- The advantage of fuzzy cognitive map is in its unique specification of integrating human knowledge and experience with computer-aided techniques so that one can achieve a human—machine synergy that is ideally suited for the complexities of medicine.
- In order to model even more complex and highly interconnected data, a new paradigm of data representation called hypergraphs will need to be implemented. A hypergraph is a graph model in which the relationships (called a hyperedge) can connect any number of nodes.
- All of these low-shot learning methodologies that do not require Big Data can be particularly useful in clinical medicine and health care.
- As the demand for computation exponentially increases in biomedicine as we head into the cognitive era of AI, quantum computing (and other types of computing such as DNA computing) will be a necessary technological tool.

IV. The future of medical intelligence

- This is a generative model (structured probabilistic graphical model to be exact) that differs from deep learning; in that, it has a scaffolding (rather than learning from scratch, or *tabula rasa*, as in deep learning).
- Siamese neural networks, called the third generation of neural networks, operate with events called spikes, which are discrete events that occur at certain points in time; it therefore is more biological than machine learning in that the neural network often relies on the timing of individual action potential of neurons.
- Swarm intelligence is intelligence derived from many individuals based on self-organizing group behavior. The collective behavior illustrates that unified systems outperform the majority of individual members.
- Transfer learning involves adaptation of a trained model to predict examples from a different dataset; this phenomenon is particularly favorable with deep learning networks since deep learning requires so much time and resources for its training.
- One key future development of big data analytics is real-time analytic processing. This is a process in which the data is captured and processed in a streaming fashion using online analytical processing and complex machine-learning algorithms.
- The aforementioned software-defined network concept has evolved into a software-defined data center (SDDC) architecture (with server virtualization, software-defined networking, and storage hypervisor) that can be entirely virtualized so that all of the infrastructure and the management can be automated by software.
- By embedding intelligence into all aspects of medical data from graph database and meta-database management system to customized cloud infrastructure and to SDDC and virtualization, the aforementioned strategies can accelerate this transformation in biomedicine from fragmented and unstructured data sets to cohesive and agile information imbued with medical intelligence.
- Medical data and analytics need to be transformed from a "vertically integrated, closed, proprietary, slow innovating" data system to that of a "horizontally integrated, open interfaces, and rapid innovation" data ecosystem.
- The future of AI in medicine is extremely propitious with a myriad of advanced AI techniques such as deep reinforcement learning, one-shot learning, and cognitive architecture that will need to be in synergy with clinicians with cognitive skills to allow data to be an enabler of new knowledge and intelligence in biomedicine and health care.
- For us to fulfill our vision of precision medicine and population health, we need to change the paradigm of evidence-based medicine to that of a data science-driven intelligence-based medicine.
- The final stage of human-to-machine relationship is convolution, a term that describes the sinuous ridges of the brain in biological terms but conveniently a third function that is promulgated from two existing functions in mathematical terms. In our lifetime, we will observe human and machine intelligences be harmoniously intertwined and difficult to discern which contributed to what aspect of this endeavor. We may even stop calling anything "artificial" intelligence.

References

[1] Etzioni O. How to regulate artificial intelligence. *New York Times* September 1, 2017.

[2] Knight W. Forget killer robots—bias is the real AI danger. MIT Technol Rev, October 3, 2017;.

[3] Hsu W, Elmore J. Shining light into the black box of machine learning. JNCI—J Natl Cancer Inst 2019;111 (9):877−9.

[4] Char DS, Shah NH, Magnus D. Implementing machine learning in health care—addressing ethical challenges. N Engl J Med 2018;378(11):981−3.

[5] One Hundred Year Study on Artificial Intelligence (AI100).<https://ai100.stanford.edu>.

[6] McKewon N. Making SDNs Work. YouTube April 2012.

Conclusion

This esprit de corps of the artificial intelligence (AI) in medicine community reminds us how essential relationship and collaboration is in this domain. While we discuss "machines" and "artificial" intelligence, there are humans behind these entities (at least for now). Relationship needs to be fostered in AI in medicine and health care at all levels and across all dimensions. Often, the best strategy entails forming dyadic relationships or hybrids:

First and foremost, it needs to be between human and machine. There are numerous published manuscripts in which there are no clinicians among the authors as if data science in medicine can be an entirely isolated discipline and science. This hubris, albeit unintentional, is unacceptable. On the other hand, clinicians need to be much more knowledgeable and accommodating of this new AI paradigm in medicine.

Second, the relationship needs to be between human and human. The clinician and data scientist dyad needs to be engendered and much more effective in creating an important interface between data science and clinical medicine and health care. The universal theme that prevailed around the world is empowering patients and enabling clinicians to use AI to be innovative and transformative. From a plastic surgeon in Toronto who leveraged deep learning (DL) in medical images to improve her care of burn patients to the data scientist from Senegal who is working on reinforcement learning models in health care, these global AI in medicine citizens continue the mission of creating an AI in medicine and health-care ecosystem and community to affect transformative change.

Third, the relationship needs to be between machine and machine. The advent of Internet of Things is leading to the need for AI to be embedded in wearable technology. Thus the future will be machine collaborating with machines to become Internet of Everything. In addition, DL algorithms will need to be in synergy with those principles of cognitive architecture to maximize the yield of AI in medicine and health care.

Even though AI programs such as *AlphaZero* have now heralded the advent of high-level, self-learning AI that relies more on cognition and algorithms than brute pattern recognition of data [1], AI remains at least partly a human intelligence imprinted into machines that reflect human thinking. The clear message from recent events is that we need to continually define and clarify the human-to-machine relationship. Interestingly, there are biological terms for this anthropocentric relationship. The first stage of this special human–machine relationship is symbiosis, or the living together of two dissimilar organisms. We are mostly at this stage currently. The second stage is synergy in which the total combined effect is usually greater than the sum of the parts. Muscles and nerves are often "synergistic." We already have the early underpinnings for this level of relationship.

We need to reach the final stage of human-to-machine relationship: convolution. This is a term that describes the sinuous ridges of the brain in biological terms but conveniently a third function that is promulgated from two existing functions in mathematical terms.

In our lifetime, we will observe human and machine intelligences be harmoniously intertwined and difficult to discern, which contributed to what aspect of this endeavor. We may even stop calling anything "artificial" intelligence.

Much has happened even in the few years since the inception of this book and *AlphaGo* versus Lee Sedol matches in 2016. *AlphaGo Zero*, the follow-up and much more capable version of *AlphaGo*, learned to play Go without the benefit of learning from human-played Go games in the past and was able to defeat its predecessor *AlphaGo* 100 games to 0. It was evident that *AlphaGo Zero* (named for its *tabula rasa* inception) played the game Go with innovative moves not previously seen among human Go players (so "new intelligence"). What was absolutely astounding was that *AlphaGo Zero* learned to play Go in 40 days and surpassed all human players with about 2500 years of learned history. *AlphaZero*, the congener of *AlphaGo Zero*, is also a self-learning program but can play more two-player games (specifically chess, shogi, and Go) and play with a style not observed with human players before, and easily win with an uncanny flexibility and adaptation. Lastly, *AlphaStar* has very recently soundly defeated human champions in yet another game that was deemed nearly impossible for AI to conquer: the real-time strategy (RTS) game of *StarCraft II*. This was perhaps an even more impressive an AI feat as the previous programs since RTS games have additional challenges compared to the game Go: imperfect information, long-term planning, real-time, and large action space. These AI feats have very significant implications for clinical medicine and health care as clinicians face very similar situations as complex RTS games on a daily basis.

What was enlightening about the human versus *AlphaGo* competition discussed at the beginning of this book was not only the brilliant and innovative 37th move made by *AlphaGo* during the second game, but also the 78th move of the meaningless fourth game (as *AlphaGo* had already won three games in the five-game match) in which Lee Sedol made an equally brilliant and creative move. This move was not widely publicized as the 37th move but clearly demonstrated that the computer had a positive influence on man and a man-to-computer synergy would yield the best result for AI, perhaps especially in clinical medicine and health care. The biologist E.O. Wilson would remind us that the ultimate union of sciences, including biological sciences and AI, is with the humanities in the form of consilience. Perhaps the best reward of AI and its essence in our lives, akin to children, is a more in depth understanding and appreciation of ourselves as human beings.

Machines have computation, instructions, and objectivity, while humans have purpose, creativity, and passion. With AI in clinical medicine and health care and intelligence-based medicine, it should never be human *versus* machine, but always human *and* machine.

Major takeaways in AI in clinical medicine and health care

After this long decades long journey, and meeting/discussing AI in clinical medicine and health care at many international venues and with thousands of clinicians, data and computer scientists, informatics experts, AI experts, educators, researchers, investors, hospital administrators, IT specialists, and students, these are the major takeaways (not in any particular order and a few principles may be duplicated from above but worth repeating):

DL is very good for selected situations. As impressive as DL is for medical image interpretation and complicated nonlinear situations with large volume of data, machine learning (ML) may be better suited for simpler models with less data. In addition, DL faces significant challenges for projects such as electronic health record (EHR). Lastly, robotic process automation (RPA) is adequate for tasks that have repeated steps that would benefit from automation.

Cognitive elements in AI will be needed. Clinicians will need to be much more engaged with the next wave of AI in medicine projects that will be more focus on cognitive architecture. *AlphaGo Zero* from Google DeepMind was much more about algorithms than big data. Important cognitive elements include memory, relationships, imagination, creativity, and abstract concepts.

Sharing of data will be essential. Collaboration among centers and pooling of data on a global scale will contribute significantly to the overall AI in medicine effort. Data in health care fragmented and disorganized to begin with, but some of this deficiency can be neutralized with pooling of health care data into data reservoirs, perhaps with blockchain and AI for privacy can security.

A human-to-machine convolution, not synergy, will advance biomedicine. We need to balance machine intelligence with human cognition so that a third realm, "medical intelligence," can result from this hybrid process. A good algorithm improves human decisions but a human can improve the algorithm by creatively solving its failures. Human and machine can be intertwined to convolve.

We overestimate AI in the short term (Amara's Law). There is some hype about AI in biomedicine especially given the performance of DL and convolutional neural network (CNN) in medical image interpretation. We need to be careful with the AI hype as we can easily create an artificial AI in medicine winter, which may exhaust the valuable resources we need for a long-term investment in AI.

But we will underestimate AI for the long term. The longer term gain is very high so we need to "pace" ourselves for the longer journey in AI in medicine. In the coming decades, a myriad of new discoveries in the areas of disease knowledge, drug discovery, and protein folding will lead to new heights of understanding that will be impressive to even scientists.

Data overload is complicated by system complexity. The data conundrum in health care is a delicate balance: while more good (accurately labeled) data will be what is needed for the DL methodologies to yield good-to-excellent performances of predictions, the system complexity of biological systems and biomedical scenarios is an additional daunting challenge to overcome.

We need to push the AI in medicine agenda together. The two main domains (clinical medicine and data science) need to understand each other's cultures and domains; the more we can collaborate and trust each other, the better. In these days of incessant texting and emails, we are hyperconnected but undercommunicating.

We need a major educational effort in AI in health care. A group of champions needs to facilitate an interface between the clinician and data scientist groups for educational and training curricula for both domains. The chasm between clinical medicine and data science is huge but surmountable if we remain patient and put aside any hubris.

Big data in health care is limited. Although the clinicians need to work collaboratively to gather as much as the biomedical data as possible (national and international levels), our data scientist colleagues need to appreciate that a large volume of good data is not possible sometimes, especially with rare diseases. Innovative solutions are needed to get around the data conundrum.

The AI technology is surging forward but other dimensions are not keeping up. The AI technology is rapidly improving, but regulation, ethics, and laws are not keeping up with this exponential trajectory. Regulatory agencies, ethics groups, and legal entities will all need to appreciate that AI is an ultrafast moving methodology and traditional strategies and timelines are simply out-of-date.

Performance of a prediction tool should not be only area under the curve (AUC) of receiver operating curve (ROC). Almost never are other important indices about performance discussed: AUC of the precision-recall curve and the F_1 score (see text). The reason is that if the population is unbalanced (usually a large number of true negatives), metrics can appear to be superior. Performance also does not correlate with outcome.

Data and databases in health care need a major makeover. Most of health care data and databases are still in relational database structure and will be much better in the more flexible and dynamic graph or hypergraph format for the future. This change is especially necessary for DL and its variants as well as cognitive architecture tools to better accommodate health-care data.

AI and neurosciences will need to be in more synergy. The recent successes with Google DeepMind projects are simply astounding, and these self-learning general AI tools are likely to be influential in medicine. These tools are the "smart" AI tools (with intuition, creativity, and "common sense") that everyone has been waiting for, but perhaps with some appropriate trepidation.

It is more about AI interpretability, not explainability. Much has been written about AI explainability, and work is being done to have the best of both worlds (explainability and performance). It is more about interpretability, however, which is being able to observe a cause and effect without intricate knowledge of all the details inside the technology or device.

Clinicians will need to participate in this major paradigm shift in biomedicine. Clinicians will need time and resources to explore this new resource of AI and to observe clinical validation studies to be entirely convinced of the value of this new paradigm. Physicians and their clinical wisdom alone is valuable and they do not need to be facile with data science to be part of this movement.

Wearable and implantable technology will need embedded AI. The upcoming data "tsunami" from the panoply of wearable and implantable medical and health devices will be generating a large volume of continuous and real-time physiologic data; all of this data will need to be organized, stored, and analyzed peripherally in the form of peripheral or embedded AI.

Future of AI will be more about synthesis of data, not big data. There is already a trend to synthesis data and not have over reliance on available labeled data for learning. There simply will not be enough human labeled data, especially in rare diseases and small patient populations. The generative AI tools like generative adversarial networks will help to navigate around this limitation in medicine.

Relationships will become increasingly important in AI models. Often in clinical situations, features that have a relationship are sometimes not associated in ML/DL models. This aspect of clinical medicine is a dimension will need to be duplicated or mimicked by neural AI models (such as recursive cortical networks) in the near future to increase relevance in medicine.

Medicine needs real-time AI tools. While medical imaging has been successful coupled with AI, subspecialties with more focus on real-time decision-making need AI tools that can accommodate these time-series data and decisions that are also often based on too little information that constantly change. There is hope that self-learning deep reinforcement learning (DRL) can provide the AI resource that is vital.

AI can be the equalizer in clinical medicine and health care. The present health-care ecosystem fosters a mindset for hospitals to compete and be part of ranking systems (akin to college sports teams). With strategic use of AI, the concept of venerable experts only at the premiere institutions will be neutralized so that this special AI-generated expertise can be democratized.

AI needs to be in the minds of many, not in the hands of a few. Just as personal computers became ubiquitous, AI will need to have the same level of adoption in clinical medicine. In the future, programming will not be needed, making AI available will elevate primary care physicians into mini subspecialists and subspecialists into multidimensional experts all over the world.

DL has performed well with medical images, but other applications will be more challenging. The much more complex nature of unstructured EHR (compared to medical images and its pixels/voxels) and other data in biomedicine and health care render this realm significantly more challenging for DL even though its nonlinear capability is a good strategy.

We need to learn from the earlier AI in medicine era and remember these lessons. Good old fashioned AI (GOFAI) (in the form of expert systems) failed to be widely adopted in the earlier era not only because it was too slow and cumbersome and not useful for new and/or complex cases, but also it failed to meet lofty expectations. We need to incorporate the strengths of GOFAI into current data science and AI.

AI can be used for most if not all subspecialties and areas in medicine. The image-focused subspecialties (radiology, ophthalmology, pathology, cardiology, etc.) are among the first adopters because of the impressive performance of the CNN work. For the next wave of subspecialties to fully adopt AI, real-time decision support tools and other applications will need to mature.

Despite all the advances of AI, medicine and health care are behind other sectors. Due to intrinsic challenges (such as data access and knowledge deficits), AI has not been widely adopted by clinical medicine and health care; activity and interest, however, are increasing substantially in certain subspecialties (especially ones that are image focused) and areas around the world.

The recent surge of AI in medicine is a derivative of the AI triad. The three elements of AI: sophisticated algorithms (especially DL), big data, and computational power/cloud storage have promulgated the recent AI revolution. The relative success of DL and specifically CNN in medical image interpretation has launched AI in medicine in this era.

AI is much more than just the well-publicized tools. Despite IBM Watson and its publicity (both positive and negative) as well as CNN and its early successes in medical image analytics, AI is a full "symphony" of "instruments": natural language processing (NLP) for language, RPA for automation of repetitive tasks, and many types of machine and DL (such as self-learning DRL).

Medicine is not a dichotomous or categorical science. While the experience thus far with CNN and medical images have been relatively straightforward, much of clinical medicine is not based on dichotomy (such as alive–dead). Many cases of diseases are subclinical (with a time element) and/or have gradations of disease; this level of complexity and nuance creates a challenge for ML/DL.

There is information in what we do not see in EHR. In clinical medicine, often the absence of a significant positive piece of evidence is sometimes paradoxically the best evidence for a diagnosis; this aspect may or may not be adequately captured in an ML/DL model unless domain expertise in disease processes are directly involved in the model.

Resist the temptation to design AI tools before defining the problem. Problems in need of a solution is pervasive in health care. If one were to wear an "AI lens," some of these problems can be readily solved with an AI solution. This approach is aligned with the design thinking process in which the initial steps of problem-solving are to empathize with people and to define the problem.

AI tools can rehumanize medicine. There is an undercurrent of AI dehumanizing medicine but AI can actually "rehumanize" medicine. If clinicians are partly relieved of their burden with data and information from EHR with AI strategies (such as DL information tools embedded into the EHR or NLP-directed dictation tools), they will have more time with their patients.

Learning about AI enables us to learn about ourselves. As we delve into ethical and philosophical issues about AI and human elements of thinking, creativity, imagination, consciousness, and ethics, we are in essence of bringing clarification (or at least more discussions and intellectualization) into what humans need to define going into the next century with our AI partners.

When AI works well and humans trust it, we will no longer call it AI. With increasing trust and familiarity with AI tools, these tools will become commonplace and we will gradually cease calling it AI but rather simply accept these tools as part of our devices or practices. This will be particularly evident with wearable devices as well as hospital monitors that will be smart with embedded AI.

We humans sometimes have an unfair expectation of our machine partner. When we critique an AI tool with an AUROC of 0.83, we need to be reminded that humans usually perform much lower (the few accidents with autonomous vehicles are in the headlines even though sometimes humans are partly to blame for not being vigilant). This performance expectation should be symmetric.

There is a tendency to trust electronic data more than one should. There are many humans involved with the EHR in the data inputting so it can be full of errors. It is perhaps more accurate if we remove at least some if not all the humans from the loop of entering or validating various types of data in the EHR. Real-time analytics in the form of data correction should be in place.

An AI hybrid or even ensemble with two or more methodologies can be a winning strategy. There are several examples of a hybridization of two or more methodologies, such as DRL. Other examples include semisupervised learning and CRNN. In the future, it will be routine to have several specialized AI tools together (similar to instrument sections of an orchestra).

It is useful to look at AI tools in other sectors and adopt these tools for health care. AI is deployed in many sectors and some of the innovations in AI are not known to health-care AI stakeholders. DeepMind's work on *AlphaStar* and *AlphaZero* and Event Horizon Telescope work on the image of the black hole have implications for many situations in biomedicine and health care.

The future role of randomized controlled trials needs to be reconfigured. The future of clinical research will be more data-driven and bottom-up rather than top-down designed. Recruitment of patients will also need to be enabled by AI and not be human-driven as this process is too tedious and too slow without AI.

A dually educated cohort of clinician—data scientists can be the catalyst for this domain. A small cohort of dually educated clinician—data scientists can help solidify this interface between the two domains and this expertise and perspective result in a geometric (not additive) gain in both insight and knowledge. This can be analogous to a musician—composer in music industry.

AI in medicine is in its very early era. If AI in medicine is compared to music, we are presently in the equivalent of the Renaissance period (after the Medieval period) and about to enter the Baroque era. If we all remain dedicated and patient, we can see the wondrous AI in medicine accomplishments in the Classical (Mozart) and Romantic (Chopin) periods to follow.

AI is the exit velocity we need to leave the present health-care conundrum. It is not possible to escape the gravitational forces of the present set of weighty issues in medicine and health care. AI coupled with emerging technologies can be the potent force we need in order to have the escape velocity to leave the present milieu.

The future CIO will be Chief Intelligence Officer. As AI becomes more pervasive and less enigmatic in health care, organizations will need a cohort of experts in the realm of AI in medicine and health care. In addition to the new type of CIO, there will also be new related careers such as a health care AI architect or data strategist to accommodate the demand of more sophisticated AI.

AI-related stakeholders need to spend time with clinicians. There are many examples of startup companies failing to follow this strategy and end up with an AI service that is adequate or even good but not congruent with what the clinicians actually need in their practice. This data science-to-clinical medicine misalignment is not infrequent.

Superior prediction does not mean better outcomes. It is important to understand that a higher AUC of the ROC does not automatically translate to better outcome, and a diagnosis is only the beginning of an opportunity to improve survival or outcome. There are many necessary steps in this health-care chain that will need to be executed from diagnosis of a medical image.

AI and its tools do not change human behavior. Some AI tools can lead to better prediction of disease (such as 36.8% chance of diabetes in the next 5 years), but there are only a few AI tools that are coupled to an effort to change human behavior (such as weight management to reduce the likelihood of diabetes).

And finally,

The human-to-human is more essential than ever before. The best moments during this long journey have been meaningful social interactions with many special people in this domain (always without machines) who took the time: an intimate dinner in Vienna with data science colleagues, coffee in Boston with a young aspiring surgeon, late evening soiree in London with technophiles, and tea in Hangzhou with a startup group.

Reference

[1] Silver D, Schrittwieser J, Simonyan K, et al. Mastering the game of go without human knowledge. Nature 2017;550(7676):354—9.

Artificial intelligence in medicine compendium

	Topic	Score	Notes
		2 points per correct answer (10 questions × 5 answers for total of 100 points)	
AI	Timing	Name the reasons why AI is so relevant now for medicine and health care: • *Big data/volume of data* • *Maturity of algorithms/improved methodology* • *Computing power (GPUs)* • *Cloud capability/cloud storage* • *Performance of DL/CNN*	
Data	Big data	List the "Vs" of Big Data: • *Velocity* • *Volume* • *Variety* • *Veracity* • *Value/visualization/variability*	
AI	Components	What are components of AI? • *Machine learning* • *Deep learning* • *Cognitive computing* • *Natural language processing* • *Robotics* • *Expert systems*	
AI	Types	Name different types of AI: • *Weak versus strong* • *Specific versus broad* • *Narrow versus general* • *Assisted/augmented/autonomous* • *Descriptive/predictive/prescriptive (analytics)*	
Data	Conundrum	List issues with health care data: • *Format: mostly unstructured* • *Location: in many locations* • *Integrity: much data missing* • *Consistency: definitions inconsistent* • *Size: escalating in size*	

Data science	Supervised learning	List the five common supervised learning methodologies: • *Support vector machines* • *k-Nearest neighbor* • *Naive Bayes* • *Decision trees* • *Logistic regression* • *Linear regression*
AI	Neural networks	Determine if the following are correct in methodology and use: • *CNN—MRI interpretation* • *RNN—ICU data analytics* • *Machine learning—patient clustering* • *Cognitive computing—EKG analysis* • *Natural language processing—EHR*
AI	Subspecialties	Name five subspecialties that are most active in the area of AI applications: • *Radiology* • *Cardiology* • *Critical care medicine* • *Ophthalmology* • *Dermatology* • *Pathology* • *Surgery*
AI	Obstacles	What are the main obstacles for AI adoption? • *Trust* • *Explainability* • *Data access and security* • *Domain knowledge* • *Workflow*
AI	History	Name significant events in history of AI: • *Enigma* • *Dartmouth conference* • *Deep Blue and chess* • *Watson and Jeopardy!* • *AlphaGo and Go* • *Many others*
AI	History	(BONUS) Name the AI methodology associated with the event: • MYCIN: *expert system* • Deep Blue and chess: *expert system* • Jeopardy: *IBM Watson/cognitive computing* • Go match: *reinforcement learning/machine learning/deep learning* • ImageNet: *DL/CNN*

AI, Artificial intelligence; *CNN*, convolutional neural network; *DL*, deep learning; *EHR*, electronic health record; *EKG*, electrocardiogram; *GPUs*, graphical processing units; *ICU*, intensive care unit; *MRI*, magnetic resonance imaging; *RNN*, recurrent neural network.

Glossary

This glossary contains both biomedical terms as well as data science vocabulary often observed in biomedical data science and artificial intelligence in biomedicine. This glossary consists of words collected over two decades; while a few of the terminologies may be dated, these are nevertheless important words in the evolution of data science and artificial intelligence.

5G (or 5G NR) the cellular network technology's fifth generation that is a significant improvement in speed (the International Mobile Telecommunications guidelines stipulate that 5G should have a peak data rate of 20 Gb/s for downlink and 10 Gb/s for uplink) and in reliability in mobile Internet connectivity

501(k) submission a premarket submission to the FDA with the caveat that the new device is "substantially equivalent to a device already placed into one the three device classifications before it is marketed"

A

Accountable Care Organization (ACO) creation of health-care entities to accommodate three aims: better care for individuals, better health for populations, and slower growth in costs through improvements in care coordination

Accuracy in the confusion matrix, true positives + true negatives/total as a reflection of how often the classifier is correct

Activation function an activation function converts an input signal of a node into an output signal and therefore enables an artificial neural network to "learn" with nonlinear properties (types include sigmoid function, hyperbolic tangent function, and ReLU or rectified linear units)

AdaBoost (see Adaptive Boosting)

Adaptive Boosting (or AdaBoost) a metaalgorithm used in machine learning to improve the performance of a set of relatively weak classifiers into a stronger one by a general ensemble method (final classifier is linear combination of weak classifiers). Adaboost is a very good strategy to boost performance of decision trees in a binary classification problem

Adoption Model for Analytics Maturity (AMAM) a HIMSS proposal for a systems framework and strategic roadmap for analytics in a health-care organization that ranges from stage 0 to 7 (0 being fragmented point solutions to 7 being personalized medicine and prescriptive analytics)

Affordable Care Act (ACA) the comprehensive health reform signed in March 2010 by President Obama into law to render preventive care more accessible and affordable starting in January 2014

Agent (see intelligent agent)

AI winters time periods (1974–80 and then 1987–93) during which funding for AI research and projects was decreased due to less optimism about AI

Algorithm term describing the computer process of following a well-defined list of instructions with historical origin traced to Al Khwarizmi and later popularized by Leonardo Fibonacci

Algorithmic Accountability Act proposed act to require large companies to assess and manage the risks associated with automated decision systems

AlphaGo Google's DeepMind group developed this computer program to play the game Go that utilizes a Monte Carlo tree search with value and policy networks enabled by deep learning technology

AlphaGo Zero an improved version of AlphaGo that learned to play the game Go without any human-created data and was able to soundly defeat its progenitor AlphaGo

AlphaStar the AI system from Google's DeepMind that was able to defeat human champions in a real-time strategy game (*StarCraft II*) and thus reached a new AI milestone

AlphaZero a game-playing reinforcement learning system that taught itself how to play multiple games (chess, shogi, and Go) and was able to defeat the human champions in each aforementioned game

AMAM (see Adoption Model for Analytics Maturity)

American Standard Code for Information Interchange (ASCII) a character-encoding scheme to represent common text in computers and related equipment or devices that display text

API (see application programming interface)

Application programming interface (API) a set of programming instructions and standards (commands, functions, protocols, and objects) for accessing a web-based software application or web tool

Application-specific integrated circuit (ASIC) an integrated circuit that is designed for a specific use

Area under the curve (AUC) in receiver operating characteristic, or ROC, where true positive rate (sensitivity) is plotted versus false-positive rate (100 − specificity), AUC is used to reflect performance of a classifier to separate two classes (usually condition vs no condition). A value of 1 is perfect as a classifier and 0.5 is random chance

Artificial conversational entity (see chatbot)

Artificial general intelligence (AGI) also known as strong AI, this is an AI that is at least or more skillful or flexible as a human with machine intelligence that can think and function similar to humans

Artificial intelligence (AI) the science and engineering of making intelligent machines, especially intelligent computer programs (John McCarthy, Stanford)

Artificial neural network (ANN) a computational model inspired by natural neurons with communication channels between neurons with signals that can be weighted (positive or negative)

ASCII (see American Standard Code for Information Interchange)

Association analysis data mining methodology that is useful for discovering interesting relationships hidden in large data sets that can be represented in the form of association rules (sets of frequent items)

AUC (see area under the curve)

Augmented intelligence (or intelligence augmentation) use of technology in a supportive role to enhance human intelligence (vs assistive and automated intelligence). This term is sometimes used in place of artificial intelligence for reassurance

Augmented reality (AR) a technology in which 3D virtual objects are integrated into a 3D real environment in real time as a form of advanced computer-assisted navigation or visualization technology

AUROC (see ROC and AUC)

Autoencoder a neural network with two networks (encoder and decoder network) that is used in an unsupervised way for a compressed low dimensional representation of the input (for dimension reduction or data denoising)

Automated reasoning a subdiscipline of AI that is an intersection of computer science, cognitive science, and logic that focuses on the process of reasoning

Automation bias a group of errors that humans have a tendency to make in highly automated decision-making situations. Elements of automation bias include errors of omission and commission, overreliance, and over compliance

Autonomous intelligence automated decisions made by the adaptive intelligent systems autonomously with little or no human oversight

Autonomous systems robots and any other device that operates with a level of independence that humans grant for execution of a task. Also the convergence of robotics, mechatronics, and AI

Availability heuristic an intellectual shortcut that relies on immediate recall of one's intrinsic knowledge when evaluating a situation

Avatar a graphical representation of oneself in the virtual world

B

Backpropagation (or back propagation)   short for backward propagation of errors, this is a process in which the "error" of the network from its guess about the data is back transmitted to adjust the network's parameters

Backward chaining (or reasoning)   an inference method that is described as working backward from the goal to determine if there is data to support these outcomes

Bag of words (BoW)   a model used in NLP to represent text data in machine learning that involves disregarding the order or structure of words in the document

Bagging   creation of many duplicates or different sets of training data followed by training with the same model that increases the classification accuracy; random forest is an example

Bayesian belief network (BBN) (also Bayes network or belief network)   a graphical model (directed acyclic graph) that encodes the probabilistic relationships or dependencies among the variables of interest and thereby can be used to learn causal relationships (such as disease and symptoms). In this model, nodes are features while edges represent the relationship between features

Belief network   (see Bayesian belief network)

Bias   how high the weighted sum of the input values needs to be before a node is to be activated

Big data   recent paradigm describing the coupling of the massive amount of data with sophisticated data analytics to acquire new knowledge or insight. The three "Vs" of big data are volume, variety, and velocity

Biocybernetics (also biocybernetic system)   science of information flow and processing to regulate the processes in entire living organism with an interdisciplinary approach that involves laboratory data using biomodeling and computational elements

Bioinformatics   an interdisciplinary study of storage, retrieval, organization, and analysis of biological data and information (genomic sequence alignment, molecular dynamics, and gene expression data) that utilizes computer science, biology, statistics, and mathematics

Biomedical informatics   an interdisciplinary field of quantitative and computational methods for biomedical data and information designed to solve problems across the entire spectrum from biology to medicine and includes the disciplines of bioinformatics and medical or clinical informatics

Biomedical signal analysis   a visualization and interpretation method in biology and medicine for detection, storage, transmission, analysis, and display of images

Biometrics   use of biological markers (such as genetic information, fingerprints, voice patterns, facial characteristics, or eyes) for identification or quantification purposes of humans with the aid of technology (statistical analysis and measurement)

Biomimicry   an innovative methodology that is inspired by nature for design and production of materials or systems

Bionics   body parts that are made stronger or more capable by special electronic or electromechanical devices; use of a biological entity and electronic devices that are put together to create the required implant

Bitmap (BMP)   a binary representation in which a bit (or a set of bits) corresponds to part of an object (usually as an image). In graphics, bitmap is an image format that creates and stores the computer graphics but considered to be inferior to JPEG or vector

Black box   the perception of AI, deep learning in particular, that the methodology and its steps or process cannot be easily explained and perhaps trusted

Blockchain (or block chain)   an innovative secure database that is shared by all parties in a distributed network of computers so that an anonymous exchange of digital assets can occur, and the data are resistant to modification. Blockchain was initially used in bitcoin, and it is based on a distributed ledger principle

Blog a truncation of "web" and "log," it is a virtual informational site that consist of users placing posts (or entries)

Boltzmann machine (see restricted Boltzmann machine)

Boolean algebra (or binary algebra) a branch of algebra in which the values of the variables are the truth values true and false (denoted by binary numbers 1 and 0, respectively)

Boosting (see Adaptive Boosting)

Bots (see chatterbot)

Brain-based device (BBD) a synthetic neural model with behavioral tasks embodied in robotic phenotypes, or essentially a neurally controlled cognitive robot

Brain—computer interface (BCI) [also mind—machine interface (MMI), neural—control interface (NCI), or brain—machine interface (BMI)] this is a bidirectional communication between the brain and computer or device to enhance or augment either mental or physical capability of the human

Butterfly effect a significant part of chaos theory in which a very small change in one state of a complex nonlinear system can lead to a large difference in the system or the observation that these systems are very sensitive to initial conditions

C

Capsule neural network (also Capsule Network or CapsNet) a relatively new deep learning concept promulgated by Geoffrey Hinton that is based on human brain modules called "capsules" that are good for routing visual images to the appropriate capsule for improved hierarchical relationships

Case-based reasoning this is an artificial intelligence technique of utilization of former experiences to comprehend and solve new problems, and the four-step cycle of retrieve, reuse, revise, and retain is used for this methodology

Causality (or cause and effect) the principle is which a state or a process, called a cause, leads to another state or process termed an effect, and this effect is dependent on the cause. For this relationship to be have causality (and not merely correlation), the relationship between the cause and the effect cannot be explained by a third variable

Central processing unit (CPU) the electronic circuitry in the computer (control unit, arithmetic logic unit, and immediate access store) that performs the basic functions and operations (such as arithmetic, logical, control, and input/output) specified by the instructions. The four primary functions of the CPU are fetch, decode, execute, and store

Chaos theory the study of nonlinear dynamics in mathematics in which seemingly random events are predictable from simple deterministic equations in complex systems, such as human physiology and pathophysiology (see also butterfly effect)

Chatbot (also chatterbot or bot) an AI-driven (with NLP) software application that performs a conversation with auditory or written texts to simulate a human and the thinking process

Chunk (or data chunk or chunking) a performance improvement process to organize a large multidimensional data set for fast and flexible data access by storing these data sets in multidimensional rectangular chunks to speed up slow access

Circos plot a visualization tool using circular ideogram layout (with the 22 pairs of chromosomes in circular orientation) to identify similarities and differences between genomes

Classification a supervised learning methodology in which the output variable is a category (binary or multiclass) with the help of a model that separates the data into these categories; examples of classification models include support vector machine, decision tree, and naive Bayes

Classification trees (see decision trees)

Classifier (or classification model) an algorithm (or mathematical function) that implements a classification (binary or more) in machine learning such as naive Bayes, random forest, boosted trees, or support vector machine

Clinical decision support system (CDSS) an information system that entails tools (such as computerized alerts, clinical guidelines, and condition-specific order sets) that provide the clinical staff data and information that can improve health care

Clinical Document Architecture (CDA) a markup standard based on XML from HL7 that defines the medical record structures for certain documents (such as discharge summaries and progress notes but also images and other types of media) to facilitate exchange of information between providers and patients

Cloning creation of an organism that is an exact genetic copy of another with every DNA material being identical

Cloud computing (also cloud) a jargon term to describe the workload shift via the Internet from local computers to a remote network of computers (infrastructure, platform, and software as a service, or IaaS, PaaS, and SaaS) with data storage and compute power being main benefits

Cluster analysis an unsupervised machine learning and data science methodology of dividing data or objects into groups or clusters that exude the same or similar characteristics; an example is k-means clustering

CNTK (or The Microsoft Cognitive Toolkit) this is an open-source production-grade toolkit for deep learning that is fast and flexible as well as easy to use

Codon a three-nucleotide sequence of DNA or RNA that specifies a single amino acid

Cognitive analytics highest level of analytics using reinforcement learning and cognitive computing to achieve human-like decision support (What is the best that could happen?)

Cognitive architecture the third and upcoming wave of AI with attention to more interobject relationships akin to how humans think

Cognitive computing (or cognitive system) an AI system (such as IBM Watson) that simulates the thinking process in a human with the intersection of three disciplines: neuroscience, cognitive science, and computer science with algorithms that use machine learning, pattern recognition, and NLP

Cohen's kappa (or Cohen's kappa coefficient, k) a measurement of the agreement between two raters or a measure of the classifier's performance versus performance by chance

Combinatorics (or combinatorial mathematics) field of mathematics that is concerned with problems of selection, arrangement, and operation within a discrete system; this branch of mathematics includes graph theory

Comma-separated value (CSV) a plain-text file format that stores data in with commas separating the numbers and with an extension .csv

Complementary-DNA (c-DNA) a DNA molecule synthesized by an RNA-dependent DNA polymerase from an RNA template

Complementary learning system the theory that intelligent agents need to possess two learning systems as it exists in the brain: neocortex (with structured knowledge using parametric representations) and hippocampus (individual experiences using nonparametric representations). Through a process called replay, information from the first system is passed to the second system

Complexity theory (see chaos theory)

Computer-aided detection (CADe) a technology involved in noting conspicuous structures and sections; these systems are geared for location of lesions in medical images

Computer-aided diagnosis (CADx) a technology involved in evaluating the conspicuous structures; these systems are geared for to perform characterization of the lesions

Computer-assisted design (CAD) (also computer-aided design and drafting, or CADD) the utilization of computer systems and software to aid the creation and optimization as well as storage of a particular design (not to be confused with CADe or CADx)

Computer-generated imagery (CGI) three-dimensional computer graphics used for creating scenes or special effects

Computer vision (see machine vision)

Concatenation the linkage of two or more character strings (words) or separated things end-to-end to be addressed as a single item (e.g., air and line can be concatenated into airline)

Confirmation bias   tendency to search for information that confirms one's preexisting hypothesis

Confusion matrix   a table that is used to delineate the performance of a classification model on a set of test data

Continuity of care document (CCD)   an XML-based markup standard designed for a patient summary clinical document that can be shared by all computer applications; it uses the HL7 CDA elements

Convolution   a mathematical function derived from two given functions by integration to derive a third function

Convolution filter (also convolution kernel or kernel matrix)   tool used in CNN for processing the input image data and converting the image data into a feature map

Convolutional neural network (CNN)   a special type of forward-feeding artificial neural network with a convolution layer at the beginning used for medical image detection and interpretation in which the neurons are inspired by the cortex of the brain with overlapping influence

Copy number variation (CNV)   variation from one person to the next in the number of copies of a particular gene or DNA sequence

Correlation   measurement of the relationship between two variables

Cost function   this is the optimization theory concept of an average of loss functions of an entire training set (while loss function is the error for a single training example)

CPU   (see central processing unit)

Cross-validation method   in the evaluation of a model the process in which the original data set is divided into k equal-sized subsets called folds that are used as training data set

Crowdsourcing   distributive problem solving for services or ideas by distributing tasks to a large online virtual community to mine collective intelligence

CURES Act (or the 21st Century CURES Act)   legislation, passed in 2016, designed to accelerate medical product development (drugs and devices) in order to advance innovations

Current Procedural Terminology (CPT)   list of medical, surgical, and diagnostic services organized by the AMA in five-digit codes

Curse of dimensionality   phenomena (usually unfavorable) that occur with learning algorithms in high-dimensional spaces that are not observed in lower dimension spaces

Cybersecurity   the technologies and practices that are designed for the protection of computer systems and information from theft and disclosure

Cyborg   from "cybernetic" and "organism," a being with biological and artificial parts but a combination of living organism and machine (vs robots which are mainly machine)

Cyc   an AI project started by computer scientist Douglas Lenat in 1984 that focused on using human-like reasoning in collecting ontology and knowledge base concepts; considered by some the precursor to IBM Watson

D

Data   signals, facts, statistics, and items of information with little meaning without context

Data analytics   process of transforming and modeling data sets for discovery of useful information

Data-driven medicine   data-centric approach promulgated from Stanford that compute on massive amounts of data to discover previously unrecognized patterns and to make clinical relevant predictions to improve health care

Data Engineer   someone who is mainly focused on processing and managing large data sets with software engineering to accommodate the data science team; also known as database administrators or data architects

Data exhaust   data generated as information byproducts (resulting from all of digital as well as online activities) and consist of storable choices, actions, and preferences; this type of data can be very revealing about an individual or group

Data frame a table or two-dimensional array as a storage mechanism for data tables that is more general than a matrix (in which columns have the same data type)

Data hub (or enterprise data hub) a collection of data from a variety of sources organized for sharing or distributing that is different than a data warehouse (unintegrated) or a data lake (raw data)

Data Intelligence Continuum data to higher levels of this continuum: information, knowledge, and intelligence with final level being wisdom

Data lake a storage mechanism for all data within a system in its natural or raw format; usually favored by data scientists (as opposed to data warehouses being favored by business intelligence experts)

Data leakage data transferred electronically or physically from an organization of individual to another entity that is not authorized to be transmitted. Data leakage in the context of machine learning is when data outside the training data set are used for the model generation

Data loss prevention (DLP) processes and tools designed to prevent data, especially sensitive data, to be lost or stolen by unauthorized users

Data mart a subset of data warehouse that focus on a particular subject or department

Data mining the process of automatically discovering useful information or patterns in large data repositories (using machine learning or statistical models); part of knowledge discovery in databases

Data reservoir a managed and organized version of data lake in which there are data platforms (such as Apache Hadoop and NoSQL databases as well as relational database servers) to serve both the data scientist as well as the business users

Data scientist (or data science) someone who works in the interdisciplinary field (statistics, machine learning, data mining, predictive analytics, and mathematics) focused on strategies to extract knowledge from data in structured and unstructured forms

Datathon a venue for data scientists (and other disciplines depending on the theme) to work closely and intensely together (in the spirit of hackathons) on biomedical data

Data visualization visual representation of data with statistical graphics, plots and information graphics to maximize data communication with clarity

Data warehouse (DW) a database or collection of databases for reporting and data analysis to be used for management decision-making

Decision theory identification of value and uncertainty relevant to a decision and process of determining the optimal solution

Decision trees decision support mechanism to display an algorithm that uses a tree-figured graph or model of decisions (at structures called decision nodes) with consequences

Deductive reasoning (or deductive logic) general theory or observations that support a process of reasoning to reach a logical conclusion (also termed "top-down logic")

Deep belief network (or deep belief nets) a class of deep neural network with multiple layers or a generative graphical model in machine learning. This network is essentially a graphical model that learns to extract a deep hierarchical representation of the data

Deep Blue the IBM supercomputer that defeated the human chess champion in chess with search algorithms and GOFAI

Deep learning an artificial intelligence technique that extends the traditional machine learning techniques with multiple layers of neural networks that are capable of nonlinearity

DeepMind (Google) London-based computer scientist and neuroscientist group headed by Demis Hassabis that developed the *AlphaGo* software as well as a myriad of others that defeated the human champions in games

DeepQA project a project aimed at illustrating how the advancement and integration of natural-language processing, information retrieval, machine learning, and knowledge representation to accommodate open-domain questions as observed with the IBM Watson supercomputer

Deep Q-network model that leverages deep neural network to estimate the Q-value function and has the synergistic advantage of both reinforcement learning as well as deep learning

Defense Advanced Research Projects Agency (DARPA)   a federal agency, established in 1958, to prevent strategic surprise from negatively impacting the US national security by maintaining the technological superiority of the US military in many disciplines; the innovative engine for the Department of Defense

Dendral expert system   an early AI project at Stanford in which a heuristic system is designed to be a chemical analysis expert system

Descriptive analytics   statistical methodologies as well as data mining and visualization for mostly a reporting function (What happened?)

Determinism   the belief that events are determined by existing conditions or causes. A deterministic algorithm will always give the same output given the same input

DGX-1 (NVIDIA)   an integrated and robust system for deep learning that features eight Tesla P100 accelerators connected though NVLink, the NVIDIA high-performance GPY interconnect

Diagnostic analytics   queries and root cause analysis using statistical analysis with focus on insight (Why did it happen?)

Differential privacy   the cryptographical process of maximizing accuracy of queries from statistical databases while minimizing chances of identifying its records

Digital Imaging and Communications in Medicine (DICOM)   an international standard for storing, displaying, processing, and transmitting medical images for integration of medical image devices so that images are interoperable

Digital medicine   use of digital tools in medicine to record clinical data and generate medical knowledge that is more precise, more effective, more experimental, and more widely distributed

Dimensionality reduction (also dimension reduction or generalization)   a methodology of reducing variables by feature selection and/or extraction, so there are only principal variables; this process reduces storage space and speeds computation as well as improves model performance (by reducing multicollinearity)

Distributed computing   study of distributed systems in which components are in different networked computers. A computer program within the system is called a distributed program

Distributed ledger (or distributed ledger technology, DLT)   independent computers in a network that are able to record and share transactions and data in individual ledgers; blockchain is a type of distributed ledger with a specific set of features

DNA computing (or Molecular computing)   computing technology that incorporates DNA and RNA and molecular biology hardware instead of silicon-based technology

DNA microarray (or DNA chip)   a system in which many probes with known identity are fixed on a solid support with slots that can be DNA, cDNA, or oligonucleotides

Drone (also unmanned aerial vehicle, UAV)   a pilotless air vehicle that is operated by remote control; AI will provide real-time machine learning technology to enable drones individually or in groups to perceive their surroundings

Drug discovery   process in which novel candidate medications for diseases are discovered though computational means

Dual process thinking   the process in which both System 1 (fast and experiential) and System 2 (slow and analytical) thinking are in balance

E

Edge computing   the constellation of data storage and computing power being closer to the user and away from a central location so that response times and transfer rates are ideal

Edge detection   the process of locating boundaries via image brightness discontinuities so that image segmentation and data extraction can be performed

eHealth (or e-Health) health care accompanied by electronic processes to enhance the communication and quality of delivery

Eigenvector a vector that does not change direction when a linear transformation is performed; this is used to reduce noise in data as well as in dynamic problems. Eigenvectors are paired with eigenvalues, which are numbers (with all the eigenvectors associated with an eigenvalue being and eigenspace)

Electronic health (or medical) record (EHR or EMR) digital record of a patient's paper health (comprehensive) or medical (narrower) status that is available for authorized users for reading and decision-making

Embedded AI (eAI) the application of AI in small devices to push machine and deep learning tools away from only central sources

EMR Adoption Model (EMRAM) the HIMSS model to track health institutions and adoption of EMR in stages (0–7) with stage 7 adoption being no use of paper for medical records

EMRAM (EMR Adoption Model)

Encyclopedia of DNA Elements (ENCODE) an international collaboration of research groups funded by the National Human Genome Research Institute (NHGRI) to identify all functional elements in the human genome

Enigma machine a cipher machine invented by German engineer Arthur Scherbius at the end of World War I to be used for encipher/decipher messages. Its code was eventually deciphered by the British mathematician Alan Turing

Ensemble learning machine learning strategy that involves training a large number of models that together will surpass the performance of a single model; a metamodel that includes bagging, boosting, and stacking

Enterprise data warehouse (EDW) (see data warehouse)

Enterprise data warehouse (EDW) (or data warehouse) a system for reporting and analyzing data in the form of business intelligence for an organization

Enterprise resource planning (ERP) a software that integrates various functions of an organization such as CRM, human resources, accounting, and inventory and order management with a shared central database that can support these functions

Entity Relation (ER) Model or Diagram a graphical representation of the entities and the relationships to illustrate data and database infrastructure

Epigenetics the study of the interactions of chemicals and genes and the factors that influence these interactions such as DNA methylation, histone modification, and nucleosome location

ETL (see extract, transform and load)

Evidence-based medicine an interdisciplinary strategy to learn by making decisions on research studies that are carefully selected; the five-step process includes ask–acquire–appraise–apply–analyze

Evolutionary algorithms (see genetic algorithms)

Exabyte (EB) one quintillion or 10^{18} B (or 1 billion gigabytes) which is enough storage capacity to store 100,000 times all printed material

Exome the part of the genome that contains the DNA record for the protein coding part of the genome

Exon a sequence of DNA that codes information for protein synthesis that is transcribed to messenger RNA

Exoskeleton (or powered exoskeleton) a powered suit to assist the wearer to increase strength and/or endurance with health-care uses, including lifting of heavy patients and allowing disabled patients to walk

Expert systems an artificial intelligence methodology, among the first successful techniques used in artificial intelligence, in which a computer system emulates the decision-making process of a human expert

Explainable AI (also XAI, interpretable AI or transparent AI) methodologies of AI that can be trusted and understood by stakeholders so that the "black box" perception is no longer present. Interpretability has lesser expectation that explainability in that the former mandates cause-and-effect and not necessarily all the details of the methodology

Extract, transform and load (ETL) a process that occurs in a data repository in which data is extracted out of the source and organized into a data warehouse or other data structure

Extensible markup language (XML) a markup language that defines a set of rules for encoding documents for storage and transport in a human- and machine-readable format

F

F score (also F1 score or F measure) a measure of the test's accuracy that is defined as the weighted harmonic mean of the precision and recall of the test

Facebook an online social networking site with potential problems including exposure to inappropriate materials, cyberbullying, sexting, and even Facebook depression

Fast Healthcare Interoperability Resources (FHIR) (pronounced "fire") an interoperability standard for electronic exchange of health-care information developed by HL7 to provide a framework for exchange to support clinical practice

Feature an individual measurable element of a phenomenon being observed (number, string, or graph) for use in machine learning algorithm for pattern recognition, classification, and regression

Feature engineering the process of feature extraction with added domain expertise

Feature extraction the process of taking existing features from feature selection and transforming this group of features into one that will be good for the model

Feedforward neural network an ANN that is simplistic with only connections between nodes and the information moving in one direction; this ANN is without a cycle (as in RNN)

FHIR (see Fast Healthcare Interoperability Resources)

Field-programmable gate array (FPGA) these are integrated semiconductor devices that utilize the concept of configurable logic blocks so that these are programmable after manufacturing

File Transfer Protocol (FTP) a standard Internet client–server protocol designed to transmit files between computers over TCP/IP connections

Floating point operations per second (FLOPS) a computer performance measurement benchmark for rating the speed of microprocessors. For example, one teraFLOPS is equal to one trillion FLOPS

Forward chaining a process in expert systems in which inference rules are used to extract more data until goal is reached

Forward propagation calculation flow of the neural network in the forward direction from the input to the output

Fuzzy cognitive maps (FCM) fuzzy logic combined with neural networks to form these cognitive maps, that is, good for modeling complex systems

Fuzzy logic or reasoning it is a problem-solving control system that accommodates degrees or graded truth rather than absolute (true or false) Boolean logic

G

Gamification the application of digital game design techniques and leveraging of technology and psychology to nongame contexts such as social impact or health-care challenges

Gated recurrent unit (GRU) a variation of LSTM that is structurally similar (but simpler with two rather than three gates) and without internal memory

GDPR (see General Data Protection Regulation)

Gene therapy techniques that involve the use of genes to treat or prevent disease: replacing or inactivating a mutated gene or introducing a new gene

General AI (see artificial general intelligence, AGI)

General Data Protection Regulation (GDPR) policy approved by the EU Parliament in 2016 to harmonize data privacy laws across Europe for personal data and protect and empower EU citizens data privacy; noncompliance is accompanied by heavy penalties

Generalization (see dimensionality reduction)

Generative adversarial network (GAN) neural networks that are composed of two separate deep neural networks (called the generator and the other, the discriminator) that are in essence competing each other but can recursively create and improve new content

Genetic algorithms (GA) an artificial intelligence technique that mimics biological evolution by representing the solution to the problem as a genome and applying genetic operators to evolve the eventual best solution

Genetic engineering technologies that can modify the genetic makeup of cells and can involve highly sophisticated manipulations of genetic material, also called genetic modification

Genetic Information Nondiscrimination Act (GINA) signed by President Bush in 2008, the act protects Americans against discrimination based on their genetic information in situations that involve health insurance and job employment

Genetically modified organism (GMO) a plant or animal that has been genetically engineered with DNA from a source such as bacteria, virus, or another plant or animal

Genetics, Robotics, Internet, and Nanotechnology (GRIN) Technologies emerging technologies that follow a curve of exponential change. Another related term is nanotechnology, biotechnology, information technology, and cognitive science (NBIC)

Genome project the quest to sequence all 3 billion base pairs of the human genome that was led by the National Institute of Health and completed in 2003 with discovery of more than 1800 disease genes

Genome-wide association study (GWAS) an approach used in genetics research to look for associations between many (typically hundreds of thousands) specific genetic variations (most commonly single-nucleotide polymorphisms) and particular diseases

Genomics genomics is the study of functions and interactions of all the genes in the genome, not just of single genes as in genetics

GitHub (Microsoft) an open-source Git repository hosting service with collaboration features for stored source codes from individuals (a cloud platform for code)

Global Innovation Index (GII) a valuable benchmarking tool of innovation based on national economy pillars (such as institutions, human capital and research, infrastructure, market sophistication, and business sophistication) as well as innovation output pillars: knowledge and technology and creative outputs

GOFAI (see Good Old-Fashioned AI)

Good Old-Fashioned AI (GOFAI) also known as symbolic AI, GOFAI was the first wave of AI (1950–80s) that described intelligence in symbolic terms with basis on symbolic representations of problems, logic, and search

Google a search engine company with the mission "to organize the world's information and make it universally accessible and useful" by using "Googlebot" to crawl and search via algorithms. Active health science entities in Google include Verily Life Sciences and DeepMind Health

Googlebot the search algorithm bot (also called a "spider") that crawls and collects documents from the web for the Google search engine

Google Glass a wearable personal computer with optical head-mounted display with capability to communicate with the Internet with natural-language commands with potential for application in medicine and surgery

GPU (see graphical processing unit)

Gradient descent this is an optimization algorithm that iteratively finds the optimal parameters (weights and biases) that will minimize the cost function; essentially this is how machine learning algorithms "learn"

Graph database or graph theory a NoSQL type of database that uses graph structures for semantic queries with nodes and edges to represent and store data

Graphene a material that is composed of a one-atom-thick layer of carbon that is purported to be much stronger than steel while able to conduct electricity; potential applications include screens and displays, biomedical devices and sensors, and memory chips and electronic processors

Graphical processing unit (GPU) a single-chip processor with highly parallel structure used to manage and boost the performance of video and graphics as well as algorithms for large blocks of data in order to lessen the burden of the CPU; also known as visual processing unit (VPU)

Graphical user interface (GUI) an interface that allows the user to be able to communicate with the electronic device via graphical icons and visual indicators

Ground Truth a term to describe a reality check or checking for accuracy for machine learning algorithms against the real world

GRU (see gated recurrent unit)

H

Hackathon an event in which computer and data scientists gather to collaborate on coding intensely in a very short time frame (usually days or weekends)

Hadoop (Apache) software library that serves as a framework for distributed processing of large data sets across clusters of computers using relatively simple programming models; one of the leading technologies for big institutions to mine big data

Haplotype a set of DNA variations or polymorphisms that tend to be inherited together

Haptic technology (or haptics) the science of understanding and improving human interaction with the physical world through the sense of touch

Health 2.0 (or Medicine 2.0) combination of health care with the concept of Web 2.0 with an additional assumption of patient empowerment

Healthcare Effectiveness Data and Information Set (HEDIS) an information tool used by health plans in order to measure 81 care and service performance metrics in five domains

Healthcare Information and Management Systems Society (HIMSS) an international not-for-profit organization focused on better health though information and technology. The HIMSS EMR Adoption Model (EMRAM) tracks health institutions and adoption of EMR in stages (0–7)

Health Information Technology for Economic and Clinical Health (HITECH) Act signed into law in 2009 by President Obama, HITECH was designed to encourage the adoption and implementation of EHR and its supporting technologies for meaningful use

Health Insurance Portability and Accountability Act (HIPPA) the 1996 act, endorsed by the US Congress, provided regulations for use and disclosure of an individual's health information. HIPPA is now often discussed in the context of data privacy and security

Health Level 7 (HL7) a not-for-profit standards developing organization that is dedicated to provide a comprehensive framework for the exchange and integration of electronic health information

Hebbian theory or learning a neuroscientific theory of how neuronal connections can be enforced (the strength of the connection) and is the basis for weight selection in artificial neural networks as part of learning in these networks

Heuristic (also heuristic technique or method) a technique that is designed for solving a problem more quickly when traditional methods are too slow or simply, a mental shortcut to decrease cognitive load

Hidden layer the layer between the input and output layers in a neural network that performs computations received from the input layer, and then the results are passed to the output layer

Hidden Markov model (HMM) a popular statistical methodology for modeling a wide range of time series data as well as signal processing, particularly speech processing and applications include reinforcement learning and temporal pattern recognition. HMM can be equivalent to a simple dynamic Bayesian network

HIPPA (see Health Insurance Portability and Accountability Act)

HL7 (see Health Level 7)

Holdout method in evaluation of a model the master data set is divided into two or three subsets: training set, validation set, and test set

Holography (also hologram) 3D image of an object that is projected and captured on a 2D surface formed by a split laser beam

HTML (see Hypertext Markup Language)

Hughes phenomenon the observation that as the number of features increases, the classifier's performance increases until the optimal number of features is reached

Human Brain Project a large and collaborative research project that started in 2013 using an ICT-based infrastructure to advance knowledge of neuroscience and computing

Human−computer interaction (HCI) the interdisciplinary field of study that involves interaction between human users and computers; computer science, behavioral science, design science, cognitive psychology, and communication theory are all involved in this field

Hybrid Assistive Limb (HAL) a cyborg-type robot with a voluntary control system designed to expand or improve the user's physical capability by receiving biosignals from the skin

Hypergraphs a special kind of graph structure in which an edge (called hyperedges) can join any number of vertices

Hyperparameter hyperparameter is a parameter that has a set value prior to the learning process (value cannot be estimated from the data and therefore set manually) and is external to the model. Examples include the k in k-nearest neighbor, the learning rate in training a neural network, and the C and sigma hyperparameters in SVM

Hyperplane a decision surface in a high-dimensional space that represents the largest separation between two classes that is used in support vector machines

Hypertext Markup Language (HTML) the standard markup language for creating web pages

I

Image classification supervised learning using labeled examples then training a model to recognize objects in images

ImageNet large visual image database with over 14 million images (with about 1000 images per synset) for research in visual recognition that spawned the recent AI research, especially deep learning

Imputation the statistical method of replacing missing data with substituted values to reduce bias and inefficiency in data; the many kinds of imputation include mean or mode imputation, predictive mean imputation, and hot deck nearest neighbor imputation

Inductive logic programming this is a subfield of AI that is the intersection between machine learning and logic programming and can lead to automated learning of logic rules from examples and background knowledge

Inductive reasoning starting with data and specific observations and ending with general principles (also termed "bottom-up logic")

Inference engine a component of the knowledge base of the expert system that makes logical deductions about knowledge assets; the inference engine also is coupled with the user interface

Informatics for Integrating Biology and the Bedside (i2b2) an NIH-funded center for developing a scalable informatics framework that will enable clinical research with the existing clinical data for discovery research

Information data in a more structured as well as meaningful context

Information communications technology (ICT) the infrastructures and components needed for modern computing: cloud computing, software, hardware, transactions, communications technology, data, and Internet access

Information technology (IT)   the use of computers and other devices to create and store as well as secure all forms of electronic data

In-frame exon skipping   the skipping of an exon that contains a multiple of three nucleotides during splicing of pre-mRNA, resulting in the preservation of the reading frame for translation

Innovation   defined as the act of introducing something new or different that creates value. Innovation in medicine and health care is becoming more intertwined with emerging technologies, especially AI

Intelligence   ability to acquire and apply knowledge to achieve goals

Intelligent agent (or agent)   a unit that can perceive its environment via sensors and act upon the environment via effectors

Intelligent automation   combining AI and automation to lead to systems that can sense information and automate entire processes or workflows while learning and adapting during this process

Interaction networks   a model that can reason about how objects in a complex system interact so that dynamic prediction can be made

Internet of Everything (IoE)   the intelligent connection (embedded AI) of people, process, data, and things to make networked connections from IoT (see next) more relevant and valuable

Internet of Things (IoT)   the infrastructure of the information society that links together networks, devices, and data for the purpose of collecting and exchanging data

Internet Protocol (IP)   protocol by which data are sent from one computer to another on the Internet; each computer has a unique IP address

Interoperability   according to HIMSS, the ability of different information systems, devices, or applications to connect, in a coordinated manner, within and across organizational boundaries to access, exchange, and cooperatively use data among stakeholders with the goal of optimizing the health of individuals and populations

Intron   a noncoding segment of DNA between exons that can interrupt a gene-coding sequence

IP   (see Internet Protocol)

Isabel   a web-based diagnostic checklist and decision support tool to help clinicians broaden their differential diagnosis and recognize a disease at the point of care

J

JavaScript Object Notation (JSON)   an open-standard data format that is used for asynchronous browser/server communication with human-readable text to transmit data objects

JPEG (Joint Photographic Experts Group)   a data reduction (usually 10:1) methodology and standard with lossy (irreversible) compression for images

JSON   (see JavaScript Object Notation)

Julia   a high-level and high-performance fast programming language (with syntax similar to Python) that is designed to enable high-performance numerical analyses and computational science

Jupyter (or Jupyter notebook)   an open-source project (named after three common programming languages: Julia, Python, and R) rooted in Python (as IPython notebook) that is used by data scientists and engineers as a tool for collaboration

K

Kaggle   a crowdsourcing approach to a platform for predictive modeling and analytics competitions for the best predictive model that is transparent. It has now evolved into a cloud-based workbench for data science and AI

Keras   a high-level neural network API in Python for fast experimentation and especially designed for TensorFlow, CNTK, or Theano

Kernel   a filter used in CNN and also a mathematical function used in SVM (the latter can be in linear or radial form)

k-Means clustering   an unsupervised learning methodology for cluster analysis in data mining by which an unlabeled data set is partitioned (with iterative refinement) into small number of clusters (by minimizing the distance between each data point and the center of the cluster, called centroid, that the point belongs to)

k-Nearest neighbor (kNN)   a supervised machine learning algorithm that is a nonparametric method for classification and regression; the underlying principle is calculation of distances and choosing the k value

Knowledge   more contextual form of information and can be explicit or tacit

Knowledge-based systems (KBS)   artificial intelligence techniques that include rule-based learning, case-based learning, and model-based learning; the other school is intelligent computing method

Knowledge discovery in databases (KDD)   data mining for searching hidden knowledge in a large amount of data with data preparation, selection, cleansing, and interpretation. Data mining is the pattern extraction phase of KDD, so the latter is a broader process

Knowledge Graph (Google)   the system of gathering facts, people, and places to create an interconnected search results that would be accurate

Knowledge representation   an artificial intelligence technique to represent knowledge in symbols in order to derive at conclusions from these elements

Kolmogorov−Smirnov Test   this is a nonparametric statistics test that compares a known hypothetical probability distribution to the distribution generated by one's data

L

Lab-on-a-chip (or microfluidics)   performing laboratory operations on a small scale (to tens of micrometers) using miniaturized microfluidic devices

Legacy systems   an older technology or system related to a previous (usually outdated) computer system

Linear discriminate analysis (or normal discriminant analysis)   a supervised linear transformation technique that is used for dimensionality reduction

Linear regression   the relationship between a dependent and an independent variable to predict future values of the dependent variable

Lisp (programming language)   lisp, or LISP, is a high-level programming language initially specified in 1958 as part of the Dartmouth Summer Research Project on Artificial Intelligence by John McCarthy

Logical Observation Identifiers Names and Codes (LOINC)   a database and international standard for medical laboratory observations and measurements

Logistic regression   a regression model where the dependent variable is categorical and is a supervised learning methodology used for classification (despite its nomenclature)

LOINC   (see Logical Observation Identifiers Names and Codes)

Long short-term memory (LSTM)   a simple recurrent neural network that can be built into a bigger recurrent neural network

Loss function   a method of assessing how well the algorithm models the data set (the higher the number, the worse the performance) by measuring the absolute difference between the prediction and actual value. Gradient descent is often used to find the minimum point of this loss function

Lossless compression   data compression algorithm that renders the original data to be reconstructed from the compressed data (picture quality remains the same as the original)

Lossy compression   data compression algorithm that renders the original data to be partly reconstructed

Low-shot (or few-shot) learning   the strategy of suppling a learning model very small amount of training data. The advantages of this methodology include overcoming scarcity of supervised data and the labeling burden of huge data sets

LSTM   (see long short-term memory)

M

Machine learning   a branch of artificial intelligence with statistical models, which allows computers to automatically learn from data without programming

Machine to machine (M2M)   communication network that entails wireless devices forming an ecosystem that minimizes human interaction

Machine vision   technology and methodologies used to provide image-based analytics for inspection or control as well as robotic guidance in various industries

MapReduce (Apache)   a programming paradigm that accommodates massive scalability of unstructured data across thousands of commodity cluster servers

Markov model   a stochastic model used to accommodate changing systems in such a way that future states depend only on current state (but not on states prior to the current state). This is called Markov property

Mashup   a web application or derivative work that consists of a combination of data and information of various sources

Massive open online course (MOOC)   online distance education course and connection that is meant for a large-scale participatory audience via web access. Examples are Coursera, edX, and Udacity

MATLAB (matrix laboratory)   a high level fourth generation language and interactive environment for numerical computation, visualization, and programming

Matrix (or Data Matrix)   a collection of data elements in a two dimensional rectangular table in rows and columns (a vector is a row or column of a matrix)

Max pooling   this is a process in which the input representation of an image or output matrix is down-sampled with reduced dimensionality with features intact

Meaningful use   in the use of EHR, meaningful use is for improving quality and care coordination as well as engaging patients, and maintaining privacy to lead to better clinical outcomes

Mechanical Turk   a fake mechanical chess playing automaton in the 18th century that was in fact manned by a human (also an Amazon website to find and accept assignments using crowdsourcing)

Medical decision support   (see clinical decision support)

Medical image processing or analysis   use of machine intelligence to perform quantitative analytics of medical imaging modalities such as CT, MRI, PET, or microscopy

Medicare Access and CHIP Reauthorization Act (MACRA)   a bipartisan legislation signed into law in 2015 that changes the way Medicare rewards clinicians for value over volume

Medicine 2.0   (see Health 2.0)

Metabolomics   the comprehensive characterization of small molecule metabolites in biological systems

Metadata   defined as data about data or how data are collected or formatted with relevance to data warehouses

mHealth (also m-health or mobile health)   practice of medicine and public health supported by mobile devices such as mobile phones or tablet computers

Microbiome   a discipline to examine how bacteria interact with each other and the human body to cause or prevent disease

Microelectromechanical systems (MEMS)   technology of microscopic devices, especially ones with moving parts

Microfluidics   (see lab-on-a-Chip)

MicroRNA (miRNA)   a short regulatory form of RNA that binds to a target RNA molecule and generally suppresses its translation by ribosomes

Misclassification rate (or error rate)   the false positives and negatives together over the entire population

Mixed reality   a convergence of the physical with the digital world in a virtual continuum

Model-based reasoning   an artificial intelligence methodology in which an inference method is used based on a model of the physical world

Modified National Institute of Standards and Technology (MNIST) a large database of handwritten digits that is utilized as an image processing system training data set

Modifier genes genes that have a relatively small quantitative effect on the expression of another gene

MongoDB a free and open-source platform that is a document-oriented NoSQL database program with scalability and flexibility

Monte Carlo tree search (MCTS) a heuristic search algorithm that is a combination of classic tree search and reinforcement learning designed for certain types of decision processes (most recently publicized as part of the *AlphaGo* AI program)

Moore's Law observation by Gordon Moore, founder of Intel, that the number of transistors per square inch on integrated circuits double every 2 years since the circuit was invented

Moravec's paradox observation by Hans Moravec, an AI robotics expert, that machines can solve things that are difficult for humans (such as geometrical problems), but at the same time, tasks that are easy for humans are often difficult for machines

Multiparameter Intelligent Monitoring in Intensive Care (MIMIC) database that contains physiologic and pathophysiologic signals and vital sign time series from monitors as well as comprehensive clinical data for analytics and research. This resource is often used for modeling biomedical data

Multiprotocol Label Switching (MPLS) a methodology used for WAN connectivity in order to route traffic inside a telecommunications network

Munging when referring to data, the process of cleaning and shaping the data prior to data mining; similar to data wrangling

MYCIN an early expert system AI program that originated at Stanford University (with Ted Shortliffe) for treating blood infections

N

Naive Bayes classifier a supervise learning methodology based on Bayesian theorem of independence between features and is well suited for data with high input dimensionality

Nanobots (or nanorobots) devices ranging in size from 0.1 to 10 micrometers and made of nanoscale or molecular components with promising use in biomedical technology especially coupled with AI technology

Nanomedicine use of nanotechnology and nanomaterials for clinical applications such as in vivo contrast agents, drug carriers, or diagnostic devices

Nanotechnology the intentional design, characterization, production, and applications of materials, structures, devices, and systems by controlling their size and shape in the nanoscale range (1–100 nm)

Nanotechnology, biotechnology, information technology, and cognitive science (NBIC) emerging technologies that follow a curve of exponential change. Another term is Genetics, Robotics, Internet, and Nanotechnology (GRIN) Technologies

Natural-language generation (NLG) AI technology that transforms data into language akin to a human analyst with relevance, intuition, and expediency

Natural-language processing (NLP) a field in artificial intelligence and computer science to study the interaction between human natural language and computers

Natural-language understanding (NLU) (or natural-language interpretation, NLI) an AI technology focused on machine reading comprehension and is a postprocessing part of NLP

Neat (vs Scruffy) AI the two philosophies of AI research: Neats prefer solutions that are elegant and clear, whereas scruffies believe that intelligence is very complex; this underlying difference may be rooted in the symbolic versus connectionist dichotomy

Neo4j a graph database management system with native graph storage and processing that is presently the most popular graph database

Net neutrality the basic principle that disallows Internet service providers from altering any content by speeding up/slowing down or by blocking any content or applications one like to use; the neutrality keeps Internet free and open

Network File System (NFS) a file system that enables storage and retrieval of data from files from multiple sources

Neural lace (Neuralink) Elon Musk's concept of a brain–computer system what would link human brains with a computer interface to achieve a human–machine symbiosis

Neural network (or Nets) (see artificial neural network)

Neural Turing machine (NTM) a RNN that extends the neural network concept and couple this with logical flow and external memory sources

Neuromorphic Chip silicon chips that are modeled on biological brains designed to process sensory data as a human brain without specific programming

Neuromorphic computing design of computers that will possess three characteristics that brains have (but computers do not): low power consumption, fault tolerance, lack of need to be programmed

Neuromorphic engineering (see neuromorphic computing)

NewSQL a class of RDMBS that has the scalability of NoSQL but combines ACID (atomicity, consistency, isolation, and durability) guarantee of RDBMS

Next-generation sequencing (NGS) the inexpensive production of large volumes of sequence data that holds an advantage over the first-generation automated Sanger sequencing technique

NLP (see natural-language processing)

Nodes the decision points in decision trees that are interconnected by branches

Normalization (data or database normalization) reconfiguration of data so that there is no redundancy of data and that data dependencies are logical; this strategy is designed to increase performance and decrease storage

Not Only SQL (NoSQL) databases that are characterized by large data volumes, scalable replication and distribution, and efficient queries

O

Object detection algorithms using extracted features in the process of detecting instances of objects

One-shot learning a machine learning strategy promulgated by Stanford machine learning expert Li Fei Fei to learn information about object categories from one or very few training images (instead of large amounts of such data)

Online analytical processing (OLAP) an approach used in business intelligence to answer analytical query that is multidimensional (relational database, reports, and data mining)

Online transaction processing (OLTP) a class of information systems that deal with small and interactive transactions that require very low response times (as opposed to batch processing)

Ontology in information science, ontology entails representation and definition of concepts, data, and entities and their relationships

Open Neural Network Exchange (ONNX) an open-source artificial intelligence ecosystem with machine and deep learning models

Optical character recognition (OCR) electronic conversion and information entry as well as recognition of images of text (typed or handwritten) to a machine-encoded text

Optimization algorithm an iterative procedure for the goal of achieving an optimum or best solution

Organ printing biomedical application of rapid prototyping, or additive layer-by-layer biomanufacturing, is an emerging in situ transforming technology that is new compared to traditional solid scaffold-based tissue engineering

Overfitting a type of modeling error in which the function or learning system is too closely fitting a training data set as to not able to accurately predict outcomes of the untrained data set (describes the noise instead of the underlying concept)

P

P4 (predictive, personalized, preventive, and participatory) medicine philosophy to allow biotechnology to manage an individual's health and wellness instead of disease in a personalized approach

Parallel algorithms an algorithm that is capable of being executed simultaneously on different processing devices and then the results combined to yield the final group result

Parameters values derived from training that are characteristic of the population and are internal to the model. Parameters of a neural network are the weights of the connections

Parsing an analysis used in NLP of a string of symbols in natural or computer language following the rules of grammar

Part-of-speech tagging (POST) the natural-language process in which the words in a sentence are marked to correspond to part of speech

Patient-Centered Outcomes Research Institute (PCORI) organization that helps to make informed health-care decisions and better outcomes by promoting evidence-based information

Perceptron a machine learning algorithm for supervised learning of binary classifiers initially conceived in 1957 by Frank Rosenblatt; the basis for the modern artificial neural network

Personalized medicine a more precise and customized extension of traditional approaches to understanding and treating disease that has advanced due to wide availability of genetic information

Petabyte one quadrillion or 10^{15} B which is enough storage capacity to store the DNA of all Americans (1000 TB = 1 PB and 1000 PB = 1 EB)

Pharmacogenetics the study of how the actions of and reactions to drugs are dependent on the variations of an individual's genes and metabolic pathways

Pharmacogenomics a term at times confused with pharmacogenetics, pharmacogenomics is the genomic discipline to study genes in the entire genome across the population that influence drug response (the premise for personalized medicine)

Phenomics use of large-scale, high-throughput assays and bioinformatics approaches to investigate how genetic instructions actually translate into tangible phenotypic traits

Picture archiving and communication system (PACS) medical imaging technology for storing as well as presenting medical images by a myriad of imaging modalities (MRI, X-Ray, CT, and ultrasound)

Pixel the smallest element or a sample of a picture in digital imaging

Pluripotency (also induced pluripotent stem cells) the potential of a cell to develop into more than one type of mature cell, usually any of the three germ layers (endoderm, mesoderm, or ectoderm)

Podcast a multimedia digital file that is accessible on the Internet and downloadable to a computer or other device with the potential for medical education and patient education

Polanyi's paradox the observation that our intuitive tacit knowledge of how the world works (bottom of iceberg) often exceeds our explicit understanding (tip of the iceberg)

Policy in reinforcement learning, a function that maximizes the reward in a long-term setting; reward maximization

Policy network a component of deep reinforcement learning strategy in *AlphaGo* that evaluates the current situation and predicts the next step

Posterior probability in Bayes' rule, it is a conditional probability of a given event that is computed after observing a second event

Precision the degree in which repeated measurements show the same results; reproducibility. Also called positive predictive value

Precision medicine the coupling of clinicopathological profiles with molecular profiles to tailor the diagnostic and therapeutic regimen to the individual

Precision−Recall curve a curve plotting precision and recall which can neutralize the imbalanced classification problem

Predictive analytics   the capability to predict future human behavior or events by data mining and basic machine learning techniques (What will happen?)

Predictive modeling   this strategy uses statistics to predict outcomes via one or more classifiers

Preimplantation genetic diagnosis (PGD) (or preimplantation genetic screening, PGS)   the genetic profiling of the embryo prior to implantation and as part of in vitro fertilization

Prescriptive analytics   machine and deep learning techniques for predictive modeling and with focus on optimization (What should I do?)

Principal component analysis (PCA)   a simple but powerful statistical technique that is useful in extracting relevant information from confusing and/or large data sets to result in dimensionality reduction

Probabilistic reasoning   usage of past situations and examples and apply statistics to predict a likely outcome

Protected health information (PHI)   information regarding health status and care as well as payment that can be associated with an individual

Proteomics   study of structure and function of proteins

Publication bias   tendency in biomedicine to publish only studies with a positive diagnostic or therapeutic result

Python   a powerful (multiparadigm) dynamic programming language that is used in a myriad of application domains with certain advantages for research: language interoperability, data structures, available libraries, and balance of high- and low-level programming

PyTorch (Python)   a Python-based library used as a deep learning developmental framework for an efficient experimentation

Q

Q-learning (quality learning)   a reinforcement learning algorithm that has the goal of finding the best action with the highest reward given a current state, and does so with explorative—exploitative actions that are outside the current policy (off-policy)

Quantifiable self   self-knowledge of medical conditions or disease through self-tracking with devices and biomarkers

Quantum bit (or qubit)   the basic unit of information in quantum computing that is analogous to the bit in classical computing

Quantum computing   computing that will harness the power of atoms and molecules to perform memory and processing tasks with quantum bits, or qubits

Quantum dot   materials that consist of a core-and-shell structure (e.g., CdSe coated with zinc and sulfide with a stabilizing molecule and a polymer layer coated with a protein)

Quick Response (QR) code   a matrix barcode that is attached to an item which has information related to the item

R

R   a computer programming language and environment for statistical computing and high-quality graphics particularly well suited for biomedical data science

Radio-frequency identification (RFID)   usage of radio waves to communicate between the reader and an electronic tag that is affixed to an object or person

Radiomics   the methodology of extracting features from high volume medical images and converting these features into mineable and analyzable data for precision medicine

Random forest   an ensemble of decorrelated decision trees (with each tree constructed by using a random subset of training data) that will output a prediction value

Randomized controlled trial (RCT)   a study design involving treated and untreated individuals in a randomized assignment format

Recall percentage of total relevant results correctly classified by the model; also known as sensitivity

Receiver operating characteristic (ROC) a graphical plot of true positive rate (y-axis) versus false-positive rate (x-axis) at various thresholds with the area under the curve reflecting the performance of the model

Rectified linear unit (ReLU) (or rectifier) ReLU is the most used activation function in deep learning as it is easy to train and performs relatively well due to its ability to avoid the vanishing gradient problem

Recurrent neural network (RNN) (or recurrent nets) a type of deep learning with a loop that is useful in sequential data such as text and speech as well as time series data

Recursive Cortical Network (RCN) a generative object-based model that deviates from deep learning in that it begins with a scaffold (rather than *tabula rasa*) and therefore is much more data efficient

Regenerative medicine field of biomedical science that utilizes technology to use stem cells to rejuvenate, replace, and regenerate body cells, tissue, and organs

Region of interest (ROI) image samples identified within the data set for a particular purpose

Regression statistical methodology to estimate the relationship between dependent and independent variables

Regularization the process of constraining the coefficient estimates toward 0 as to discourage learning a complex model and thus avoiding the risk of overfitting

Reinforcement learning machine learning that is inspired by behavioral psychology to maximize cumulative reward in a balanced state of exploration and exploitation. Deep reinforcement learning is combining this learning with deep learning and has been used for multiple AI strategies in game playing by DeepMind of Google

Relational database a database with a relational model of data (in tables) and the software systems that maintain these databases are called relational database management system (RDBMS)

ReLU (see rectified linear unit)

ResearchKit (Apple) an open-source framework introduced by Apple that provides an opportunity for researchers and developers to create apps for medical research

Residual neural network (ResNet) a type of artificial neural network that has skip connections (with biological origin in pyramidal cells in the cerebral cortex) which connect output of previous layers with the output of new layers

Resource Description Framework (RDF) a framework for describing resources on the web that is designed to be understood by computers

Restricted Boltzmann machine (RBM) a generative stochastic and shallow (two layers being visible and hidden layers) artificial neural network with the learning based on statistical methods; the restriction in these RBMs is that there is no intralayer communication

Rich Site Summary (RSS) (also Really Simple Syndication) a format for delivery of regularly changing web content and a family of web feed formats to gather updated information, including blog entries, audio and video information, and news articles, and to present a summarized text for an update

Robotic process automation (RPA) intelligent software robots deployed to automate repetitive activities with the user interface of a computer system

Robotic surgery robotic systems (the Intuitive Surgical da Vinci Surgical System and the Zeus MicroWrist Surgical System) used to assist in the performance of certain surgical procedures, especially laparoscopic procedures

Robots technology that entails design, construction, maintenance, and application of robots with their computer environment; health-care applications include rehabilitation, exoskeletons, and virtual visits

Rule-based reasoning an artificial intelligence technique (under the knowledge-based systems or rule-based expert systems) that involves rules, database, and interpreter for the rules

Rules engine part of the expert system that is connected to the knowledge base with if-then rules stored within it

S

Scalar   a number, whereas a vector is a list of numbers (in a row or column)

Scruffy AI   (see neat and scruffy AI)

Segmentation   the process of dividing or partitioning a medical image in computer vision into many segments (called pixels) for analysis

Self-organizing map (SOM)   a data visualization technique to display an artificial neural network that uses unsupervised learning to produce a low-dimensional representation

Semantic net (or network)   a knowledge representation methodology used for propositional information (or mathematically a labeled directed graph)

Semantic Web   extension of the World Wide Web with metadata that will allow users to share content beyond the traditional boundaries of applications and websites and for computers to "talk" to each other; also termed "Web 3.0"

Semisupervised learning   a hybrid of supervised and unsupervised learning that utilizes a small amount of labeled data and then a relatively large amount of unlabeled data

Sentient AI   AI that possesses properties of self-awareness

Sentiment analysis   extraction of subjective information with NLP from mining of data to determine the individual's feeling about certain issues

Signal processing   technology that uses mathematical and computational representation for transferring information contained in various formats

Simple Protocol and RDF Query Language (SPARQL)   a semantic query language for databases stored in RDF format

Single (simple) nucleotide polymorphisms (SNPs)   a single-nucleotide variation in a genetic sequence that is a common form of variation in the human genome

Singularity   this refers to the point in time (estimated to be around the year 2045) during which technological intelligence supercedes human intelligence. This is a concept initially attributed to mathematician John von Neumann but popularized by the science fiction writer Vernor Vinge and the futurist Ray Kurzweil

Smart wearable systems (SWS)   devices ranging from sensors and actuators to other monitoring devices for management of patients' health status

SNOMED CT   (see Systemized Nomenclature of Medicine—Clinical Terms)

Social media   the creation, sharing, and exchange of information and ideas in the virtual community; includes tools such as Twitter, Facebook, YouTube, Foursquare, and other social tools

Softmax   an activation function that changes numbers into a vector of probabilities

Software agent   a computer program capable of performing without direct supervision and is the computer analog of an autonomous robot

Software as a service (SaaS)   a distribution model in which software is made available to the client via a provider hosts applications over the Internet

Spark (Apache)   an open-source big data processing engine with the ecosystem consisting of structured data, streaming analytics, machine learning, and graph computation

Spiking neural network (SNN)   a third-generation neural network that bridges the gap between machine learning and neuroscience by using neuronal properties. SNN uses discrete events rather than continuous values that conventional machine or deep learning use

SQL   (see Structured Query Language)

Stem cell (or embryonic stem cell)   unspecialized cells capable of renewing via cell division and also capable of being induced to become tissue specific cells

String   a data type as a sequence of characters in computer programming, usually represented in quotation marks

Strong AI   (see artificial general intelligence)

Structured Query Language (SQL)   a special-purpose query programming language for accessing and managing databases organized in a relational database management system (RDBMS)

Superintelligence   a form of intelligence that supersedes the intelligence of humans and is associated with technological singularity

Supervised learning   machine learning of predicting from labeled training data with each example being an input object and a desired output

Support vector machines (SVM)   a machine learning technique with support vectors in which supervised learning models with learning algorithms are used that analyze data for classification and regression analysis

Swarm intelligence   collective intelligence of a decentralized but self-organizing system that was introduced in the context of cellular robotic systems

Syllogism   a form of reasoning (deductive) in which a conclusion is drawn from two given or assumed propositions

System 1 (and System 2) Thinking   Nobel laureate Daniel Kahneman's work on the brain having System 1 (fast and experiential) and System 2 (slow and analytical) thinking

Systemized Nomenclature of Medicine—Clinical Terms (SNOMED CT)   a suite of designated standards of clinical terminology used by clinicians and others in electronic health and medical records

Systems biology   integration of biology and medicine along with technology and computation as a discipline to study biological components from molecules to organisms or entire species

T

Tag cloud   (see word cloud)

Telehealth   health-care services (with a wider spectrum than telemedicine) being delivered via telecommunication with promise to increase access to specialized services

Telemedicine   the use of medical information exchanged from one site to another site via electronic communications

Temporal convolutional network (TCN)   a variation of CNN for sequence modeling tasks that can be faster and more accurate than RNNs, LSTM, and GRUs

TensorFlow (Google)   an open-source software library for machine intelligence developed by Google Brain team for numerical computation using dataflow graphs as well as deep learning

Tensor processing unit (TPU) (Google)   machine learning chips, more powerful than CPUs or GPUs, designed specifically to increase the speed of machine learning tasks

Terabyte (TB)   computer memory storage capacity that is a trillion bytes or a thousand gigabytes; 10 TB can hold the entire literary collection of the Library of Congress (1000 TB = 1 PB and 1000 PB = 1 EB)

Test data   data used for assessing the generalization error of the final chose model and is the smaller subset of data compared to training data

Text mining (text analytics or text data mining)   the process of extracting high-quality information from natural-language text

Theano   a Python library for fast computation of deep learning for creating deep learning models

Tissue engineering   (see regenerative medicine)

Tokenization   the process of breaking up a stream of text into words or phrases called tokens that can be used to secure information for encryption

Training data   data used to fit the model or algorithm and is usually a larger data set than the test data

Transfer learning   machine learning type of learning that focuses on storing knowledge gained from solving one problem and then applying this knowledge to solve another problem

Transhumanism   movement with philosophy of improving humans with available technological modifications from intellectual and physical perspectives

Transmission Control Protocol

Turing test a test of artificial intelligence devised by Alan Turing, the famous mathematician, wherein a human interrogator is given the task of distinguishing between a human and a computer based on the replies to questions

Twitter an online social networking and microblogging service that utilizes "tweets" or 140-character text messages as its main mode of communication

U

Underfitting phenomenon when a statistical model or machine learning algorithm fails to fit the data sufficiently so that the model is excessively simple

Unified Medical Language System (UMLS) key terminology and standards from the NIH to provide interoperable biomedical information systems a common dictionary

Unmanned aerial vehicle (UAV) (see drone)

Unsupervised learning type of machine learning in which a function is predicted based on unlabeled data so there is no error/reward signal for the predicted solution

User experience (UX) the optimization of the usability of the product or service from the user's perspective (human first) and is more analytical and technical than UI (something that looks great but not feel good to use is good UI but not good UX)

User interface (UI) the graphic design of the product or service and at times confused with UX (something that feels good to use but not good to look at is good UX but not good UI)

V

Validation data data used to estimate the prediction error of the model and about the same size as the test data

Value network the component of deep reinforcement learning of *AlphaGo* that evaluates the situation and predicts which side will win

Vanishing gradient (problem) with increasing number of layers and accompanying activation functions to the neural network, the loss function gradients can approach zero and thus rendering the training and learning of the network increasingly more difficult

Variance the difference in fits between data sets as higher complexity models have a higher variance

Variants of uncertain significance (VUS) (or allelic variant of unknown significance) an alteration in the normal sequence of a gene that has unknown clinical significance and disease risk

Variational autoencoder an autoencoder that not only compresses the data but is also generative as it synthesizes new similar data to the observed data

Vector a row or column of numbers whereas scalar is a number and matrix is an array of numbers (one or more rows and one or more columns)

Virtual assistant (VA) a sentient digital assistant that uses artificial intelligence for facilitating one's digital life or other activities and tasks (e.g., Apple's Siri, Microsoft's Cortana, Amazon's Echo, Facebook's M, or Google's Now)

Virtual private networks (VPN) a private network extension to connect a shared or public network to enable organizations and individuals to transmit data between computers

Virtual reality (VR) technology that uses images and sounds in order to simulate a user's physical presence in a virtual environment

Visual analytics the science of analytical reasoning supported by interactive visual interfaces and consists of an integral approach combining visualization, human factors, and data analysis

Voxel a value in three-dimensional space (as opposed to pixel in two-dimensional picture)

W

Watson (IBM) a supercomputer that utilizes a portfolio of natural-language processing, information retrieval, knowledge representation, and machine learning with 4 TB of disk storage to be able to read close to 100 million pages per second and defeat human contestants on the quiz show *Jeopardy!*

Weak AI (or narrow AI) AI or machine intelligence that is based on a single-focused task (such as chess or Go playing)

Wearable technology or devices devices that can monitor vital signs (such as heart rate, blood pressure, or pulse oximetry) and electrocardiogram (for waveform analysis and heart rhythm assessment)

Web 2.0 a term describing new collaborative Internet applications with key elements include RSS, blogs, wikis, and podcasts

Web 3.0 (see Semantic Web) a term to describe the evolution of the web in finding and organizing new information beyond the boundaries of websites

Weights the modifying factor for inputs to lead to a final output of a neural network; these can be adjusted to lead to different outputs

Wide area network (WAN) a computer or telecommunications network over a large geographical area

Wiki a website or online resource that allows certain users to add and edit medical information as a collective group

Wireless sensor networks (WSN) a network of autonomous sensors for monitoring conditions in an environment but has potential health-care applications

Wisdom an understanding of principles derived from intelligence and values and beliefs with self-reflection and future vision

Wolfram Alpha (or WolframAlpha) a powerful computational knowledge engine that uses natural-language processing and answers questions (similar to DeepQA project by IBM)

Word cloud (or tag cloud) a method of representing text data to visualize keyword metadata on websites (by size of font and by color)

Wrangling (data wrangling) as in data munging, the process of transforming data from a raw form into a more refined form for analytics. ETL is different than data wrangling in that ETL works with more structured data and is focused on IT as the end user

X

XAI (see Explainable AI)

XML (see Extensible markup language)

Y

YouTube an online video-sharing website with video clips uploaded by individuals or by media corporations and even hospitals

Z

Zero-shot learning (ZSL) the process that a machine learns how to recognize an image with no labeled image as its classification training data

Zettabytes (ZB) one sextillion or 10^{21} B (or 1 billion terabytes) with the total amount of global data around 3 zettabytes

Key references

* Recommended

** Highly recommended

Books on data science, artificial intelligence, and human cognition

** Agarwal A, et al. Prediction machines: the simple economics of artificial intelligence. Boston, MA: Harvard Business Review Press; 2018.

—*One of the top books on AI from the economic perspective with easy to understand but essential concepts in AI.*

* Armstrong, S. Smarter than us: the rise of machine intelligence. Berkeley, CA: Machine Intelligence Research Institute; 2014.

—*A short treatise on AI with a well-rounded overview of the relevant current issues.*

Boden MA. AI: its nature and future. New York: Oxford University Press; 2016.

** Bostrom N. Superintelligence: paths, dangers, strategies. Oxford: Oxford University Press; 2014.

—*An outstanding perspective on the nuances of AI in the current era and in the future.*

* Brockman J. Thinking; the new science of decision-making, problem-solving, and prediction. New York: HarperCollins Publishers; 2013.

—*A thought-provoking and insightful collection of essays on topics such as rational thought, decision-making, intuition, and prediction.*

* Brockman J. Possible minds: 25 ways of looking at AI. New York: Penguin Press; 2019.

—*Although the names of AI dignitaries are not as well recognized as the other compendium by Martin Ford, there are some excellent chapters.*

** Broussard M. Artificial unintelligence: how computers misunderstand the world. Cambridge, MA: MIT Press; 2018.

—*An insightful look into the world of AI from someone who is a computer scientist and a writer so the dual perspective is unique.*

Christian B, Griffiths T. Algorithms to live by: the computer science of human decisions. New York: Henry Holt and Company LLC; 2016.

** Paul D, James WH. Human + machine: reimagining work in the age of AI. Boston, MA: Harvard Business Review Press; 2018.

—*These two Accenture technology leaders delineate the world of AI and the human machine relationship better than any other book.*

* Pedro D. The master algorithm; how the quest for the ultimate learning machine will remake our world. New York: Basic Books; 2015.

*—*A very good book on machine learning for anyone who would like to know more about machine learning beyond the average media publication but would not like to be buried by the esoteric allusions of computer and data science.*

** Ford M. Architects of Intelligence: The Truth About AI From the People Building It. Packt Publishing, 2018.

—*The best collection of interviews with the hall of fame AI experts by the New York Times futurist Martin Ford. With very few exceptions, these interviews are worth reading twice.*

** Gerrish S. How smart machines think. Cambridge, MA: MIT Press; 2018.

—*Excellent insight into machines such as deep neural networks and natural language processing and explained in such a way that anyone can really appreciate the intricacies of these technological marvels.*

* Hawkins J. On intelligence. How a new understanding of the brain will lead to the creation of truly intelligent machines. Times Books; 2004.

—*An excellent book on how to relate the brain and neuroscience to computers in an innovative way by the founder of palm computing.*

* Kaplan J. Artificial intelligence: what everyone needs to know. Oxford: Oxford University Press; 2016.

—*Good to excellent primer on the relevant topics on AI, including law, human labor, social equity, and the future.*

* Kasparov G. Deep thinking: where machine intelligence ends and human creativity begins. New York: Perseus Books; 2017.

—*An amazingly insightful book by the former world's best chess player reflecting on machine intelligence after his famous defeat by Big Blue.*

** Knuth DE. The art of computer programming, vols. 1–4. Boston, MA: Addison Wesley; 1997.

—*This is the pinnacle of not only computer programming books but arguably scientific books in general with its large four volumes and in its 39th printing (it is even dedicated to a computer, the Type 650 computer).*

* Ray K. How to create a mind: the secret of human thought revealed. New York: Penguin Books; 2012.

—*A fascinating look at the mind and future of artificial intelligence from the futurist Ray Kurzweil.*

* Thomas M. Superminds: the surprising power of people and computers thinking together. New York: Hachette Book Group; 2018.

*—*An enlightening book on not only synergy between human and machine but the collective wisdom of the crowd as well.*

** Marcus G, Freeman J. The future of the brain: essays by the world's leading neuroscientists. Princeton, NJ: Princeton University Press; 2015.

—*A must read for anyone who would like to stay ahead in understanding the future underpinnings of artificial intelligence as a cognitive science.*

* McAfee A, Brynjolfsson E. Machine, platform, crowd: harnessing our digital future. New York: W.W. Norton and Company; 2017.

—*Very good update from the authors of The Second Machine Age on the digital revolution, including many references on AI and its impact in our society.*

** Minsky, M, Papert SA. Perceptrons: an introduction to computational geometry. Boston, MA: Massachusetts Institute of Technology; 1960.

—*An excellent historical treatise on the perceptron and its evolution as the precursor of deep learning with only one caveat: the math is very esoteric but the story itself is compelling.*

* Motyl P. Labyrinth: the art of decision making. Vancouver, Canada: Page Two Books; 2019.

—*Good readable book on how nuances in decision making can be understood with many good points for biomedicine.*

** Sejnowski TJ. The deep learning revolution. Cambridge, MA: MIT Press; 2018.

—*A very readable personal perspective on deep learning that can still be understood by anyone who is not a data scientist.*

* Tegmark M. Life 3.0: being human in the age of artificial intelligence. New York: Penguin Random House LLC; 2017.

—*A lengthy treatise on AI and its impact on the future of life in all dimensions by an MIT physicist.*

Textbooks on data science and artificial intelligence

** Hastie T, Tibshirani R, Friedman J. The elements of statistical learning: data mining, inference, and prediction. New York: Springer; 2013.

*—*Well-written like its more introductory book (see James et al., 2013) but very heavy on the mathematics and not easy for most mortals.*

* Howard RA, Abbas AE. Foundations of decision analysis. Upper Saddle River, NJ: Pearson Education Inc.; 2016.

—*Outstanding textbook on the entire process of decision making and its analysis as a science.*

** James G, Witten D, Hastie T, et al. An introduction to statistical learning with applications in R. New York: Springer; 2013.

—*This textbook and its more advanced sister book (see Hastie et al., 2013) are hands-down the best books I have seen during my years in school studying biomedical data science.*

Lucci S, Kopec D. Artificial intelligence in the 21st century. Dulles, VA: Mercury Learning and Information; 2013.

** Russell S, Norvig P. Artificial intelligence: a modern approach. 3rd ed. Upper Saddle River, NJ: Pearson Education, Inc.; 2010.

—*The most comprehensive and authoritative textbook on artificial intelligence with an incredible historical and futuristic perspective as well as amazing depth and breadth of AI methodologies.*

* Tan PN, Steinbach M, Kumar V. Introduction to data mining. New York: Addison-Wesley Publishing; 2006.

*—*My personal favorite textbook on data mining covering classification, association, clustering, and anomaly detection in a clear and comprehensive manner.*

Books on data science, artificial intelligence, and human cognition in biomedicine

* Agah A. Medical applications of artificial intelligence. Boca Raton, FL: CRC Press; 2014.

—*A comprehensive and current textbook on medical applications of AI with more emphasis on the computer and data science aspects.*

* Beckerman AP, Petchey OL. Getting started with R: an introduction for biologists. Oxford: Oxford University Press; 2012.

—*A must read as an introduction to the world of R as it relates to biology and biomedicine.*

* Chettipally, UK. Punish the machine! The promise of artificial intelligence in health care. Charleston, SC: Advantage Press; 2018.

—*A personal perspective on the promise of AI in medicine and health care authored by one of the passionate advocates of AI and application.*

* Clancey WJ, Shortliffe EH. Readings in medical artificial intelligence: the first decade. Addison-Wesley Publishing; 1984.

—*Good collection of essays during the first decade of AI in medicine with predominantly works in knowledge-based area.*

Ceophas TJ, Zwinderman AH. Machine learning in medicine. New York: Springer; 2013.

* Consoli S, Recupero DR, Petkovic M. Data science for health care. Cham, Switzerland: Springer; 2019.

—*Good reference but mainly designed for data scientists and engineers. The clinician can perhaps glean something from the first few chapters.*

Dua S, Chowriappa P. Data mining for bioinformatics. Boca Raton, FL: CRC Press; 2014.

* Giabbanelli PJ, Mago VK, Papageorgiou EI. Advanced data analytics in health. Cham, Switzerland: Springer; 2018.

—*Good collection of topics covering data exploration and visualization to machine learning and modeling.*

** Groopman, J. How doctors think. Boston, MA: Houghton Mifflin Company; 2007.

—*Timeless but excellent read on how clinicians think and how clinicians can improve this complicated process.*

Liebowitz J, Dawson A. Actionable intelligence in health care. Boca Raton, FL: CRC Press; 2017.

*Lu L, Zheng YF, Carneiro G, et al. Deep learning and convolutional neural networks for medical image computing. Cham, Switzerland: Springer; 2017.

—*Good textbook on this topic but designed for the more advanced reader in computational science and computer vision.*

Luxton DD. Artificial intelligence in behavioral and mental health care. Elsevier, Academic Press: London; 2016.

—*Good collection of implementation of AI in aspects of mental health care and is well ahead of its time as a book.*

Mahajan P. Artificial intelligence in health care. Self-published; 2018.

Natarajan P, Frenzel JC, Smaltz DH. Demystifying big data and machine learning for health care. Boca Raton, FL: CRC Press; 2017.

* Ranschaert ER, Morozov S, Algra PR. Artificial intelligence in medical imaging. Switzerland: Springer; 2019.

—*One of the better references on AI in medical imaging with some chapters better than others.*

* Reddy CK, Aggarwal CC. Health care data analytics. Boca Raton, FL: CRC Press; 2015.

—*A comprehensive book on data analytics in health care with broad range of topics including natural language processing, visual analytics, clinical decision support systems, computer-assisted medical image analysis systems, and information retrieval.*

** Scarlet, A. A machine intelligence primer for clinicians. San Bernardino, CA; 2019.

—*An excellent primer on machine learning for the clinician who has a basic understanding of data science but would like to understand applications of machine learning.*

** Shortliffe EH, Cimino JJ. Biomedical informatics: computer applications in health care and biomedicine (health informatics). 4th ed. London: Springer; 2014.

—*An outstanding timeless textbook on health informatics that provides an essential framework for any reading of AI in biomedicine and health care.*

* Szolovits P. Artificial intelligence in medicine. In: AAAS selected symposia series, vol. 51. Boulder, CO: Westview Press Inc.; 1982.

—*A very good historical perspective on the state of the art during this early period of artificial intelligence in medicine with emphasis on expert systems.*

** Ten Teije, A, Popow C, Holmes JH. Artificial intelligence in medicine: 16th conference on artificial intelligence in medicine, AIME 2017, Vienna, Austria, Proceedings. New York: Springer; 2017.

—*Excellent academic content of all aspects of AI in medicine published by the leadership of AIME for its biennial meeting in Europe.*

* Topol, E. Deep medicine: how artificial intelligence can make health care human again. New York: Basic Books; 2019.

—*A summary of recent AI activities in medicine and health care with the thesis of utilizing AI for improvement of health care for everyone, including the clinicians.*

Wachter R. The digital doctor: hope, hype, and harm at the dawn of medicine's computer age. New York: McGraw Hill; 2015.

Yang H, Lee EK. Health care analytics: from data to knowledge to health care improvement. Hoboken, NJ: John Wiley and Sons; 2016.

Journals of interest

AI Magazine (AAAI)
Editor: Ashol Goel (AAAI, Palo Alto, CA)
Artificial Intelligence in Medicine (AIMed)
Editors: Anthony Chang and Freddy White (AIMed, London and Orange, CA)
Artificial Intelligence in Medicine (AIIM)
Editor: Carlo Combi (Elsevier, Boston, FL)
Journal of American Medical Informatics Association (JAMIA)
Editor: Lucila Ohno-Machado (Oxford University Press, Oxford)
Journal of Medical Artificial Intelligence (JMAI)
Editor: Jia Chang (AME Publishing, Shatin)
MIT Technology Review
Editor: David Rotman (MIT Press, Boston, MA)
Nature and Nature Machine Intelligence/Nature Medicine
Editor: Magdalena Skipper (Springer, New York)
Wired
Editor: Nicholas Thompson (Wired Media Group, New York)
Top 100 articles (and more) on artificial intelligence and artificial intelligence in medicine
(In the spirit of crowd wisdom and "swarm intelligence," no author has more than one article listed.)

Group papers

Artificial intelligence and life in 2030: one hundred year study on artificial intelligence, <https://ai100.stanford.edu>.

European Group on Ethics in Science and Technologies. Statement on artificial intelligence, robotics, and autonomous systems; 2018.

Executive Office of the President: National Science and Technology Council and Committee on Technology. Preparing for the future of artificial intelligence; 2016.

FDA. Proposed regulatory framework for modifications to artificial intelligence/machine learning (AI/ML)-based software as a medical device (SaMD): discussion paper and request for feedback. <regulations.gov>; 2019.

Authors papers

Abramoff MD, Lavin PT, Brich M, et al. Pivotal trial of an autonomous AI-based diagnostic system for detection of diabetic retinopathy in primary care offices. NPJ Digit Med 2018;1:39.

Alagappan M, Glissen Brown JR, Mori Y, et al. Artificial intelligence in gastrointestinal endoscopy: the future is almost here. World J Gastrointest Endosc 2018;10(10):239−49.

Altman R. AI in medicine: the spectrum of challenges from managed care to molecular medicine. AI Mag 1999;20 (30):67−77.

Amirkhani A, Papageorgiou EI, Mohseni A, et al. A review of fuzzy cognitive maps in medicine: taxonomy, methods, and applications. Comput Methods Programs Biomed 2017;142:129−45.

Angermueller C, et al. Deep learning for computational biology. Mol Syst Biol 2016;12(878):1−16.

Balayla J, Shrem G. Use of artificial intelligence (AI) in the interpretation of intrapartum fetal heart rate (FHR) tracings: a systematic review and meta-analysis. Archives of Gynecology and Obstetrics 2019;300.

Banaee H, Ahmed MU, Loutfi A. Data mining for wearable sensors in health monitoring systems: a review of recent trends and challenges. Sensors (Basel) 2013;13(12):17472−500.

Barry DT. Adaptation, artificial intelligence, and physical medicine and rehabilitation. Phys Med Rehabil 2018; S131−4.

Beam AL, Kohane IS. Big data and machine learning in health care. JAMA 2018;319(13):1317−18.

Bejnordi BE, Veta M, Van Diest PJ, et al. Diagnostic assessment of deep learning algorithms for detection of lymph node metastases in women with breast cancer. JAMA 2017;318(22):2199−210.

Benjamins JW, Hendriks T, Knuuti J, et al. A primer in artificial intelligence in cardiovascular medicine. Neth Heart J 2019;1−9.

Benke K, Benke G. Artificial intelligence and big data in public health. Int J Environ Res Public Health 2018;15:2796−805.

Bennett TD, Callahan TJ, Feinstein JA, et al. Data science for child health. J Pediatr 2018;208:12−22.

Bur AM, Shew M, New J. Artificial intelligence for the otolaryngologist: a state of the art review. Otolaryngol Head Neck Surg 2019;160(4):603−11.

Cabitza F, Locoro A, Banfi G. Machine learning in orthopedics: a literature review. Front Bioeng Biotechnol 2018;6:75.

Chang AC. Precision intensive care: a real-time artificial intelligence strategy for the future. Pediatr Crit Care Med 2019;20(2):194−5.

Chang HY, Jung CK, Woo JI, et al. Artificial intelligence in pathology. J Pathol Transl Med 2019;53:1−12.

Char DS, Shah NH, Magnus D. Implementing machine learning in health care—addressing ethical challenges. N Engl J Med 2018;378(11):981−3.

Chartrand G, Cheng PM, Vorontsov E, et al. Deep learning: a primer for radiologists. RadioGraphics 2017;37 (7):2113−31.

Chen Y, Argentinis E, Weber G. IBM Watson: how cognitive computing can be applied to big data challenges in life sciences research. Clin Ther 2016;38(4):688−701.

Ching T, Himmelstein DS, Beaulieu-Jones BK, Kalinin AA, Do BT, Way GP, et al. Opportunities and obstacles for deep learning in biology and medicine. J R Soc Interface 2018;15:20170387.

Coiera EW. Artificial intelligence in medicine: the challenges ahead. J Am Med Inform Assoc 1996;3(6):363−6.

Connor CW. Artificial intelligence and machine learning in anesthesiology. Anesthesiology 2019;131.

Darcy AM, Louie AK, Roberts LW. Machine learning and the profession of medicine. JAMA 2016;315(6):551−2.

Deo RC. Machine learning in medicine. Circulation 2015;132:1920−30.

Desai GS. Artificial intelligence: the future of obstetrics and gynecology. J Obstet Gynecol India 2018;68(4):326−7.

Dey D, Slomka PJ, Leeson P, et al. Artificial intelligence in cardiovascular imaging: JACC state-of-the-art review. J Am Coll Cardiol 2019;73(11):1317−35.

Dimitrov D. Medical internet of things and big data in health care. Healthc Inform Res 2016;22(3):156−63.

Ekins S. The next era: deep learning in pharmaceutical research. Pharm Res 2016;33(11):2594−603.

Esteva A, Robicquet A, Ramsundar B, et al. A guide to deep learning in health care. Nat Med 2009;25:24−9.

Farley T, Kiefer J, Lee P, et al. The biointelligence framework: a new computational platform for biomedical knowledge computing. J Am Med Inform Assoc 2013;20(1):128−33.

Ferrucci D, Brown E, Chu-Carroll J, et al. Building Watson: an overview of the DeepQA project. AI Mag 2010;31 (3):59−79.

Fogel AL, Kvedar JC. Perspective: artificial intelligence powers digital medicine. NPJ Digit Med 2018;1:5−8.

Ganapathy K, Abdul SS, Nursetyo AA. Artificial intelligence in neurosciences: a clinician's perspective. Neurol India 2018;66:934−9.

Gawehn E, Hiss JA, Schneider G. Deep learning in drug discovery. Mol Inform 2016;35(1):3−14.

Ghassemi M, Celi LA, Stone DJ. State of the art review: the data revolution in critical care. Crit Care 2015;19:118−27.

Goodfellow IJ, Pouget-Abadie J, Mirza M, et al. Generative adversarial networks. arXiv:1406.2661.

Grapov D, Fahrmann J, Wanichthanarak K, et al. Rise of deep learning for genomic, proteomic, and metabolomic data integration in precision medicine. OMICS 2018;22(10):630−6.

Greenhalgh T, Howick J, Maskrey N. Evidence based medicine: a movement in crisis? Br Med J 2014;348:g3725.

Greenspan H, van Ginneken B, Summers RM. Guest editorial/deep learning in medical imaging: overview and future promise of an exciting new technique. IEEE Trans Med Imaging 2016;35(5):1153−9.

Griebel L, Prokosch HU, Kopcke F, et al. A scoping review of cloud computing in health care. BMC Med Inform Decis Mak 2015;15:17.

Gubbi S, Hamet P, Tremblay J, et al. Artificial intelligence and machine learning in endocrinology and metabolism: the dawn of a new era. Front Endocrinol 2019. Available from: https://doi.org/10.3389/fendo.2019.00185.

Gulshan V, Peng L, Coram M, et al. Development and validation of a deep learning algorithm for detection of diabetic retinopathy in retinal fundus photographs. JAMA 2016;316:2402−10.

Hanson WC, Marshall BE. Artificial intelligence applications in the intensive care unit. Crit Care Med 2001;29 (2):427−35.

Hashimoto DA, Rosman G, Rus D, et al. Artificial intelligence in surgery: promises and perils. Ann Surg 2018;268 (1):70−6.

Hassabis D, Kumaran D, Summerfield C, et al. Neuroscience-inspired artificial intelligence. Neuron Rev 2017;95 (2):245−58.

He J, Baxter SL, Xu J, et al. The practical implementation of artificial intelligence technologies in medicine. Nat Med 2019;25:30−6.

Hinton G. Deep learning—a technology with the potential to transform health care. JAMA 2018;320(11):1101−2.

Hosny A, Parmar C, Quackenbush J, et al. Artificial intelligence in radiology. Nat Rev Cancer 2018;18(8):500−10.

Hueso M, Vellido A, Montero N, et al. Artificial intelligence for the artificial kidney: pointers to the future of a personalized hemodialysis therapy. Kidney Dis 2018;4(1):1−9.

Iniesta R, Stahl D, McGuffin P. Machine learning, statistical learning, and the future of biological research in psychiatry. Psychol Med 2016;46:2455−65.

Jha S, Topol EJ. Adapting to artificial intelligence: radiologists and pathologists as information specialists. JAMA 2016;316(22):2353−4.

Jiang F, Jiang Y, Zhi H, et al. Artificial intelligence in health care: past, present, and future. Stroke Vascu Neurol 2017;2:e000101.

Johnson AE, Pollard TJ, Shen L, et al. MIMIC-III, a freely accessible critical care database. Sci Data 2016;3:160036.

Johnson KW, et al. Artificial intelligence in cardiology. J Am Coll Cardiol 2018;71(23):2668−79.

Kapoor R, Walters SP, Al-Aswad LA. The current state of artificial intelligence in ophthalmology. Surv Ophthalmol 2019;64:233−40.

Kim YJ, Kelley BP, Nasser JS, et al. Implementing precision medicine and artificial intelligence in plastic surgery: concepts and future prospects. Plast Reconstr Surg Glob Open 2019;7:e2113.

Klein JG. Five pitfalls in decisions about diagnosis and prescribing. Br Med J 2005;330:781−3.

Komorowski M, Celi LA, Badawi O, et al. The artificial intelligence clinician learns optimal treatment strategies for sepsis in intensive care. Nat Med 2018;24(11):1716−20.

Krittanawong C, et al. Artificial intelligence in precision cardiovascular medicine. Journal of American College of Cardiology 2017;69(21):2657−64.

Lamanna C, Byrne L. Should artificial intelligence augment medical decision making? The case for an autonomy algorithm. AMA J Ethics 2018;20(9):E902−10.

LeCun Y, Bengio Y, Hinton G. Deep learning. Nature 2015;521:436−44.

Lee S, Mohr NM, Street WN, et al. Machine learning in relation to emergency medicine clinical and operational scenarios: an overview. West J Emerg Med 2019;20(2):219−27.

Lin SY, Mahoney MR, Sinsky CA. Ten ways artificial intelligence will transform primary care. J Gen Intern Med 2019;34:1−5.

Londhe VY, Bhasin B. Artificial intelligence and its potential in oncology. Drug Discov Today 2019;24(1):228−32.

Lynn LA. Artificial intelligence systems for complex decision-making in acute care medicine: a review. Patient Safety Surg 2019;13:6−14.

Mathur P, Burns ML. Artificial intelligence in critical care. Int Anesth Clin 2019;57(2):89−102.

Middleton B, Sittig DF, Wright A. Clinical decision support: a 25 year retrospective and a 25 year vision. Yearb Med Inform 2016;(Suppl. 1):S103−16.

Miller DD, Brown EW. Artificial intelligence in medical practice: the question to the answer? Am J Med 2018;131:129−33.

Miller PL. The evaluation of artificial intelligence systems in medicine. Comput Methods Programs Biomed 1986;22:5−11.

Miotto R, Wang F, Wang S, et al. Deep learning for healthcare: review, opportunities and challenges. Brief Bioinform 2017;19:1−11.

Mnih V, Kavukcuoglu K, Silver D, et al. Human-level control through deep reinforcement learning. Nature 2015;518:529–33.

Morgan DJ, Bame B, Zimand P, et al. Assessment of machine learning vs standard prediction rules for predicting hospital readmissions. JAMA Netw Open 2019;2(3):e190348.

Mousses S, Kiefer J, Von Hoff D, et al. Using biointelligence to search the cancer genome: an epistemological perspective on knowledge recovery strategies to enable precision medical genomics. Oncogene 2008;27:S58–66.

Naugler C, Church DL. Automation and artificial intelligence in the clinical laboratory. Crit Rev Clin Lab Sci 2019;56(2):98–110.

Niazi MKK, Parwani AV, Gurcan MN. Digital pathology and artificial intelligence. Lancet Oncol 2019;20(5): e253–61.

Nichols JA, Herbert Chan HW, Baker MAB. Machine learning: applications of artificial intelligence to imaging and diagnosis. Biophys Rev 2019;11(1):111–18.

Norman GR, Monteiro SD, Sherbino J, et al. The causes of errors in clinical reasoning: cognitive biases, knowledge deficits, and dual process thinking. Acad Med 2017;92(1):23–30.

Nsoesie EO. Evaluating artificial intelligence applications in clinical settings. JAMA Netw Open 2018;1(5):e182658.

Obermeyer Z, Emanuel EJ. Predicting the future—big data, machine learning, and clinical medicine. N Eng J Med 2016;375:13–16.

Patel V, Shortliffe EH, Stefanelli M, et al. The coming of age of artificial intelligence in medicine. Artif Intell Med 2009;46:5–17.

Peek N, Combi C, Marin R, et al. Thirty years of artificial intelligence in medicine (AIME) conferences: a review of research themes. Artif Intell Med 2015;65(1):61–73.

Rajkomar A, Dean J, Kohane I. Machine learning in medicine. N Eng J Med 2019;380:1347–58.

Rajpurkar P, Irvin J, Zhu K, et al. CheXNet: radiologist-level pneumonia detection on chest X-rays with deep learning. arXiv 2017; arXiv:1711.05225.

Ramesh AN, Kambhampati C, Monson JR, et al. Artificial intelligence in medicine. Ann R Coll Surg Engl 2004;86 (5):334–8.

Ravi D, et al. Deep learning for health informatics. IEEE J Biomed Health Inform 2017;21(1):4–21.

Reddy S, Fox J, Purohit MP. Artificial intelligence-enabled health care delivery. J R Soc Med 2019;112(1):22–8.

Rosenblatt F. The perceptron: a probabilistic model for information storage and organization in the brain. Psychol Rev 1958;65(6):386–408.

Rusk N. Deep learning. Nat Methods 2016;13(1):35.

Russell S, Hauert S, Altman R, et al. Robotics: ethics of artificial intelligence. Nature 2015;521(7553):415–18.

Sacchi L, Holmes JH. Progress in biomedical knowledge discovery: a 25-year retrospective. Yearb Med Inform 2016;S117–29.

Saito T, Rehmsmeier M. The precision-recall plot is more informative than the ROC plot when evaluating binary classifiers on imbalanced datasets. PLoS One 2015;10(3):e0118432.

Schwartz WB. Medicine and the computer: the promise and problems of change. N Engl J Med 1970;283:1257–64.

Senders JT, Arnaout O, Karhade AV, et al. Natural and artificial intelligence in neurosurgery: a systematic review. Neurosurgery 2018;83:181–92.

Sheridan TB. Human-robot interaction: status and challenges. Hum Factors 2016;58(4):525–32.

Shortliffe, E.H. Artificial intelligence in medicine: weighing the accomplishments, hype, and promise. IMIA Yearb Med Inform 2019; 28.

Shortliffe EH, David R, Axline SG, et al. Computer-based consultations in clinical therapeutics: explanation and rule acquisition capabilities of the MYCIN system. Comput Biomed Res 1975;8(4):303–20.

Shu LQ, Sun YK, Tan LH, et al. Application of artificial intelligence in pediatrics: past, present, and future. World J Pediatr 2019;15(2):105–8.

Sidey-Gibbons JAM, Sidey-Gibbons CJ. Machine learning in medicine: a practical introduction. BMC Med Res Methodol 2019;19:64.

Silver D, Huang A, Maddison CJ, et al. Mastering the game of go with deep neural networks and tree search. Nature 2016;529:484–9.

Stewart J, Sprivulis P, Dwivedi G. Artificial intelligence and machine learning in emergency medicine. Emerg Med Australas 2018. Available from: https://doi.org/10.1111/1742-6723.13145 [Epub ahead of print].

Szolovits P, Patil RS, Schwartz W. Artificial intelligence in medical diagnosis. Ann Intern Med 1988;108:80–7.

Thrall JH, Li X, Li Q, et al. Artificial intelligence and machine learning in radiology: opportunities, challenges, pitfalls, and criteria for success. J Am Coll Radiol 2018;15(3):504−8.

Thukral S, Singh Bal J. Medical applications on fuzzy logic inference system: a review. Int J Adv Networking Appl 2019;10(4):3944−50.

Topol EJ. Hi-performance medicine: the convergence of human and artificial intelligence. Nat Med 2019;25:44−56.

Tseng HH, Wei L, Cui S, et al. Machine learning and imaging informatics in oncology. Oncology 2018. Available from: https://doi.org/10.1159/000493575.

Vellido A. Societal issues concerning the application of artificial intelligence in medicine. Kidney Dis 2019;5:11−17.

Wahl B, Cossy-Gantner A, Germann S, et al. Artificial intelligence and global health: how can AI contribute to health in resource-poor settings? BMJ Global Health 2018;3:e000798.

Wang R, Pan W, Jin L, et al. Artificial intelligence in reproductive medicine. Soc Reprod Fertil 2019;R139−54.

Wartman SA, Combs CD. Reimagining medical education in the age of AI. AMA J Ethics 2019;21:146−52.

Williams AM, Liu Y, Regner KR, et al. Artificial intelligence, physiological genomics, and precision medicine. Physiol Genomics 2018;50:237−43.

Wong ZAY, Zhou J, Zhang Q. Artificial intelligence for infectious disease big data analytics. Infect Dis Health 2019;24:44−8.

Yamashita R, Nishio M, Do RKG, et al. Convolutional neural networks: an overview and application in radiology. Insights Imaging 2018;9:611−29.

Yang YJ, Bang CS. Application of artificial intelligence in gastroenterology. World J Gastroenterol 2019;25(14):1666−83.

Yu KH, Beam AL, Kohane IS. Artificial intelligence in health care. Nat Biomed Eng 2018;2:719−31.

Websites of interest

American Medical Informatics Association
(AMIA.org)
Artificial Intelligence in Medicine (AIMed)
(Ai-Med.io)
Artificial Intelligence in Medicine
(www.journals.elsevier.com/artificial-intelligence-in-medicine)
Association for the Advancement of Artificial Intelligence (AAAI)
(www.aaai.org)
Data Science Institute (DSI) at American College of Radiology (ACR)
(www.acrdsi.org)
Machine Learning for Health Care (MLHC)
(www.mlforhc.org)
MIMIC III
(MIMIC.physionet.org)

Videos of interest
Introductory
How Researchers are Teaching AI to Learn Like a Child (AAAS)
(https://www.youtube.com/watch?v = 79zHbBuFHmw)
AlphaGo (Documentary, 2017)
Do You Trust This Computer? (Documentary 2018)
Artificial Intelligence: Mankind's Last Invention
(https://www.youtube.com/watch?v = cM8Nk7b_X1o)
Three Principles for Creating Safer AI (TED2017/Stuart Russell)
(https://www.ted.com/talks/stuart_russell_3_principles_for_creating_safer_ai?language = en)
What Happens When Our Computers Get Smarter Than We Are? (TED2015/Nick Bostrom)
(https://www.ted.com/talks/nick_bostrom_what_happens_when_our_computers_get_smarter_than_we_are?language = en)
IBM Watson: The Science Behind an Answer

(https://www.youtube.com/watch?v = DywO4zksfXw)
A Gentle Introduction to Machine Learning
(https://www.youtube.com/watch?v = Gv9_4yMHFhI)

More advanced
Machine Learning Fundamentals: Bias and Variance
(https://www.youtube.com/watch?v = EuBBz3bI-aA)
The Rise of Artificial Intelligence Through Deep Learning (TEDxMontreal/Yoshua Bengio)
(https://www.bing.com/videos/search?q = artificial + intelligence + and + TED&qpvt = artificial + intelligence
    + and + TED&view = detail&mid = C0477628C68142A174BFC0477628C68142A174BF&&FORM = VDRVRV)
But What Is a Neural Network? Deep Learning, Chapter 1 (3Blue1Brown series)
(https://www.youtube.com/watch?v = aircAruvnKk)
Gradient Descent, How Neural Networks Learn. Deep Learning, Chapter 2 (3Blue1Brown series)
(https://www.3blue1brown.com/videos-blog/2017/10/16/gradient-descent-how-neural-networks-learn-deep-
    learning-chapter-2)
What is Backpropagation Doing Really? Deep Learning, Chapter 3 (3Blue1Brown series)
(https://www.3blue1brown.com/videos-blog/2017/11/3/what-is-backpropagation-really-doing-deep-learning-
    chapter-3)
100 + Companies to Know in AI in Medicine and Health Care
There are now over 200 companies in the AI in health-care domain, and listed below are the companies that are
    important to know. (Note. The largest companies and their activities in health care are described in the various
    sections of the book rather than listed here.)
The information provided here include name of company, funding amount, current address of company, found-
    ing year, estimated number of employees, area of focus, and specific services. (Note. Sources include company
    websites, various lists, CBInsights, and Crunchbase.)

Company	AI in health-care focus	Specific service	URL
3Scan ($21M) San Francisco (CA) (2011) [10−50]	Digital and computational pathology	Knife edge scanning microscope (KESM)	www.3scan.com
Ada ($60M) London (UK) (2011) [NA]	AI-powered doctor app and telemedicine service	Personal health guide	www.ada.com
AiCure ($27M) New York (NY) (2010) [10−50]	AI-powered platform for medication adherence and treatment efficacy	IMA (interactive medical assistant)	www.aicure.com
AIDoc ($42M) Tel Aviv (IL) [10−50]	AI platform for radiology image interpretation and workflow	Platform for interpretation	www.aidoc.com

(Continued)

(Continued)

Company	AI in health-care focus	Specific service	URL
AIMed (NA) London (UK) (2016) [10−50]	Multimedia platform for AI in medicine and health care	AIMed meetings AIMed academic magazine	www.ai-med.io
Alignment Health ($240M) Orange (CA) (2013) [250−500]	Health-care analytics for population health management	Advanced clinical model	www.alignmenthealthcare.com
AliveCor ($63M) Mountain View (CA) (2010) [10−50]	AI-enabled methodology for EKG monitoring	Kardia mobile	www.alivecor.com
Amara Health Analytics ($75K) San Diego (CA) (2010) [10−50]	Real-time predictive analytics	Clinical vigilance for sepsis software	www.amarahealthanalytics.com
Apixio ($36M) San Mateo (CA) (2009) [50−100]	Decision support analytics platform for risk adjustment	HCC profiler cognitive computing solution	www.apixio.com
Arterys ($44M) San Francisco (CA) (2011) [50−100]	Radiology and imaging analytics platform	Cardiac MRI analytics	www.arterys.com
Atomwise ($51M) San Francisco (CA) (2012) [10−50]	Drug development based on AI tools	AtomNet	www.atomwise.com

(*Continued*)

(Continued)

Company	AI in health-care focus	Specific service	URL
Ayasdi ($106M) Menlo Park (CA) [50–100]	Complex data analytics platform for enterprise	Symphony AyasdiAI	www.ayasdi.com
Babylon Health ($85M) London (UK) (2013) [1000–5000]	AI-powered system for primary care queries	Health check	www.babylonhealth.com
BaseHealth ($18M) Sunnyvale (CA) (2008) [10–50]	Predictive analytics for population health management and risk adjustment	BaseHealth Engine	www.basehealth.com
Bay Labs ($5M) San Francisco (CA) (2013) [10–50]	AI-powered echocardiography	EchoMD Auto EF	www.baylabs.io
BenevolentAI ($202M) London (UK) (2013) [50–100]	Accelerated drug development with knowledge graph	Benevolent platform	www.benevolent.ai
BioXcel Therapeutics (IPO) Branford (CT) (2017) [10–50]	AI-enabled drug discovery for neuroscience and immunooncology	InveniAI	www.bioxceltherapeutics.com
Bot MD ($2M) Palo Alto (CA) (2018) [1–10]	Smartphone-based AI assistant for clinicians	Chatbot for clinicians	www.botmd.io
Buoy Health ($9M) Boston (MA) (2014) [1–10]	Diagnostic algorithm and analytics platform	Diagnostic platform	www.buoyhealth.com

(Continued)

(Continued)

Company	AI in health-care focus	Specific service	URL
Butterfly Network ($350M) Guilford (CT) (2011) [100−250]	Medical imaging device that reduces cost with AI medical image potential	Butterfly iQ	www.butterflynetwork.com
Cardiogram ($2M) San Francisco (CA) (2016) [10−50]	Analytics on heart rate and its variability	Cardiogram app DeepHeart	www.cardiogr.am
Clarify Health Solutions ($63M) San Francisco (CA) (2015) [10−50]	Digital technology combined with analytics for efficient care	Analytics platform with real-time insights	www.clarifyhealth.com
Clinithink (Series B) London (UK) (2009) [10−50]	Data query NLP platform for unstructured data	CLiX ENRIC CLiX CNLP	www.clinithink.com
CloudMedx ($5M) Palo Alto (CA) (2015) [10−50]	AI with ML and NLP for EHR and data-driven decisions	AI Assistant	www.cloudmedxhealth.com
Cognoa ($20M) Palo Alto (CA) (2014) [10−50]	Machine learning for diagnostic device for autism	Cognoa Child Development app	www.cognoa.com
CureMetrix (Series A) La Jolla (CA) (2014) [10−50]	Medical image analysis for mammography	cmTriage	www.curemetrix.com
Deep 6 (Seed) Pasadena (CA) (2015) [10−50]	AI-enabled methodologies to locate patients for clinical trials	AI-enabled patient-trial matching	www.deep6.ai

(Continued)

(Continued)

Company	AI in health-care focus	Specific service	URL
Deep Genomics ($17M) Toronto (CA) (2014) [10–50]	Intersection of genomics and AI leading to drug development	Project Saturn	www.deepgenomics.com
Doc.ai ($12M) Palo Alto (CA) (2016) [1–10]	Blockchain-based AI for predictive analytics with data in one location	Health data app	www.doc.ai
Doctor Evidence ($2M) Santa Monica (CA) (2004) [50–100]	Analytics for evidence analysis using published data and patient outcomes	DRE AI	www.drevidence.com
DreaMed Diabetes ($5M) Petah Tiqva (IL) (2014) [1–10]	Closed-loop solution for patients with diabetes	DreaMed Advisor Pro	www.dreamed-diabetes.com
Droice Labs ($440K) New York (NY) (2016) [10–50]	Health analytics with NLU with real-world clinical data	Healthcare analytics platform	www.droicelabs.com
Enlitic ($30M) San Francisco (CA) (2014) [10–50]	Radiology and deep learning for medical image interpretation and workflow	Radiology workflow and early detection	www.enlitic.com
FDNA (NA) Boston (MA) (2011) [10–50]	AI-enabled tool for facial dysmorphology analysis with genetic search	Face2Gene	www.fdna.com
Flatiron Health (Acquired by Roche) New York (NY) (2012) [250–500]	Data analytics with platform for oncology	OncoCloud OncoEMR OncoBilling	www.flatiron.com

(*Continued*)

(Continued)

Company	AI in health-care focus	Specific service	URL
Freenome ($78M) San Francisco (CA) (2014) [10–50]	Noninvasive disease screening with AI focused on cancer	IMPACT	www.freenome.com
Gauss Surgical ($52M) Los Altos (CA) (2011) [10–50]	AI-enabled platform for real-time measurement of surgical blood loss	Triton for real-time monitoring of surgical blood loss	www.gausssurgical.com
Ginger.io ($28M) San Francisco (CA) (2011) [50–100]	AI-enabled coaching and behavioral health support	Behavioral health coaching Teletherapy	www.ginger.io
Glooko ($71M) Mountain View (CA) (2010) [100–250]	Universal diabetes platform for wearable devices and glucose levels	Glooko app	www.glooko.com
GNS Health Care ($77M) Cambridge (MA) (2000) [100–250]	Health-care analytics focused on applying big data to empower stakeholders	Platform using clinical trial and real world data	www.gnshealthcare.com
H2O.ai ($74M) Mountain View (CA) (2012) [10–50]	AI solutions for health care based on algorithms and applications	Open source ML platform	www.h2o.ai
Health Catalyst ($377M) Salt Lake City (UT) (2008) [250–500]	Analytics and risk prediction for improvement	Catalyst.ai tools	www.healthcatalyst.com
Health Fidelity ($19M) San Mateo (CA) (2011) [50–100]	Risk adjustment solutions with analytics and NLP for organizations	HCC Scout	www.healthfidelity.com

(Continued)

(Continued)

Company	AI in health-care focus	Specific service	URL
HealthTap ($88M) San Francisco (CA) (2010) [100−250]	Virtual care with immediate access to clinicians	Healthtap app	www.healthtap.com
iCarbonX ($600M) Shenzhen (CN) (2015) [100−250]	Digitalization of life and health	Meum DigitalMe	www.iCarbonx.com
IDx ($52M) Coralville (IA) (2010) [10−50]	Autonomous and clinically aligned diagnostic algorithms for disease detection	IDx-DR	www.eyediagnosis. net
Imagen Technologies ($60M) New York (NY) (2016) [10−50]	AI for medical images for error reduction	Medical image platform	www.imagen.ai
Infermedica ($4M) Wroclaw (PL) (2012) [50−100]	AI technology to reduce medical care waste with triage tools	Symptom Checker Call Center Triage	www.infermedica. com
Infervision ($71M) Beijing (CN) (2015) [250−500]	Computer vision with AI for diagnosis of cancer	Medical image platform	www.infervision.com
Jvion ($9M) Duluth (GA) (2011) [100−250]	Cognitive clinical science for prediction	Eigen Sphere Machine	www.jvion.com
KenSci ($30M) Seattle (WA) (2015) [50−100]	AI for care variation and hospital operations	Predictive analytics	www.kensci.com

(Continued)

(Continued)

Company	AI in health-care focus	Specific service	URL
K Health ($44M) New York (NY) (2016) [10−50]	Insights from health-care users with data analytics	AI health app	www.khealth.ai
Lark ($46M) Mountain View (CA) (2011)[50−100]	AI platform for chronic disease management and prevention	Diabetes prevention and hypertension program	www.lark.com
Livongo Health ($235M) Mountain View (CA) (2014) [250−500]	Data analytics for chronic disease management	Applied Health Signal Engine	www.livongo.com
Lumiata ($31M) San Mateo (CA) (2013) [10−50]	Data analytics and risk prediction and quality of care	Risk Matrix	www.lumiata.com
Lunit ($21M) Seoul (KR) (2013) [50−100]	AI-enabled multidimensional medical image interpretation platform	Lunit Insight	www.lunit.io
Medalogix ($5M) Nashville (TN) (2009) [10−50]	Predictive analytics solutions for risk prediction	Touch Bridge Nurture Nurture	www.medilogix.com
MedAware ($10M) Ra'anana (IL) (2012) [10−50]	Data analytics for reducing prescription error and improving patient safety	MedAS MedRIM MedQC MedRAF	www.medaware.com
Medopad ($29M) London (UK) (2011) [100−250]	Connected digital health ecosystem enabled by AI lab	Ecosystem for rare diseases and complex conditions	www.medopad.com

(Continued)

Company	AI in health-care focus	Specific service	URL
MedWhat ($3M) San Francisco (CA) (2010) [1−10]	Medical AI with cognitive computing, deep learning, and NLP	Human-level AI for medicine platform	www.medwhat.com
MI10 ($500K) Irvine (CA) (2019) [10−50]	AI strategy platform for health-care organizations and companies	Medical Intelligence Quotient (MIQ)	www.MI10.ai
Numedii ($6M) Menlo Park (CA) (2008) [10−50]	Big Data technology for drug discovery and precision therapeutics	Artificial Intelligence for Drug Discovery (AIDD)	www.numedii.com
Numerate ($17M) San Bruno (CA) (2007) [10−50]	Computational platform for drug design and development	Biological modeling for drugs	www.numerate.com
Nuritas ($50M) Dublin (IE) (2014) [10−50]	Computational approach to nutrition and health along with genomics	Discovering bioactive peptides	www.nuritas.com
Nutrino ($12M) Tel Aviv (IL) (2011) [10−50]	Personalized nutrition platform with analytics and wearable technology	FoodPrint	www.nutrinohealth.com
Olive ($75M) Columbus (OH) (2012) [100−250]	AI-powered solutions for health-care administration	AI-as-a-Service (RPA)	www.oliveai.com
Oncora Medical ($3M) Philadelphia (PA) (2014) [10−50]	AI methodology for radiation oncology	Oncora Patient Care	www.oncoramedical.com

(*Continued*)

(Continued)

Company	AI in health-care focus	Specific service	URL
OrCam Technologies ($86M) Jerusalem (IL) (2010) [50−100]	AI-inspired artificial vision with wearable platform and simple interface	OrCam MyEye 2	www.orcam.com
Paige.AI ($25M) New York (NY) (2018) [NA]	Computational pathology for AI-enabled support	Pathology AI Guidance Engine	www.paigeai.com
PathAI ($75M) Boston (MA) (2016) [50−100]	AI-powered pathology	The PathAI Solution	www.pathai.com
Pearl ($11M) Los Angeles (CA) (2019) [10−50]	AI portfolio of solutions for dental professionals	Second Opinion Practice Intelligence Smart Margin	www.hellopearl.com
PhysIQ ($19M) Naperville (IL) (2013) [10−50]	Analytic platform for health indicators for chronic diseases	Personalized physiology analytics platform (PPA)	www.physiq.com
Picwell ($11M) Philadelphia (PA) (2012) [10−50]	Benefits purchasing ecosystem with enterprise AI software	Plan selection and decision support	www.picwell.com
Prognos ($43M) New York (NY) (2010) [50−100]	Largest registry of clinical diagnostic information with algorithms	AI analytics for early tracking and prediction of disease	www.prognos.ai
Proscia ($12M) Philadelphia (PA) (2014) [10−50]	Digital pathology platform for AI-powered pathology	Concentriq digital pathology platform	www.proscia.com

(Continued)

(Continued)

Company	AI in health-care focus	Specific service	URL
Pure Storage ($531M) Mountain View (CA) (2009) [1000−5000]	Data storage with cost efficiency with new class of enterprise storage	Pure1 cloud-based management platform	www.purestorage. com
Qventus ($45M) Mountain View (CA) (2012) [10−50]	Real-time predictive operations management platform for hospitals	Platform for operations	www.qventus.com
Recursion Pharmaceuticals ($226M) Salt Lake City (UT) (2013) [50−100]	AI and automation combined with experimental biology for precision medicine	Platform for rug repurposing	www. recursionpharma. com
Renalytix AI ($29M) Penarth (UK) (2018) [10−50]	Analytics platform with EHR, genomic profiles, and biomarkers	KidneyIntelX	www.renalytixai.com
Saykara ($8M) Seattle (WA) (2015) [10−50]	AI-powered healthcare virtual assistant for documentation	Documentation tool	www.saykara.com
Sensely ($12M) San Francisco (CA) (2013) [10−50]	Avatar-based platform as virtual assistance for stakeholders	Virtual Nurse Molly	www.sensely.com
Sight Diagnostics ($52M) Tel Aviv (IL) (2011) [10−50]	AI-driven platform for blood analysis and infectious disease diagnostics	Digital fluorescent microscopy for malaria OLO POC	www.sightdx.com
SkinVision ($12M) Amsterdam (NL) (2012) [10−50]	AI-enabled skin lesion diagnosis	Platform to make skin lesion diagnosis	www.skinvision.com

(Continued)

(Continued)

Company	AI in health-care focus	Specific service	URL
SOPHiA Genetics ($140M) Lausanne (CH) (2011) [250–500]	Data computing and life sciences combined for data analytics solutions	SOPHiA AI	www.sophiagenetics. com
Spring Health ($8M) New York (NY) (2016) [10–50]	AI-enabled personalized mental health care	Precision Mental Health Care	www.springhealth. com
Synyi ($81M) Shanghai (CN) (2016) [50–100]	Big data-driven AI applications for research and services	Semantic analysis system	www.synyi.com
Systems Oncology (NA) Scottsdale (AZ) (2015) [10–50]	Cognitive computing and multi-scalar approach to drug discovery	Cognitive platform for drug discovery	www. systemsoncology.com
Turbine (Seed) Budapest (HU) (2015) [50–100]	AI-driven cancer treatment with new drug discoveries	The Turbine Stack	www.turbine.ai
twoXAR ($14M) San Francisco (CA) (2014) [10–50]	AI-driven drug discovery with de-risk strategy for studies	Computational platform for drug discovery	www.twoxar.com
Wellframe ($25M) Boston (MA) (2011) [50–100]	AI-centric mobile strategy for care management and patient engagement	Health plan solutions	www.wellframe.com
Welltok (250M) Denver (CO) (2009) [250–500]	Analytics for personalized health action plans	CafeWell Health Optimization Platform	www.welltok.com

(Continued)

(Continued)

Company	AI in health-care focus	Specific service	URL
Univfy ($6M) Los Alsots (CA) (2009) [10–50]	AI-supported program for maximizing success of IVF for women	PreIVF Report IVF Refund Program	www.univfy.com
Verge Genomics ($36M) San Francisco (CA) (2015) [10–50]	AI-enabled drug discovery for neuroscience	Platform for neurodegenerative diseases	www.vergegenomics. com
VIDA ($9M) Coralville (IA) (2004) [10–50]	Quantitative analysis and AI tools for lung imaging in pulmonary diseases	LungPrint	www.vidalung.ai
Virta ($82M) San Francisco (CA) (2014) [50–100]	Online medical specialty clinic for type 2 diabetes	Medical supervision Personal health coach	www.virtahealth.com
VisualDx (NA) Rochester (NY) (1999) [10–50]	Diagnosis with photographs and symptoms	Visual CDSS	www.visualdx.com
Viz.ai ($30 M) San Francisco (CA) (2016) [1–10]	AI platform for stroke patients	VizLVO VizCTP	www.viz.ai
Woebot Labs ($8M) San Francisco (CA) (2017) [NA]	AI-enabled conversational agent for cognitive behavioral therapy	Chatbot for therapy	www.woebot.io
Your.MD ($17M) London (UK) (2012) [10–50]	AI health information service in partnership with NHS	Symptom checker	www.your.md

(Continued)

(Continued)

Company	AI in health-care focus	Specific service	URL
Zebra Medical Vision ($20M) Shefayim (IL) (2014) [10–50]	Imaging analytics and AI tools for radiology	Imaging analytics	www.zebra-med.com
Zephyr Health ($33M) San Francisco (CA) (2011) [50–100]	Insight as a service for biomedicine	Zephyr Illuminate	zephyrhealth.com

Index